THE PORPHYRINS

Volume II

Structure and Synthesis, Part B

Edited by

DAVID DOLPHIN

Department of Chemistry
University of British Columbia
Vancouver, British Columbia, Canada

ACADEMIC PRESS   New York   San Francisco   London   1978

A Subsidiary of Harcourt Brace Jovanovich, Publishers

ACADEMIC PRESS, INC.
111 Fifth Avenue, New York, New York 10003

United Kingdom Edition published by
ACADEMIC PRESS, INC. (LONDON) LTD.
24/28 Oval Road, London NW1

Library of Congress Cataloging in Publication Data

Main entry under title:

The Porphyrins.

Includes bibliographies and indexes.
CONTENTS: v. 2
pt. B.   Structure and synthesis.—
v. 5.   pt. c.   Physical chemistry.—
1.   Porphyrin and porphyrin compounds.   I.   Dolphin,
David.   [DNLM: 1.   Porphyrins.   WH190 P837]
QD401.P825   547'.593   77-14197
ISBN 0−12−220102−7

Contents

1   Synthesis and Stereochemistry of Hydroporphyrins

HUGO SCHEER

2   Hydroporphyrins: Reactivity, Spectroscopy, and Hydroporphyrin Analogues

HUGO SCHEER AND HANS HERLOFF INHOFFEN

3   The Porphyrinogens

D. MAUZERALL

4   Oxophlorins (Oxyporphyrins)

P. S. CLEZY

5   Irreversible Reactions on the Porphyrin Periphery (Excluding Oxidations, Reductions, and Photochemical Reactions)

J.-H. FUHRHOP

6   Chemical Transformations Involving Photoexcited Porphyrins and Metalloporphyrins

FREDERICK R. HOPF AND DAVID G. WHITTEN

7   Linear Polypyrrolic Compounds

ALBERT GOSSAUER AND JÜRGEN ENGEL

8 Metal Complexes of Open-Chain Tetrapyrrole Pigments

J. Subramanian and J.-H. Fuhrhop

9 Stereochemistry and Absolute Configuration of Chlorophylls and Linear Tetrapyrroles

Hans Brockmann, Jr.

10 Pyrrolic Macrocycles Other than Porphyrins

R. Grigg

List of Contributors

Numbers in parentheses indicate the pages on which the authors' contributions begin.

HANS BROCKMANN, JR. (287), Gesellschaft für Molekular-biologische Forschung mbH, Braunschweig-Stöckheim, West Germany, and Fakultät für Chemie, Universität Bielefeld, Bielefeld, West Germany

P. S. CLEZY (103), Department of Organic Chemistry, The University of New South Wales, Kensington, New South Wales, Australia

JÜRGEN ENGEL (197), Institut für Organische Chemie der Technischen Universität, Braunschweig, West Germany

J.-H. FUHRHOP (131, 255), Gesellschaft für Biotechnologische Forschung, Braunschweig-Stöckheim, West Germany, and Institut für Organische Chemie der Technischen Universität, Braunschweig, West Germany

ALBERT GOSSAUER (197), Institut für Organische Chemie der Technischen Universität, Braunschweig, West Germany

R. GRIGG (327), Queens University, Belfast, Northern Ireland

FREDERICK R. HOPF (161), Department of Chemistry, University of North Carolina, Chapel Hill, North Carolina

HANS HERLOFF INHOFFEN (45), Institut für Organische Chemie der Technischen Universität, Braunschweig, West Germany

D. MAUZERALL (91), The Rockefeller University, New York, New York

HUGO SCHEER (1, 45), Institut für Botanik der Universität, Munich, West Germany

J. SUBRAMANIAN (255), Gesellschaft für Biotechnologische Forschung mbH, Braunschweig-Stöckheim, West Germany

DAVID G. WHITTEN (161), Department of Chemistry, University of North Carolina, Chapel Hill, North Carolina

General Preface

Man cannot give a true reason for the grass under his feet
why it should be green rather than red or any other color.

Sir Walter Raleigh
History of the World: Preface (1614)

Just over two centuries after these words of Raleigh, Verdeil in 1844 converted chlorophyll to a red pigment which prompted him to suggest a structural relationship between chlorophyll and heme. Shortly thereafter, Hoppe-Seyler, in 1880, strengthened this hypothesis by showing the spectral resemblances between hematoporphyrin and an acid degradation product of chlorophyll. The final steps in these structural elucidations were initiated by Willstätter, and culminated in the heroic work of Hans Fischer who showed that but for two hydrogen atoms grass would indeed be red and that only two more hydrogen atoms would have ensured that Raleigh and his countrymen would indeed have been blue-blooded Englishmen.

The close structural similarity between the porphyrins and chlorins gives little measure of the relationships among and the diversity of their numerous and important biochemical functions. All life on this planet relies directly on the central role of the chlorophylls and cytochromes in photosynthesis by means of which photonic energy is converted and stored as chemical energy. It is likely that long before oxygen was abundant in the Earth's atmosphere the cytochromes were responsible for respiration. With the advent of photosynthesis the oxygen produced is the terminal electron acceptor for all aerobic respiration. For many organisms the

xi

means by which oxygen is transported, stored, reduced, and activated are frequently mediated by heme proteins. In mammals, oxygen is transported by the cooperative tetrameric protein hemoglobin and stored by monomeric myoglobin. When oxygen is reduced to water, in the terminal step of respiration, four electrons are transported via a series of cytochromes to cytochrome oxidase. Cytochrome oxidase contains two iron porphyrins and two copper atoms. In addition, nature also brings about one- and two-electron reductions to superoxide and peroxide. Both the decomposition and further activation of hydrogen peroxide are mediated by the heme proteins catalase and peroxidase. Furthermore, heme proteins function as both mono- and dioxygenases, and recently cytochrome P-450, which functions as a monooxygenase by combining properties of both oxygen binding and electron transport, has been shown to be important in a wide variety of biological hydroxylations.

This brief insight into a few of the many central roles played by metalloporphyrins in nature plus the challenges that porphyrins present to the inorganic, organic, physical, and biochemist suggest the wealth of knowledge that is documented in these areas. It is the objective of "The Porphyrins" to present a full and critical coverage of all the major fields relating to porphyrins, their precursors, catabolic derivatives, and related systems in a manner that we trust will be useful to those in physics, chemistry, biochemistry, and medicine.

The treatise consists of seven volumes. Volumes I and II (Structure and Synthesis, Parts A and B) cover nomenclature, history, geochemistry, synthesis, purification, and structural determination of porphyrins, metalloporphyrins, and mono- and polypyrrolic compounds and related systems. Volumes III, IV, and V (Physical Chemistry, Parts A, B, and C) cover electronic structure and spectroscopy including uv-vis, ORD, CD, MCD, mass, ir, resonance Raman, Mössbauer, Zeeman, NMR (diamagnetic, paramagnetic), ESR, and X-ray crystallography. In addition, redox chemistry, electron transfer, aggregation, oxygenation, and solid state phenomena are included. Volumes VI and VII (Biochemistry, Parts A and B) cover the biosynthesis and enzymatic synthesis of porphyrins, chlorophylls and their precursors, and the chemistry and biochemistry of the bile pigments and the roles of porphyrins and bile pigments in clinical chemistry. The structure and function of the major hemoproteins are also covered.

It remains for me to thank my colleagues and co-workers for their support and assistance. A special debt of gratitude goes to my mentors: Alan Johnson who introduced me to these areas and who taught me why chlorophyll is green, and Bob Woodward who showed the world how to make chlorophyll and taught me why.

DAVID DOLPHIN

Preface

Volume II (Structure and Synthesis, Part B) contains chapters on porphyrins reversibly modified at the periphery by either oxidation (oxophlorins) or reduction (hydroporphyrins, porphyrinogens) and by irreversible reactions at the periphery. The synthesis (including photoexcited porphyrins and metalloporphyrins), properties, and stereochemistry of these systems as well as those of the chlorophylls and linear polypyrroles are covered. In addition, chapters on linear polypyrroles, their metal complexes, and macrocycles other than porphyrins are included. The volume complements Volume I (Structure and Synthesis, Part A) which includes chapters on nomenclature, historic aspects, and geochemistry, as well as methods for the synthesis, characterization, and purification of porphyrins and metalloporphyrins.

The final result is an up-to-date and critical review of the areas described above. This treatise provides, for the first time, a complete and comprehensive review of all the major aspects of porphyrin chemistry and biochemistry.

I wish to take this opportunity to thank the contributors to this volume. For those who completed their chapters on time, I give my thanks for their patience during the period between submission of their manuscript and the publication of this book. Of those who were not so prompt I ask that they understand my impatience.

DAVID DOLPHIN

xiii

Contents of Other Volumes

VOLUME I   STRUCTURE AND SYNTHESIS, PART A

VOLUME V   PHYSICAL CHEMISTRY, PART C

VOLUME VI   BIOCHEMISTRY, PART A

VOLUME VII   BIOCHEMISTRY, PART B

1

Synthesis and Stereochemistry of Hydroporphyrins

HUGO SCHEER

I. INTRODUCTION

The fully unsaturated porphyrin macrocycle (**1**) (see Scheme 1, structures **1–13**) contains 11 conjugated double bonds. Strictly speaking, the term "porphyrin" applies only to this system, but a variety of compounds are known in which the macrocyclic porphyrin skeleton is retained, while one or more of the double bonds are removed. These compounds are formally derived from porphyrins by hydrogenation, and are, therefore, commonly termed hydroporphyrins. Reduction is probably the most common pathway to hydroporphyrins, and the known redox reactions and isomerizations of true hydroporphyrins are summarized in the reaction scheme (Scheme 1). However, compounds that contain the chromophore of hydroporphyrins are accessible, too, by oxidation of porphyrins yielding the oxy and oxo analogues of the hydroporphyrins (see Chapter 2).

1

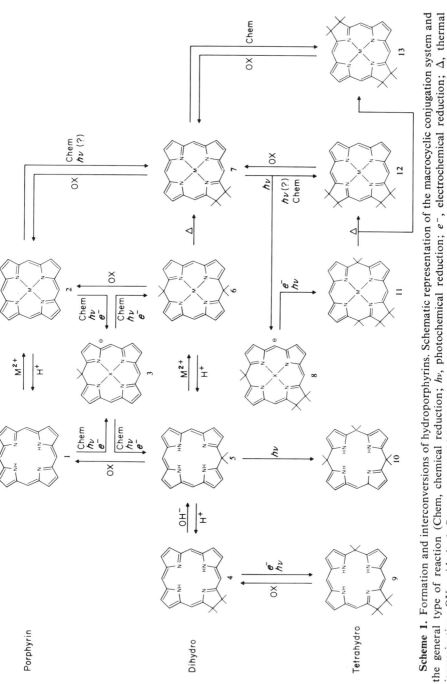

Scheme 1. Formation and interconversions of hydroporphyrins. Schematic representation of the macrocyclic conjugation system and the general type of reaction (Chem, chemical reaction; $h\nu$, photochemical reduction; e^-, electrochemical reduction; Δ, thermal isomerization; OX, oxidation). Structures at a higher hydrogenation level than 4 have not been considered. For the porphyrinogens, see Chapter 3.

Some of the hydroporphyrins are of central biological importance. Most chlorophylls are Mg complexes of chlorins (4) or bacteriochlorins (13), in which one or two of the peripheral double bonds are reduced, the sirohemes found in nitrate and sulfate reductases[1a] contain the "cis" bacteriochlorin chromophore (12),[1b] porphyrinogens (16) and porphomethenes (10)[1c] are intermediates in porphyrin biosynthesis (Volume VI, Chapters 1 and 2), and oxophlorins (18) are discussed as first intermediates in the degradation of porphyrins into bile pigments (Volume VI, Chapter 5). In addition to these naturally occurring hydroporphyrins, a variety of synthetic systems has been characterized in the past, which yielded a detailed insight into the reactivity and electronic structure of the porphyrin macrocycle.

The hydroporphyrins can be arbitrarily divided into two classes, depending on whether or not the macrocyclic conjugation is retained. In chlorins (4), bacteriochlorins (12, 13), dihydrobacteriochlorins (14),[1] and corphin dications (15),[2] one or more of the macrocyclic peripheral double bonds are reduced

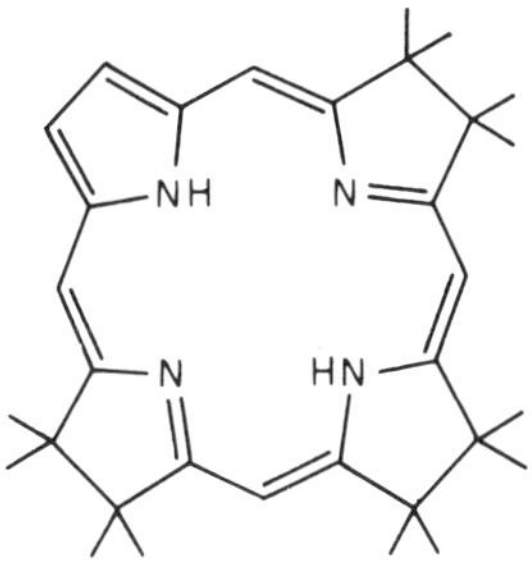

14

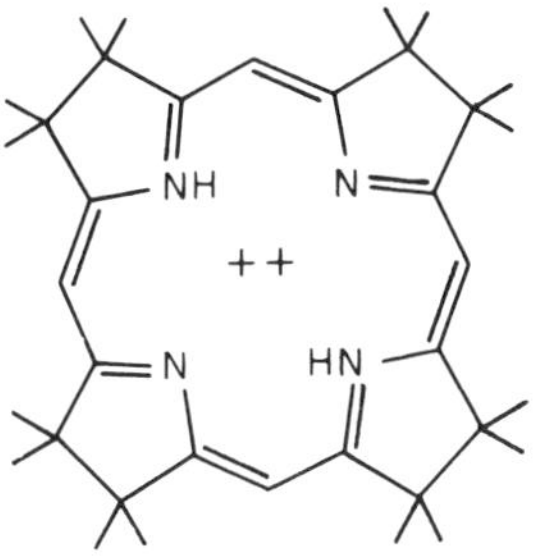

15

without loss of the macrocyclic conjugation as evidenced, for example, by the aromatic (4, 12, 13, 14) and antiaromatic (15) ring-current effects, respectively, in their nuclear magnetic resonance (nmr) spectra.[3] In the nonaromatic hydroporphyrins, the cyclic conjugation is usually interrupted at one or more of the bridging methine positions, leaving one to four conjugated pyrrole

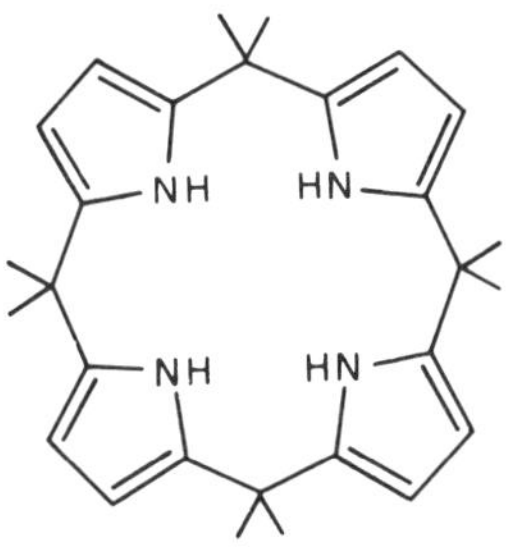

16

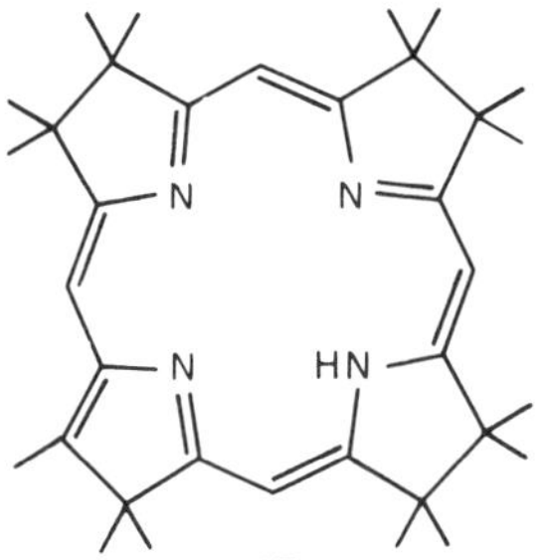

17

subunits (structures **5, 6, 9, 10, 16**). In the corphins (**17**),[4,5] the cyclic conjugation is interrupted in an essentially different manner at one α-pyrrolic carbon atom and one N atom. Some related structures in which the conjugation is interrupted by other means are discussed elsewhere. In most cases, the respective carbon atoms become sp^3 hybridized by either hydrogenation of or addition of new C–C bonds. However, exocyclic oxohydroporphyrins with exocyclic C=O bonds are widely investigated due to their greater stability toward oxidation (see Chapter 2).

II. SYNTHESIS OF HYDROPORPHYRINS

A. Chemical Reduction

1. CHEMICAL REDUCTION TO CHLORINS

To date no synthetic method has been devised in which the chlorin, and almost any other hydroporphyrin macrocycle,[2,6] is built in the rational step-by-step fashion now commonly employed in porphyrin synthesis. Chlorins are usually found as the undesired by-products in *meso*-tetraarylporphyrins,[7,8] and a wide variety of chlorins is accessible by suitable modification of chlorophylls.[9–12a] The recent progress in the chemical modification of chlorophyll derivatives is summarized in the appendix to this chapter. The general synthetic approach to chlorins, however, involves first the synthesis of the respective porphyrin and then its subsequent reduction to chlorin. (It is noteworthy that the same approach is used in the biosynthesis of chlorophylls, too.[13])

There are 26 isomeric neutral dihydroporphyrin structures possible[14] for symmetric porphyrins that are chemically reasonable. Most of these structures are still theoretical, and, indeed, only three of them have been prepared as defined compounds and characterized in detail: the chlorins (**4**), the phlorins (**5**), and the porphodimethenes (**6**). Nonetheless, the synthetic problem is complicated by the possibility of stereoisomers, and, when unsymmetrically substituted porphyrins are to be reduced, of structural isomers. The selective reduction of porphyrins is then a major problem, and, for example, the regio- and stereoselective reduction of the 7,8-double bond was one of the key reactions in Woodward's chlorophyll synthesis.[15] In spite of the straightforward reactions now available to prepare the common model chlorins like $H_2(OEC)$ and $H_2(TPC)$, there is still no generally applicable set of conditions to obtain chlorins with sensitive substituents.

The best approach to the etio-type chlorins is the treatment of porphyrins with reagents typical for the hydrogenation of isolated double bonds. Reactions of this type support an electronic structure of the porphyrin macrocycle in which at least two of the peripheral double bonds do not fully participate with the 18π aromatic conjugation system. Catalytic hydrogenation[15a-17] or reduction with diimide[18,19] yields exclusively the *cis*-chlorins. Borohydration yields a 5:1 mixture of the cis and trans isomers,[17,20] possibly due to two competing pathways in which *trans*-OEC is formed via a phlorin intermediate, rather than by a direct attack of B_2H_6 on the peripheral double bond. *trans*-OEC is best prepared by reduction of Fe(III)(OEP)Cl with sodium in amyl alcohol, and subsequent demetalation of the formed Fe(OEC).[19,21-22c] The Sn(IV) and Mn(III) complexes of OEP can be reduced in comparable yield to the metallochlorin,[21-24] but demetalation is less satisfactory in this case for preparative purposes.[24]

Considerable attention has been given to the influence of the central metal on the reduction of metalloporphyrins. Central metals generally reduce the possibility of side reactions. Phlorin formation is precluded at low pH and in neutral solutions,[25] although at higher pH phlorin (= porphodimethene) anions (3) can be formed.[26-29] The preferential reduction of metalloporphyrins with highly charged central metals can be rationalized in first order by an electrostatic model, which was introduced to explain their one-electron redox potentials.[30] The latter ones are primarily determined by the electron density at the periphery, which, in turn, can be controlled by the electronegativity of the central metal. Thus, Fe(III), Mn(III), and Sn(IV) porphyrins are easily reduced both chemically[21-22c] and photochemically.[23,24] On the other hand, Mg and Zn porphyrins are more difficult to reduce. However, Zn is exceptional in stabilizing chlorins and/or in promoting the rearrangement of *meso*-hydrogenated by-products into chlorins.[26,28,31-31b] Such subtle influences of the central metal have found some attention, but they cannot yet be rationalized. Finally, the demetalation properties of the formed metallochlorins are an important synthetic aspect. Fe(III), Mn(III), and Sn(IV) complexes can be demetalated only under harsh conditions not suitable for chlorins that are easily oxidized,[17,20] or which bear delicate substituents, while metals of intermediate stability like Zn seem to be again the best choice.

The synthesis of *meso*-tetraarylporphyrins from pyrroles and benzaldehydes[32,33] is always accompanied by the formation of hydroporphyrins, especially chlorins.[7,8] These are usually undesired contaminants of the porphyrins, and methods have been devised for their removal by selective oxidation to porphyrins.[33] If the chlorins are the desired products, they can be separated by fractionated crystallization (which leaves the chlorins enriched in the mother liquor), by fractionated acid extraction (in which the lower basicity[34] of the chlorins is used[9,19]), and, only with difficulty and only if small amounts have to be separated, by chromatography. The mechanism of hydroporphyrin formation during the synthesis of TPP has been studied by several groups.[7,8,35,36] It could be shown that the initially formed

TPP-porphinogen is oxidized in an acid-catalyzed reaction to the porphyrin via a porphodimethene[37] intermediate.[8] The chlorin is then formed by the subsequent re-reduction of the porphyrin.[8] The best synthetic approach to TPC is the diimide reduction method of Whitlock et al.,[19] which yields about 50% of the crude product.

Chlorins obtained by the above procedures are always accompanied by bacteriochlorins, which can be enriched and isolated if the reaction conditions are carefully controlled. Both isomers, namely, the true bacteriochlorins (13) and the isobacteriochlorins (12), are formed, the latter ones usually in higher yield. True bacteriochlorins are obtained in highest yield under the conditions of the Wolff–Kishner reduction,[20,38] and there seems to be a tendency that central metals stabilize the isobacteriochlorins.[19,20,22a,22b,25,38,39] If chlorins are the desired reaction products, the yield can be increased by selective reoxidation of the bacteriochlorin by-products to the chlorin level.[19,40,41]

In summary, chlorins derived from the highly symmetric model porphyrins like OEP and TPP can be prepared in high yields by chemical reduction, but less symmetric chlorins like the ones related to chlorophylls are not accessible in this way. Surprisingly, the formation of a mono-*meso*-substituted chlorin more closely related to the chlorophylls was reported as early as 1929.[42] Treatment of the phylloporphyrin **18** with sodium ethoxide at 185°C produced in good yield (20%) a chlorin to which structure **19** was assigned on the basis

                    18   19

of its uv-vis spectrum. This structure was proved correct 40 years later by X-ray crystallography.[43] Although the mechanism of its formation is not known, the driving force seems to be the same steric hindrance between the peripheral substituents on which Woodward's approach[15] to chlorophyll *a* (Chl *a*) was based. The outline and the details of this synthesis are reviewed earlier.[15,44] Here we want to discuss briefly only the final and crucial conversion steps leading to the 7,8-*trans*-chlorin **27**. The principal concept of the selective reduction at C-7 and C-8 was based on the hypothesis of the "overcrowded periphery." In porphyrins possessing neighboring *meso-* and β-pyrrole substituents, there exists considerable steric hindrance between these coplanar groups. This interaction provides a driving force for the selective sp^3 hybridization of either one of the substituent sites in the "over-crowded" region, by which the respective substituent(s) are positioned out of plane, and, thus, the steric hindrance is relieved. There are now numerous examples of regio- and stereoselective reductions of porphyrins to support this principle (see also Section III), which will here be exemplified by the respective reactions in the original work, the last steps of the Chl *a* synthesis leading to chlorin-e$_6$ trimethyl ester (**20**).

Double condensation of the two dipyrromethanes **21** and **22** (via a Schiff base intermediate) with 12 *N* HCl in MeOH leads not to a porphyrin, but to

20   21   22

the phlorin **23** instead. Compound **23** was the first example of this class of compounds, and its formation is obviously favored (as compared to the porphyrin formation) by the steric relief due to the out-of-plane positioning of the γ-substituent. Compound **23** can be oxidized, e.g., with iodine, to the corresponding γ-substitued porphyrin **24**. The latter is not only easily re-reduced back to the phlorin **23**, but also undergoes a remarkable acid-catalyzed isomerization to the phlorin **25**. Thus, the steric hindrance due to the γ-substituent in **24** must be large enough to overcome the resonance stabiliza-tion of the cyclic conjugated system, and, in acetic acid, an equilibrium is established with about equal amounts of **24** and the γ-allylphlorin **25**. No similar equilibrium of **24** was found with the desired chlorin, but this goal

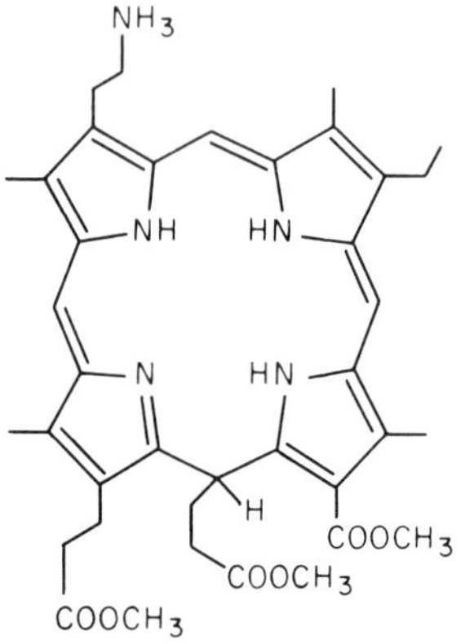

23

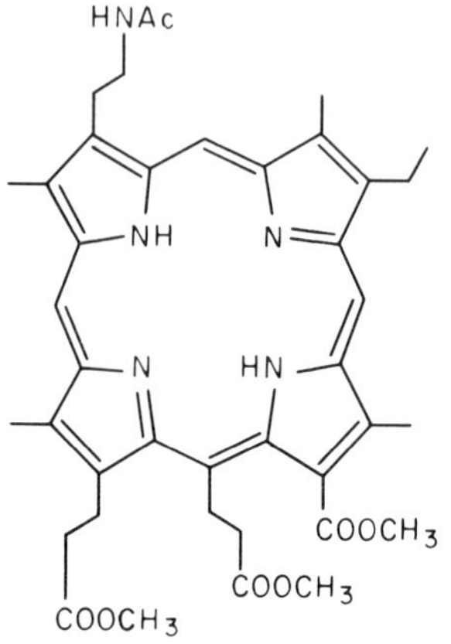

24

was achieved by an additional oxidation of **25** to the γ-allylporphyrin **26**. The latter compound is no more capable of undergoing a similar isomerization (**24** $\rightarrow$ **25**) due to the double bond in the γ-substituent. Instead, in this case, the steric strain is relieved by isomerization of the porphyrin **26** into the

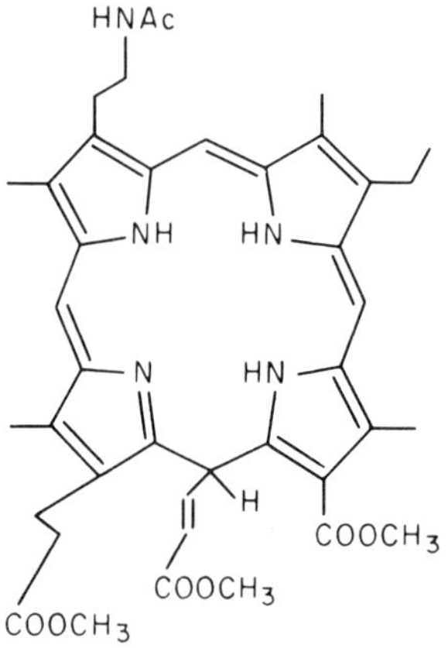

25

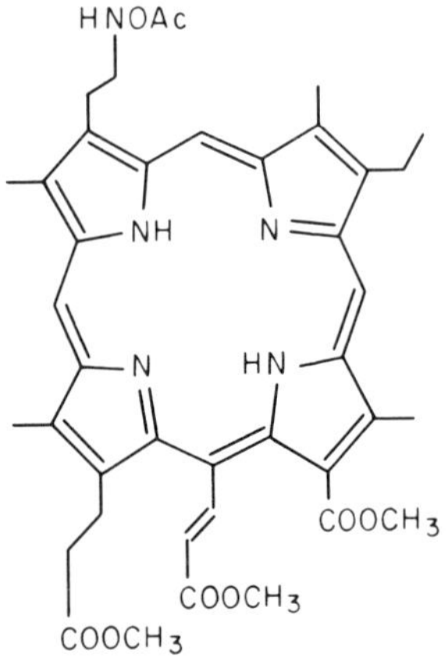

26

chlorin (= purpurin) **27**.* This step is both stereo- and regioselective. Not only is the desired 7,8-chlorin preferentially formed, due to the larger steric hindrance of the 6-$COOCH_3$ versus the 7-propionic side chain, but also the right stereochemistry at ring IV is obtained by the transoid arrangement of

---

* Recently, the need for an electrophylic β-pyrrole substituent for this isomerization has been invoked from model studies [L. Witte and J.-H. Fuhrhop, *Angew. Chem.* **87**, 387 (1975)].

the substituents with the largest effective volume, namely, the 7-propionester and the 8-CH$_3$ group. By this means, the essential 7,8-*trans*-chlorin structure has been formed in high yields in a rational approach. The subsequent conversions of **27** and resolution of the racemate then lead to chlorin-e$_6$ trimethyl ester (**20**) and finally to chlorophyll *a* (**28a**) by methods devised

(a) R = C$_{20}$H$_{39}$ (phytyl)
(b) R = H

already by the groups of Willstätter and Stoll, and of Fischer.[9,45] The functionalization of the 3-methyl group of **20** leading to the chlorophyll *b* series has been described by Inhoffen *et al.*[45a] (Chapter 2).

2. CHEMICAL REDUCTION TO INTERRUPTED SYSTEMS

The most widely investigated chemical reduction that leads to interrupted systems is the treatment of porphyrins (**1**) with alkali metals,[46–50] or, better, with aromatic radical anions[26,28] to yield the porphyrin dianions, and the subsequent addition of protic solvent (reductive protonation) or alkylating reagents like alkyliodides (reductive alkylation). The reaction of metalloporphyrins with aromatic radical anions was first investigated by Closs and Closs.[26] Two subsequent[50,51] one-electron reductions yielding the dianion can be followed by uv-vis and esr spectroscopy.[26,47] Addition of 1 mole of methanol to the dianion yields the anion of type **3**, which can be considered as the conjugated base of either a phlorin or a porphodimethene. Addition of a second mole of methanol then results (with Zn as central metal) in an isomerization of the porphodimethene (**6**) into a chlorin (**4**).[26,28] The reductive protonation was studied in the OEP series for a variety of central metals by

Buchler *et al.*[28,52,53] and can be summarized as follows: (a) the reduction step leads normally to the dianion, although with some transition metals (Co, Cu, Ni, Cr) only the monoanion is accessible; (b) upon protonation, four reaction products are generally observed: the metal complexes of the porphodimethene (**6**), the chlorin (**7**), the isobacteriochlorin (**12**), as well as the metal-free porphyrin. The chlorin yield (isomerization of **6** into **7**) is best for the Zn complexes; (c) there are two yet to be identified types of porphodimethenes (?) with absorptions at about 455 and 520 nm, respectively; (d) the protonation is accompanied by disproportionation reactions, as evidenced by the formation of the porphodimethene from monoanions, as well as by the formation of isobacteriochlorins from the dianions.

In an alternative reaction, the dianion can be treated with alkylating electrophilic agents instead of methanol. This reductive alkylation allows an insight into the reaction mechanism and the electron density of the dianion, even in cases in which the ultimately isolated reaction product is the re-oxidized porphyrin. With CH_3I as methylation reagent, and subsequent oxidation, the α,γ-methylated OEP **29a** was obtained from the dianion of

(a) R = Et, M = H₂
(b) R = Me, R = Zn
29

(a) M = H₂
(b) M = metal-ion
30

OEP in 60% yield, thus confirming a high electron density of the dianion at opposite methine positions.[53]

The uv-vis spectra of the dimethylporphyrins **29** show considerable bathochromic shifts (as compared to the educt porphyrins), which are characteristic for oligopyrrole pigments that are twisted due to steric hindrance (Brunings–Corwin effect[17,31-31b]; for a critical study see ref. 55). In analogy with Woodward's considerations concerning the effects of an overcrowded porphyrin periphery,[15] one would expect the α,γ-dimethyl-porphyrins to be destabilized, while the reduced dimethylporphodimethenes should be stabilized. In agreement with this concept, the free porphomethene **30a** and a series of its metal complexes could be isolated.[28,52,53] It is further proved by the direct comparison of the two isomeric porphodimethenes obtained by photoreduction of the α,γ-dimethylporphyrin **29b**.[56] One of them, the α,γ-dimethyl-β,δ-porphodimethene **31a**, is again stabilized for steric reasons. The other product is spectroscopically very similar, but is rapidly oxidized to the educt. Therefore, the isomeric structure **31b** was proposed,

(a) R = Me, R' = H
(b) R = H, R' = Me

31   32

which is not sterically stabilized against rearomatization. In an interesting variant of the reductive alkylation, the diethylchlorin **32** was recently obtained by alkylation of the OMP-dianion.[57] In this reaction, two ethyl groups have seemingly migrated from methine to β-pyrrole positions.

B. Photochemical Reduction

Life on earth depends on photosynthesis, and the conversion step of light into chemical energy involves chlorophylls in all known photosynthetic organisms. Photoreactions of chlorophylls, which are usually either chlorins or bacteriochlorins,[58] as well as photoreactions of porphyrins in general are, therefore, extensively studied.[59] At least two* photochemical conversions of chlorophylls *in vivo*[59a] are well characterized: the photoreduction of proto-chlorophyllide *a* (**33**) to chlorophyllide *a* (**28b**) as one of the last steps of Chl *a* biosynthesis in higher plants and some algae,[13] and the reversible oxidation of reaction center chlorophyll as the primary reaction of photosystem I (for leading references, see refs. 60; Volume IV, Chapter 3; and Volume V, Chapter 9). Comparably little is known about a third one, the primary reaction of photosystem II, except that chlorophyll *a* is involved as well.[61,62]

Although chlorophyll was used as the red photosensitizer for photographic emulsions as early as the beginning of this century,[63,64] the first defined

---

* For the photochemistry of chenopodium chlorophyll protein, c.f. T. Oku and G. Tomita, *Plant Cell Physiol.* **16**, 1009 (1975); and *Photochem. Photobiol.* **25**, 199 (1977).

reversible *in vitro* photoreaction of chlorophyll was not described before 1948 by Krasnovskii.[65] In this reaction, chlorophyll *a* in pyridine is reversibly reduced by absorbic acid. First, a transient yellow product (λ_{max} = 470, 340 nm) can be observed, followed by a fairly stable pink product (λ_{max} = 515 nm). The latter pink Krasnovskii product was recently characterized as the α,γ-porphodimethene **34**.[66] During the Krasnovskii reaction reductive

33   34

capacity is generated, therefore, the reaction was extensively studied as a possible model for the photosynthetic conversion of light into chemical energy. It could be shown that the reaction is quite general for porphyrins,[67–70] and that various reductants can be used, including hydrazines, sulfhydryl compounds, hydroquinones, and stannous chloride. With ascorbic acid as the reducing agent, the presence of a base is always required, and the reaction occurs via electron transfer from the ascorbate anion, and subsequent proton transfer from the protonated base[70] present in the ion pair. On the other hand, reduction with $SnCl_2$ is quantitative in strongly acidic solutions.[23,71] Several investigations are focused on the influence of pH[71,72,72a] and solvent system[25,31–31b,72b,72c,73] on the reaction.

The original work of Krasnovskii deals with the reduction of chlorophyll *a* (**28**), and, due to its potential implications for photosynthesis, a large amount of work has been focused on this molecule, a metallochlorin. Most of the recent studies are concerned with other porphyrin type compounds, however, be it for use of model systems or for the interest as such in these structures. Although some details of the Krasnovskii reductions are still to be clarified, the general scope of the reaction does strongly resemble the scheme outlined for the chemical reduction in Section II, A, 2. It can be summarized as follows. (a) Metalloporphyrins (**2**) are reduced subsequently to metalloporphodimethenes (**6**), metalloisobacteriochlorins (**11**), and metallohypobacteriochlorins (**14**). (b) Porphyrins (**1**) are reduced subsequently to phlorins (**5**) and porphomethenes (**10**). (c) The reduction of chlorins yields the corresponding structures with one peripheral double bond removed, namely, chlorinphlorins (**9**) are obtained from chlorins (**4**), and metallochlorinporphodimethenes (**11**) are obtained from metallochlorins (**7**). Structures of chlorin reduction products beyond the dihydrochlorin level are yet to be characterized. (d) Phlorins (**5**) can isomerize to chlorins (**4**) (see Section II, B), or they can be metalated to metalloporphodimethenes (**6**).[74] (e) Metalloporphodimethenes (**6**) can isomerize to metallochlorins (**7**). They can be demetalated to porphodimethenes, and certain results indicate that they can be demetalated to phlorins as well. (f) Chlorinphlorins can probably isomerize to porphomethenes.

The chromophore in some of the products with retained macrocyclic conjugation (chlorins, bacteriochlorins) is relatively safely identifiable by means of uv-vis spectroscopy. However, reduced forms with interrupted cyclic conjugation, but conjugation of more than one pyrrolic subunit, usually have absorptions close to 500 nm. Early additional techniques involved acid–base and redox titrations for the characterization of these types of compounds[14]; in addition, the usefulness of infrared spectroscopy (ir) and especially nmr has been demonstrated recently.[3,31b,56,66,71,75–79] Thus, at least four different compounds with an absorption at $\lambda_{max} = 500$ nm have been characterized as reduction products: porphomethenes (**10**),[14,79] chlorinphlorins (**9**),[78,80] chlorinporphodimethenes (**11**),[66] and porphodimethenes (**6**).[31b,56] All these products contain the common dypyrromethene chromophor, which gives rise to the observed absorption.[81]

1. METALLOPORPHYRINS TO METALLOCHLORINS

The photoreduction of protochlorophyllide *a* (**33**) to chlorophyllide *a* (**28b**) is, in higher plants and some algae, one of the last steps in the biosynthesis of chlorophyll *a*.[13] This step is both regio- and stereospecific *in vivo*: only the 7,8-double bond in (**33**) is reduced, and the "extra" hydrogens are transoid to each other. The thermodynamic stability of this structure as compared to other isomers was first discussed by Woodward and proved by the approach of the Harvard group[15] to the synthesis of chlorophyll *a* (see Section II, A, 1).

In spite of the thermodynamic stability of chlorophyll *a*, however, attempts to stimulate the photoreduction of protochlorophyll *a* and related Zn complexes *in vitro* have been only partially successful. The reduction yields the 7,8-*cis* rather than the *trans* isomers,[31–31b,82–83a] and there are indications that the reaction proceeds via a porphodimethene intermediate, which subsequently isomerizes to the chlorin.[31–31b,82]

Two products can be identified spectrophotometrically during the photo-reduction of etio-type Zn porphyrins and Zn pheoporphyrins: a product with an absorption band at $\lambda_{max} \approx 515$ nm, and the chlorin ($\lambda_{max} \approx 650$ nm). (In the case of TPP, a primary intermediate with $\lambda_{max} = 448$ nm is observed instead.[84])

After a possible induction period due to the presence of oxygen,[25,85] the 515-nm product is formed almost quantitatively in the beginning of the reduction (color changes from purple to yellow-orange). In the dark and under the exclusion of oxygen, it is converted rapidly into the chlorin, and more slowly, back into the porphyrin[31a,31b,82,86] educt. For the photoreduction of Zn porphin, the second step has also been shown to be inducible by light.[25] These results of the photoreduction of Zn phylloerythrin (35) were interpreted from by-product analysis and ^{2}H incorporation studies[31b,82] as the primary formation of two isomeric porphodimethenes (36a,b) ($\lambda_{max} = 515$ nm). These can subsequently isomerize in a dark (or light[25]) reaction into the *cis*-chlorin (37a), or are partially reoxidized into the porphyrin (35). The isomer 36a

35   36a

also undergoes a reaction in which ring E is opened to form a rhodochlorin[31b] (Section II, B, 2). Based on quantum yield studies, an intermediate reduced below the dihydroporphyrin level was proposed during the rearrangement to the chlorin.[25]

Yet, a third reaction product ($\lambda_{max} = 470$ nm) is spectrophotometrically identified by Suboch *et al.*[83] A band in this region was already earlier described,[86] but without a clear kinetic separation from the 515-nm band. It should be mentioned that a similar product is observed, as well, during the chemical reduction of porphyrins, which was interpreted as arising from an isomeric porphodimethene.[28] Based on this third product, Suboch *et al.*[83,83a] discuss a different pathway in which the chlorin and the porphodimethene ($\lambda_{max} = 470$ nm) are formed independently. The chlorin is then further

36b

37a

reduced to the pink Krasnovskii product[65] (chlorin porphodimethene,[66] $\lambda_{max} = 515$ nm). Although the above discussed kinetics seem to disfavor this interpretation, at least for the photoreduction of **35**, the question remains open as long as the structure of the intermediates is not determined in a more direct approach (namely, by *in situ* [1]H nmr.[66,80]). Formation of two isomeric porphodimethenes (**36a,b**), or another type of "isomerism" (namely, ionic versus neutral) is conceivable, which is suggested by the similar spectroscopic properties of the porphodimethene isomers obtained by reductive alkylation of symmetric porphyrins.[28]

To stress the similarity between the chemical and photochemical reduction once more, Sn(IV)(OEC) was obtained in excellent yield from Sn(IV)(OEP) by photoreduction with tetramethylethylenediamine.[24] Chemical reduction of Sn(IV) complexes is well known and has been interpreted by an electrostatic model with increased macrocycle electron density.[29] As a preparative method, the reduction became attractive after a method had been developed to demetalate the Sn(IV) chlorins.[24] Unfortunately, only porphyrins without functional groups like C=O are accessible by this method.

Spectral differences between the chlorins produced by photoreduction of Zn pheophorphyrins, and the corresponding chlorins delivered from natural sources, were first observed in 1970 by Krasnovskii[86] and by Scheer.[82] As demonstrated independently by two groups,[31–31b,82–83a] these differences are due to the *cisoid* configuration of the extra hydrogen atoms in the synthetic photoproducts, as compared to the *transoid* configuration in their natural counterparts (see Section III, C).

The above discussed results, namely, the formation of chlorins via isomerization of primary photoreduction products (e.g., **36**) and the "wrong" stereochemistry of the chlorins at C-7 and C-8, present two problems: They render the *in vitro* reaction only to be a limping model of the biosynthetic reduction step of protochlorophyllide *a* **33** → **22b**, and they present an interesting problem, as this reaction sequence produces identical structures (with respect to the 7,8-stereochemistry) under various conditions.*[31–31b,83,83a]

2. METALLOPORPHYRINS TO PORPHODIMETHENES

Metalloporphodimethenes (**6**) are generally the first long-lived intermediates in the photoreduction of metalloporphyrins (**2**). Although a dipyrromethene chromophore was concluded earlier for these products, direct proof was not obtained until recently.[56] Photoreduction of the α,γ-dimethylporphyrin **29b** yielded two spectroscopically similar isomers (**31a, b**). One of them (**31b**) is easily reoxidized to the starting material, while the other isomer could be isolated and identified by ^{1}H nmr as the β,γ-porphodimethene **31a**. It is suggested that both the α,γ- (**31b**) and the β,δ-isomer (**31a**) were formed, but that only the latter one is stablized for steric reasons.[56] The formation of two isomeric porphodimethenes from Zn phylloerythrin methyl ester (**25**) as well is suggested from ^{1}H/^{2}H exchange experiments and from the careful analysis of the reoxidation products.[31b] In the β,δ-porphodimethene, the isocyclic ring is considerably stressed. This is manifested by the isolation of the rhodo-chlorin **37b** among the products in about 5% yield, which results from rupture of the C-γ to C-10 bond instead of a C–H bond.

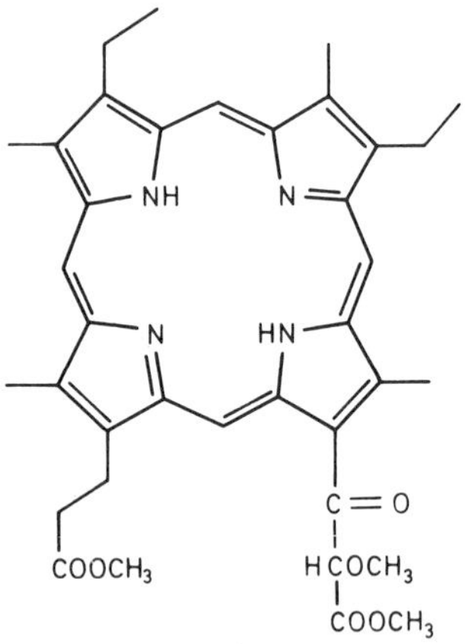

37b

The photoreduction of Zn-etio has been investigated at low temperature.[27] At first a phlorin anion (= porphodimethene anion **3**) is formed, which upon

---

* Nothing is known as yet on the stereochemistry of the products obtained by reduction of Sn(IV) (OEP).[23,24]

warming is protonated to the porphodimethene. Although the products are only characterized by their uv-vis spectra, the characteristic absorptions of the products and the striking analogy to the reductive protonation[26,28] make these assignments of the structure tempting. The most interesting finding is the light-induced deprotonation of the porphodimethene back to the anion, which is again thermally reprotonated. This process stabilizes the anion in the light, allowing it to thus react separately from the dimethene.

3. METALLOCHLORINS TO METALLOCHLORINPORPHODIMETHENES

The photoreduction of metallochlorins is principally very similar to that of metalloporphyrins. The hydrogenation of the 7,8-double bond is, with respect to the photochemistry, probably only a minor perturbation. In its original work,[65] Krasnovskii reports the formation of a pink product ($\lambda_{max} = 523$, 411 nm) upon illumination of a pyridine solution of Chl a in the presence of ascorbic acid. The assignment of both absorption bands to one product is probable from recent ${}^1H$ nmr experiments, which showed formation of a single product only ($\geq 90\%$).[66] This product is probably formed by the following sequence.[69,87]

(a) Electron transfer from the ascorbate anion[70] and formation of a short-lived ($< 10^{-2}$ sec) Chl-radical anion* with $\lambda_{max} = 475$ and 745 nm.[73,93-97]

(b) Proton transfer from the protonated base[70] and formation of a monohydrochlorophyll with $\lambda_{max} = 480$ nm.†[93,98,98a,99]

(c) Repetition of processes (a) and (b)[100] to form the pink product. This latter process is favored by kinetic data,[100] but there are indications of a disproportionation step of the monohydrochlorophyll[94] similar to the one observed with phlorins.[101] Even in the absence of oxygen, this reduction is reversible in the dark, and the latter product could not yet be isolated from the reaction product in pure form. The proposed β,δ-dihydrochlorophyll structure[88] (34 = 7,8-chlorin-α,γ-porphodimethene) was proved, however, by carrying out the reduction directly with H_2S in the nmr tube.[66]

(d) The reduction requires water or other proton donors.[72c,73,102]

(e) Due to the lability of magnesium in the dihydroproduct, pheophytinization is a frequent side reaction,[102-104] and, in fact, photopheophytinization is believed to always involve intermediate photoreduction.[87,105]

---

* A direct observation of the anion radical of the pigments[88] during the photoreduction was not yet possible, while signals of the reducing agents, ascorbic acid,[89] and hydroquinone[90] are well studied.[91,92]

† In analogy to the intermediates observed during the reductions with aromatic radical anions,[26,28] and the photochemical reduction of porphyrins,[27] a phlorin–anion structure 3 is probable for the intermediate with $\lambda_{max} = 475$ and 745 nm in the Krasnovskii reduction as well.

4. PORPHYRINS TO PHLORINS AND PORPHOMETHENES

The reduction sequence porphyrin (1) → phlorin (5) → porphomethene (10) has been observed under various reaction conditions and over a wide pH range from neutral[14] to highly acidic.[71,106] The reaction gives usually high spectroscopic yields, and the only major side reaction is formation of iso-bacteriochlorins.[106] Porphomethenes can be isolated in pure form and were studied directly.[71]

The phlorin structure for the first long-lived intermediate in the photo-reduction of porphyrin-free bases was first established by Mauzerall.[14] This conclusion was drawn from redox titrations, and from the characteristic[107] pH-dependent spectral changes of the formed dihydro product. Recent $^1H/^2H$ exchange experiments[75,76] were interpreted on the same basis, involving an intermediate formation of a phlorin. For the primary reduction product of the monoazaporphyrin **38a**,[108] the γ-phlorin structure **38b** has been

38a   38b

proposed.[76] By kinetic analysis, it has been demonstrated recently[108a] that the phlorin radical formed as the initial product of TPP photoreduction undergoes a light-dependent dimerization. The dimer is suggested as the immediate precursor of TPC (and TPP), which is obtained in the TPP series instead of the phlorin. The preparation and chemistry of phlorins is discussed in detail in Chapters 3 and 4.

38c   38d

Upon longer reaction, porphyrin-free bases are reduced beyond the phlorin stage to porphomethenes. The structure was again established first by a combination of redox and acid–base titrations,[14] and recently supported by isotope exchange experiments[75,76] and by direct ^1H nmr and ir measurements.[71] The reduction of the monoazaporphyrin 38a yields two isomeric porphomethenes[108] for which the α-methene (38c, $\lambda_{max} = 545$ nm) and the β-methene structure (38d, $\lambda_{max} = 510$ nm), respectively, have been proposed on the basis of $^1H/^2H$ exchange studies.[76]

5. PHOTOREDUCTION OF CHLORINS

The photoreduction of chlorin-free bases gives rise to two principal products. In a comparative study of various chlorins, these primary products, as well as some of their reaction products, were characterized by their absorption spectra.[109] Chlorins without an additional isocyclic five-membered ring yield one product only. It shows an absorption at 525 nm in neutral solution and at about 610 nm in acidic medium. Upon admission of air, products with bacteriochlorin-type absorptions are obtained. The product obtained from chlorin-e_6 trimethyl ester 20 could recently be isolated and characterized by ^1H nmr as the 7,8-chlorin-γ-phlorin 39.[78] This isomer is

39

probably again stabilized against oxidation for the same steric reasons discussed for the porphodimethenes. It is noteworthy, however, that electrochemical reduction of the same chlorin 20 yields solely an isomer of 39, which is reduced at the β-position (see Sections II, C and III) and which was shown to be stabilized for steric reasons, too.[110] In the case of pheophorbides (namely, chlorins containing an isocyclic five-membered ring), the chlorin-phlorin is no longer stabilized and could only be observed spectroscopically. In addition to it, a second, yet uncharacterized, product is formed with a broad absorption peaking at 580–600 nm.[109] The reduction of chlorins in acidic media was studied spectrophotometrically for a variety of educts.[72a,74] Among other structures, the formation of isomeric chlorin porphodimethene dications (chlorinphlorin dications) was discussed. The isomerization of reduced chlorins to porphomethenes (9 → 10) leads to porphyrins in the reoxidized reaction mixture.[14]

6. FORMATION OF BACTERIOCHLORINS

Besides the chlorins, bacteriochlorins are the second major group of naturally occurring hydroporphyrins. The last steps in bacteriochlorophyll biosynthesis are not yet well understood, but there is no evidence that light-induced reactions are involved. Bacteriochlorin type pigments are observed, however, as by-products in many *in vitro* photoreductions, and in at least one case, the main product belongs to this class of compounds.

Bacteriochlorins, strictly speaking, have two peripheral double bonds reduced at opposite pyrrole rings (13). The parent compounds of this class are the bacteriochlorophylls a[111] and b.[41] The isomeric isobacteriochlorins ("cis"-bacteriochlorins, 12) are often encountered as synthetic products, and recently this chromophore has been found in the sirohemes,[1b] which are present in sulfate and nitrate reductases.[1a] The identification of bacteriochlorins in complex reaction mixtures is facilitated by their characteristic long wavelength absorption, while isobacteriochlorin pigments ($\lambda_{max} \approx$ 620 nm) are likely to be obscured by other bands.

Reduction products with absorption spectra of true bacteriochlorins were first observed by Byteva et al.[112] and by Krasnovskii and Voinovskaya.[113] Similar by-products of photoreductions were occasionally observed by other investigators,[79] but to our knowledge none of them has been analyzed for their molecular structure. The formation of bacteriochlorin pigments from several Mg-containing chlorophyll derivatives was studied in some detail.[114] Photoreduction in the presence of air leads to the subsequent formation of two pigments of this type. The results indicate that both of the pigments are formed indirectly by reoxidation of colorless intermediates at a lower oxidation level.

The formation of isobacteriochlorins seems to be favored considerably with respect to the isomeric bacteriochlorins. Products of this type have been consistently reported for the photoreduction of porphyrins[1,23,25,115–117] (for the spectral properties, see Seely and Inhoffen et al.[115,117]).

The most carefully investigated reduction to isobacteriochlorins is the formation of hypochlorophyll a (40) from chlorophyll a 24.[115] The reduction proceeds in high yield, and the structure 40 for hypochlorophyll was suggested mainly on the basis of absorption spectra and redox titrations. However,

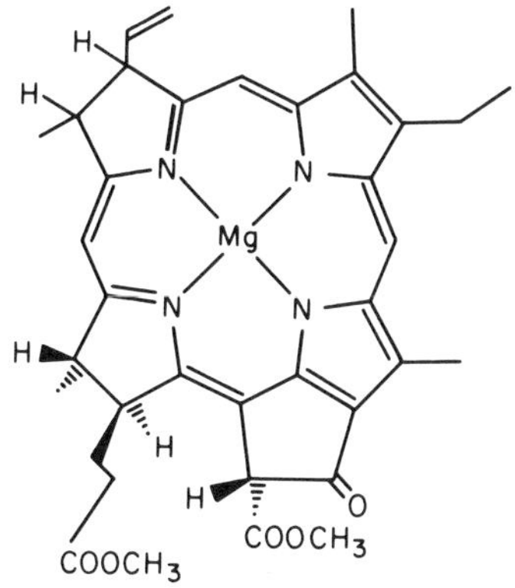

40

proof of the structure by more detailed methods like ^{1}H nmr is desirable. Of special interest would be the stereochemistry of the additional hydrogens at ring A, as 7,8-*cis*-chlorins are formed under similar conditions from metalloporphyrins.[31a,82] Chlorophylls without the 2-vinyl group do not undergo the hypochlorophyll reduction, and attempts to obtain the free base hypochlorin by demetalation were unsuccessful.[115]

7. FORMATION OF HYPOBACTERIOCHLORINS

The formation of hexahydro compounds with the hypobacteriochlorin chromophore (14) was suggested by Seely and Calvin.[1] In the reduction of Zn-TPP with photoactivated benzoin, formation of a product with a chlorin-type absorption ($\lambda_{max} = 642, 425$ nm) was observed. It forms upon prolonged irradiation of the isobacteriochlorin (12), and can be reoxidized in high yield to the latter. The structure proposed was based on the absorption spectrum,

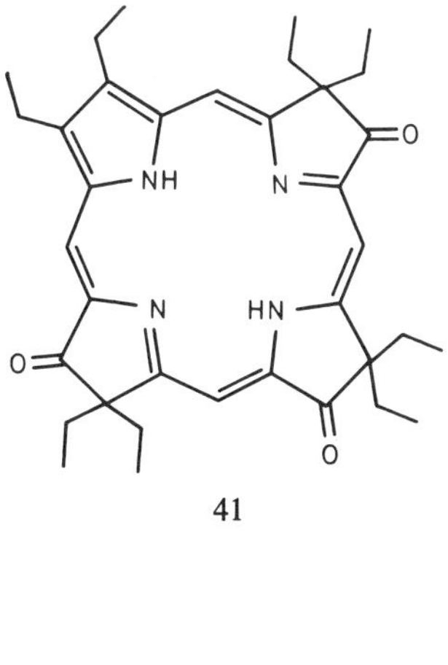

41

and this assignment is supported by the chlorin-type spectra observed for geminiporphin triketones, namely, **41**.[118] Similar products are encountered in other studies on the photoreduction of metalloporphyrins, but their instability has precluded thus far a more detailed investigation.

C. Electrochemical Reduction

In the polarographic reduction of porphyrins, three distinct steps have been well established, and there is possibly a fourth one close to the dissociation voltage of the solvent. The first step yields, in a one-electron reduction in aprotic solvents, the radical anion.[29,48,49,119,120,121] In protic solvents, this electron transfer is thought to be followed rapidly by a proton transfer, thus yielding a neutral radical.[48,49] This hypothetical intermediate is unstable, however, and the first detectable reduction product is a phlorin.* Thus, the first polarographic step in protic solvents is a two-electron reduction.[122-124] It has been shown that porphyrin-free radicals generated photochemically disproportionate rapidly into porphyrin and phlorin.[101] A similar disproportion close to the electrode would account for the polarographic result. A subsequent polarographic step has been characterized in protic solvents,[124] but the structure of the product is not known. The anion radicals of chlorophylls and pheophytins have been obtained by electrochemical reduction of the parent pigments.[88,120,121]

Phlorins can be produced electrochemically in good yield by reduction at a controlled potential.[48,49] They are easily reoxidized to the parent porphyrins, but their structure has been established[48,49] from their characteristic[10,14] uv-vis absorption spectrum, from coulometric studies, and from $^1H/^2H$ exchange experiments. In a similar reaction, chlorinphlorins are produced from chlorins. The required potential is about 100 mV less negative† for the chlorin ($E_{1/2} \approx -0.5$ V) than for the porphyrins ($E_{1/2} \approx \leq 0.6$ V for etiotype porphyrins), but it varies considerably with the peripheral substitution.[110] Stability and structure of the chlorinphlorins show large variations, too, depending on the structure of the chlorin educt.[110] Sterically nonhindered chlorins without an isocyclic ring form a mixture of isomeric chlorinphlorins with a half-life on the order of minutes to hours, while sterically hindered chlorins like chlorin-e_6 trimethyl ester (**20**) form exclusively 7,8-chlorin-β-phlorins (namely, **42**), which are stable for days. Selectivity and stability in the latter case can be rationalized by the model of the overcrowded periphery.[14,107] However, 7,8-chlorin-γ-phlorins are produced photochemically with a similar degree of selectivity, and their similar high stability has been accounted for by the same model.[78] The choice between the two attacks leading to products of comparable

---

* For the reactivity of phlorins, as studied by electrochemical methods, see Chapter 2.
† Polarographic half-wave potentials in methanolic HCl buffered with aniline.

42

stability, namely, at C-β and C-γ, are probably governed by the electronic structure of the anion radical. In the electrochemical reduction, this primary product is formed in the ground state, in the photoreduction it is produced in the excited state. In the chlorins, the relative electron densities vary considerably between the ground[125-128] and first excited singlet and triplet states,[127,128] but no comparable calculations have been made for the anion.

Very short-lived chlorin phlorins are produced from pheophorbides,[110] namely, chlorins with an isocyclic ring between C-6 and C-γ. [1]H/^{2}H exchange experiments indicate the reduction at the position corresponding to the middle methine ^{1}H nmr signal,[110] i.e., the α-position.[129,129a]

III. STEREOCHEMISTRY

To a large extent, stereochemistry of hydroporphyrins has been the stereochemistry of chlorophylls and related structures (see also Chapter 9). The majority of stereochemical information has been drawn from degradation experiments (Chapter 9) and spectroscopic studies, especially from nmr[3] and ORD/CD[130] spectroscopy. Compared to the well over 100 X-ray analyses of true porphyrins (for recent reviews, see Hoard[131] and Fleischer,[132] and Volume III, Chapters 10 and 11), only few crystal data are available for the (less readily crystalizing) chlorins.[43,133-136] The stereochemistry of porphodimethenes has been studied by Buchler et al.,[28,137] and some attention has been focused on chlorinphlorins.[78,110] Conformation studies of β-pyrrolic substituents in solution have been reviewed recently.[3] For crystal structural studies, the reader is referred to Volume III, Chapters 10 and 11 and Hoppe et al.,[43] Hoard,[131] Fleischer,[132] Fisher,[133,134] Strouse,[135] and Kratky and Dunitz.[135a]

A. The Macrocycle

The macrocycle in the fully unsaturated porphyrins has been shown to be rather flexible. Planar structures are an exception, and the macrocycle shows generally regular deviations from planarity which are described, for example, as domed, or ruffled, and by which any steric strain is evenly distributed over the entire macrocycle.[28,131,132] The flexibility of the macrocycle is further supported by markedly different structures for a single compound in different crystal forms,[138,139] by large thermal out-of-plane motions of the atoms of the ring system[132,139] and by pronounced distortions upon local steric perturbations (for leading references, see Scheer and Katz,[3] and Strouse[135]). In contrast to the rather flexible macrocycle as a whole, however, the individual pyrrolic subunits in all porphyrins are nearly planar, thus underlining Woodward's concept of the aromatic subunits (see below).[15,107] Even in porphyrins with an "overcrowded periphery,"[15,107] the pyrrolic subunits remain essentially planar in spite of considerable local distortions.[140] This concept was the key in understanding the selective formation of 7,8-chlorins and γ-phlorins in the course of Woodwards' chlorophyll synthesis (see Section II, A, 1).[15] In the former ones, the β-pyrrole positions C-7 and C-8 are sp^3 hybridized, in the latter ones the bridging C-γ. In either case, the substituents come out of plane, and the quaternized carbon atom becomes more flexible, thus relieving the steric strain. Recent examples to underline the same principle include the stabilization of *meso*-substituted porphodimethenes[28,52,56] and chlorinphlorins.[78,110]

In the chlorins, the macrocycle is perturbed by the hydrogenation of one of the peripheral double bonds. By this means, ring D is no longer an aromatic subunit. Thus, pronounced deviations from planarity are observed in this region, not only for the substituents, but also for ring D itself. Thus far, the main interest in chlorin stereochemistry has focused on chlorins closely related to chlorophylls. A series of related metal-free chlorins[43,133,134] and more recently on their Mg complexes[135] have been investigated by X-ray analysis.

In the Fischer notation (cf. compound **28**) the chlorophylls are reduced at C-7 and C-8 (C-17 and C-18), but they also have an additional substituent at C-γ (C-15) (IUPAC notation in brackets). In all these compounds, the steric interaction of the γ-substituent with the β-pyrrole substituents at the neighboring C-6 and C-7 is, therefore, an additional strong perturbation, which impedes inferences on the steric consequences of the reduced double bond. Substitution at C-γ leads to a pronounced dissymmetry in the region of the γ-7a bond. The γ-substituent is forced out of plane to one side, the 7-carbon to the opposite side of the macrocycle (Fig. 1). As all natural chlorins are chiral, the twist of the γ-7a double bond is chiral, too. Due to the

C-8   C-7   C-10

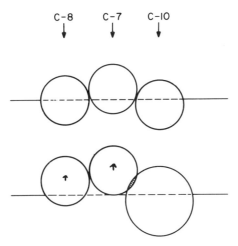

Fig. 1. Steric hindrance of a bulky substituent at C-γ with ring D substituents in chlorins. Schematic drawing of a chlorophyllide (cf. **37**), and of a strongly hindered chlorin like the peripheral metal complex **43**. For numbering see formula **28**.

bulky propion ester side chain situated "above" the plane (7S configuration),[141–143] C-7 is pushed "up," too, while C-γ comes below the plane.[133] The twist of the 7a-γ bond and of the adjacent bonds render the macrocycle inherently dissymmetric (Fig. 2). This is indicated from the chiroptic properties of pheophorbides,[130] and reflected in the X-ray analysis.[43,133–135] The basic features of the ORD/CD spectra are determined by the chirality of C-7, while substituents and chirality of the asymmetric C-8, 9, and 10 yield incremental contributions superimposed on the spectrum of the inherently dissymmetric chromophore.[130]

The increased flexibility of the reduced ring D was first demonstrated by ¹H nmr studies of δ-substituted chlorins.[129a,143a,144] Introduction of a bulky substituent into this position in pheophorbides leads to pronounced incremental chemical shifts of proton resonances arising from ring D substituents,

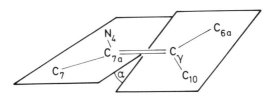

Fig. 2. Twist of the C-γ, C-17 double bond due to the steric hindrance of the "substituents" C-10 and C-7. Schematic drawing of a partial structure; for numbering see formula **28**.

which have been interpreted as arising from a conformational change of ring D. Similar induced conformational changes have recently been observed as the result of increased steric interaction with the γ-substituents in peripheral metal complexes of porphyrins.[145] Stability differences in these enolic structures seem to be mainly due to a varying degree of steric hindrance. The complex **43** of methylpheophorbide *a* is stable in the absence of strongly

43

competing ligands for magnesium.[145] Under the conditions where equilibrium amounts of the free ligand methylpheophorbide *a* are below 2%, the strongly hindered pheoporphyrin complex **44** exists only in a 1:4 equilibrium with the free ligand.[124] On the other hand, the s-*cis* β-diketone **45** is enolized even in the absence of chelating metal ions.[188] Spectroscopic differences observed between the natural 7,8-*transoid* chlorins and their cisoid isomers are a further

44   45

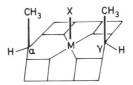

Fig. 3. Stereochemistry of α,γ-dimethyloctaethyl-β,δ-porphodimethenes and their metal complexes. Schematic drawing according to Buchler *et al.*[137]

indication of the enhanced flexibility of ring D.[31a,130] [1]H nmr signals originating from protons in ring D may show pronounced differences, while other signals are shifted uniformly and to a lesser degree.[31a] All available X-ray data support this interpretation of a more flexible ring D. They show pronounced deviations from planarity in ring D, allowing a local relief of steric strain.[43,133–135] In contrast, ring D is planar in the corresponding pheophorphyrins, and the steric strain is dissipated over the entire macrocyclic system.[146]

The sterochemistry of dimethylporphodimethenes has been investigated in some detail by Buchler *et al.*[28,52,147] α,γ-Dimethyl-β,δ-porphodimethenes are stabilized against reoxidation (rearomatization) by steric interaction of the *meso*-methyl groups with the β-pyrrole substituents in the parent porphyrins. This is yet another example of Woodward's thesis of the "overcrowded periphery,"[15] which was invoked from ^{1}H nmr data,[28,52] from a comparative study between two isomeric dimethylporphodimethenes,[56] and which has recently been proved by X-ray analysis.[136] The macrocycle of the two porphodimethenes of type **30b** (M = Ni, TiO, respectively) is folded like a gabled roof along the two reduced bridging C atoms. The two methyl groups are cisoid to each other and in "chimney"[137] (exo) positions, and the central metal with the fifth ligand occupies a chimney position as well.[136] The details of this structure are then mainly determined by the size and preferential coordination of the central metal[28] (Fig. 3).

Some aspects of the stereochemistry of chlorin phlorins were inferred from the selective formation of chlorin-β-phlorins (namely, **41**) during electrolytic reduction of chlorin-e_6 and related structures. Space-filling models show that, by this means, the steric relief at the periphery is probably even larger than by sp^3 hybridization of C-γ.[110] However, a respective γ-phlorin (**39**) has been obtained with a similar degree of selectivity by a photochemical reduction.[78] The steric concept is supported by the formation of isomeric chlorinphlorin mixtures upon reduction of chlorins lacking the 6- or γ-substituent.

B. Metallochlorins

Insertion of a metal into porphyrins gives rise to two additional stereochemical aspects; the metal ion can be out of the plane of the macrocycle,

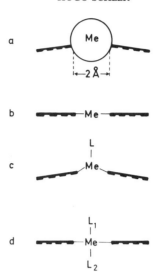

Fig. 4. Stereochemistry of metallochlorins. Metal ions with a diameter considerably larger than 2.01 Å, which do not fit into the central 4-N cavity (a), metal ions with a diameter ≤ 2.01 Å and a coordination number of four (b), five (c), and six (d). In porphyrins without σ_h plane (cf. in all chlorophylls), isomers are possible in the cases a and c, and in case d with different axial ligands, depending on the orientation of the metal with respect to the two faces of the macrocycle.

and it can be ligated in various distinct ways. In chlorins (and other hydropor-phyrins), the possible nonequivalency of the two sides of the macrocycle gives rise to a third problem (see Fig. 4). If the metal is out of plane, or if it is ligated with two different extra ligands, isomeric structures will be formed depending upon their orientation with respect to the two nonequivalent sides of the macrocycle. In the crystal structure of both chlorophyll a and b, the central Mg is five coordinate and displaced "below" the macrocycle if written in the normal way (i.e., on the same side as the 7-H, the 8-CH_3 and the 10-$COOCH_3$ groups).[135] No example of this kind of isomerism is known as yet in solution, although it may play a key role in chlorophyll aggregation[148] (see Volume V, Chapter 9).

Metals with an ionic radius not significantly larger than 2.01 Å fit principally without distortions into the central cavity of the porphyrin macrocycle.[131] All metals commonly encountered in chlorins, especially Mg being present in all chlorophylls, belong to this class. In this case, the type of ligation essentially determines whether the metal is in plane or out of plane (Fig. 4). In a symmetric ligand field (no or two similar extra ligands), the metal is in plane. With only one extra ligand, or with two different ligands, the metal is usually positioned out of plane. For five coordinate Mg chlorins with pyridine as the axial ligand, a displacement of 0.7–0.8 Å from the mean plane of the macrocycle has been estimated by ^{1}H nmr.[149] X-Ray analysis of methylchlorophyllides *a* and *b* yield similar structures that have the Mg about 0.45 Å out of the mean plane.[135] In addition, they show a regular, bowl-shaped deformation of the macrocycle, with the four nonbonding N orbitals directed toward the metal. This type of macrocyclic distortion (Fig. 4) is commonly encountered in metalloporphyrins with the metal situated out of plane.

The coordination type of Mg chlorins was first investigated by Miller and Dorough[150] and in some detail by Fried and Sancier.[151] The equilibrium between five and six coordination is temperature dependent.[151] Katz *et al.*[152] derived from ^{1}H nmr data equilibrium constants for the formation of the monosolvate of chlorophyll *a* with alcohols, which are about 100 times larger than for the disolvate. The preferential five coordination of chlorophylls is also manifested in chlorophyll aggregates, which arise from ligation of the central Mg of one molecule, preferentially with the peripheral 9-carbonyl group in another molecule[3,148] (for details, see Volume V, Chapter 9). Recently, a shift of the orange band in the uv-vis spectrum of bacteriochlorophyll *a* from 580 to 610 nm could be related to the change from coordination number five to six,[153] and related spectral shifts have been observed as well for chlorophylls *a* and *b*.[153]

C. Asymmetric Centers

In many cases, the formation of hydroporphyrins is accompanied by the formation of asymmetric C atoms. Research in this field was focused again predominantly on chlorophylls, which can have asymmetric C atoms at C-2a, C-3, C-4, C-7, C-8, and C-10* (Fig. 5). In addition, a variety of modified

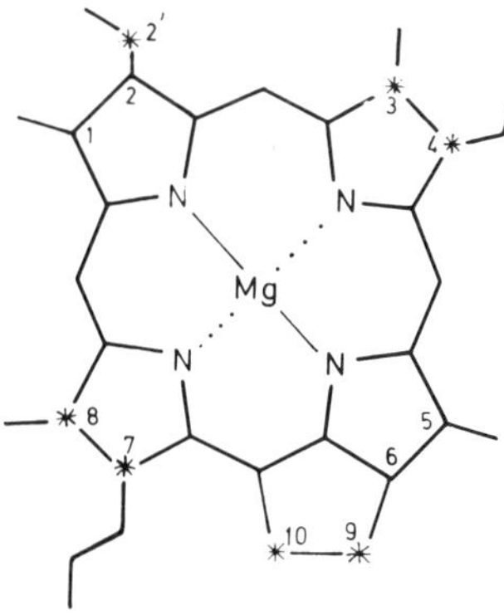

Fig. 5. Location of asymmetric C-atoms (*) studied in chlorophylls and their derivatives.

* Phytol has two additional asymmetric centers at C-P7 and C-P11.[154]

or synthetic chlorins with nonnatural configurations or additional asymmetric C atoms have been investigated in the past decade.[3,130]

C-7, C-8. Chlorophylls of the chlorin series (chlorophylls *a*, *b*, and *d*) contain two asymmetric centers at C-7 and C-8, and usually* a third one at C-10. All natural chlorins, as well as those obtained under equilibrating conditions, have the thermodynamically favored[15,19,157,158] 7,8-trans configuration.† The (relative) trans configuration of C-7 and C-8 was first suggested from racemization experiment.[157,158] It was supported by the outcome of the synthesis of Chl *a*, in which the natural isomer (as racemic mixture) was formed under equilibrating conditions,[15] by the small ¹H nmr coupling constant of the "extra" hydrogens at C-7 and C-8,[159] and by the induced conformational changes of ring D upon substitution introduction of bulky γ-[145] or δ-substituents.[143,143a] The chemical proof of the 7,8-transoid configuration was achieved by chromic acid degradation[141] to trans-disubstituted succinic acid derivatives. The absolute configuration at C-7 and C-8 has been determined by the same oxidative degradation method. By correlation with hematinic acid derived from (−)-α-santonine,[142] and with hydrocarbons of known absolute configuration,[143] the chlorin C atoms were shown to be 7*S*, 8*S* configurated. Recently, the same conclusion has been drawn[160] by applying a modification of Horeau's[161] method on 9-hydroxypheophorbides of the a series.[162,163]

7,8-*cis*-Chlorins are obtained from the corresponding (metallo) porphyrins by chemical reduction with typical cis-reducing agents as well as by photochemical hydrogenation (Section II, B, 1). *cis*-Chlorins are thermodynamically less stable than the corresponding trans isomers. This is due to steric hindrance of the cisoid substituents and is manifested by a bathochromic shift of the uv-vis spectrum,[17,19,20,31-31b] a more pronounced dehydrogenation in the mass spectrometer,[20,31-31b] and by a more ready oxidation with quinones.[17,20] The ¹H nmr spectra of *cis*-chlorins show generally a slight reduction of the aromatic ring current as compared to the trans isomers, but the spectral differences are mostly dominated by neighboring group effects in the vicinity of the reduced pyrroline ring.[31a] The ORD/CD spectra of pheophorbides are comparably insensitive to the relative configuration of C-7 and C-8, but the sign of the major π–π* transitions is determined by the absolute configuration at C-7. This striking feature has been interpreted in terms of an inherently dissymmetric chromophore[130] (see above).

---

* The bacteriochlorophylls *c*, *d*, and *e* have an asymmetric C-2a instead.[155,156]

† In the following the shorter but less precise terms "cis" and "trans" are used instead of "cisoid" and "transoid," respectively.

C-9, C-10.   The trans configuration of C-7 and C-10 was deduced from a combined ¹H nmr and ORD/CD investigation of C-10 mono- and disubstituted methylpheophorbides.[164,165] From systematic variations of the C-10 substituents in pheophorbides of the a[164] and b series,[165] characteristic increments of the neighboring proton chemical shifts and the ORD/CD amplitudes have been derived. Both methods can be used independently and were recently reviewed.[3,130]

The same methods that have been used to link the stereochemistry at C-10 with that at C-7 and C-8 can be used to link the stereochemistry of C-9 with that at C-10. In the chlorophylls, C-9 is sp^2 hybridized and, thus, achiral. Reduction of the 9-CO group to a CHOH group yields C-9 stereoisomers, however, which have been studied in some detail.[160,162,163] The increments of the 9-OH group to the ORD/CD spectra is much less pronounced than that of C-10 substituents, but the configuration at C-9 can be deduced by ¹H nmr and ir spectroscopy.[162,163] In all the C-9 alcohols, H-bonds are observed between the 9-OH and the C-10 carbomethoxy, the C-10 alkoxy, or the C-7c propionester group, respectively (Fig. 6). Identification of the bridgehead substituents can then be used as additional aid for the configuration assignment, besides the chemical shift argument.

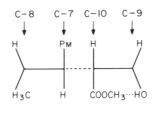

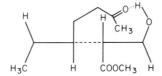

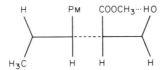

Fig. 6. Hydrogen bonds between substituents at C-7, C-9, and C-10 in 9-desoxo-9-hydroxypheophorbides. Schematic drawing of the periphery, viewed parallel to the macrocyclic plane, and parallel to the axis through C-α and C-γ.

D. Configurational Stability

The formulation of possible chlorophyll structures containing asymmetric C atoms[166] focused some interest on their optical activity. In the first study, Stoll and Wiedemann[167] did demonstrate chlorophylls to be optically active, but they also noticed an extremely easy racemization of their samples. The subsequent investigations of Fischer and Stern[168] did support the optical activity, but they found it to persist over extended times and during extensive transformations of the molecule. From the compounds studied, it became clear that the optical activity was due to C-7 and C-8 and that their configuration was stable against refluxing in mineral acidic solutions (transesterification of the propionic acid), heat (pyrolysis to pyropheophorbides), and (in some cases) reduction to porphinogens. However, epimerization is possible under alkaline conditions. In strongly basic solutions, all benzylic protons are exchangeable,[12a] which leads in the case of the "chlorin hydrogens" to configuration equilibration.[157] Only *trans*-chlorins are formed under these equilibration conditions. The loosening of the benzylic "chlorin hydrogens" with strong bases is further demonstrated by the chlorin to phlorin isomerization at high pH.[169]

The fate of the asymmetric C-7 and C-8 upon reduction to the porphinogens, and their subsequent reoxidation, is still not completely understood. Most leuco compounds investigated by Fischer[168] were found to be inactive, although optical activity was clearly demonstrated in at least one case (leucomesopurpurin-7-trimethyl ester). With the early analytical equipment, no decision could be made as to whether the optical activity was retained in the mesopheophorbides obtained by catalytic hydrogenation and subsequent reoxidation of pheophorbides.[170] Recently, the configurational stability of C-7 and C-8 has been demonstrated for mesopheophorbides obtained by this procedure (as well as for pheophorbides isolated as by-products from the HI-isomerization).[31a,31b,82] The results obviously prove the presence of "chlorinogens" with asymmetric C-7 and C-8 as intermediates. It is not clear, however, if they convert to porphinogens, and both are oxidized separately to the respective chlorins and porphyrins, or whether a common "chlorinogen" is oxidized to a different degree. The results also indicate that re-reduction of the porphyrin during reoxidation is not important in this system. Reactions of this type have been demonstrated[8] to account for the formation of tetraphenylchlorin during the Rothemund reaction.[32]

C-10. When chlorophyll *a* is chromatographed, it is always accompanied by a minor band of a pigment that has very similar spectral properties and which was, therefore, termed Chl *a'*.[171] Both Chl *a* and *a'* are rapidly interconvertible to an equilibrium mixture of about 7:1, and they have been identified by ^{1}H nmr as the 10(S) and 10(R) epimers,[172]* respectively. Similar configurationally labile epimers have been found for other chlorophylls and pheophorbides containing an enolizable β-ketoester system, and even in cases where a chromatographic separation is not possible, satellite lines in the nmr spectra indicate the presence of prime pigments.[165,172] In most chlorophylls (except for the bacteriochlorophylls c,d, and e[155,156]) C-10 is asymmetric and part of an enolizable β-ketoester system. In neutral solutions, the equilibrium is almost entirely on the side of the diketo form. Even under these conditions, however, C-10 epimerizes at room temperature within hours, and the equilibration has been followed by ^{1}H nmr.[172] The epimerization is greatly enhanced in the presence of base, by which means the equilibrium concentrations can be changed as well. In view of the easy epimerization it is surprising, however, that the C-10 epimers of chlorophylls can be separated by chromatography on powdered sugar, and the reepimerization followed by ^{1}H nmr. The isolation of an optically active protochlorophyll a (33) has been reported by Sauer et al.[175] In 33, C-10 is the only asymmetric center, thus the racemization must be slow under the workup conditions. This indicates a considerably enhanced stability of C-10 in 33 (a porphyrin) as compared to 28 (a chlorin), which is probably due to the steric destabilization of the enol intermediate. This conclusion is supported by a similar destabilization of the peripheral complexes of protopheophytin a (44)[174] (see above).

The configuration at C-10 in pheophorbides can be stabilized by removal of either one of the carbonyl groups, or by substitution of the enolizable proton at C-10.[130] 10-Alkoxymethylpheophorbides are configurationally stable under most conditions. They were, therefore, investigated in some detail to evaluate the chiroptical and ^{1}H nmr properties of the pure C-10 epimers.[163–165] In strongly acidic solution the configuration is unstable, however, and, from isotope labeling experiments, an S_N2 type inversion has been suggested[163] for the acidic alcoholysis. 10-Alkoxypyropheophorbides, in which the 10-carbomethoxy group is removed, can be oxidized to porphyrins with retention of the C-10 configuration.[31a] Epimerization in the presence of base is probable, however, because the 10-protons in pyromethylpheophorbide a are easily exchangeable in refluxing pyridine.[39,176] Upon reduction of the 9-carbonyl group, the configuration at C-10 is stabilized toward bases, but is again unstable in strongly acidic media.[162,163]

---

* Recently, an enolic chelate structure was proposed for the primed pigments on the basis of uv-vis spectra.[173] This structure seems unlikely both on the basis on the nmr evidence for the 10-epimer formulation, and in view of the dramatically different spectra of chlorophyll enolates[10] and peripheral Mg chelates.[145,174]

The C-9 configuration of 9-hydroxypheophorbides is stable against acid, but labile against base.[162,163] Under the (basic) reaction conditions of the NaBH$_4$ reduction of the 9-CO group, epimerization at C-9 of the kinetically to the thermodynamically favored epimers has been observed. During this reduction, pheophorbides of unnatural C-10 configuration are formed from methylpheophorbide a. This is a clear chemical indication for the presence of minor amounts of the prime pigment in solution, but it is not clear if the observed 10(S):10(R) ratio of 1:11 in the reduction product does represent the equilibrium concentrations.

IV. CONCLUSIONS

Reductions are an important aspect of porphyrin chemistry both for hydroporphyrin synthesis and for the understanding of the reactivity of the porphyrin macrocyclic system. Hydroporphyrins derived from the principal model porphyrins OEP and TPP are accessible in moderate to good yields by selective chemical reduction. Selective hydrogenation of less symmetrically substituted or more sensitive porphyrins remains difficult. Photochemical and electrochemical methods have been suitable for this purpose, but the wealth of compounds available through photoreduction is still only partly charac-terized. A detailed product analysis beyond the chromophoric system (by uv-vis spectroscopy) has been possible recently for some systems by the application or ir and especially of nmr spectroscopy, by which structural and stereo-isomers can be distinguished. The importance of stereochemical factors, and to some extent of electronic factors, has been emphasized to account for the products, but these rationalizations are often ambiguous and do not yet permit the prediction of products in a reliable way. Further research is desirable to link the selectivity of, and the relations between, the chemical, photochemical, and electrochemical reductions of porphyrins in a more defined model.

The conformational mobility of the porphyrin, and especially of the hydroporphyrin macrocycle, has evolved as a new stereochemical concept. The macrocyclic conformation is determined on one side by the peripheral substitution pattern, the central metal, and the "extra ligands" at the latter, on the other side by the tendency to distribute strain evenly on the π-system. These effects have been studied both in the crystal and in solution. Probably the most exciting example in the past has been the application to the coopera-tivity in hemoglobin–oxygen binding, and there is accumulating evidence that stereochemistry plays an equally fundamental role in the primary events of photosynthesis.

V. APPENDIX: RECENT STUDIES IN
CHLOROPHYLL CHEMISTRY

The last review on chlorophyll chemistry appeared in 1966,[10] and some more recent results have been incorporated in reviews of porphyrin chemistry.[11,12] This appendix is a brief summary on the chemistry of chlorophylls of the past decade. Most of the work has been carried out on the more stable, more readily accessible, but from the biochemical point of view less interesting metal-free pheophorbides. However, the chlorophyllides (= Mg complexes) are now accessible from almost any pheophorbide by the metalation procedures described in Chapter 2, Section I, A, 3. For the numbering scheme of chlorophylls, see structure **28**.

Reactions at C-2.   The 2-vinyl has been degraded to the 2-formyl group with KMnO$_4$.[177] 2-Acetyl chl *a* and related products are obtained by oxidation of Bchl *a* with quinones,[40] or by acid catalized isomerization of Bchl *b*.[41]

Functionalization of 3-CH$_3$ and 4-CH$_2$CH$_3$.   The 3-CH$_3$ and 4-CH$_2$CH$_3$ groups can be functionalized[45a] via β-oxophlorins[110] by the methods described in Chapter 2, Section II, B, 2. Compounds with 4-CHOH-CH$_3$ groups, or functional groups derived thereof, are accessible, too, from Bchl *b*.[41]

7,8-Stereoisomers.   Racemization at C-7,8 occurs under the conditions of the Wolf-Kishner reaction.[157,158] The 7,8-*cis*-pheophorbides are available via oxidation of *trans*-pheophorbides to the respective porphyrins, and photoreduction of the Zn-complexes.[31,31a,82,83]

Oxidation at C-9 and C-10.   Oxidation at C-10 of methylpheophorbides of the *a* and *b* series with quinones like chloranil leads to 10-alkoxymethylpheophorbides, which can be pyrolized to 10-alkoxypyropheophorbides.[164,165] The latter are available directly from pyropheophorbides via oxidation with thallium-trifluoroacetate.[178] Prolonged treatment with the latter reagent leads further to 9,10-dioxo derivatives.[178] C-9 alcohols are oxidized with CrO$_3$ to the 9-CO derivatives.[163] Oxidation at ring E leading to the allomerization products 10-hydroxypheophorbides, δ-lactones, anhydrides and the like have been reinvestigated[9,10,179] and the products characterized.[180,181]

Reduction of Carbonyl Groups.   Selective reduction of the 3-CHO group in derivatives of the Chl *b* series is possible with NaBH$_4$.[182] Prolonged reaction leads to a simultaneous reduction of the 9-carbonyl group.[160,162,163,]

[182,184] With LiAlH$_4$, both the benzylic 3-OH and 9-OH groups are hydrogenolized to the 3-CH$_3$ and 9-CH$_2$ derivatives.[160,183,184] Under these conditions, the 7d-ester (and other C=O groups) are simultaneously reduced to alcohols. Tosylation of the latter, and, subsequently, treatment again with LiAlH$_4$ leads to a 7-propyl substituent.[183,184] 3- and 9-Desoxo compounds are also available from the respective alcohols via catalytic hydrogenolysis,[39,82] or with HI.[185] The stereochemistry of 9-hydroxy derivatives of the *a* series,[162,163] and of polyols derived from pheophorbides of the *a* and *b* series,[160,184] has been studied in some detail[130] (see Sections III, C and D). The aggregation[3,186] and stereochemistry[187] of 2[α-hydroxy]ethylpheophorbides has been investigated (see Chapter 9).

Peripheral Complexes.   Peripheral complexes with Mg or Zn bound to the β-ketoester system of ring E are formed in anhydrous solvents with the anhydrous metal salts.[145] Diketo derivatives like **45** form more stable complexes of the same type.[188] The latter diketones are obtained by intramolecular Dieckmann condensation of the 7-propionic ester with the 10-ester. A Δ-9,10 structure has been suggested, too, for the product obtained by Vilsmeier-Haak formylation of Cu-pyromethylpheophorbide *a*.[189]

Dimerization.   The dimerizations reported so far[190,191] involve condensation via the 7-propionic ester side chains. Transesterification with diols (glycols) leads to half-esters, which can then be condensed with a pheophorbide free acid by catalysis with dicyclohexylcarbodiimide,[190] phosgene,[191] or the like.

Isotope ($^1H/^2H$) Exchange.   The methine protons of pheophorbides next to a reduced ring are exchangeable with acid.[155,192] In chlorophylls, they are already (partially) exchangeable in protic solvents like methanol.[192] The β- and δ-protons are exchangeable via reduction to the Krasnovskii product and reoxidation.[66] The δ, α, and to considerably lesser extent the β-methine protons in pheophorbides not containing the enolizable β-ketoester system are exchanged[12a] during metalation with Mg-perchlorate.[193] All benzylic protons are exchangeable with base in the order[12a] 10-H[192] ≫ 5-CH$_3$[11,39] ≫ 7,8-H[157,158] > 3-CH$_3$ > 1-CH$_3$ ≈ 2-CH$_2$ ≈ 4-CH$_2$. Exchange of other than 10-H$_{(2)}$ is best carried out in the pyropheophorbides due to loss of the 10-COOCH$_3$ group under the reaction conditions.[176] The 5-CH$_3$ protons can be exchanged selectively by first exchanging both the 10-H$_{(2)}$ and 5-CH$_3$ in refluxing pyridine,[11,39] and then reexchanging the former[12a] at the conditions of the Wolff-Kischner reduction.[157,158] No similarly selective methods have been developed for labeling either one of these positions. The latter reaction also leads to racemization at C-7 and C-8 by proton exchange and also to reduction of the 2-vinyl group.[158] In addition, base catalysis leads to the exchange of enolic protons, viz., the 7b-protons in 7c-esters,[12a] or the 2a-protons of 2-acetylchlorins.[178] The facile exchange of the 10-protons even in pyro derivatives[176] is due to the combined effect of enolic and benzylic exchange.

Miscellaneous. Pheophorbides substituted at C-δ (e.g., Cl) are accessible by electrophilic substitution of free base pheophorbides,[125,129,180,189] while metallopheophorbides seem to be less suitable for this purpose.[189] Mono-, di-, and tri-N-methylchlorins have been prepared by methylation with methyl fluorosulfonate and methyl iodide, respectively.[194] The oxidation of pheophorbides to porphyrins, with[180] and without[31a,164] opening of the isocyclic ring, has been studied by several groups. In the porphyrin series, formation of ring E from 6-(β-dicarbonyl)porphyrins,[195,196] as well as the inverse reaction,[31b] have been investigated. Aminolysis of pheophorbides leads to rupture of ring E.[197] The "prime" chlorophylls (e.g., chl *a*′) have been shown to be the C-10 epimers of the respective chlorophylls.[172] Separation of chlorophyll homologues has been achieved by charge-transfer chromatography.[198] For redox reactions of chlorins and bacteriochlorins, see this chapter, and Chapter 2. For the introduction of Mg, see Chapter 2, Section I, A, 3. The nmr spectroscopy of chlorophylls and their derivatives has been reviewed[3] (see also Chapter 9 and Volume V, Chapter 9). The chiroptic properties of chlorophylls and pheophorbides have been reviewed.[130]

ACKNOWLEDGMENTS

Professor H. H. Inhoffen and Dr. J. J. Katz are thanked for their generous support of this work and for helpful discussions.

REFERENCES

1a. J. M. Vega, R. H. Garett, and L. M. Siegel, *J. Biol. Chem.* **250**, 7890 (1975).

1b. M. J. Murphy, L. M. Siegel, H. Kamin, and D. Rosenthal, *J. Biol. Chem.* **248**, 2801 (1973).

1c. R. Poulson, J. Boon, and W. J. Polglase, *Can. J. Chem.* **52**, 21 (1974).

1. G. R. Seely and M. Calvin, *J. Chem. Phys.* **23**, 1068 (1955).

2. P. M. Müller, S. Farooq, B. Hardegger, W. S. Salmond, and A. Eschenmoser, *Angew. Chem.* **85**, 954 (1973); *Angew. Chem., Int. Ed. Engl.* **12**, 914 (1973).

3. H. Scheer and J. J. Katz, *in* "Porphyrins and Metalloporphyrins" (K. M. Smith, ed.), 2nd ed. p. 399. Elsevier, Amsterdam, 1975.

4. A. P. Johnson, P. Wehrli, R. Fletcher, and A. Eschenmoser, *Angew. Chem.* **80**, 622 (1968); *Angew. Chem., Int. Ed. Engl.* **7**, 623 (1968).

5. H. H. Inhoffen and N. Müller, *Tetrahedron Lett.* p. 3209 (1969).

6. B. Franck and C. Wegner, *Angew. Chem.* **87**, 419 (1975); *Angew. Chem., Int. Ed. Engl.* **14**, 423 (1975).

7. A. D. Adler, L. Sklar, F. R. Longo, and J. D. Finarelli, *J. Heterocycl. Chem.* **5**, 669 (1968).
8. D. Dolphin, *J. Heterocycl. Chem.* **7**, 275 (1970).
9. H. Fischer and H. Orth, "Die Chemie des Pyrrols," Vol. II, Parts 1 and 2. Akad. Verlagsges., Leipzig, 1940.
10. G. R. Seely, *in* "The Chlorophylls" (L. P. Vernon and G. R. Seely, eds.), Chapter 3. Academic Press, New York, 1966.
11. H. H. Inhoffen, J. W. Buchler, and P. Jäger, *Fortsch. Chem. Org. Naturst.* **26**, 785 (1968).
12. K. M. Smith, *Q. Rev., Chem. Soc.* **25**, 31 (1971).
12a. H. Scheer, J. R. Norris, and J. J. Katz, *J. Am. Chem. Soc.* (in press).
13. J. B. Wolff and L. Price, *Arch. Biochem. Biophys.* **72**, 293 (1957); K. Shibata, *J. Biochem. (Tokyo)* **44**, 147 (1957); see also L. Bogorad, *in* "Chemistry and Biochemistry of Plant Pigments" (T. W. Goodwin, ed.), p. 29. Academic Press, New York, 1965; and Vol. VI, Chapter 3.
14. D. Mauzerall, *J. Am. Chem. Soc.* **84**, 2437 (1962).
15. R. B. Woodward, *Angew. Chem.* **72**, 651 (1960); *Pure Appl. Chem.* **2**, 283 (1961); R. B. Woodward, W. A. Ayer, J. M. Beaton, F. Bickelhaupt, R. Bonnet, P. Buchschacher, G. L. Closs, H. Dutler, W. Leimgruber, J. Hannah, F. P. Hauck, S. Itô, A. Langemann, E. Le Goff, W. Lwowski, J. Sauer, Z. Valenta, and H. Volz, *J. Am. Chem. Soc.* **82**, 3800 (1960).
15a. G. E. Ficken, R. P. Linstead, E. Stephen, and M. Whalley, *J. Chem. Soc.* p. 3879 (1958).
16. M. Whalley, *Chem. Soc., Spec. Publ.* **3**, 83 (1955).
17. H. H. Inhoffen, J. W. Buchler, and R. Thomas, *Tetrahedron Lett.* p. 1145 (1969).
18. H. Parnemann, Ph. D. Thesis, Technische Hochschule, Brauschweig, West Germany (1964).
19. H. W. Whitlock, Jr., R. Hanauer, M. Y. Oester, and B. K. Bower, *J. Am. Chem. Soc.* **91**, 7485 (1969).
20. R. Thomas, Ph. D. Thesis, Technische Hochschule, Braunschweig, West Germany (1967).
21. H. Fischer and H. Helberger, *Justus Liebigs Ann. Chem.* **471**, 285 (1929); H. Fischer and K. Herrle, *ibid.* **530**, 230 (1937); H. Fischer and F. Balaz; *ibid.* **553**, 166 (1942).
22. W. Schlesinger, A. H. Corwin, and L. J. Sargent, *J. Am. Chem. Soc.* **72**, 2867 (1950).
22a. U. Eisner, *J. Chem. Soc.* p. 3461 (1957).
22b. U. Eisner, A. Lichtarowicz, and R. P. Linstead, *J. Chem. Soc.* p. 733 (1957).
22c. A. H. Corwin and O. D. Collins, *J. Org. Chem.* **27**, 3060 (1962).
23. D. G. Whitten and J. C. N. Yau, *Tetrahedron Lett.* p. 3077 (1969); D. G. Whitten, J. C. N. Yau, and F. A. Carrol, *J. Am. Chem. Soc.* **93**, 2291 (1971).
24. J. H. Fuhrhop, T. Lumbantobing, and J. Ulrich, *Tetrahedron Lett.* p. 3771 (1970).
25. G. R. Seely and K. Talmadge, *Photochem. Photobiol.* **10**, 195 (1964).
26. G. L. Closs and L. E. Closs, *J. Am. Chem. Soc.* **85**, 818 (1963).
27. A. N. Sidorov, *Zh. Strukt. Khim.* **14**, 255 (1973).
28. J. W. Buchler, L. Puppe, K. Rohbock, and H. H. Schneehage, *Ann. N. Y. Acad. Sci.* **206**, 116 (1973).
29. J. H. Fuhrhop, *Struct. Bonding (Berlin)* **18**, 1 (1974).
30. D. Mauzerall and J. H. Fuhrhop, *J. Am. Chem. Soc.* **91**, 4174 (1969); J. H. Fuhrhop, *Z. Naturforsch., Teil B* **25**, 255 (1970).
31. H. Wolf and H. Scheer, *Tetrahedron Lett.* pp. 1111 and 1115 (1972).

31a. H. Wolf and H. Scheer, *Justus Liebigs Ann. Chem.* **1973**, 1710 (1973).

31b. H. Scheer and H. Wolf, *Justus Liebigs Ann. Chem.* **1973**, 1741 (1973).

32. P. Rothemund, *J. Am. Chem. Soc.* **57**, 2010 (1935).

33. N. Datta-Gupta and G. E. Williams, *J. Org. Chem.* **36**, 2019 (1971); G. H. Barnett, M. F. Hudson, and K. M. Smith, *J. Chem. Soc. Perkin Trans. I* p. 1401 (1975); K. Rousseau and D. Dolphin, *Tetrahedron Lett.* p. 4251 (1974).

34. A. Neuberger and J. J. Scott, *Proc. R. Soc. London, Ser. A* **213**, 307 (1952).

35. A. D. Adler, F. R. Longo, J. D. Finarelli, J. Goldmacher, J. Assour, and L. Korsakoff, *J. Org. Chem.* **32**, 476 (1967); J. B. Kim, J. J. Leonard, and F. R. Longo, *J. Am. Chem. Soc.* **94**, 3986 (1972); A. Treibs and N. Häberle, *Justus Liebigs Ann. Chem.* **718**, 183 (1968).

36. G. M. Badger, R. A. Jones, and R. L. Laslett, *Aust. J. Chem.* **17**, 1028 (1964).

37. D. Mauzerall and S. Granick, *J. Biol. Chem.* **232**, 1141 (1958).

38. A. N. Sidorov, *Russ. Chem. Rev. (Engl. Transl.)* **35**, 152 (1966).

39. C. D. Mengler, Ph. D. Thesis, Technische Hochschule, Braunschweig, West Germany (1967).

40. J. R. L. Smith and M. Calvin, *J. Am. Chem. Soc.* **88**, 4500 (1966).

41. H. Scheer, W. A. Svec, B. T. Cope, M. H. Studier, R. G. Scott, and J. J. Katz, *J. Am. Chem. Soc.* **96**, 3714 (1974).

42. A. Treibs and E. Weidemann, *Justus Liebigs Ann. Chem.* **471**, 146 (1929).

43. W. Hoppe, G. Will, J. Gassmann, and H. Weichselgartner, *Z. Kristallogr., Kristallgeom, Kristallphys., Kristallchem.* **128**, 18 (1969).

44. W. Lwowski, *in* "The Chlorophylls" (L. P. Vernon and G. R. Seely, eds.), Chapter 5. Academic Press, New York, 1966.

45. H. Fischer and W. Lautsch, *Justus Liebigs Ann. Chem.* **528**, 265 (1937); H. Fischer and A. Spielgelberger, *ibid.* **515**, 130 (1935); R. Willstätter and A. Stoll, "Untersuchungen über Chlorophyll," p. 188ff. Springer-Verlag, Berlin and New York, 1913.

45a. H. H. Inhoffen, P. Jäger, and R. Mählhop, *Justus Liebigs Ann. Chem.* **749**, 109 (1971).

46. A. V. Shablya and A. N. Terenin, *Opt. Spektrosk.* **9**, 533 (1960).

47. V. E. Kholmogorov and A. V. Shablya, *Opt. Spektrosk.* **17**, 298 (1964).

48. J. W. Dodd and N. S. Hush, *J. Chem. Soc.* p. 4607 (1964).

49. N. S. Hush and J. R. Rowlands, *J. Am. Chem. Soc.* **89**, 2976 (1967).

50. D. W. Clack and N. S. Hush, *J. Am. Chem. Soc.* **87**, 4238 (1965).

51. R. H. Felton and H. Linschitz, *J. Am. Chem. Soc.* **88**, 1113 (1966).

52. J. W. Buchler and L. Puppe, *Justus Liebigs Ann. Chem.* **1974**, 1046 (1974).

53. J. W. Buchler and L. Puppe, *Justus Liebigs Ann. Chem.* **740**, 142 (1970).

54. K. J. Brunings and A. H. Corwin, *J. Am. Chem. Soc.* **64**, 593 (1942).

55. H. Falk and O. Hofer, *Monatsh. Chem.* **106**, 115 (1975).

56. A. M. Shul'ga, G. N. Sinyakov, V. P. Suboch, G. P. Gurinovich, Yu. V. Glazkov, A. G. Zhuravlev, and A. N. Sevchenko, *Dokl. Adad. Nauk SSSR* **207**, 457 (1972).

57. G. N. Sinyakov, V. P. Suboch, A. M. Shul'ga, and G. P. Gurinovich, *Dokl. Adad. Nauk Beloruss. SSR* **17**, 660 (1973).

58. With the exception of the chlorophylls *c*: R. C. Dougherty, H. H. Strain, W. A. Svec, R. A. Uphaus, and J. J. Katz, *J. Am. Chem. Soc.* **92**, 2826 (1970); H. H. Strain, B. T. Cope, G. N. McDonald, W. A. Svec, and J. J. Katz, *Phytochemistry* **10**, 1109 (1971).

59. For a review of the literature until 1966, see L. P. Vernon and B. Ke in "The Chlorophylls" (L. P. Vernon and G. R. Seely, eds.), Chapter 18. Academic Press, New York, 1966.

59a. For a review of the literature until 1966, see G. R. Seely, in "The Chlorophylls" (L. P. Vernon and G. R. Seely, eds.), Chapter 17. Academic Press, New York, 1966.

60. J. R. Norris, H. Scheer, and J. J. Katz, *Ann. N. Y. Acad. Sci.* **244**, 260 (1975); G. Feher, A. J. Hoff, R. A. Isaacson, and L. C. Ackerson, *ibid.*, p. 239; J. J. Katz and J. R. Norris, *Curr. Top. Bioenerg.* **5**, 41 (1973); W. W. Parson and R. J. Cogdell, *Biochim, Biophys. Acta* **416**, 105 (1974); A. J. Bearden and R. Malkin, *Q. Rev. Biophys.* **7**, 131 (1975); R. K. Clayton, *Ann. Rev. Biophys. Bioeng.* **2**, 131 (1973); J. R. Bolton and J. T. Warden, *Ann. Rev. Plant Physiol.* **27**, 375 (1976).

61. W. L. Butler, *Acc. Chem. Res.* **6**, 177 (1973); for leading references, see also B. Ke, S. Sahu, E. Shaw, and H. Beinert, *Biochim. Biophys. Acta* **347**, 36 (1974).

62. G. Döring, G. Stiehl, and H. T. Witt, *Z. Naturforsch., Teil B* **22**, 639 (1967); G. Döring, G. Renger, J. Vater, and H. T. Witt, *ibid.* **24**, 1139 (1969).

63. J. Böhi, *Helv. Chim. Acta* **12**, 121 (1929).

64. J. C. Ghosh and S. B. B. Gupta, *J. Indian Chem. Soc.* **11**, 65 (1934).

65. A. A. Krasnovskii, *Dokl. Akad. Nauk SSSR* **60**, 421 (1948).

66. H. Scheer and J. J. Katz, *Proc. Natl. Acad. Sci. U.S.A.* **71**, 1626 (1974).

67. A. A. Krasnovskii, *Usp. Khim.* **29**, 736 (1960); *Russ. Chem. Rev. (Engl. Transl.)* **29**, 344 (1960).

68. A. N. Sidorov, *in* "Elementary Photoprocesses in Molecules" (B. S. Neporent, ed.), p. 201. Plenum, New York, 1968.

69. A. A. Krasnovskii, *Prog. Photosynth. Res. Proc. Int. Congr.* [*1st*]; *1968* Vol. 2, p. 709 (1969).

70. G. R. Seely and A. Folkmanis, *J. Am. Chem. Soc.* **86**, 2763 (1964).

71. V. P. Suboch and G. P. Gurinovich, *Dokl. Akad. Nauk Beloruss. SSR* **14**, 112 (1970).

72. V. B. Evstigneev, A. A. Kazakova, and B. A. Kiselev, *Biofizika* **18**, 53 (1973).

72a. V. B. Evstigneev and V. S. Chudar, *Biofizika* **19**, 425 (1974).

72b. V. B. Evstigneev and O. D. Bekasova, *Dokl. Akad. Nauk. SSSR* **154**, 946 (1964).

72c. T. T. Bannister, *Plant Physiol.* **34**, 246 (1959).

73. G. Zieger and H. T. Witt, *Z. Phys. Chem.* **28**, 286 (1961).

74. D. A. Savel'ev, A. N. Sidorov, R. P. Evstigneeva, and G. V. Ponomarev, *Dokl. Akad. Nauk SSSR* **167**, 135 (1966).

75. A. M. Shul'ga, I. F. Gurinovich, Yu. V. Glazkov, and G. P. Gurinovich, *Zh. Prikl. Spektrosk.* **15**, 671 (1971).

76. A. M. Shul'ga, G. P. Gurinovich, and I. F. Gurinovich, *Biofizika* **18**, 32 (1973).

77. G. N. Sinyakov, V. P. Suboch, A. M. Shul'ga, and G. P. Gurinovich, *Dokl. Akad. Beloruss. SSR* **17**, 660 (1973).

78. V. P. Suboch, A. M. Shul'ga, G. P. Gurinovich, Yu. V. Glazkov, A. G. Zhuravlev, and A. N. Sevchenko, *Dokl. Akad. Nauk SSSR* **204**, 404 (1972).

79. A. M. Shul'ga, I. F. Gurinovich, G. P. Gurinovich, Yu. V. Glazkov, and A. G. Zhuravlev, *Zh. Prikl. Spektrosk.* **15**, 106 (1971).

80. H. H. Inhoffen, P. Jäger, R. Mählhop, and C. D. Mengler, *Justus Liebigs Ann. Chem.* **704**, 188 (1967).

81. H. Falk and O. Hofer, *Monatsh. Chem.* **106**, 97 (1975).

82. H. Scheer, Ph. D. Thesis, Technische Universität, Braunschweig, West Germany, 1970.

83. V. P. Suboch, A. P. Losev, and G. P. Gurinovich, *Photochem. Photobiol.* **20**, 183 (1974).

83a. G. P. Gurinovich, A. P. Losev, and V. P. Suboch, *Photosynth., Two Centuries After Its Discovery, Proc. Int. Congr. Photosynth. Res., 2nd, 1971* Vol. 1, p. 299 (1972); A. P. Losev, V. P. Suboch, and G. P. Gurinovich, *Dokl. Akad. Nauk Beloruss. SSR* **15**, 550 (1971).

84. A. N. Sidorov, *Dokl. Akad. Nauk SSSR* **158**, 4 (1964); D. A. Savel'ev, A. N. Sidorov, R. P. Evstigneeva, and G. V. Ponomarev, *ibid.* **167**, 135 (1966).
85. I. M. Byteva, O. M. Petsol'd, and V. M. Poddubnaya, *Biofizika* **15**, 977 (1970).
86. A. A. Krasnovskii, M. I. Bystrova, and F. Lang, *Dokl. Akad. Nauk SSSR* **194**, 1441 (1970).
87. V. E. Kholmogorov, *Biofizika* **15**, 983 (1970).
88. J. Fajer, D. C. Borg, A. Forman, D. Dolphin, and R. H. Felton, *J. Am. Chem. Soc.* **95**, 2739 (1973).
89. B. H. J. Bielski and A. O. Allen, *J. Am. Chem. Soc.* **92**, 3793 (1970). B. H. J. Bielski and A. O. Allen, *J. Am. Chem. Soc.* **92**, 3793 (1970). B. H. J. Bielski, D. A. Comstock, and R. A. Bowen, *J. Am. Chem. Soc.* **93**, 5624 (1971).
90. R. A. White and G. Tollin, *Photochem. Photobiol.* **14**, 43 (1971).
91. G. Tollin, *J. Bioenerg.* **6**, 69 (1974).
92. A. V. Umrikhina, N. V. Bublichenko, and A. A. Krasnovskii, *Biofizika* **18**, 565 (1973).
93. V. B. Evstigneev and V. A. Gavrilova, *Dokl. Akad. Nauk SSSR* **114**, 1066 (1957).
94. A. P. Losev, I. M. Byteva, B. M. T. Dzhagarov, and G. P. Gurinovich, *Biofizika* **17**, 213 (1972).
95. Yu. N. Kozlov, B. A. Kiselev, and V. B. Evstigneev, *Biofizika* **18**, 59 (1973).
96. A. K. Chibisov, A. V. Karyakin, and M. E. Zubrilina, *Dokl. Akad. Nauk SSSR* **161**, 483 (1965).
97. B. I. Barashkov and A. K. Chibisov, *Biofizika* **17**, 764 (1972).
98. V. B. Evstigneev and V. A. Gavrilova, *Dokl. Akad. Nauk SSSR* **115**, 530 (1957).
98a. V. B. Evstigneev, *Prog. Photobiol., Proc. Int. Congr., 3rd, 1960* p. 559 (1961); A. A. Krasnovskii, *ibid.* p. 561.
99. R. Livingston and P. J. McCartin, *J. Am. Chem. Soc.* **85**, 1571 (1963).
100. G. P. Gurinovich and I. M. Byteva, *Biofizika* **15**, 602 (1970).
101. D. Mauzerall and G. Feher, *Biochim. Biophys. Acta* **88**, 658 (1964).
102. T. T. Bannister, "Photochemistry in the Liquid and Solid State." Academic Press, New York, 1960.
103. V. B. Evstigneev and V. A. Gavrilova, *Dokl. Akad. Nauk SSSR* **91**, 89 (1953).
104. A. A. Krasnovskii and G. P. Brin, *Dokl. Akad. Nauk SSSR* **73**, 1239 (1950).
105. E. V. Pakshina and A. A. Krasnovskii, *Biokhimiya* **29**, 1132 (1964).
106. A. V. Umrikhina, G. A. Yusupova, and A. A. Krasnovskii, *Dokl. Akad. Nauk SSSR* **175**, 1400 (1967).
107. R. B. Woodward, *Ind. Chim. Belge*, p. 1293 (1962).
108. A. M. Shul'ga, G. P. Gurinovich, and I. F. Gurinovich, *Biofizika* **17**, 986 (1972).
108a. Y. Harel, J. Manassen, and H. Levanon, *Photochem. Photobiol.* **23**, 337 (1976).
109. A. M. Shul'ga and V. P. Suboch, *Biofizika* **16**, 214 (1971).
110. R. Mählhop, Ph. D. Thesis, Technische Hochschule, Braunschweig, West Germany (1966); H. H. Inhoffen and R. Mählhop, *Tetrahedron Lett.* p. 4283 (1966); H. H. Inhoffen, P. Jäger, R. Mählhop, and C. D. Mengler, *Justus Liebigs Ann. Chem.* **704**, 188 (1967).
111. H. Fischer, R. Lambrecht, and H. Mittenzwei, *Hoppe Seyler's Z. Physiol. Chem.* **253**, 1 (1938).
112. I. M. Byteva, A. P. Losev, and G. P. Gurinovich, *Biofizika* **10**, 953 (1965).
113. A. A. Krasnovskii and K. K. Voinovskaya, *Dokl. Akad. Nauk SSSR* **96**, 1209 (1954).
114. M. V. Sarzhevskaja and A. P. Losev, *Biokhimiya* **35**, 681 (1970).
115. G. R. Seely, *J. Am. Chem. Soc.* **88**, 3417 (1966).
116. V. P. Suboch and A. M. Shul'ga, *Biofizika* **16**, 603 (1971).

117. H. H. Inhoffen, J. W. Buchler, and R. Thomas, *Tetrahedron Lett.* p. 1141 (1969).
118. H. H. Ihnoffen and W. Nolte, *Tetrahedron Lett.* p. 2185 (1963); *Justus Liebigs Ann. Chem.* **725**, 167 (1969).
119. G. S. Wilson and B. P. Neri, *Ann. N.Y. Acad. Sci.* **206**, 568 (1973).
120. R. H. Felton, G. M. Sherman, and H. Linschitz, *Nature (London)* **203**, 637 (1964); H. Berg and K. Kramarczyk, *Biochim. Biophys. Acta* **131**, 141 (1967).
121. J. Fajer, D. C. Borg, A. Forman, D. Dolphin, and R. H. Felton, *J. Am. Chem. Soc.* **95**, 3234 (1973).
122. P. Jäger, Ph. D. Thesis, Technische Hochschule, Braunschweig, West Germany (1965).
123. H. H. Inhoffen and P. Jäger, *Tetrahedron Lett.* p. 1317 (1964).
124. A. Ricci, S. Pinamonti, and V. Bellavita, *Ric. Sci.* **30**, 2497 (1960).
125. R. B. Woodward and V. Skarii, *J. Am. Chem. Soc.* **83**, 4676 (1961).
126. A. E. Pullman, *J. Am. Chem. Soc.* **85**, 366 (1963).
127. P. S. Song, T. A. Moore, and M. Sun, in "The Chemistry of Plant Pigments" (C. O. Chichester, ed.), p. 33. Academic Press, New York, 1972.
128. H. A. Otten, *Photochem. Photobiol.* **14**, 589 (1971); C. Weiss, *J. Mol. Spectrosc.* **44**, 37 (1972), J. V. Knop and J. H. Fuhrhop, *Z. Naturforsch. B* **25**, 729 (1970); C. Weiss, H. Kobayashi, and M. Gouterman, *J. Mol. Spectrosc.* **16**, 415 (1965).
129. H. H. Inhoffen, G. Klotmann, and G. Jeckel, *Justus Liebigs Ann. Chem.* **695**, 112 (1966); G. Jeckel, Ph.D. Thesis, Technische Hochschule Braunschweig, West Germany (1967).
129a. W. Trowitzsch. *Org. Magn. Res.* **8**, 59 (1976).
130. H. Wolf and H. Scheer, *Ann. N.Y. Acad. Sci.* **206**, 549 (1973).
131. J. L. Hoard, *Science* **174**, 1295 (1971).
132. E. B. Fleischer, *Acc. Chem. Res.* **3**, 105 (1970).
133. M. S. Fischer, Ph. D. Thesis (UCRL 19524). University of California, Berkeley (1969); M. S. Fisher, D. H. Templeton, A. Zalkin, and M. Calvin, *J. Am. Chem. Soc.* **94**, 3613 (1972).
134. J. Gassmann, I. Strell, F. Brandl, M. Sturm, and W. Hoppe, *Tetrahedron Lett.* p. 4609 (1971).
135. C. E. Strouse, *Proc. Natl. Acad. Sci. U.S.A.* **71**, 325 (1974); H.-C. Chow, R. Serlin, and C. E. Strouse, *J. Am. Chem. Soc.* **97**, 7230 (1975); R. Serlin, H.-C. Chow, and C. E. Strouse, *ibid.*, p. 7237;
135a. L. Kratky and J. D. Dunitz, *Acta Crystallogr.* **31B**, 1586 (1975).
136. P. N. Dwyer, J. W. Buchler, and W. R. Scheidt, *J. Am. Chem. Soc.* **96**, 2789 (1974).
137. J. W. Buchler, L. Puppe, K. Rohbock, and H. H. Schneehage, *Ann. N.Y. Acad. Sci.* **206**, 116 (1973).
138. S. J. Silvers and A. Tulinksy, *J. Am. Chem. Soc.* **86**, 927 (1964).
139. M. J. Hamor, T. A. Hamor, and J. L. Hoard, *J. Am. Chem. Soc.* **86**, 1938 (1964).
140. M. B. Hursthouse and S. Neidle, *J. Chem. Soc., Chem. Commun.* p. 449 (1972).
141. G. E. Ficken, R. B. Johns, and R. P. Linstead, *J. Chem. Soc.* p. 2272 (1956).
142. I. Fleming, *Nature (London)* **216**, 151 (1967).
143. H. Brockmann, Jr., *Justus Liebigs Ann. Chem.* **754**, 139 (1971).
143a. G. Brockmann and H. Brockmann, Jr., *IIT NMR Newslett.* pp. 117–162 (1968).
144. D. N. Lincoln, V. Wray, H. Brockmann, Jr., and W. Trowitzsch, *J. Chem. Soc. Perkin Trans. II* p. 1920 (1974).
145. H. Scheer and J. J. Katz, *J. Am. Chem. Soc.* **97**, 3273 (1975).
146. R. C. Pettersen and L. E. Alexander, *J. Am. Chem. Soc.* **90**, 3873 (1968); R. C. Pettersen and L. E. Alexander, *J. Am. Chem. Soc.* **93**, 5629 (1971).

147. J. W. Buchler and K. L. Lay, *Z. Naturforsch.* **30b**, 385 (1975).
148. J. J. Katz, *in* "Bioinorganic Chemistry" (G. Eichhorn, ed.), Chapter 29. Am. Elsevier, New York, 1972.
149. C. B. Storm, *J. Am. Chem. Soc.* **92**, 1423 (1970).
150. J. R. Miller and G. D. Dorough, *J. Am. Chem. Soc.* **74**, 3977 (1952).
151. S. Fried and K. M. Sancier, *J. Am. Chem. Soc.* **76**, 198 (1954).
152. J. J. Katz, H. H. Strain, D. L. Leussing, and R. C. Dougherty, *J. Am. Chem. Soc.* **90**, 784 (1968).
153. T. A. Evans and J. J. Katz, *Biochim. Biophys. Acta* **396**, 414 (1975).
154. J. W. K. Burrel, L. M. Jackman, and B. C. L. Weedon, *Proc. Chem. Soc., London* p. 263 (1959).
155. J. W. Mathewson, W. R. Richards, and H. Rapoport, *J. Am. Chem. Soc.* **85**, 364 (1963); *Biochem. Biophys. Res. Commun.* **13**, 1 (1963); A. S. Holt, D. W. Hughes, H. J. Kende, and J. W. Purdie, *J. Am. Chem. Soc.* **84**, 2835 (1962); K. M. Smith and J. F. Unsworth, *Tetrahedron* **31**, 367 (1975).
156. A. Gloe, N. Pfennig, H. Brockmann, Jr., and W. Trowitzsch, *Arch. Mikrobiol.* **102**, 103 (1975); W. Trowitzsch, Ph. D. Thesis, Technische Universität, Branuschweig, West Germany (1974).
157. H. Fischer and H. Gibian, *Justus Liebigs Ann. Chem.* **550**, 208 (1942).
158. H. Fischer and H. Gibian, *Justus Liebigs Ann. Chem.* **552**, 153 (1942).
159. G. L. Closs, J. J. Katz, F. C. Pennington, M. R. Thomas, and H. H. Strain, *J. Am. Chem. Soc.* **85**, 3809 (1963).
160. H. Brockmann, Jr., and J. Bode, *Justus Liebigs Ann. Chem.* **1974**, 2431 (1974).
161. A. Horeau and H. Kagan, *Tetrahedron* **20**, 2431 (1963); A. Marquet and A. Horeau, *Bull. Soc. Chim. Fr.* p. 128 (1967).
162. H. Wolf, H. Brockmann, Jr., H. Biere, and H. H. Inhoffen, *Justus Liebigs Ann. Chem.* **704**, 208 (1967); H. Wolf, H. Brockmann, Jr., I. Richter, C.-D. Mengler, and H. H. Inhoffen, *ibid.* **718**, 162 (1968).
163. H. Wolf, I. Richter, and H. H. Inhoffen, *Justus Liebigs Ann. Chem.* **725**, 177 (1969).
164. H. Wolf and H. Scheer, *Justus Liebigs Ann. Chem.* **745**, 87 (1971).
165. H. Wolf and H. Scheer, *Tetrahedron* **28**, 5839 (1972).
166. J. B. Conant, E. M. Dietz, C. F. Bailey, and S. E. Kamerling, *J. Am. Chem. Soc.* **53**, 2383 (1931); J. B. Conant, E. M. Dietz, and T. Werner, *ibid.* p. 4436; H. Fischer, F. Broich, S. Breitner, and L. Nüssler, *Justus Liebigs Ann. Chem.* **498**, 228 (1932); H. Fischer, O. Süs, and G. Klebs, *ibid.* **490**, 38 (1931).
167. A. Stoll and E. Wiedemann, *Helv. Chim. Acta* **16**, 307 (1933).
168. H. Fischer and A. Stern, *Justus Liebigs Ann. Chem.* **519**, 58 (1935); **520**, 88 (1935); see also H. Fischer and H. Orth, "Die Chemie des Pyrrols," Vol. II, part 2, p. 353. Akad. Verlagsges., Leipzig, 1940.
169. H. W. Whitlock and M. Y. Oester, *J. Am. Chem. Soc.* **95**, 5738 (1973).
170. H. Fischer and K. Bub, *Justus Liebigs Ann. Chem.* **530**, 213 (1937).
171. H. H. Strain and W. M. Manning, *J. Biol. Chem.* **146**, 275 (1942).
172. J. J. Katz, G. D. Norman, W. A. Svec, and H. H. Strain, *J. Am. Chem. Soc.* **90**, 6841 (1968).
173. P. H. Hynninen, *Acta Chem. Scand.* **27**, 1487 (1974).
174. H. Scheer and J. J. Katz, unpublished results.
175. C. Houssier and K. Sauer, *J. Am. Chem. Soc.* **92**, 779 (1970); *Biochim. Biophys. Acta* **172**, 492 (1969).
176. F. C. Pennington, H. H. Strain, W. A. Svec, and J. J. Katz, *J. Am. Chem. Soc.* **86**, 1418 (1964).

177. A. S. Holt and H. V. Morley, *Can. J. Chem.* **37**, 507 (1959).
178. H. Scheer, unpublished results.
179. A. S. Holt, *Can. J. Biochem. Physiol.* **36**, 439 (1958).
180. G. W. Kenner, S. W. McCombie, and K. M. Smith, *J. Chem. Soc. Perkin Trans. I* p. 2517 (1973).
181. F. C. Pennington, H. H. Strain, W. A. Svec, and J. J. Katz, *J. Am. Chem. Soc.* **89**, 3875 (1967).
182. A. S. Holt, *Plant Physiol.* **34**, 310 (1959).
183. D. Müller-Enoch, Ph.D. Thesis, Technische Universität, Braunschweig, West Germany (1971).
184. J. Bode, Ph. D. Thesis, Technische Universität, Braunschweig, West Germany (1971).
185. M. Strell and A. Kalojanoff, *Justus Liebigs Ann. Chem.* **652**, 218 (1962).
186. H. Brockmann, Jr., W. Trowitzsch, and V. Vray, *Org. Magn. Res.* **8**, 380 (1976).
187. H. Brockmann, Jr., *Phil. Trans. R. Soc. Lond.* **273**, 277 (1976).
188. H. Falk, G. Hoornaert, H.-P. Isenring, and A. Eschenmooser, *Helv. Chim. Acta* **68**, 2347 (1975).
189. W. Trowitzsch, Ph.D. Thesis, Technische Universität, Braunschweig, West Germany (1974).
190. S. G. Boxer and G. L. Closs, *J. Am. Chem. Soc.* **98**, 5406 (1976).
191. M. R. Wasielewski, M. H. Studier, and J. J. Katz, *Proc. Natl. Acad. Sci. U.S.A.* **73**, 4282 (1976).
192. R. C. Dougherty, H. H. Strain, and J. J. Katz. *J. Am. Chem. Soc.* **87**, 104 (1965).
193. S. J. Baum, B. F. Burnham, and R. A. Plane, *Proc. Natl. Acad. Sci. U.S.A.* **52**, 1439 (1964).
194. R. Grigg, G. Shelton, A. Sweeney, and A. W. Johnson, *J. Chem. Soc. Perkin Trans. I* **1972**, 1789 (1972).
195. M. T. Cox, T. T. Howarth, A. H. Jackson, and G. W. Kenner, *J. Am. Chem. Soc.* **91**, 1232 (1969).
196. G. W. Kenner, S. W. McCombie, and K. M. Smith, *J. Chem. Soc. Perkin Trans. I*, **1974**, 527 (1974).
197. F. C. Pennington, S. D. Boyd, H. Horton, S. W. Taylor, D. G. Wulf, J. J. Katz, and H. H. Strain, *J. Am. Chem. Soc.* **89**, 387 (1967).
198. T. Kemmer, Ph. D. Thesis, Technische Universitat, Braunschweig, West Germany (1976).
199. W. Oettmeier, T. R. Janson, M. C. Thurnauer, L. L. Shipman, and J. J. Katz, *J. Phys. Chem.* **81**, 339 (1977).

2

Hydroporphyrins: Reactivity, Spectroscopy, and Hydroporphyrin Analogues

HUGO SCHEER and HANS HERLOFF INHOFFEN

I. REACTIVITY OF HYDROPORPHYRINS

A. Chlorins and Bacteriochlorins

The chlorin macrocyclic system, as compared to that of porphyrins, shows two distinct differences in reactivity. The methine positions next to the pyrroline ring have a higher electron density and are, thus, susceptible to electrophilic attack, and chlorins are more easily oxidized both by one- and two-electron oxidants. Both features have been linked to the reactivity of chlorophylls, which are, with the exception of the chlorophylls *c*, either chlorins or bacteriochlorins. The easy isotope exchange of the δ-H in chlorophyll *a* (**1**) is a direct result of the increased electron density at this position, and especially the photooxidation to the cation radical is of eminent biological importance as the genuine light conversion step in photosystem I (see Volume II, Chapter 6; Volume IV, Chapters 3–5; and Volume V, Chapters 2 and 9).

However, chlorophyll chemistry is determined not only by the central Mg ion, but by the functional groups of the peripheral substituents. The most prominent one is probably the enolizable β-ketoester grouping at the isocyclic ring E, which is (together with the central Mg) responsible for the strong and specific self-aggregation of chlorophylls[1,2] and for which the involvement in photosystem II oxygen evolution has been discussed.[3,4] Chlorophyll chemistry and chlorophyll aggregation have been reviewed thoroughly[2,5–8] (see also Volume V, Chapter 9). Besides the basic features of the chlorin nucleus, only one special aspect of chlorophyll chemistry shall be discussed here, namely, the metalation or the free phorbin ligand with Mg. A brief summary of the recent research in chlorophyll chemistry is provided in the appendix of Chapter 1.

1. ELECTROPHILIC SUBSTITUTION

The increased electron density at the methine positions next to the pyrroline ring in chlorins was first concluded by Woodward and Skarič[9] from reactivity studies. It could be shown that the product obtained on reaction of chlorins with HCl/H_2O_2 was not a dioxychlorin, as originally proposed,[10] but rather a δ-chloro-7,8-chlorin. The formation of this *meso*-substituted chlorin by a selective electrophilic attack was rationalized in a new formulation of the

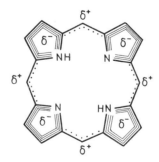

Fig. 1. Schematic representation of the porphyrin π-electron charge densities, according to Woodward.[11]

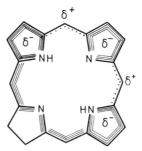

Fig. 2. Schematic representation of the chlorin π-electron charge densities, according to Woodward.[11]

porphyrin macrocycle,[11] in which the four pyrrolic subunits retain, to a certain degree, their aromaticity ($= 6\pi$ systems) by borrowing electron density from the neighboring methine bridges (Fig. 1). In the chlorins, the reduced pyrroline ring can no longer be such an aromatic subunit and, thus, the electron density at the neighboring methine positions is increased (Fig. 2). This difference between the "chlorin-type" methine protons next to the pyrroline ring (γ, δ in formula **1**), and the "porphyrin-type" methine protons

1

(α, β in **1**), respectively, was supported by MO calculations,[12-16] and is now an experimentally well established fact.

The chlorin-type methine protons next to the reduced ring are easily exchangeable against ^{2}H with acid,[9,17] while, under the same conditions, the remote porphyrin-type methine protons are nonexchangeable. This exchange is even more pronounced in some metal complexes and proceeds in the chlorophylls already in neutral solutions.[17] The differential $^1H/^2H$ exchange of a series of pheophorbides and bacteriopheophorbides has been studied[18] in an attempt to establish the methyl substitution site in the bacteriochlorophylls *d* (chlorobium chlorophylls 660), from which substitution at α or β was suggested. The results were not equivocal, however, and there is now positive proof for a δ-substituted structure instead.[19-22] The selective electrophilic attack at the chlorin-type methine positions has been demonstrated, for example, during nitration,[23] chlorination,[9,24] bromination,[20a] and formylation (of the hemins[25]). Further support for the increased electron density comes from the oxidative cleavage of $H_2(OEC)$ or $Zn(OEC)$ with thallic trifluoroacetate yielding selectively the A-dihydrobiliverdin **2** via hydrolysis

2

of a meso-trifluoroacetoxy intermediate[26] and cleavage of the oxochlorin phlorin.[27] The selective cleavage at 1-δ in δ-methylpheophorbides has been reported to occur photochemically in good yield.[22] Preparative difficulties arising in electrophilic substitution of pheophorbide metal complexes have been discussed.[20a]

Although in the latter examples, steric factors could play an additional role, the main driving force is certainly the increased electron density at the chlorin-type methine positions. In bacteriochlorins, all methine positions are adjacent to pyrroline rings, and accordingly all three methine protons in bacteriochlorophyll *a* and its pheophorbides are exchangeable.[17,18] Spectroscopic support for the increased electron density arises mainly from the characteristic high field shift of the chlorin-type methine protons in the nmr spectra of chlorins and bacteriochlorins.[28] In chlorins, the δ- and, if present, the γ-methine protons are shielded by about 1 ppm as compared to the α- and β-protons. In the bacteriochlorins, all methine protons are comparably shielded, while in the isobacteriochlorins three sets of methine resonances are observed.

While nucleophilic attack at the *meso* positions has been demonstrated in porphyrin dications[29], no similar reaction is known for chlorins. However, cation radicals of metallochlorins can be nitrated at the methine bridge next to the pyrroline ring with nitrite.[30]

2. Oxidation to Porphyrins

The conversion of chlorins to porphyrins was a major clue in establishing the relationship between heme and chlorophyll. A variety of reactions has been worked out by Fischer and his co-workers to achieve this formal oxidation of the macrocycle.[5] There are two principal modes for this reaction: the direct dehydrogenation of the reduced ring(s), or an initial reduction to colorless hydroporphyrins, followed by reoxidation beyond the chlorin level to the porphyrin.

Chlorins can be oxidized directly to the respective porphyrins with oxidants such as O_2, I_2, Fe(III), H_2O_2[5], but the reaction is best carried out with high-potential quinones. In a comparative study, o-chloranil and 2,3-dichloro-4,5-dicyano-p-benzoquinone were found most effective for the oxidation of pheophytin a to protopheophytin a.[31] With excess quinone, the oxidation is spectroscopically almost quantitative, but the yield is often reduced due to the formation of insoluble aggregates and further oxidation during workup. Quinone oxidation has been used to prepare 2-vinyl pheoporphyrins because the sensitive 2-vinyl group in chlorophylls is not attacked during the reaction,[32] and H_2(TPC) contaminations in H_2(TPC) can be removed as well by selective oxidation with quinones[33, 33a] or even with dimethylsulfoxide[34] (see Scheme 1).

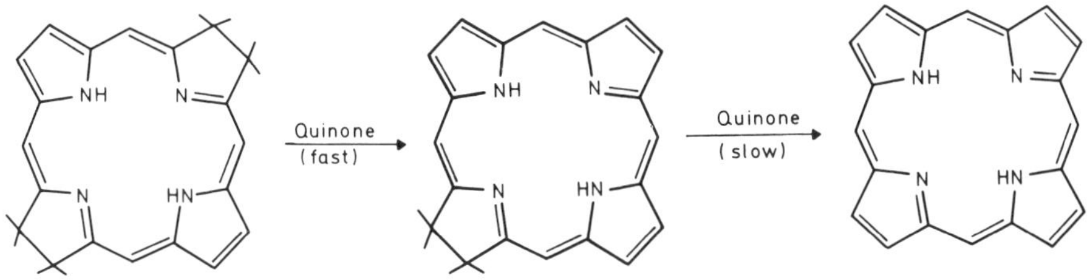

Scheme 1

Bacteriochlorins are oxidized by quinones in two kinetically well separated steps to porphyrins. The first step to the chlorin is fast and quantitative, and only an equimolar amount of the quinone is needed.[35,36] In the bacteriochlorophylls, only ring B is dehydrogenated during the first reaction (see also Fischer and Orth,[5] p. 306), this high selectivity probably being mainly due to steric factors (overcrowded periphery, Chapter 1). The subsequent dehydrogenation to the porphyrin is much slower and requires an excess of oxidant. This distinct differentiation of the two oxidation steps has been used to prepare unusually substituted chlorophylls (7,8-chlorins) from bacteriochlorophylls[35-37] and to optimize the H_2(TPC) yield in the chemical reduction of H_2(TPP) with diimide.[38] In the chlorophylls, this stability of the chlorin, as compared to the bacteriochlorin conjugation system, is probably aided by the steric strain introduced in the molecule upon oxidation of the 7,8 positions (see Chapter 1). However, symmetrically substituted bacteriochlorins follow the same reactivity pattern,[38,39] indicating only a comparably small increase in stabilization energy upon conversion of chlorins to porphyrins.

For the alternative method of converting chlorins into porphyrins by reduction with subsequent reoxidation, either Pd/H_2 or HI in acetic acid are the reagents of choice.[5] In neutral organic solvents, chlorins are reduced with Pd/H_2 to colorless "chlorinogens," which are reoxidized by air to the starting material. In acidic solutions, porphyrins are the major reoxidation products, accompanied by small amounts of the respective chlorins.[40–43] In both cases, the recovered chlorin is fully optically active.[5,44,45] Depending on structure and conditions, 2–3 moles of hydrogen are consumed during the reduction of chlorins, in excess to the 1 mole used up by the 2-vinyl group, if present. As the 7,8-stereochemistry is unchanged, the chlorinogens are, therefore, tetra- or hexahydrochlorins, respectively. It is not clear however, if the chlorinogen is converted to a porphyrinogen and both are oxidized separately, or whether a common leuco compound is reoxidized to a different degree, depending on the solvent system. It could be shown that chlorins can be formed during the oxidation of porphyrinogens by re-reduction of previously formed porphyrin under the conditions of the Rothemund $H_2(TPP)$ synthesis,[46] but an analogous sequence would not account for the optical activity in the Pd/H_2 reaction.

Catalytic hydrogenation followed by oxidation is suitable for converting sterically nonhindered chlorins in good yields to porphyrins, but, for sterically hindered pheophorbides, the reduction step is too slow and accompanied by reductive cleavage of ring E.[47] In this case, the HI "isomerization" reaction provides better yields in spite of a number of by-products.[44] HI is one of the most versatile reagents of Fischer's porphyrin chemistry, and the by-products are generally derived from attack of the 2-vinyl group. The mechanism of the isomerization is not known in detail, but uv-vis spectroscopic changes and by-product analysis indicate a porphodimethene intermediate.[47]

Besides being oxidable to porphyrins, chlorins and bacteriochlorins, as well as their metal complexes, can be oxidized under mild conditions with I_2, Fe(III), and similar one-electron oxidants to cation radicals. As chlorophyll cation radicals are generated in the light conversion step of photosystem I in photosynthesis, this reaction is of prime biological importance and widely investigated. The chemistry and spectroscopy of radical cations (and of radical anions) is beyond the scope of this Chapter, and the reader is referred to Volume II, Chapter 6; Volume IV, Chapters 3–5; and Volume VI, Chapters 2 and 9 for pertinent reviews.

3. Mg INSERTION

The binding of the central magnesium ion in chlorophylls and Mg porphyrins is thermodynamically unstable,[48] and the metal is lost very easily by acid catalysis[49] or photochemically (see Chapter 1).

By contrast, the remetalation of pheophytins with Mg to the corresponding chlorophylls has been a problem in chlorophyll chemistry until very recently. Alcohol-treated Grignard reagents,[49,50] Mg phenoxide, Mg viologen, MgI_2 hexapyridinate,[51] and anhydrous Mg salts in pyridine[52,53] have been used to metalate pheophytins in moderate to low yield. As porphyrins can be metalated in high yields under these conditions, this difference has been ascribed to their being chlorins, and has been rationalized on the basis of their lower basicity as compared to the porphyrins.[53] Recently, strongly sterically hindered phenoxides have been applied successfully in the metalation of pheophytins, including bacteriopheophytin a,* under carefully controlled conditions.[53a] This procedure is probably the method of choice for the large scale preparation of chlorophylls, since the pheophytins are usually available in large amounts and are more easily purified. The metalation of chlorins related to chlorophyll a has been investigated systematically by T. Urumov and M. Strell (*Justus Liebigs Ann. Chem.*, in press; T. Urumov, Ph.D. Thesis, Technische Universität München, West Germany, 1975). The authors report the insertion of Mg into methylpheophorbide a in refluxing DMSO, although part of the product is decarbomethoxylated. Starting with Cd-chlorins, which are easily accessible, they also obtained less readily accessible metal complexes (e.g., Mn-chlorins) by transmetalation in refluxing methanol.

$H_2(OEC)$[54] and certain pheophorbides[55a] have been metalated with anhydrous Mg $(ClO_4)_2$ in pyridine[53] in a reaction catalyzed by traces of water.[55a] By systematic variations of the substituents, it could be shown that the difficult metalation of pheophytin a (3) is not due to its chlorin structure, but rather to the presence of the enolizable β-ketoester system at ring E.[55] If this group is absent, as in pyropheophorbides, the metalation to the respective chlorophyllide usually proceeds in high yields. If the β-ketoester system is present, it competes successfully as a ligand for Mg with the central 4-N cavity, and peripheral complexes (cf. 4) are formed in which the Mg is bound to the β-dicarbonyl group. Similar complexes have recently been obtained from pheophorbides containing a rigid cis-β-diketo group,[55b] and a $\Delta9,10$-pheophorbide has been discussed.[20a] There appears to be a mutual exclusion of the two binding sites. Chlorophylls (= central Mg complexes, cf. 5) do not form peripheral complexes (cf. 6) even though the enolizable β-ketoester system is present, and the peripheral complexes of pheophorbides cannot be further metalated to the chlorophylls. Interestingly, the biosynthetic insertion of Mg takes place before the formation of the isocyclic ring bearing the β-ketoester grouping.[6,56]

This mutual exclusion seems unexpected at first, but it can be rationalized on both structural and electronic grounds. X-Ray results reveal substantial

---

* For the metalation of bacteriopheophytin a, see also M. Wasielewski, *Tetrahedron Lett.* p. 1373 (1977).

differences between chlorophyllides[57-59] and the Mg-free pheophorbides,[60,61] and similar pronounced changes are expected, too, in the rings C and E of the peripheral complexes, due to the partial double bond between C-9 and C-10. As both complexes are not very stable, the changes due to the metal at one binding site could then prevent binding of the second metal ion at the other binding site, respectively. The stability is further decreased by the electrostatic repulsion of the two proximate Mg^{2+} ions.

B. Phlorins

1. OXIDATION TO PORPHYRINS

Phlorins are generally easily oxidized to the corresponding porphyrins,[11,62-64] but they can be stabilized enough to be studied by conventional spectroscopic techniques if the parent porphyrins are sterically hindered. Thus, phlorins were first characterized during the redox reactions of a porphyrin sterically hindered by design in the course of Woodward's chlorophyll synthesis.[65,66]

The structural factors influencing the stability of phlorins were studied most systematically for a series of electrochemically generated chlorinphlorins related to chlorophylls.[63] Chlorinphlorins derived from sterically unhindered chlorins like rhodochlorin dimethyl ester (7) have a half-life (in aerobic

7

solutions) in the range of minutes. Their stability is increased by the presence of a 2-vinyl substituent, but it is strongly decreased in chlorinphlorins derived from pheophorbides. The latter compounds contain an isocyclic ring between C-6 and C-γ, by which the rigidity of the macrocyclic system is increased. On the other hand, the β-chlorinphlorins obtained by cathodic reduction of sterically hindered chlorins with substituents at C-6 and C-γ, but without the isocyclic ring (cf. 8), have a half-life in the range of hours or even days. An isomer of 8, the chlorinphlorin 9 obtained by photoreduction is the only phlorin as yet isolated.[67] Space filling models indicate for both isomers 8 and 9 a comparable relief of steric strain.

The oxidation of phlorins derived from unhindered porphyrins has been studied by a combination of rapid electrochemical techniques and absorption spectroscopy.[64,68-70] The oxidation is irreversible in neutral and alkaline solutions, with $E_{1/2} = -0.5$ to -0.7 V versus the saturated calomel electrode. Below pH 2, in aqueous HCl, the oxidation of 10 to 11 is reversible ($E_{1/2} = -0.04$ V [68]). Under these conditions, the phlorin cation 10 having an interrupted macrocyclic conjugation is in equilibrium with the cyclic-conjugated porphyrin dication 11; thus, an estimate of the porphyrin resonance energy

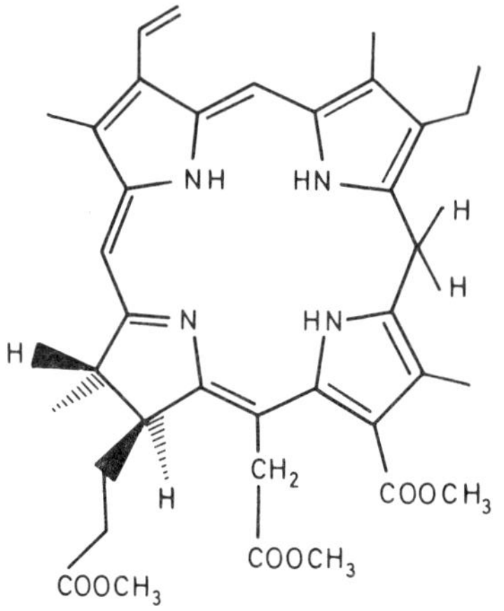

8

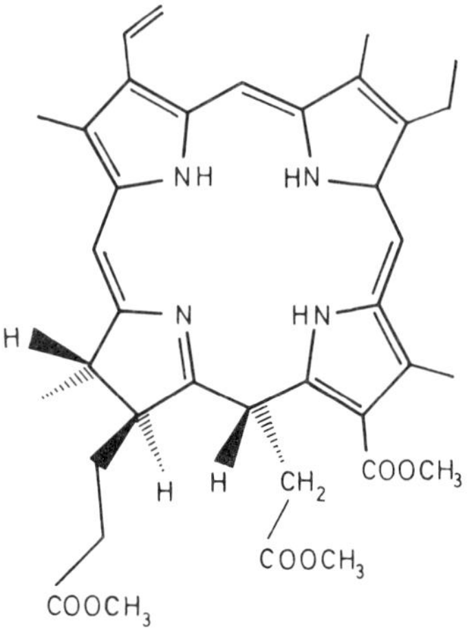

9

is possible from the electrochemical data (see below). Of particular interest is the formation and oxidation of the tetraphenylisophlorin anion, which is observed during cyclic voltametry of TPP in wet dimethylformamide.[69] In the dark, the isophlorin is oxidized to the parent porphyrin, but when

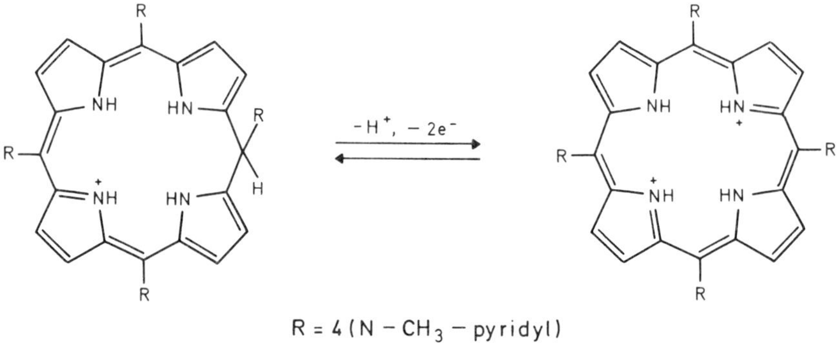

$$-H^+, -2e^-$$

$$R = 4(N-CH_3-pyridyl)$$

10   11

irradiated at its red absorption band ($\lambda_{max} = 730$ nm), the isoporphyrin[66,71,72] is formed instead, which subsequently tautomerizes to TPP. As a special case of oxidation, Mauzerall[62] has demonstrated the slow disproportionation of the urophlorin cation to porphyrin and porphomethene by a radical mechanism.

2. SYNTHESIS OF CHLOROPHYLL b

7,8-Chlorin-β-phlorins (cf. 8) are subject to a remarkable photooxidation leading to 3-methoxy-4-hydroxy- or 3,4-dioxybacteriochlorins (12), according

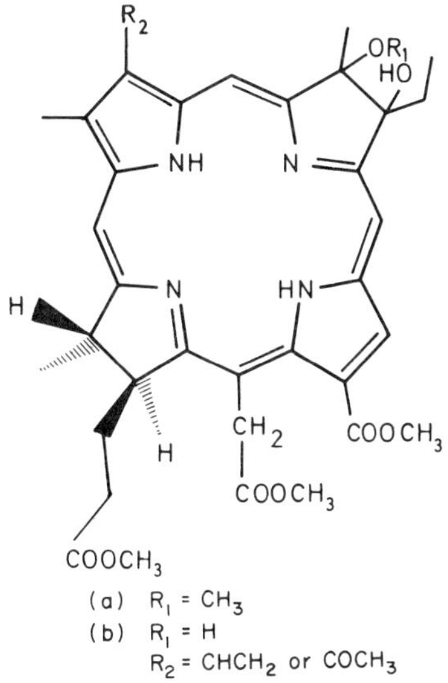

(a) R_1 = CH_3
(b) R_1 = H
     R_2 = $CHCH_2$ or $COCH_3$

12

to the solvent used (methanol or water, respectively).[63] From the latter compounds, chlorins with functionalized side chains at C-3 or C-4 are available, which open a synthetic route from the chlorophyll *a*- to the *b*-series.[63a] The β-hydrogenated structure of **8** (see above), the chlorinphlorin isomer being obtained in practically quantitative yield, could be proved by *one* cycle of electrochemical reduction in deuteromethanol and subsequent reoxidation, when only the ^1H nmr signal of the β-methine proton was decreased by exactly 50%. Photooxygenation of β-chlorinphlorins with

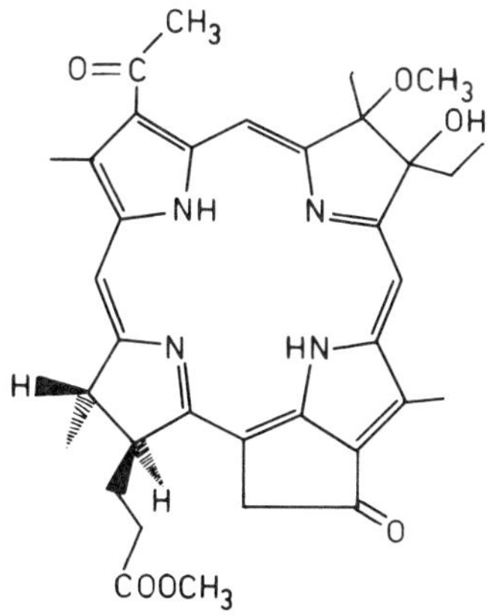

13

visible light, namely, irradiation of the chlorin-e_6 trimethyl ester-β-phlorin (**8**) in benzene/methanol in the presence of oxygen and boric acid produce, in moderate yield, two isomeric 3-methoxy-4-hydroxybacteriochlorins (**12a**) (R_2 = CHCH$_2$). The two new groups are presumably in a *transoid* orientation, thus accounting for the two isomers. The origin of the 4-*OH* from atmospheric oxygen could be demonstrated (in the bacteriopyromethyl-pheophorbide *a* series, cf. **13**) by using heavy oxygen ($^{18}O_2$) for the photo-oxidation. Mass spectroscopy then shows for the molecular ion of the dioxybacteriochlorin an increase of two mass units, but still an (M-32)$^+$ fragment originating from loss of "normal" methanol. Conversely, in view of the uptake of one carbon atom, the 3-OCH_3 group could be shown to be derived from the solvent. If mono-C-deutero-methanol (C^{1}H$_2$ ^{2}HO^{1}H) was used as solvent, the molecular ion shows *one* mass unit more, from which a 33 mass unit fragment is lost (Scheme 2).

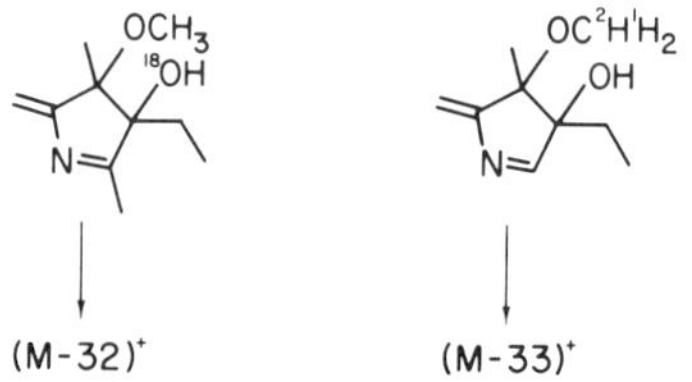

Scheme 2

Concerning the mechanism, it is probable that oxygen attacks the acidic β-phlorin hydrogen, forming a peripheral hydroperoxide. The latter loses water, producing an epoxide, which is then cleaved regioselectively by the solvent methanol. (See Scheme 3.) Accordingly, photooxidation in dioxane/water produces 3,4-diols (cf. **12b**).

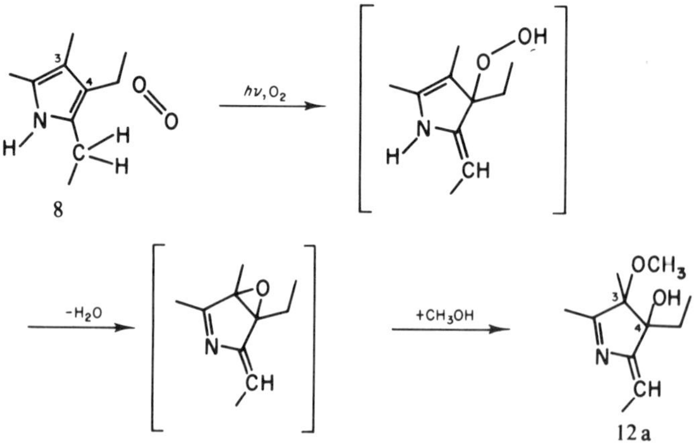

Scheme 3

Treatment of the 3-methoxy-4-hydroxybacteriochlorin **12a** with alcoholic or aqueous hydrochloric acid yielded the 4a-alkoxy- and 4a-hydroxychlorins **14a** and **14b**, respectively.[63a] The mechanism of this reaction is not known in detail, but the driving force seems to be the reformation of the 3,4-double bond. Similar products are obtained from bacteriochlorophyll *b*, which bears an ethylidene group at position 4.[36] The 4a-hydroxychlorin **14b** can be converted to the 4-vinyl or 4-acetyl derivatives **15a, b** by dehydration or oxidation, respectively.

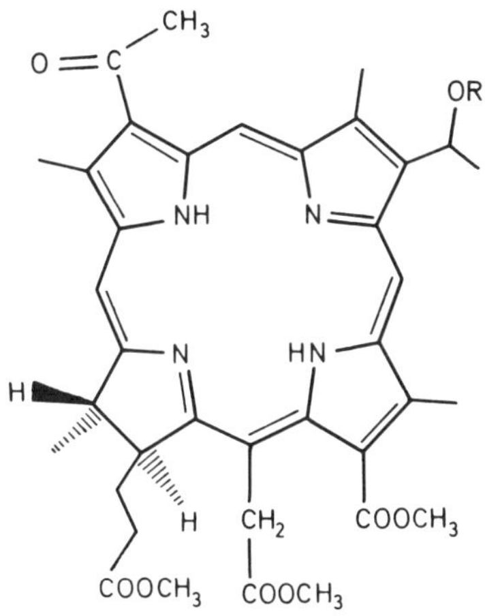

(a) R = alkyl
(b) R = H
14

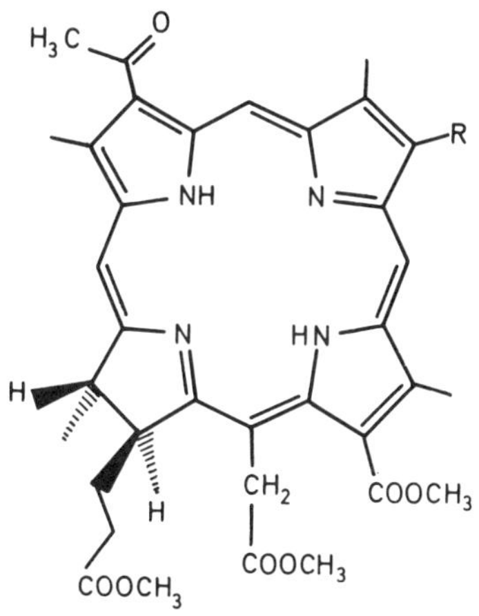

(a) R = CH=CH₂
(b) R = COCH₃
15

By the above reaction sequence, the 4-ethyl side chain in chlorophylls can be functionalized, but the selective methoxylation at C-3 prevents a similar reaction at C-3. However, if the photooxidation of **8** is carried out in aqueous dioxane, the epoxide cleavage leads to a mixture of 3,4-dihydroxybacterio-chlorins (**12b**; $R_2 = CHCH_2$), which can then be functionalized at either position 3 or 4. By this means, derivatives of the chlorophyll *b* series bearing a 3-formyl substituent are accessible from the chlorophyll *a* series.[63a]

Acid treatment of the diol **12b** ($R = CHCH_2$) leads to a mixture of the isomers **16** and **17** (Scheme 4) bearing a hydroxy group at either C-3a or C-4a. From the isomer **16**, rhodin-g_7 trimethyl ester **18a** is obtained by oxidation with dimethyl sulfoxide, as shown in the reaction Scheme 4.

In connection with the total synthesis of chlorin-e_6 trimethylester (**18b**),[65,66] (see also Chapter 1) and the conversion of rhodin-g_7 trimethyl ester (**18a**) into chlorophyll *b* worked out by the schools of Fischer and Willstätter, this

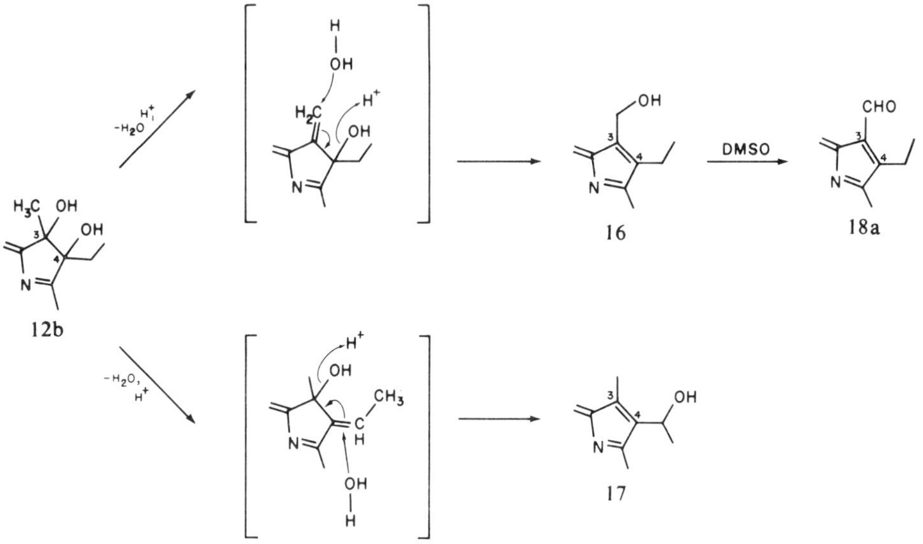

Scheme 4

partial synthesis represents the missing link in the total synthesis of chlorophyll *b*.

3. PHLORINS AND THE PORPHYRIN RESONANCE ENERGY

From stability, reactivity, and spectroscopic properties, the porphyrin macrocycle has been firmly established as being an aromatic system. The most detailed information for this statement arises probably from ^{1}H nmr data. For porphyrins and hydroporphyrins with a closed macrocyclic conjugation pathway, the ^{1}H nmr spectra are dominated by the magnetic anisitropy arising from the aromatic π-system, while this effect is absent in the ^{1}H nmr spectra of hydroporphyrins with interrupted conjugation.[28] Further evidence for the aromaticity of porphyrins arises from the equal bond lengths of the inner 16-membered conjugation pathway, which is found consistently in the X-ray analysis of metalloporphyrins (Volume III, Chapters 10 and 11), from MO calculations[13,73,75,76] (Volume III, Chapter 1), from reactivity studies[76] (Chapter 5), and from the similarity between the uv-vis spectra of porphyrins and 18π-annulenes.[77] Except for a few intermediate cases, a clear and consistent differentiation is possible by either one of these methods between the aromatic porphyrins and their nonaromatic derivatives.

In spite of this general agreement, however, the amount of stabilization is uncertain because only few quantitative experimental data are available. From the heat of combustion of several porphyrins, resonance energies

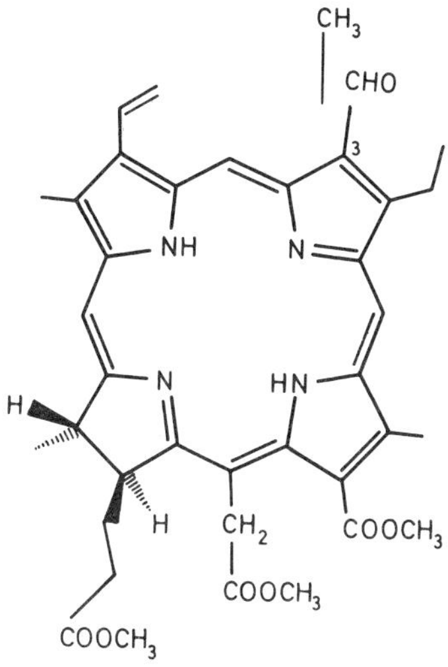

(a) R = CHO
(b) R = CH$_3$
18

between 78 and 500 kcal/mole have been calculated (see George[78]). This wide range is mainly due to the uncertainties in the bond energies, and in a critical review, using the best set of bond energies, a narrower range of 120–180 kcal/mole has been suggested recently.[78] Although in the lower range of the first estimates, this value amounts to 6.7–10 kcal/mole bond, which is comparable to that of the classical aromatic molecule, benzene (9 kcal/double bond).

A more direct approach, without the necessity to sum up a large number of bonds, would be possible by comparison with reference compounds that lack the macrocyclic conjugation, but for which the geometry, substitution pattern, etc., are comparable to the porphyrins. Phlorins and, to a lesser extent, related systems like isoporphyrins, oxophlorins, and oxaporphyrins, are useful reference compounds in this respect. So far, a direct comparison by calorimetric methods is not yet possible for the lack of stable, isolated phlorins, but equilibria between the two types have been observed in several cases: (a) between a γ-phlorinylacrylic acid and a γ-porphyrinylpropionic acid in the course of Woodward's[65] chlorophyll synthesis (see Chapter 1); (b) between the anions of octaethylchlorin and octaethylphlorin[79]; and (c) between the dication of *meso*-tetra (-4-N-methylpyridyl)porphyrin (11) and the mono-cation of the respective phlorin (10).[68] Equilibrium (a) suggests a very low resonance energy indeed, but it involves extreme structural conditions. The phlorin is not in equilibrium with a chlorin, for which a similar relief of steric hindrance was anticipated. Instead, it is in equilibrium with a porphyrin sterically hindered by design. The concept of steric hindrance as the main driving force for phlorin formation has been recently criticized.[80] On the basis of reactivity studies with OEP derivatives, the authors discuss an electronic concept instead. The second equilibrium (b) is observed at high pH and elevated temperatures, and a correlation with normal conditions is, therefore, difficult. So far the best model is the equilibrium (c) between compounds **10** and **11**. From the redox potential of -0.041 V versus the Ag/AgCl electrode, a resonance energy of 103 kcal/mole has been calculated by George,[78] which is in the lower range of estimates obtained from heat of combustion experiments. It should be noted, however, that the phlorin cation has not been investigated well enough, by the methods mentioned above, to exclude its partial conjugation. In particular, phlorin monocations have been suggested to be somewhat intermediate between the aromatic porphyrins and the nonaromatic phlorins on the basis of ^{1}H nmr data,[81] and a similar situation in **10** would then render the resonance energy too low.

4. ISOMERIZATION AND SUBSTITUTION

Phlorins and phlorin anions are important intermediates in porphyrin and metalloporphyrin reductions because they cannot only undergo further redox reactions but can also isomerize to chlorins or porphodimethenes (see also Chapter 1). The anion of Zn tetraphenylphlorin, which has been obtained by partial protonation of the Zn(TPP) dianion, undergoes (spectroscopically) quantitative isomerization to Zn(TPC).[81a] For the photoreduction of Sn(IV)-OEP(OAc)$_2$ to the respective chlorin, a phlorin intermediate has also been observed.[82] While in both cases the isomerization is irreversible, phlorin and chlorin anions are in equilibrium at high pH and elevated temperatures.[79] From the uv-vis data reported by Savel'ev et al.[83] it is likely that phlorins isomerize during metalation to porphodimethenes, and, yet, a third isomerization (isophlorin → phlorin) has been observed by Peychal-Heiling and Wilson,[69] without specification of the isophlorin structure.

The methine proton opposite to the methylene bridge is easily exchangeable in phlorins, and deprotonation of both phlorins and (metallo) porphodimethenes yields mono-*meso*- protonated anions (see Chapter 1), as evidenced by their characteristic phlorin-type spectra. Therefore, phlorins and porphodimethenes appear to be in tautomeric equilibrium at less basic conditions than required for the phlorin–chlorin equilibrium.

C. Porphodimethenes and Others

Like the isomeric (or possibly tautomeric) phlorins, porphodimethenes are easily oxidized to porphyrins and unstable toward air, but considerably stabilized if the parent porphyrins are sterically hindered. The chemistry of the thus stabilized α,γ-dimethyl octaethyl-β,δ-porphodimethene (**19a**) and its metal complexes has been studied in some detail by Buchler *et al.*[84–86] The stability of the metal complexes toward oxidation indicates a similar dependence on the central metal as has been observed by Fuhrhop[76] for metalloporphyrins, and rationalized by an electrostatic model. Thus, the Zn complex **19b** is easily oxidized, and the stability increases in the order **19a**, **19c**, and **19d**. However, only the free base **19a** yields the expected porphyrin, while **19b**

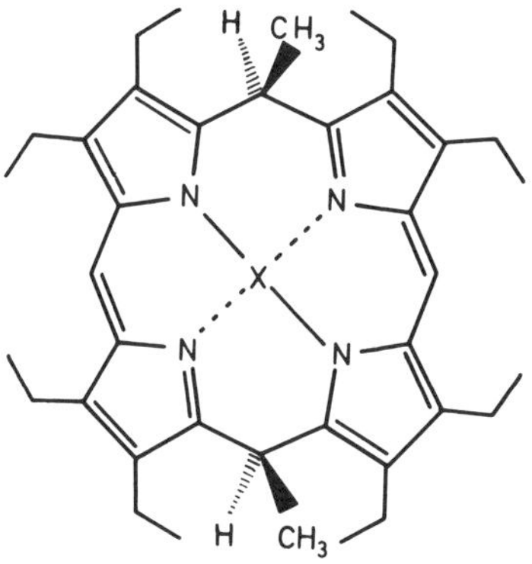

(a) X = H$_2$
(b) X = Zn
(c) X = Ni
(d) X = Cu
(e) X = Fe(III)B

19

yields what might be bile pigments.[84] The stability of the metal complexes toward demetalation follows the general pattern observed for porphyrins, but unlike the unsubstituted octaethyl derivatives, the extremes are more pronounced. Neither the porphodimethene (**19a**) nor its metal complexes isomerize to chlorins or phlorins, but the formation of 7,8-diethyl octamethyl-chlorin by reductive ethylation has been reported by Sinyakov *et al.*[87] For the Fe(III) complex **19e**, the monomer-μ-oxo-dimer equilibrium and ligand exchange reactions have been studied recently.[86] The direct oxidation of *meso*-tetraphenyl-α,γ-porphodimethene to H$_2$(TPP) has been studied by Dolphin.[46] Porphodimethenes and chlorinporphodimethenes formed in the Krasnovskii reduction of Zn pheoporphyrins or chlorophylls, respectively, are oxidized back to the parent compounds in the dark.[47,88,88a] For the Krasnovskii product of chlorophyll *a*, a half-life of about 1 hour under anaerobic conditions has been determined by ^{1}H nmr.[88]

The redox chemistry and the protonation–deprotonation equilibria of uroporphomethene have been studied by Mauzerall.[62] The porphomethene is more difficult to oxidize to urophlorin than the latter to uroporphyrin. With sulfite and dithionite, uroporphomethene is reduced to colorless product(s).

II. SPECTROSCOPY OF HYDROPORPHYRINS

A review on hydrophorphyrins would be incomplete without a discussion of their spectra. However, many aspects of hydroporphyrin spectroscopy are treated in the pertinent chapters on porphyrin spectroscopy, on the analysis of the porphyrin π-electron system, and on *in vivo* and *in vitro* studies of chlorophylls. Therefore, only a short summary of the characteristic features of hydroporphyrin spectroscopy is given here.

A. UV-VIS Spectra

Electron excitation ($S_0 \rightarrow S_1$) or uv-vis spectroscopy is by far the most widely applied spectroscopic method in hydroporphyrin chemistry and biochemistry. Due to the characteristic and intense absorptions of many hydroporphyrins and the large number of known spectra, the method is sensitive and selective. For a given chromophore, large shifts can be observed. These have been correlated in many cases to distinct molecular changes in substitution pattern, solvent system, aggregation, and the like by systematic structural variations, and by simultaneous investigation of defined model systems with more informative methods such as ^1H nmr spectroscopy. Considerable effort has also gone into a theoretical interpretation of the uv-vis spectra of hydroporphyrins.

Among the hydroporphyrins, the chlorins, bacteriochlorins, and phlorins, as well as their metal complexes, have characteristic absorption bands in the ranges between 350 and 450 nm (Soret or B-band),* and 600–900 nm (red or Q-band). In these cases, the assignment of a certain chromophoric system by uv-vis measurements is relatively safe even in reaction mixtures and biological systems. The other hydroporphyrins have less defined and less intense absorptions, around 500 nm, due to the common dipyrromethene chromophore. The assignment by uv-vis spectroscopy alone is, in these cases, ambiguous and requires support from independent techniques. In the following, the uv-vis spectroscopic features of the different hydroporphyrins

---

* It should be noted that, in most hydroporphyrins, the intensity of the "Soret" band is no longer an order of magnitude greater than the red band(s), but, rather, of comparable intensity. This is certainly due to the reduced symmetry in the hydroporphyrins and is especially pronounced for unsymmetric substitution.

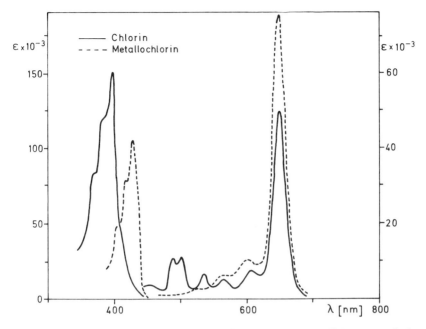

Fig. 3. Typical uv-vis spectra of pheophorbides and chlorophyllides, respectively, of the *a* series. The relative intensity of the Soret band is increased in etio-type chlorins, e.g., H₂(OEC).

are summarized. The spectra shown (Figs. 3–8) are typical for the respective etio-type compounds, if not otherwise indicated.

1. CHLORINS

Chlorins have an intense narrow red band around 660 nm ($\epsilon \approx 70,000$), and a Soret band of about threefold intensity around 400 nm (Fig. 3). A double band in the region of 500 nm ($\epsilon \approx 15,000$) is typical for free-base chlorins. Upon metalation, the disappearance of this band is the most characteristic spectral change. The red band of metallochlorins is increased in intensity, and increasingly blue-shifted with increasing electronegativity of the central metal. In the chlorophyllides (Mg chlorins), its position is almost unchanged. The Soret band is decreased in intensity and blue-shifted by about 30 nm in chlorophyllides, while the effects are less pronounced for other central metals. For the spectra of chlorophyll (Chl) *a* and *b* and their derivatives, the effects of substitution, ligation, aggregation (cf. Vernon and Seely,[6] Katz,[2] and this treatise) (Volume V, Chapter 9), and stereochemical changes [44,89] have been studied. Of some interest is the extreme red shift of the long wavelength absorption band in certain chlorophyll a hydrates ("crystalline chlorophyll"[15,58,89a,89b]), since similar molecular interactions have been proposed for the chl a "special pair"[89c] in the reaction center of photosystem I. Interestingly, a blue shift for this band has been reported recently for what has been claimed to be true chl a hydrate single crystals.[90] The electronic structure of chlorins has been investigated by MO calculations (for leading references, see 12–16). Recently, it has been suggested that the "isolated" C-3,C-4 double bond and the 9-carbonyl group are an integral part of the π-system and not locally excited.[90a]

2. BACTERIOCHLORINS

Bacteriochlorins have a narrow absorption ($\epsilon \sim 80,000$) at about 750 nm, a split Soret band, and an absorption of intermediate intensity at about 540 nm (Fig. 4). As compared to the free base, the spectrum of the metal complexes is red-shifted. As in chlorins, the intensity of the red band increases and that of the Soret band decreases upon metalations. The structural[36] and environmental influences on the uv-vis spectra of bacteriochlorophylls have been studied in $vitro$ in some detail,[2,90b] and the π electron structure has been investigated by MO calculations.[14,16,91] The uv-vis spectra of isobacteriochlorins (Fig. 5)[89,92,92a] are in the red region similar to those of the chlorins, but blue-shifted by about 30 nm for similarly substituted compounds. The Soret band of isobacteriochlorins is split as in bacteriochlorins.

3. HYPOBACTERIOCHLORINS, CORPHINS

The uv-vis spectrum of tetraphenylhypobacteriochlorin is of the chlorin type, though considerably blue-shifted, in spite of the formal interruption of the macrocycle conjugation.[93] The hypobacteriochlorin structure had been originally assigned from mechanistic considerations and MO calculations,[93] the latter indicating a participation of the nonbonding N-electron pair as in pyrrole. Support comes from the similar uv-vis spectra of the well-characterized trioxo derivatives[94] (see Section III, B). In the corphins, the macrocyclic conjugation is efficiently interrupted by cross conjugation of one pyrrolic $C\alpha$-$C\beta$ double bond.[95,96] Corphins have a broad absorption around 500 nm, and two bands ($\epsilon \sim 40,000$) between 400 and 300 nm (Fig. 5). In the corphin cation, the macrocyclic conjugation is restored, as evidenced by the reappearance of the Soret band.

4. PHLORINS

In contrast to the preceding hydroporphyrins with closed macrocyclic conjugation, the phlorins have broadened uv-vis spectra with less fine

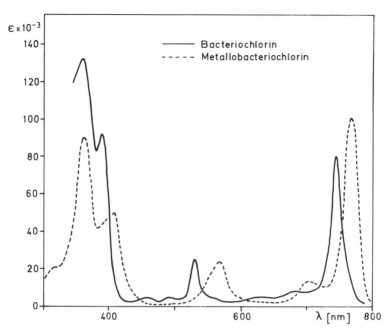

Fig. 4. Typical uv-vis spectra of bacteriopheophorbides and bacteriochlorophyllides, respectively, of the *a* or *b* series. The relative intensity of the Soret band is increased in etio-type bacteriochlorins.

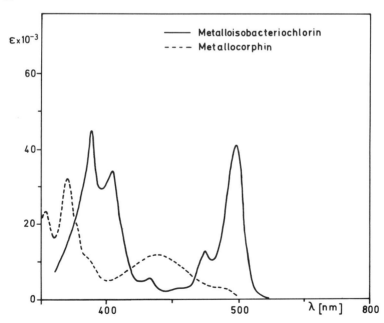

Fig. 5. Typical uv-vis spectra of metalloisobacteriochlorins (e.g., see Mengler[92]) and metallocorphins (e.g., see Johnson *et al.*[95]).

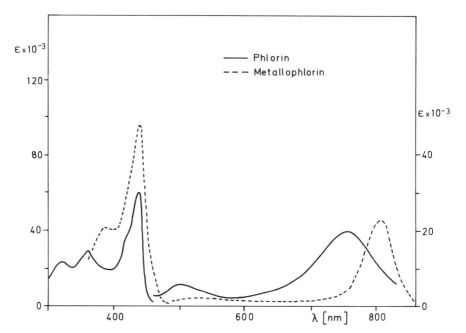

Fig. 6. The uv-vis spectra of tetraphenylphlorin and Zn tetraphenylphlorin anion (e.g., see Lanese and Wilson[70]).

structure (Fig. 6). The extinction coefficients of both the red band and the Soret band are considerably reduced, although the oscillator strength of the red band is comparable to that of the former compounds. The red band of free-base phlorins occurs around 650 nm, and a shift to 750 nm upon cation formation is characteristic for the phlorin chromophore.[62] Metallophlorins, which are probably always anions, have a considerably more intense Soret band, and the red band is shifted to the 900-nm range.[1,70,76] The spectra of chlorinphlorins (Fig. 7) are shifted about 150 nm to the blue, as compared to that of phlorins.[63] For a series of chlorinphlorins related to chlorophylls, a single, broad band between 550 and 600 nm ($\epsilon \sim 30,000$) has been observed.[63]

5. PORPHODIMETHENES AND PORPHOMETHENES

Porphodimethenes generally have a weak absorption around 500 nm and an intense one around 430 nm, the double band system probably arising from the two interacting parallel dipyrromethene chromophores. This spectral type has been observed in various metalloporphodimethenes.[47,84–86,88,97–101] However, on the basis of relative intensity variations, the possibility of the two bands arising from chemically different species has been discussed.[97,102] In the case of the Krasnovskii photoreduction product of chlorophyll *a* with

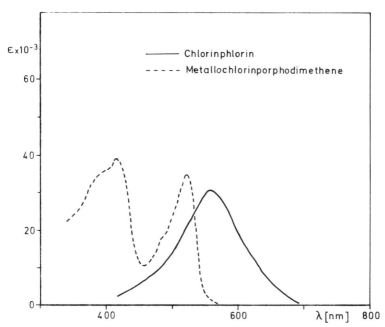

Fig. 7. The uv-vis spectra of chlorinphlorins and of the Krasnovskii photoreduction product of chlorophyll a,[88] a chlorinporphodimethene.

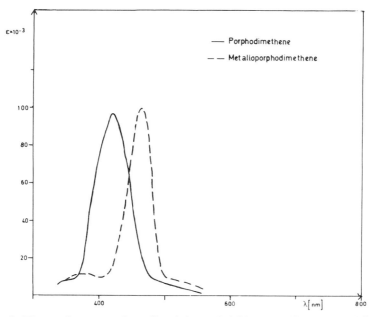

Fig. 8. The uv-vis spectra of α,γ-dimethyloctaethyl-β,δ-porphodimethene and its Zn complex (e.g., see Buchler and Puppe[84]).

H_2S, the origin of the bands from a single species has been demonstrated by ^{1}H nmr studies.[88] In the latter compound, a Mg chlorinporphodimethene, the two bands are of comparable intensity (Fig. 7), due to the lower symmetry. The 500-nm band is decreased to a shoulder in the dimethylporphodimethene **19a** and its metal complexes (Fig. 8).[84-86] The rooflike structure possibly favors an interaction of the two chromophores. Having only one dipyrromethene chromophore, porphodimethenes show a single intense absorption around 500 nm followed by a weak uv band.[62]

B. Fluorescence Spectra

Monomolecular chlorins and bacteriochlorins, as well as the chlorophylls, are highly fluorescent. The strong $0 \rightarrow 0$ band is often accompanied by a red shoulder, which is possibly related to aggregation. The fluorescence is efficiently quenched in chlorophyll aggregates (Volume V, Chapter 9), which can be used to study the state of chlorophylls *in vivo*.[103] The loss of fluorescence has also been observed in the phase test intermediates of chlorophylls (enolate anions), and in the peripheral Mg complexes of pheophorbides,[55] and related diketones[55b] and is probably due to chelation.[104] Delayed fluorescence of chlorophylls, which arises from population of S_1 states in the photosynthetic reaction centers through dark channels, has been observed in photosynthetic systems.[105,106] The *in vitro* photochemiluminescence of chlorophylls has been studied and reviewed recently by Krasnovskii and Litvin.[107] The intrinsic lifetimes of the fluorescence of chlorophyll *a* and bacteriochlorophyll *d* have been estimated to be 16 and 18 nsec, respectively, by Festisova and Borisov,[108] based on a critical examination of the absorption spectra.

C. Triplet Spectra of Chlorophylls

Due to the key role of chlorophylls in the light conversion reaction of photosynthesis, there is growing interest in other excited states besides the first excited singlet (S_1). This is especially true for the lowest triplet states (T_1), which have been discussed as possible intermediates in photosynthesis. After earlier optical studies, the detailed knowledge of the triplet state has been extended recently by the application of magnetic resonance and double resonance techniques (Volume IV, Chapter 5). Triplet–triplet absorption spectra of chlorophyll *a* and *b* and their pheophytins were first obtained by Livingston *et al.*[109-111] by flash techniques, and the influences of solvent and aggregation state have especially been studied.[112-116] The triplet spectra of the chlorophylls and the respective pheophytins in disaggregated solutions are similar. They consist of three broad and overlapping bands centered around

420 ($\epsilon \sim 6 \times 10^4$), 500 ($\epsilon \sim 2 \times 10^4$) and 670 nm ($\epsilon \sim 5 \times 10^3$)[116] with a shoulder extending above 1000 nm. For two of the bands, in-plane polarization of the absorption dipoles has been suggested from dicroic spectra.[117] Even in disaggregated (= fluorescent) solution, the quantum yield of singlet–triplet conversion (intersystem crossing) is high (chlorophyll a:$\Phi = 0.49$;[117] pheophytin a: $\Phi = 0.95$).[116] The phosphorescence (= triplet emission) of chlorophylls has first been detected by Kautsky and Hirsch[118] in deoxygenated aggregates. Due to efficient radiationless decay by vibronic coupling, the phosphorescence is very weak. Recent data determined directly are $\lambda_{max} = 985$ nm for chlorophyll a, and $\lambda_{max} = 940$ nm for pheophytin a.[119] Both the relative order and the energy range have been supported[120] by an nmr technique,[121] in which the selective broadening of resonances arising from that component of a mixture with the lowest triplet energy is monitored. The recent success in investigating the triplet state of chlorophylls[121a] and related compounds by esr techniques is reviewed elsewhere in this treatise (Volume IV, Chapter 5).

D. Vibrational Spectra

An ir band at about 1615 cm^{-1}, which was first reported by H. Golden *et al.*,[122] occurs in chlorins, but not in porphyrins and bacteriochlorins. This "chlorin band" had been tentatively assigned to a skeleton vibration, but, from Raman spectroscopic studies, a C=N vibration has been suggested.[123] The aggregation of chlorophylls *in vitro* has been studied by ir spectroscopy, mainly by analysis of the C=O region[124-126] (see Volume V, Chapter 9). These studies have been recently extended by resonance Raman spectroscopy.[127] The latter technique is both sensitive and selective and, thus, principally suited for *in vivo* studies. Due to fluorescence and apparative limitations, resonance Raman spectra have been obtained so far only by excitation in a narrow range at the long wavelength side of the Soret band. From intensity changes, depending on the excitation wavelength, a separation of the contributions from chlorophyll *a* and *b*, as well as of the different chlorophyll aggregates observed *in vivo*, seems possible.

E. Polarimetric Spectra

The ORD/CD spectra of a series of chlorins related to the chlorophylls *a* and *b* have been studied and reviewed.[45] Most of the cotton effects are related to $\pi \rightarrow \pi^*$ absorptions of the aromatic macrocycle. From systematic variations of the peripheral substituents, the macrocycle has been proposed to be inherently dissymmetric. This dissymmetry arises from the steric hindrance between the substituents at C-7 and C-γ, and is thus determined by the configuration at C-7[44,45] (for the stereochemistry of chlorins, see also

Chapters 1 and 9 and Scheer and Katz[28]). Superimposed on this basic ORD/CD spectrum are incremental spectra that reflect the configurations at C-10, and, to a lesser extent, at C-8 and C-9. The configuration determination at C-7 and C-10 is possible by ORD/CD alone, while the increments from asymmetric substitution at C-8 and C-9 are too small and require independent support (^{1}H nmr, ir) for the assignment.[44,45,128,129] In the ORD/CD spectra of pheophorbides bearing the 10-COOCH$_3$ group characteristic for most chlorophylls, an additional strong cotton effect has been observed which is not related to aromatic $\pi \rightarrow \pi^*$ transitions. Instead, it arises from the β,γ-unsaturated ester system. As only certain conformations are possible for the 10-COOCH$_3$ group, this cotton effect allows an independent determination of the configuration at C-10.[44,45,130]

The ORD/CD spectra of chlorophyll aggregates have been studied *in vivo* and *in vitro* and interpreted by an exciton model.[131,132] From correlation with transition moment calculations, a noncoplanar geometry of chlorophyll aggregates has been concluded[133,134] in agreement with results obtained by nmr spectroscopy.[135] The magnetic CD spectra of a series of chlorins related to chlorophyll *a* have been reported by Briat *et al.*[136] The red band has been analyzed in terms of the molecular symmetry, while an analysis of the Soret region is complicated by the complex structure of this band system.

F. Nuclear Magnetic Resonance Spectra

As in the ^{1}H nmr spectra of porphyrins, the spectra of the chlorins and bacteriochlorins are dominated by the ring-current-induced shifts (RIS)[28] of the aromatic macrocycle. This effect is absent in the spectra of hydroporphyrins with interrupted macrocyclic conjugation, and the spectra are, therefore, similar to those of the linear tetrapyrrolic bile pigments. (For recent reviews, see Scheer and Katz,[28] and Janson *et al.*, this treatise.)

Chlorins and bacteriochlorins are the hydroporphyrins which have been most thoroughly investigated by ^{1}H nmr. As compared to porphyrins, the most distinct differences are (a) slight reduction of the aromatic ring-current; (b) a pronounced high-field shift of the methine proton(s) next to the reduced ring(s), which is due to the increased electron density at these positions (see below); (c) decreased RIS and increased spin–spin couplings for the substituents at the reduced "chlorin" positions which both complicate a complete analysis (for details see Scheer and Katz[28]).

The ^{13}C nmr spectra of several chlorins related to chlorophylls have been reported recently,[20,21,137–139] culminating in the complete analysis of chlorophyll *a* and methylpheophorbide *a*[140]. On the other hand, the common model systems have received little attention. Ring-current and electron density (diamagnetic) contributions are less important in ^{13}C nmr spectroscopy.

Although the effects listed above for ^1H nmr are also operative,[27] the state of hybridization is much more important. For example, the chlorin C atoms, C-7 and C-8, resonate at a much higher field than the quaternary pyrrolic C atoms due to their sp^3 hybridization. The most important difference of chlorins (as compared to porphyrins) is the distinction between rings A and C, and B and D, respectively.[140] The α- and β-pyrrolic C atoms of the former rings correspond to those in pyridines, while the ones in ring B, and especially in ring D, are pyrrolelike. This difference between the two types of rings is even more clearly visible in the ^{15}N nmr spectra. The ^2H nmr spectrum of chlorophyll a-$^2H_{35}$ has been analyzed for isotope shifts.[141] Recently, ^{14}N and ^{25}Mg nuclear quadrupole resonance (nqr) spectra have been reported for chlorophyll a.[142]

G. Mass Spectra and X-Ray Photoelectron Spectra

The mass spectra of porphyrins are dealt with in detail elsewhere in this treatise (Volume III, Chapter 9). The main feature of chlorin spectra is the loss of the entire substituents at the chlorin C atoms by benzylic type fragmentation. In the porphyrins, a fragment smaller by 14 mass units is lost, due to the more extended aromatic systems. Chlorins are subject to dehydrogenation in the mass spectrometer (M-2 fragments), which is more pronounced in the *cis* than in the *trans* isomers.[44] The X-ray photoelectron (esca) spectra of chlorins have been studied recently by Falk *et al.*[143] As in porphyrins, two N_{1s} bands are distinguishable; these correspond to the aza and pyrrole nitrogens. Surprisingly, the difference between these two types of N atoms is decreased in the chlorins, and even smaller in the bacteriochlorins. The apparent conflict with nmr results[140] is possibly due to the much faster time scale of esca.

III. OXY AND OXO ANALOGUES OF THE HYDROPORPHYRINS

Oxidative degradation of porphyrins has been one of the early analytical tools to study their substitution pattern. While during these exhaustive oxidation reactions the macrocycle is broken down to yield cyclic imides, a variety of less vigorous methods has been developed by the schools of Fischer, Lemberg, and others, in which the macrocycle is retained in the oxidation products. Although well characterized some 40 years ago, only one of the structures could be established unambiguously at that time.[144] The course of the reactions and the structure of the products have been reinvestigated during the past decade. It was shown that oxidative attack of etio-type porphyrins usually occurs at the methine positions to yield *meso*-oxo analogues of the phlorins, porphodimethenes, porphomethenes, and porphyrinogens. However, by oxidation of the dications, and by oxidation with OsO_4, the attack is directed to the β-pyrrolic positions yielding oxy- and oxoderivatives of chlorins, bacteriochlorins, hypobacteriochlorins, and corphins.

A. *Meso*-oxygenated Hydroporphyrins

1. OXOPHLORINS

During heme catabolism, the porphyrin macrocycle is opened to bile pigments, while the α-methine C atom is lost as CO.[145,146] From studies of a model reaction, namely, the coupled oxidation of porphyrins with ascorbic acid and hydrogen peroxide,[147,148] oxidative attack, at the methine position via a *meso*-oxygenated iron porphyrin,[149] was proposed as the reaction mechanism. The *meso*-hydroxy structure **20a** or its tautomer **20b** was suggested for the demetalation product of one of the intermediates of the coupled oxidation.[150] Similar products were obtained by oxidation of hemochromes with hydrogen peroxide in pyridine, and a mono-*meso*-oxygenated structure was supported by combustion analysis[151] and benzoylation.[144] The *meso*-oxygenated structure was confirmed by a direct comparison of OEP-oxophlorin **20c** obtained by benzoylation of $H_2(OEP)$ and hydrolysis, with a synthetic product.[152]

Oxophlorins can be prepared either by synthesis, by oxidation of porphyrins or by selective reduction of more highly oxygenated porphyrins. The synthetic approach to oxophlorins has been opened by Jackson *et al.* (see Volume I, Chapter 6),[153,154] with their oxobilane route to porphyrins. Oxophlorins are intermediates in this synthesis, and the method has been used by Clezy *et al.* (see Chapter 4 for a review) to prepare a wide variety of oxophlorins.

The direct oxidation of metalloporphyrins to oxophlorins can be carried out by coupled oxidation with oxygen and ascorbic acid,[148] or by reaction with H_2O_2.[155] For the latter reaction, the dependence of the central metal has been studied.[152] Among the first-row transition metals, the oxidation is only possible if the metal ion has a readily available higher oxidation state (Fe(II), CO(II), Mn(II), (Mn(III))). The oxidation of free-base porphyrins is possible with lead tetraacetate[156] or with benzoyl peroxide,[81] and subsequent hydrolysis of the *meso*-acylated porphyrin. While these reactions proceed, probably by a radical mechanism, oxidation is possible[27] too using the electrophilic reagent thallic trifluoroacetate.[157] The latter reaction, which proceeds via a *meso*-trifluoroacetoxy intermediate, is probably the best synthetic approach to octaethyloxophlorin and similar structures for which no isomerism is possible.[27] The third pathway to oxophlorins is via the reduction of higher *meso*-oxygenated porphyrins, particularly the xanthoporphyrinogens.[144,156] These reactions have less synthetic value, but they were important chemical clues in the structure determination of the latter compounds.

The spectroscopy and chemistry of oxophlorins is discussed in detail in Chapter 4 and briefly here. The characteristic features of mono-*meso*-oxygenated porphyrins are their pH-dependent tautomerism between the hydroxyporphyrin structure (cf. **20a**) and the oxophlorin structure (cf. **20b**)

20a = IX type substitution   20b = IX type substitution
20c = octaethyl substitution   20d = octaethyl substitution

and the facile oxidation of the oxophlorins to their reactive radicals. Both have considerably complicated their structural analysis. In neutral solutions, the oxophlorin form (**20b**) is generally predominant.[81,152–154] This is evident from the ir spectra (strong CO band at about 1570 cm^{-1}, characteristic for pyrroketones), the uv-vis spectra [phlorin-type spectrum (Fig. 9)], and the ^{1}H nmr spectra (no ring-current effects). However, the *meso*-hydroxy tautomer **20c** is stabilized in *meso*-oxygenated furan and thiophen analogues,[158] and in OEP-β-hydroxygeminiporphyrin-4-ketone **21**.[159] In the former, tautomerization to the oxophlorin is precluded by the bivalent heteroatoms;

21

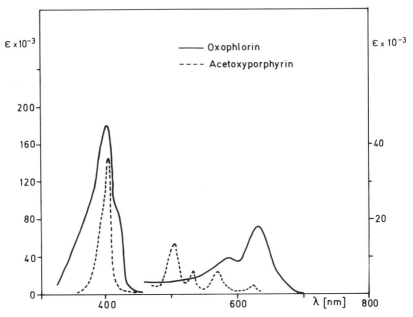

Fig. 9. The uv-vis spectra of octaethyloxophlorin and *meso*-acetoxy-OEP (e.g., see Bonnett *et al.*[81]).

in the latter the *meso*-OH is stabilized by intramolecular hydrogen bonding to the neighboring CO group.

From protonation–deprotonation studies it has been concluded that oxophlorins are both stronger acids and bases than the corresponding porphyrins.[81,154] Free-base oxophlorins are extremely sensitive to air. The redox potential of the oxophlorins is considerably lower than that of porphyrins, and the former are always contaminated with the oxophlorin radical formed by one electron oxidation and deprotonation. These paramagnetic impurities have been responsible for broadened and sometimes undetectable ^{1}H nmr spectra,[81,152,154,158] due to rapid spin exchange. The radical of OEP-oxophlorin has been obtained quantitatively by photooxidation in benzene, and its esr spectrum and reactivity have been studied. It appears that the high reactivity of oxophlorins is due primarily to the easy oxidation to and the subsequent reactions of the cation radical.[76,160]

2. DIOXOPORPHODIMETHENES

Further oxidation of porphyrins or oxophlorins[161,162] leads to α,γ-dioxo-β,δ-porphodimethenes like **229**. The best preparative procedure is again the oxidation with thallium trifluoroacetate.[161] From ^{1}H/^{2}H exchange experiments, the methine bridge opposite to the hydrogenated one in phlorins has

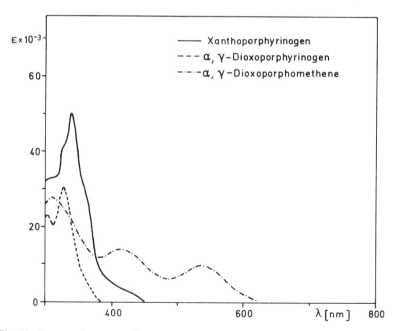

(a) R = O
(b) R = NH

22

been shown to be susceptible to electrophilic attack. This selectivity is less pronounced in the oxophlorins,[81] and α,β-dioxoporphodimethenes have been obtained as by-products of 22a.[163,164] The iminooxoporphodimethene 22b has been obtained by Fuhrhop[162] by oxidation of the *meso*-amine. The uv-vis spectra of the dioxoporphodimethenes are of the dipyrromethene type (Fig. 10), indicating an efficient separation of the two subunits by the oxo

Fig. 10. The uv-vis spectra of *meso*-oxygenated octaethyl porphodimethenes, porphomethenes, and porphyrinogens.[156]

bridges. This is supported by the ¹H nmr spectra (no ring-current effects), and by the ir spectra (strong CO, no OH).

3. OXO-PORPHOMETHENES

During the reduction studies on xanthopophyrinogens (see below), Fischer and Treibs[165] noticed the rapid oxidation of the dioxoporphyrinogen **25** to a red product. The structure of the latter was established by Inhoffen *et al.*[156] as the dioxoporphomethene **23** on the basis of ir, nmr, and ms data.

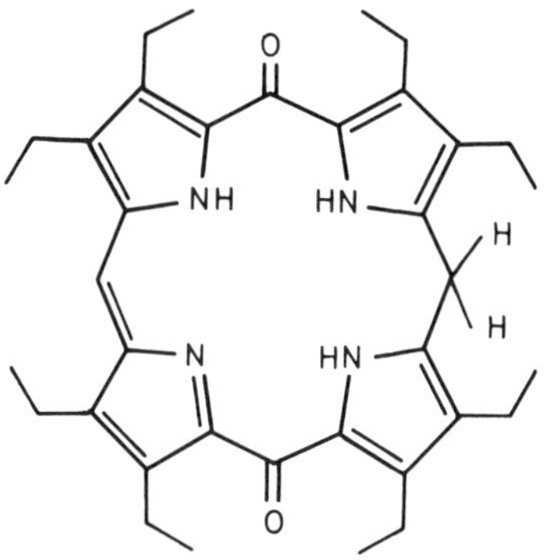

23

The opposite position of the carbonyl groups was concluded from the mass-spectroscopic fragmentation pattern.

4. OXOPORPHYRINOGENS

In 1927, in an attempt to distinguish between the indigoid and macrocyclic structures proposed for porphyrins, Fischer and Treibs[165] obtained a yellow tetraoxygenated product from the lead dioxide oxidation of porphyrins. Although Fischer and Orth[5] discussed the tetraoxoporphyrinogen **24** as a

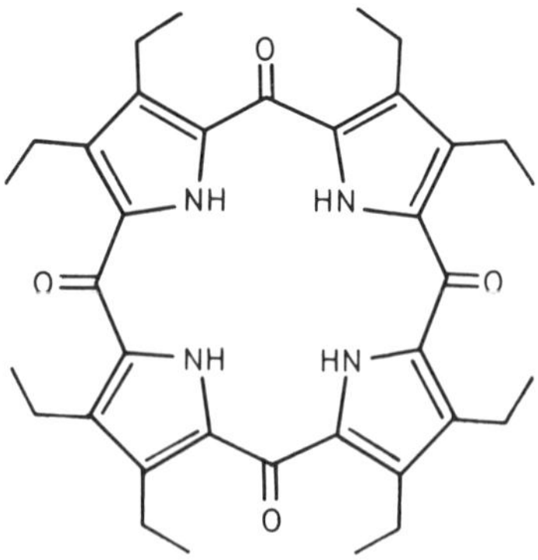

24

possible structure, this was not proved until 25 years later by Inhoffen *et al.*[156] The difficulties in assigning structure **24** were mainly due to the puzzling chemistry of the molecule: the steric hindrance of the neighboring β-pyrrolic substituents, which interferes with derivatization of the *meso*-carbonyl groups; the strong binding of two water or solvent molecules[5] in the chlathrate-type crystals[166]; and the differential reduction of two, three, or four of the carbonyl groups. The reduction with sodium amalgam[165] or lithium aluminum hydride[156] yields the porphyrins; reduction with zinc in acetic yields the α,γ-dioxoporphyrinogen **25**.[156,165] The latter compound is unstable toward

25

air[165] and oxidizes to the dioxoporphomethene **23**, which can also be obtained directly from the xanthoporphyrinogen with sodium borohydride.[156] Finally, reduction of **24** in a closed vessel with hydrogen bromide/acetic acid yields OEP-oxophlorin **20d**.[155] The reduction of **24** to **19** and subsequent benzoylation was the first definite proof of at least one oxygen of **24** in a *meso*-position.[144]

The spectra of octaethylxanthoporphyrinogen and the products derived from it by reduction have been discussed by Inhoffen *et al.*[156] (Fig. 10).

B. Hydroporphyrins Oxygenated at β-Pyrrolic Positions

Oxidation of etio-type porphyrins with hydrogen peroxide in concentrated sulfuric acid leads to oxochlorins of type **26**. Although first characterized in 1930, the proof of the structure of **26** had again to await almost 40 years.[81,94,167] The reaction can be rationalized as an electrophilic attack on the porphyrin β-pyrrolic positions, possibly leading to the β,β'-dioxychlorin **27** and subsequent pinacol–pinacolone rearrangement. Obviously, the electron density distribution is changed by dication formation to direct the electrophilic attack from the methine to β-pyrrolic positions. An alternative approach to β-hydroxylated porphyrins is by treatment with OsO_4 and subsequent

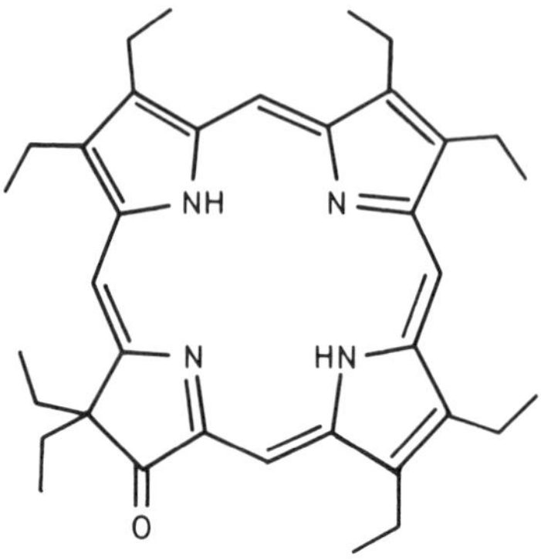

26 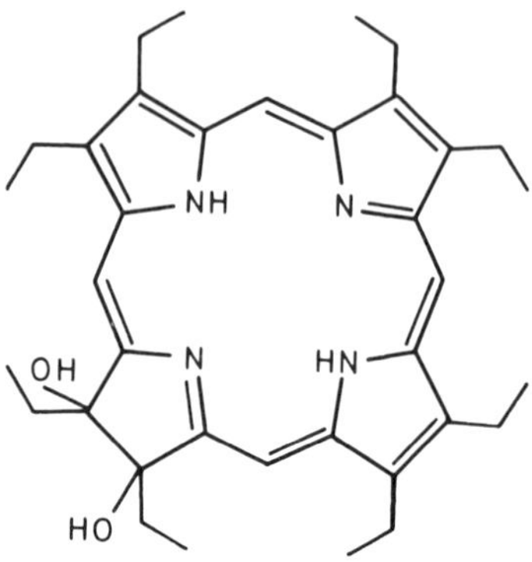   27

hydrolysis,[94,168,169] to dihydroxychlorins of type **27**. The success of this reaction again demonstrates the partial isolation of the peripheral double bond(s) from the macrocyclic conjugation pathway. The β,β'-dihydroxychlorins can be isomerized in concentrated H_2SO_4 to geminiketones, which are thus accessible by two alternate routes (Scheme 5). Both reactions have

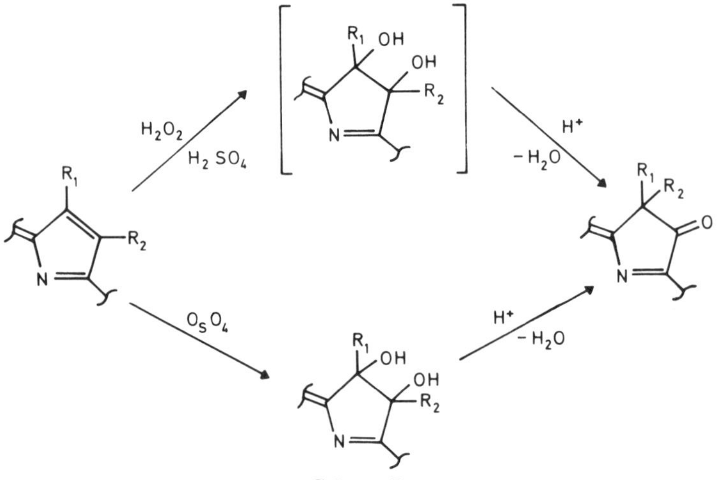

Scheme 5

been used by Inhoffen and Müller[170] in a new approach to the corphins,[95,171] which are useful model compounds for corrins, and which principally open an alternative synthetic route[171,172] to the corrin macrocyclic system. One of the major synthetic problems is posed by the quaternary β-pyrrolic C atoms in the corphyrins and corrins. The geminiporphinketones are suitable intermediates, because alkylation of the β-pyrrolic carbonyl groups is possible with lithium organic reagent.[94,170]

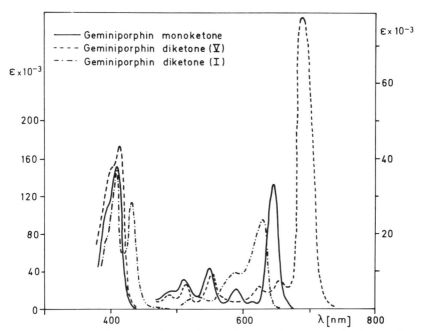

Fig. 11. The uv-vis spectra of octaethyl geminiporphyin mono- and diketones.[94]

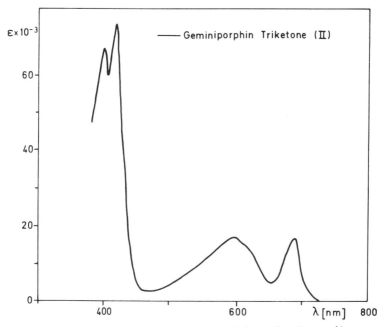

Fig. 12. The uv-vis spectrum of an octaethyl geminiporphin triketone.[94]

In exploring the H_2O_2/H_2SO_4 oxidation of $H_2(OEP)$, it was noted[81,94] that the geminiporphin monoketone **26** is always accompanied by minor amounts of di- and triketones. Starting from $H_2(OEP)$, all five geminiporphin diketones and two of the four possible triketones have been isolated and characterized.[94] The uv-vis spectra of these oxo compounds are similar to those of the respective hydroporphyrins (Figs. 5, 11 and 12), and all structures could be assigned from ^1H-nmr arguments and structural correlations between the di- and triketones. The product distribution seems to be controlled by steric hindrance: the missing triketones are the ones with two bulky geminal diethyl groups next to each other (**28, 29**), and the diketone obtained

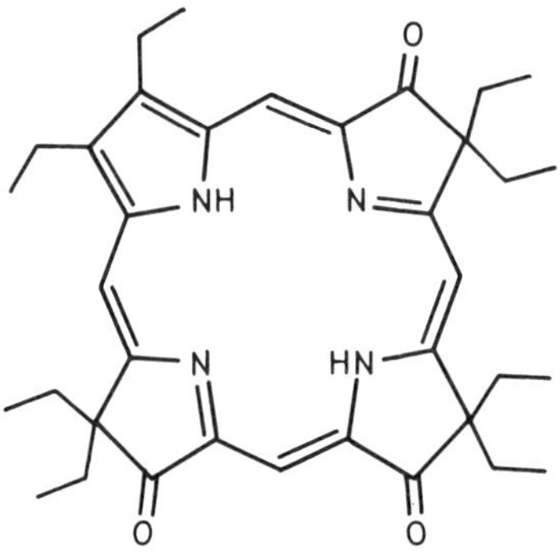

28

in lowest yield is the one with two of these groups next to an ethyl substituent (**30**). The yield of the triketones can be increased by treatment of the mono- and suitable diketones with OsO_4 and subsequent pinacol rearrangement. However, quaternization of the β-pyrrolic positions of the fourth pyrrole ring is not possible by further oxidation. Instead, the corphin macrocyclic system is accessible by primary reduction of the monoketone **26** with lithium phenyl

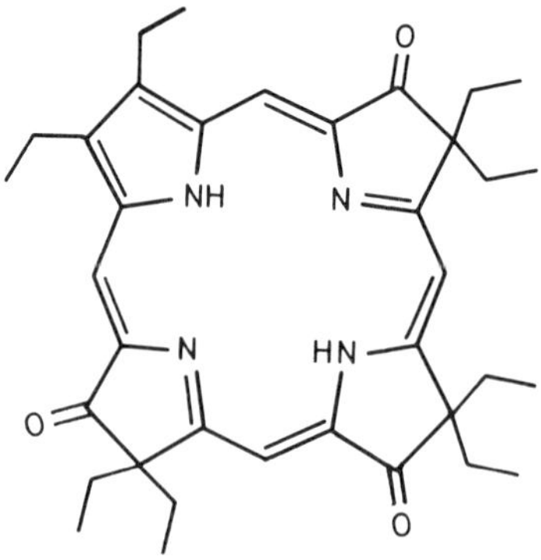

29

and dehydration of the metal complexes to salts of the type **31**, and subsequent treatment with OsO_4, followed by pinacol rearrangement to **33**. The

30   31

macrocyclic conjugation in the salts of type **31** is essentially interrupted by the cross conjugation of the 4,4a double bond, and OsO_4 hydroxylation of the three peripheral double bonds isolated from the inner olefinic conjugation system yields the hexahydroxycorphin **32**. The uv-vis spectrum of **32** is of the corphin type.[95,96] By isomerization with H_2SO_4, two of the possible eight isomeric corphin-triketones of the type **33** have been obtained in moderate yield.[170]

32   33

C. Analogues of Hydroporphyrins

In addition to the oxygenated hydroporphyrins, a variety of C analogues of chlorins and bacteriochlorins has been characterized recently which contain either exocyclic double bonds or two substituents at the reduced positions. In both cases, the macrocycle is in a higher formal oxidation state than indicated by the spectra. Isomerization to this state proceeds readily by prototropic rearrangement if an exocyclic double bond and at least one proton at the neighboring reduced position are present.[36]

The only natural C analogue known is bacteriochlorophyll b (34) bearing an ethylidene group at position 4.[36] Compound 34 is rather unstable and rearranges by acid catalysis to its endocyclic isomer, 2-desvinyl-2-acetyl-chlorophyll a (35). This isomerization is irreversible in vitro, but the possibility

34   35

of an in vivo back reaction is tempting. Ethylidene derivatives such as 34 are possible intermediates during the in vivo hydrogenation of peripheral double bonds in chlorins and porphyrins. Hudson and Smith[173] have proposed such isomerization of Mg porphyrins into 4-ethylidenechlorins as a key step in the biosynthesis of algal bile pigments. The latter are protein-bound in vivo by a thioether linkage in ring A (namely, 36), which eliminate to give 3-ethylidene-bilins in refluxing methanol.[174]

Synthetic C analogues are accessible by four general types of reactions: (a) the reduction of geminiporphinketones with lithium organic reagents,[170] (b) Woodward's[11] purpurin reaction, (c) the intramolecular cyclization of acetamidoporphyrins,[178] and (d) the addition of ethyl diazoacetate to the

36

peripheral double bond(s) of porphyrins. The latter reaction has been studied in some detail.[175-177] Starting from TPP, a variety of isomeric cyclopropanochlorins and biscyclopropanobacteriochlorins have been obtained. Due to the conformationally rigid substituents, they are suitable stereochemical models to test, for example, the magnetic anisotropy of the porphyrin macrocycle.[28] Cu(OEP) can react in a similar fashion yielding chlorins which bear four alkyl substituents at the pyrroline ring (37). As a by-product, the isomeric ethylidenechlorin 38 is obtained. It is stabilized toward rearrangement of the endocyclic isomer for the lack of a suitable proton.

The reaction of etio-type acetamidoethylporphyrins 39, under the conditions of the Vilsmeier reaction, yields methylene spiropyrrolinochlorins (40), which

can be reduced to the corresponding methyl-substituted chlorins.[178] The reaction proceeds probably via intramolecular attack at the spiro-C-to-be and prototropic shift of the peripheral to the methylene double bond.

D. Miscellaneous

Only few examples of other hydroporphyrin analogues are known.
Fuhrhop *et al.*[179] have reported the formation of the Sn(IV)-chlorodichloro-
methylchlorin **41** by reaction of Sn(IV)(OEP)(OH)$_2$ with SnBr$_4$ in chloro-

41

form. The thioacetoxyphlorin **42** is formed by reversible nucleophilic addition
of thioacetic acid,[180] a reaction certainly facilitated by the steric hindrance of
the parent porphyrin. The thiacyclic chlorin **43** (in rapid exchange equilibrium

42   43

with its mirror-image isomer) has been proposed by Clezy and Smythe[181]
because of the chlorin-type ^{1}H nmr spectrum for the product obtained by
hydrolysis of a *meso*-thioacetoxyporphyrin.

REFERENCES

1. G. L. Closs, J. J. Katz, F. C. Pennington, M. R. Thomas, and H. H. Strain, *J. Am. Chem. Soc.* **85**, 3809 (1963).
2. J. J. Katz, "Inorganic Biochemistry" (G. Eichhorn, ed.), p. 1022. Elsevier, Amsterdam, 1973.
3. J. Franck, *in* "Light and Life" (W. D. McElroy and B. Glass, eds.), p. 386. Johns Hopkins Press, Baltimore, Maryland, 1961.
4. D. Mauzerall and A. Chivvis, *J. Theor. Biol.* **42**, 387 (1973).
5. H. Fischer and H. Orth, "Die Chemie des Pyrrols," Vol. II, Part 2. Akad. Verlagsges., Leipzig, 1940 (reprinted by Johnson Reprint Organization, New York, 1968).
6. L. P. Vernon and G. R. Seely, eds., "The Chlorophylls." Academic Press, New York, 1966.
7. H. H. Inhoffen, J. W. Buchler, and P. Jäger, *Fortschr. Chem. Org. Naturst.* **26**, 785 (1968).
8. K. M. Smith, *Q. Rev., Chem. Soc.* **25**, 31 (1971).
9. R. B. Woodward and V. Skarii, *J. Am. Chem. Soc.* **83**, 4676 (1961).
10. H. Fischer and W. Lautsch, *Justus Liebigs Ann. Chem.* **528**, 247 (1937).
11. R. B. Woodward, *Ind. Chim. Belge* p. 1293 (1962).
12. A. E. Pullman, *J. Am. Chem. Soc.* **85**, 366 (1963).
13. J. V. Knop and J. H. Fuhrhop, *Z. Naturforsch., Teil B* **25**, 729 (1970).
14. P. S. Song, *in* "The Chemistry of Plant Pigments" (C. O. Chichester, ed.), p. 33. Academic Press, New York, 1972.
15. L. L. Shipman, T. M. Cotton, J. R. Norris, and J. J. Katz, *J. Am. Chem. Soc.* **98**, 8222 (1976).
16. C. Weiss, *J. Mol. Spectrosc.* **44**, 37 (1972).
17. R. C. Dougherty, H. H. Strain, and J. J. Katz, *J. Am. Chem. Soc.* **87**, 104 (1965).
18. J. H. Mathewson, W. R. Richards, and H. Rapoport, *Biochem. Biophys. Res. Commun.* **13**, 1 (1963).
19. A. S. Holt, J. W. Purdie, and J. W. F. Wasley, *Can. J. Chem.* **44**, 88 (1966).
20. D. N. Lincoln, V. Wray, H. Brockmann, Jr., and W. Trowitzsch, *J. Chem. Soc., Perkin Trans. 2* p. 1920 (1974).
20a. W. Trowitzsch, Ph.D. Thesis, Technische Universität, Braunschweig, West Germany (1974).
21. K. M. Smith and J. F. Unsworth, *Tetrahedron* **31**, 367 (1975).
22. H. Brockmann, Jr., *Phil. Trans. R. Soc. Lond.* **273**, 277 (1976).
23. R. Bonnett and G. F. Stephenson, *J. Org. Chem.* **30**, 2791 (1965).
24. R. Bonnett, I. A. D. Gale, and G. F. Stephenson, *J. Chem. Soc. C* p. 1600 (1966).
25. A. W. Nichol, *J. Chem. Soc. C* p. 903 (1970).
26. J. A. S. Cavaleiro and K. M. Smith, *J. Chem. Soc., Perkin Trans. 1* p. 2149 (1973).
27. G. H. Barnett, M. F. Hudson, S. W. McCombie, and K. M. Smith, *J. Chem. Soc., Perkin Trans. 1* p. 691 (1973).
28. H. Scheer and J. J. Katz, *in* "Porphyrins and Metalloporphyrins," 2nd ed. (K. M. Smith, ed.), p. 399. Am. Elsevier, New York, 1975.
29. D. Dolphin, Z. Muljian, K. Rousseau, D. C. Borg, J. Fajer, and R. H. Felton, *Ann. N. Y. Acad. Sci.* **206**, 177 (1973).
30. G. H. Barnett and K. M. Smith, *J. Chem. Soc. Chem. Commun.* p. 772 (1974).
31. H. Biere, Ph.D. Thesis, Technische Hochschule, Braunschweig, West Germany (1966).

32. H. Wolf, H. Brockmann, Jr., H. Biere, and H. H. Inhoffen, *Justus Liebigs Ann. Chem.* **704**, 208 (1967).
33. G. H. Barnett, M. F. Hudson, and K. M. Smith, *J. Chem. Soc., Perkin Trans. 1* p. 1401 (1975).
33a. K. Rousseau and D. Dolphin, *Tetrahedron Lett.* p. 4251 (1974).
34. N. Datta-Gupta and G. E. Williams, *J. Org. Chem.* **36**, 2019 (1971).
35. J. R. L. Smith and M. Calvin, *J. Am. Chem. Soc.* **88**, 4500 (1966).
36. H. Scheer, W. A. Svec, B. T. Cope, M. H. Studier, R. G. Scott, and J. J. Katz, *J. Am. Chem. Soc.* **96**, 3714 (1974).
37. H. Brockmann, Jr. and I. Kleber, *Tetrahedron Lett.* p. 2195 (1970).
38. H. W. Whitlock, Jr., R. Hanauer, M. Y. Oester, and B. K. Bower, *J. Am. Chem. Soc.* **91**, 7485 (1969).
39. H. H. Inhoffen, J. W. Buchler, and R. Thomas, *Tetrahedron Lett.* p. 1141 (1969a).
40. J. B. Conant and J. F. Hyde, *J. Am. Chem. Soc.* **52**, 1233 (1930).
41. H. Fischer and E. Lakatos, *Justus Liebigs Ann. Chem.* **506**, 123 (1933).
42. A. Stoll and E. Widemann, *Helv. Chim. Acta* **16**, 739 (1933).
43. E. M. Dietz and T. H. Werner, *J. Am. Chem. Soc.* **56**, 2180 (1934).
44. H. Wolf and H. Scheer, *Justus Liebigs Ann. Chem.* **1973**, 1710 (1973).
45. H. Wolf and H. Scheer, *Ann. N. Y. Acad. Sci.* **206**, 549 (1973).
46. D. Dolphin, *J. Heterocycl. Chem.* **7**, 275 (1970).
47. H. Scheer and H. Wolf, *Justus Liebigs Ann. Chem.* **1973**, p. 1741 (1973).
48. R. J. Kassner and P. S. Facuna, *Bioinorg. Chem.* **1**, 165 (1972).
49. A. H. Corwin and P. E. Wei, *J. Org. Chem.* **27**, 4285 (1962).
50. H. Fischer and S. Goebel, *Justus Liebigs Ann. Chem.* **515**, 130 (1936).
51. P. E. Wei, A. H. Corwin, and R. Arellano, *J. Org. Chem.* **27**, 4285 (1962).
52. H. Fischer, L. Filser, and E. Plötz, *Justus Liebigs Ann. Chem.* **495**, 1 (1932).
53. S. J. Baum, B. F. Burnham, and R. A. Plane, *Proc. Natl. Acad. Sci. U.S.A.* **52**, 1439 (1964).
53a. H.-P. Isenring, E. Zass, K. Smith, H. Falk, J.-L. Luisier, and A. Eschenmoser, *Helv. Chim. Acta* **58**, 2357 (1975).
54. J. H. Fuhrhop, *Z. Naturforsch., Teil B* **25**, 255 (1970).
55. H. Scheer and J. J. Katz, *J. Am. Chem. Soc.* **97**, 3273 (1975).
55a. H. Scheer, J. R. Norris, and J. J. Katz, *J. Am. Chem. Soc.* **99**, 1372 (1977).
55b. H. Falk, G. Hoornaert, H.-P. Isenring, and A. Eschenmoser, *Helv. Chim. Acta* **58**, 2347 (1975).
56. C. A. Rebeiz and P. A. Castelfranco, *Annu. Rev. Plant Physiol.* **24**, 129 (1973).
57. C. E. Strouse, *Proc. Natl. Acad. Sci. U.S.A.* **71**, 325 (1973).
58. H.-C. Chow, R. Serlin, and C. E. Strouse, *J. Am. Chem. Soc.* **97**, 7230 (1975); R. Serlin, H.-C. Chow, and C. E. Strouse, *ibid.*, p. 7237.
59. L. Kratky and J. D. Dunitz, *Acta Crystallogr., Sect. B* **31**, 1586 (1975).
60. J. Gassmann, I. Stell, F. Brandl, M. Sturm, and W. Hoppe, *Tetrahedron Lett.* p. 4609 (1971).
61. M. S. Fischer, D. H. Templeton, A. Zalkin, and M. Calvin, *J. Am. Chem. Soc.* **94**, 3613 (1972).
62. D. Mauzerall, *J. Am. Chem. Soc.* **84**, 2437 (1962).
63. H. H. Inhoffen, P. Jäger, R. Mählhop, and C. D. Mengler, *Justus Liebigs Ann. Chem.* **704**, 188 (1967).
63a. H. H. Inhoffen, P. Jäger, and R. Mählhop, *Justus Liebigs Ann. Chem.* **749**, 109 (1971).
64. G. S. Wilson and B. P. Neri, *Ann. N.Y. Acad. Sci.* **206**, 568 (1973).

65. R. B. Woodward, *J. Pure Appl. Chem.* **2**, 383 (1961).
66. R. B. Woodward, W. A. Ayer, J. M. Beaton, F. Bickelhaupt, R. Bonnett, P. Buchschacher, G. L. Closs, H. Dutler, J. Hannah, F. P. Hauck, S. Itô, A. Langemann, E. Le Goff, W. Leimgruber, W. Lwowski, J. Sauer, Z. Valenta, and H. Volz, *J. Am. Chem. Soc.* **82**, 3800 (1960).
67. V. P. Suboch, A. M. Shul'ga, G. P. Gurinovich, Yu. V. Glazkov, A. G. Zhuravlev, and A. N. Sevchenko, *Dokl. Akad. Nauk SSSR* **204**, 404 (1972).
68. B. P. Neri and G. S. Wilson, *Anal Chem.* **44**, 1002 (1972).
69. G. Peychal-Heiling and G. S. Wilson, *Anal. Chem.* **43**, 545 and 550 (1971).
70. J. G. Lanese and G. S. Wilson, *J. Electrochem. Soc.* **119**, 1040 (1972).
71. D. Dolphin, R. H. Felton, D. C. Borg, and J. Fajer, *J. Am. Chem. Soc.* **92**, 743 (1970).
72. R. Grigg, A. Sweeney, and A. W. Johnson, *Chem. Commun.* p. 1237 (1970).
73. J. Amlöf, *Int. J. Quant. Chem.* **8**, 915 (1974).
74. M. Gouterman, *J. Mol. Spectrosc.* **6**, 138 (1961).
75. C. Weiss, H. Kobayashi, and M. Gouterman, *J. Mol. Spectrosc.* **16**, 415 (1965).
76. J. H. Fuhrhop, *Angew. Chem.* **86**, 363 (1974).
77. J. F. M. Oth, H. Baumann, J. M. Gilles, and G. Schröder, *J. Am. Chem. Soc.* **94**, 3498 (1972).
78. P. George, *Chem. Rev.* **75**, 85 (1975).
79. H. W. Whitlock and M. Y. Oester, *J. Am. Chem. Soc.* **95**, 5738 (1973).
80. L. Witte and J. H. Fuhrhop, *Angew. Chem.* **87**, 387 (1975).
81. R. Bonnett, M. J. Dimsdale, and G. F. Stephenson, *J. Chem. Soc. C* p. 564 (1969).
81a. G. L. Closs and L. E. Closs, *J. Am. Chem. Soc.* **85**, 818 (1963).
82. J. H. Fuhrhop and T. Lumbantobing, *Tetrahedron Lett.* p. 2815 (1970).
83. D. A. Savel'ev, A. N. Sidorov, R. P. Evstigneeva, and G. V. Ponomarev, *Dokl. Akad. Nauk. SSSR* **167**, 135 (1966).
84. J. W. Buchler and L. Puppe, *Justus Liebigs Ann. Chem.* **740**, 142 (1970).
85. J. W. Buchler and L. Puppe, *Justus Liebigs Ann. Chem.* **1974**, 1046 (1974).
86. J. W. Buchler and K. L. Lay, *Z. Naturforsch., Teil B* **30**, 385 (1975).
87. G. N. Sinyakov, V. P. Suboch, A. M. Shul'ga, and G. P. Gurinovich, *Dokl. Akad. Nauk Beloruss. SSSR* **17**, 660 (1975).
88. H. Scheer and J. J. Katz, *Proc. Natl. Acad. Sci. U.S.A.* **71**, 1626 (1974).
88a. A. A. Krasnovskii, *Usp. Khim.* **29**, 736 (1960); *Russ. Chem. Rev.* (Engl. Transl.) **29**, 344 (1960).
89. H. H. Inhoffen, J. W. Buchler, and R. Thomas, *Tetrahedron Lett.* p. 1145 (1969).
89a. B. Ke *in* "The Chlorophylls" (L. P. Vernon and G. R. Seely, eds.), Chapter 8, p. 271. Academic Press, New York, 1966.
89b. E. E. Jacobs, A. S. Holt, R. Kromhout, and E. Rabinowitch, *Arch. Biochem. Biophys.* **72**, 495 (1957).
89c. J. R. Norris, R. A. Uphaus, H. L. Crespi, and J. J. Katz, *Proc. Natl. Acad. Sci. U.S.A.* **68**, 625 (1971); see also Volume V, Chapter 9.
90. F. Bertinelle and C. Zauli, *Mol. Cryst. Liq. Cryst.* **28**, 9 (1974).
90a. C.-A. Chin and P.-S. Song, *Int. J. Quant. Biol.* (in press).
90b. T. A. Evans and J. J. Katz, *Biochim. Biophys. Acta* **396**, 414 (1975).
91. H. Otten, *Photochem. Photobiol.* **14**, 589 (1971).
92. C. D. Mengler, Ph.D. Thesis, Technische Hochschule, Braunschweig, West Germany (1966).
92a. G. R. Seely, *J. Am. Chem. Soc.* **88**, 3417 (1966).
93. G. R. Seely and M. Calvin, *J. Chem. Phys.* **23**, 1068 (1955).

94. H. H. Inhoffen and W. Nolte, *Justus Liebigs Ann. Chem.* **725**, 167 (1969).
95. A. P. Johnson, P. Wehrli, R. Fletcher, and A. Eschenmoser, *Angew. Chem.* **80**, 622 (1968).
96. P. M. Müller, S. Farooq, B. Hardegger, W. S. Salmond, and A. Eschenmoser, *Angew. Chem.* **85**, 954 (1973).
97. J. W. Buchler and H. H. Schneehage, *Tetrahedron Lett.* p. 3805 (1972).
98. D. G. Whitten, J. C. Yau, and F. A. Carrol, *J. Am. Chem. Soc.* **93**, 2291 (1971).
99. G. R. Seely and K. Talmadge, *Photochem. Photobiol.* **3**, 195 (1964).
100. A. N. Sidorov, *in* "Elementary Photoprocesses in Molecules" (B. S. Neporent, ed.), p. 201. Plenum, New York, 1968.
101. A. M. Shul'ga, G. N. Sinyakov, V. P. Suboch, G. P. Gurinovich, Yu. V. Glazkov, A. G. Zhuravlev, and A. N. Sevchenko, *Dokl. Akad. Nauk SSSR* **207**, 457 (1972).
102. V. P. Suboch, A. P. Losev, and G. P. Gurinovich, *Photochem. Photobiol.* **20**, 183 (1974).
103. J. C. Goedheer, *in* "The Chlorophylls" (L. P. Vernon and G. R. Seely, eds.), Academic Press, New York, 1966.
104. A. A. Lamola and L. J. Sharp, *J. Phys. Chem.* **70**, 2634 (1966).
105. D. Fleischman, *Photochem. Photobiol.* **19**, 59 (1974); J. Lavorel, *Biochim. Biophys. Acta* **325**, 213 (1973).
106. G. Papageorgiou, *in* "Bioenergetics of Photosynthesis" (Govindjee, ed.), p. 320. Academic Press, New York, 1975.
107. A. A. Krasnovskii and F. F. Litvin, *Photochem. Photobiol.* **20**, 133 (1974).
108. Z. G. Fetisova and A. Yu. Borisov, *J. Photochem.* **2**, 511 (1974).
109. R. Livingston and V. A. Ryan, *J. Am. Chem. Soc.* **75**, 2176 (1953).
110. R. Livingston, G. Porter, and M. Windsor, *Nature (London)* **173**, 485 (1954).
111. R. Livingston, *J. Am. Chem. Soc.* **77**, 2179 (1955).
112. P. G. Bowers and G. Porter, *Proc. R. Soc. London, Ser. A* **296**, 435 (1967).
113. H. Linschitz and K. Sarkanen, *J. Am. Chem. Soc.* **80**, 4826 (1958).
114. S. Claesson, L. Lindquist, and B. Holmström, *Nature (London)* **183**, 661 (1958).
115. G. Zieger and H. T. Witt, *Z. Phys. Chem.* **28**, 273 (1961).
116. V. Zanker, E. Rudolph, and G. Prell, *Z. Naturforsch., Teil B* **25**, 1137 (1970).
117. B. M. Dzhagarov, E. I. Sagun, and G. P. Gurinovich, *J. Appl. Spectrosc.* **15**, 1195 (1975).
118. H. Kautsky and A. Hirsch, *Ber. Dtsch. Chem. Ges. B* **64**, 2677 (1931).
119. A. A. Krasnovskii, V. A. Romanyuk, and F. F. Litvin, *Dokl. Akad. Nauk SSSR* **209**, 51 (1973).
120. S. G. Boxer and G. L. Closs, private communication (1974).
121. S. G. Boxer and G. L. Closs, *J. Am. Chem. Soc.* **97**, 3268 (1975).
121a. J. R. Norris, *Photochem. Photobiol.* **23**, 449 (1976).
122. J. H. Golden, R. P. Linstead, and G. H. Whitham, *J. Chem. Soc.* p. 1725 (1956); H. R. Wetherell, M. J. Hendrickson, and A. R. McIntyre, *J. Am. Chem. Soc.* **81**, 4715 (1959).
123. H. Bürger, K. Burczyk, J. W. Buchler, J. H. Fuhrhop, F. Höfler, and B. Schrader, *Inorg. Nucl. Chem. Lett.* **6**, 171 (1970).
124. J. J. Katz, G. L. Closs, F. C. Pennington, M. R. Thomas, and H. H. Strain, *J. Am. Chem. Soc.* **85**, 3801 (1963).
125. J. J. Katz, R. C. Dougherty, and L. J. Boucher, *in* "The Chlorophylls" (L. P. Vernon and G. R. Seely, eds.), Chapter 7, p. 185. Academic Press, New York.
126. K. Ballschmiter and J. J. Katz, *J. Am. Chem. Soc.* **91**, 2661 (1969).
127. M. Lutz, *J. Raman Spectrosc.* **2**, 497 (1974).

128. H. Wolf and H. Scheer, *Justus Liebigs Ann. Chem.* **745**, 87 (1971).
129. H. Wolf and H. Scheer, *Tetrahedron* **28**, 5839 (1972).
130. H. Wolf, H. Brockmann, Jr., I. Richter, C. D. Mengler, and H. H. Inhoffen, *Justus Liebigs Ann. Chem.* **718**, 162 (1968).
131. E. A. Dratz, A. J. Schultz, and K. Sauer, *Brookhaven Symp. Biol.* **19**, 303 (1966).
132. D. W. Reed and B. Ke, *J. Biol. Chem.* **248**, 3048 (1973).
133. K. D. Philipson, S. C. Tsai, and K. Sauer, *J. Phys. Chem.* **75**, 1440 (1971).
134. L. Houssier and K. Sauer, *J. Am. Chem. Soc.* **92**, 779 (1970).
135. A. D. Trifunac and J. J. Katz, *J. Am. Chem. Soc.* **96**, 5233 (1974).
136. B. Briat, D. A. Schooley, R. Records, E. Bunnenberg, and C. Djerassi, *J. Am. Chem. Soc.* **89**, 6170 (1967).
137. N. A. Matwyoff and B. F. Burnham, *Ann. N. Y. Acad. Sci.* **206**, 365 (1973).
138. J. J. Katz and T. R. Janson, *Ann. N. Y. Acad. Sci.* **206**, 579 (1973).
139. R. A. Goodman, E. Oldfield, and A. Allerhand, *J. Am. Chem. Soc.* **95**, 7553 (1973).
140. S. G. Boxer, G. L. Closs, and J. J. Katz, *J. Am. Chem. Soc.* **96**, 7058 (1974).
141. R. C. Dougherty, G. D. Norman, and J. J. Katz, *J. Am. Chem. Soc.* **87**, 5801 (1965).
142. D. Lumpkin, *J. Chem. Phys.* **62**, 3281 (1975).
143. H. Falk, O. Hofer, and H. Lehner, *Monatsh. Chem.* **105**, 366 (1974).
144. E. Stier, *Hoppe Seyler's Z. Physiol. Chem.* **272**, 239 (1942).
145. T. Sjöstrand, *Scand. J. Clin. Lab. Invest.* **1**, 201 (1949).
146. R. Toxler, *Biochemistry* **11**, 4235 (1972).
147. O. Warburg and E. Negelein, *Ber. Dtsch. Chem. Ges. B* **63**, 1816 (1930).
148. R. Lemberg, *Pure Appl. Chem.* **6**, 1 (1956).
149. T. Kondo, D. Nicholson, A. H. Jackson, and G. W. Kenner, *Biochem. J.* **121**, 601 (1971).
150. R. Lemberg, B. Cortis-Jones, and M. Norrie, *Biochem. J.* **32**, 171 (1938).
151. H. Libowitzky and H. Fischer, *Hoppe-Seyler's Z. Physiol. Chem.* **255**, 209 (1938).
152. R. Bonnett and M. J. Dimsdale, *J. Chem. Soc., Perkin Trans. 1* p. 2540 (1972).
153. A. H. Jackson, G. W. Kenner, G. McGillivary, and G. S. Sach, *J. Am. Chem. Soc.* **87**, 676 (1965).
154. A. H. Jackson, G. W. Kenner, and K. M. Smith, *J. Chem. Soc. C.* p. 302 (1968).
155. H. Fischer, H. Gebhardt, and A. Rothaas, *Justus Liebigs Ann. Chem.* **482**, 1 (1930).
156. H. H. Inhoffen, J. H. Fuhrhop, and F. von der Haar, *Justus Liebigs Ann. Chem.* **700**, 92 (1966).
157. E. C. Taylor and A. McKillop, *Acc. Chem. Res.* **3**, 338 (1970).
158. P. S. Clezy and V. Diakiw, *Aust. J. Chem.* **24**, 2665 (1970).
159. H. H. Inhoffen and A. Gossauer, *Justus Liebigs Ann. Chem.* **723**, 135 (1969).
160. J. H. Fuhrhop, *in* "Porphyrins and Metalloporphyrins" (K. M. Smith, ed.), 2nd ed., p. 625. Am. Elsevier, New York, 1975.
161. K. M. Smith, *Chem. Commun.* p. 540 (1971).
162. J. H. Fuhrhop, *J. Chem. Soc. D* p. 781 (1970).
163. J. H. Fuhrhop, private communication (1975).
164. K. M. Smith, private communication (1975).
165. H. Fischer and A. Treibs, *Justus Liebigs Ann. Chem.* **451**, 209 (1927).
166. W. Sheldrick and J. H. Fuhrhop, *Angew. Chem.* **87**, 456 (1975).
167. H. H. Inhoffen and W. Nolte, *Tetrahedron Lett.* p. 2185 (1967).
168. H. Fischer and H. Pfeffer, *Justus Liebigs Ann. Chem.* **556**, 131 (1944).
169. A. W. Johnson and D. Oldfield, *J. Chem. Soc.* p. 4303 (1965).
170. H. H. Inhoffen and N. Müller, *Tetrahedron Lett.* p. 3209 (1969).
171. A. Eschenmoser, *IUPAC Congr., 23rd, 1971* Vol. 2, p. 69 (1971).

172. R. B. Woodward, *Pure Appl. Chem.* **33**, 145 (1973).
173. M. F. Hudson and K. M. Smith, *Chem. Soc. Rev.* **4**, 363 (1975).
174. W. Rüdiger, *Progr. Chem. Org. Nat. Prod.* **29**, 59 (1971).
175. H. J. Callot and A. W. Johnson, *Chem. Commun.* p. 749 (1969).
176. H. J. Callot, *Tetrahedron Lett.* p. 1011 (1971).
177. A. W. Johnson and A. Sweeney, *J. Chem. Soc., Chem. Commun.* p. 1424 (1973).
178. G. L. Collier, A. H. Jackson, and G. W. Kenner, *J. Chem. Soc. C*, p. 66 (1967).
179. J. H. Fuhrhop, T. Lumbantobing, and J. Ullrich, *Tetrahedron Lett.* p. 3771 (1970).
180. A. H. Corwin, A. B. Chivvis, R. W. Poor, D. G. Whitten, and E. W. Baker, *J. Am. Chem. Soc.* **90**, 6577 (1968).
181. P. S. Clezy and G. A. Smythe, *Chem. Commun.* p. 127 (1968).

3

The Porphyrinogens

D. MAUZERALL

I. INTRODUCTION

The porphyrinogens, which are colorless cyclic tetrapyrrylmethanes or *meso*-hexahydroporphyrins (Fig. 1), are important as precursors of the porphyrins both in biosynthesis and in the Rothemund condensation. They occur naturally in small amounts in all living cells, and are often excreted in copious quantities when porphyrin metabolism is disturbed, as in the diseases of porphyria. This subject is reviewed elsewhere in this treatise (Volume VI, Chapter 11). The porphyringens represent the most reduced state in the series of *meso*-reduced porphyrins (Fig. 1).

II. SYNTHESIS

There are essentially two routes to the porphyrinogens: reduction of porphyrins and condensation of pyrrole Mannich bases.

A. Reduction of Porphyrins

1. CHEMICAL

Fisher prepared several of the porphyrinogens in crystalline form.[1] He also developed a number of methods for the reduction of porphyrins to porphyrinogens: sodium amalgam, phosphonium iodide, and catalytic hydrogenation. In working with the naturally occurring porphyrins, we have found that 3% sodium amalgam is the most suitable reducing agent.[2] Because of the extreme efficiency of the autocatalytic porphyrin-sensitized photooxidation, the reduction is best carried out under nitrogen and in the dark. Porphyrins with reactive side groups, e.g., protoporphyrin, require very rapid reduction with a large excess of sodium amalgam.[3] The resulting porphyrinogen is a vinyl pyrrole and, thus, polymerizes extremely readily. The reactivity can be used to add sulfhydryl derivatives to the vinyl group.[4]

2. PHOTOCHEMICAL

The photoreduction of porphyrins with very mild reducing agents (EDTA and glutathione) leads to phlorins (Fig. 1), while stronger reducing agents (titannous or chromous chloride) result in the formation of porphomethenes (tetrahydroporphyrins, Fig. 1). Porphyrins are most likely photoreduced to the porphyrinogen stage with acidic stannous chloride or ascorbic acid. Whereas the quantum yield of phlorin formation[5] is near the theoretical maximum of 0.5 because of disproportionation of the primary one-electron reduced porphyrin radicals, the quantum yields of porphomethene and porphyrinogen formation are usually much lower (Volume V, Chapter 2).

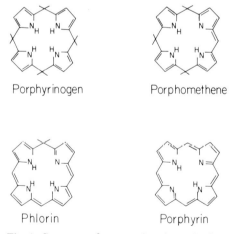

Porphyrinogen   Porphomethene

Phlorin   Porphyrin

Fig. 1. Structure of *meso*-reduced porphyrins.

This is because the most active photosensitizer is the porphyrin itself, which is rapidly reduced to the far less active phlorin,[5] thereby decreasing the photochemical rate. The phlorins are not themselves thermodynamically stable and, in the absence of air, disproportionate to porphyrins and porphomethenes.[5] The reaction is efficient but slow, with a half-time of 1 hour at 100°C or 1 day at 25°C. Further disproportionation to the porphyrin and porphyrinogen can occur, but quantitative data on this reaction have not been obtained.

B. Condensations

1. ROTHEMUND

The Rothemund condensation of aldehydes with pyrrole[6] results in the formation of porphyrinogens,[7] which are usually concomitantly autooxidized to porphyrins. Unfortunately, little work, aside from Dolphin's observations,[7] has been done to separate the condensation from the autooxidation. This condensation reaction, which is discussed in Chapter 3, Volume I, has made available a wide variety of *meso*-tetra-substituted porphyrins.

2. BIOSYNTHETIC

Interestingly, the biosynthetic pathway to porphyrins uses a similar Mannich base condensation. For a review, see Volume VI. The pyrrole porphobilinogen (Fig. 2, PBG), which is a reactive Mannich base, readily undergoes self-condensation to form uroporphyrinogen. The condensation is most rapid between pH 5 and 10. The "catch" in this condensation is that

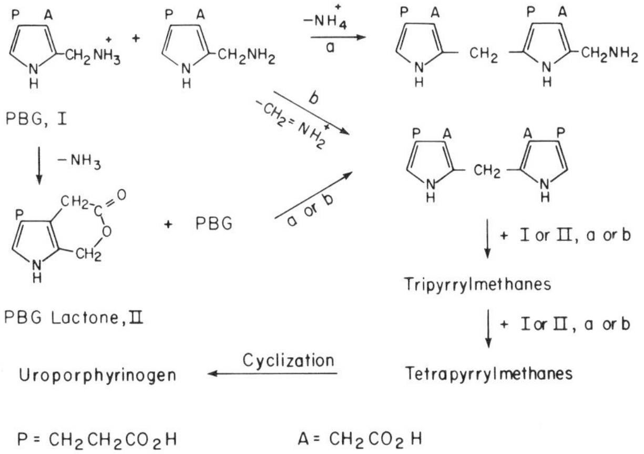

Fig. 2. Mechanism of condensation of porphobilinogen (PBG) to uroporphyrinogen.

Structure	Isomer	Probability
⌐PA-PA-PA-PA⌐	I	1/8
⌐PA-AP-PA-AP⌐	II	1/8
⌐PA-PA-PA-AP⌐	III	1/2
⌐PA-PA-AP-AP⌐	IV	1/4

Fig. 3. Structure of isomeric porphyrinogens formed from an unsymmetrically β,β'-disubstituted pyrrole and the probability of isomer formation when random.

the $-CH_2NH_2$ or $-CH_2OH$ group is almost as good a "leaving group" as is $-H$ on the pyrrole α-position. Thus, for pyrroles unequally substituted in the β-positions, four porphyrinogen isomers are possible (Fig. 3). Although the enzymatic mechanism that specifically forms uroporphyrinogen isomer III is still unknown (see Chapters 1–4 Volume VI), the chemical condensation has been shown to proceed by the mechanism[8] outlined in Fig. 2. Reaction occurs at both α-positions as shown by the formation of mixtures of isomers and by the incorporation of added ^{14}C-formaldehyde. The mixture of isomers is random in acid solution (see later) but increasingly forms isomer I towards alkaline pH (see Table 1). The isomer ratio was estimated by careful visual observation and later verified by a scanning fluorescence measurement of the chromatograms (Fig. 4), which were accurate to $\pm 2\%$. This analytical method has also been used by Kay.[9] The yield of porphyrinogen from PBG is 80% in acid solution, but it is greatly decreased in the presence of free formaldehyde

TABLE 1

Isomers and Yields of Porphyrin Formed on Condensing Isomeric Porphobilinogens [a]

Conditions	PBG			Iso-PBG		
	% yield	I	II	% yield	I	I
1 M HCl, 100°C, $\frac{1}{2}$ hr	67	0.13	0.13	67	0.14	0.10
pH 4.6, 60°C, 20 hr	55	0.21	0.11	73	0.25	0.07
pH 7.6, 60°C, 20 hr	50	0.40	0.03	56	0.42	0.03
pH 10.0, 60°C, 20 hr	53	0.49	0.01	65	0.51	< 0.01
0.1 M NaOH, 100°C, 2 hr	27	0.64	< 0.01	35	0.54	< 0.01

[a] The yield of coproporphyrin, following oxidation of uroporphyrinogen and decarboxylation of uroporphyrin, was determined by the scanning fluorescence method. The fraction of isomers I and II is given, the remainder being isomers III plus IV. (D. Mauzerall, unpublished observations, 1962.)

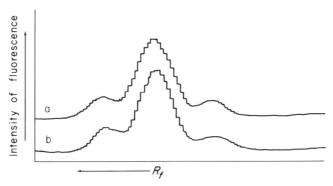

Fig. 4. The fluorescence intensity of isomeric coproporphyrins versus distance or R_f along the chromatogram. The fluorescence was excited by a low-pressure mercury lamp filtered for the 366-nm band, and measured with a photomultiplier filtered for > 600-nm emission. The chromatogram was advanced by a stepping motor. The paper was prewashed with EDTA to extract coppper ions that chelate the front running isomer and quench the fluorescence. The methodology was as described previously.[19] Curve a is a standard mixture of 1/8 I, 3/4 III + IV, and 1/8 II. Curve b is the product from PBG under acid conditions. The ratio of isomers was calculated from the areas of the respective curves. (Mauzerall, unpublished observation, 1962.)

or oxygen. The dependence on the formaldehyde:pyrrole ratio is shown in Fig. 5. In acid solution, the yield of uroporphyrinogen increases to a maximum at a molar ratio of 1, then rapidly falls. The decline in yield may be due either to the direct oxidation of pyrrylmethanes to methenes with reduction of formaldehyde, or to the condensation of formaldehyde with the intermediate polypyrrylmethanes, to form *meso*-methylenepyrrylmethanes, which can isomerize to *meso*-methylpyrrylmethenes.[10] Pyrrylmethenes are far less active

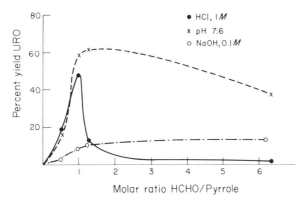

Fig. 5. The percent yield of uroporphyrin following oxidation is plotted versus the molar ratio of formaldehyde to opsopyrrole dicarboxylic acid. For details see Mauzerall.[8]

than the pyrrylmethenes in such condensations. The yield of uroporphyrin under neutral conditions decreases less drastically with an increase in the ratio of formaldehyde to pyrrole. The high yield of uroporphyrinogen from PBG may be facilitated by the formation of the lactone (Fig. 2), which effectively ties up the formaldehyde equivalent, yet activates it for reaction. Further evidence that the lactone is involved in the chemical condensation is the decrease in rate of reaction[8] of PBG at pH < 5. This lactone may also participate in the enzymic condensation.

3. ALLIED MACROCYCLES

Closely allied to the porphyrinogens are the *meso*-octamethylporphyrinogens, which are formed by condensing acetone with pyrroles.[11] A proof of structure has been given by nmr.[12] The analogous furan[13] and phenol[14] cyclic tetramers are also known. They are formed in high yield in very concentrated solutions and demonstrate the favored formation of the smallest nonhindered macrocycle by the "gem" effect,[15] that is, the favoring of cyclization by disubstitution on a carbon. This effect both favors the meeting of reactive ends by restricted rotations about the gem carbon–neighbor carbon bonds and minimizes steric hinderance in the macrocyclic form. The octamethyl porphyrinogens are stable, crystalline materials, and attempts have been made[16] to use them as host crystals to prepare dilute oriented samples of porphyrins. However, recent success[17] with simpler hosts, such as phthalic acid, may make this approach unnecessary.

III. PROPERTIES

A. Spectral

The porphyrinogens are colorless, having end absorption and a shoulder at 220 nm,[2,7] possibly an indication of interaction between the close pyrrole units. Spectral data on reduced porphyrins are collected in Table 2. The nmr spectra of tetraphenyl and octamethyl tetraphenyl[7] and of octamethyl and etioporphyrinogens[18] have been determined.

B. Autooxidation

The most obvious property of the porphyrinogens is their almost perverse ability to autooxidize. They share this property with many leucobases, and also the property of being more resistant to oxidation with increasing purity.[1] Autooxidation is photosensitized by the product, porphyrin, and is, therefore, strongly autocatalytic in the light[2] (Fig. 6). The autooxidation is

TABLE 2

Spectral Properties of Reduced Porphyrins[a]

Compound	λ_{max}	$\epsilon(\times 10^{-4})$	FWHM	pK	Ref.
Urophlorin					
Acid	735	2.2	120		
	440	6.2	36	9.3	5
Alkaline	650	2.0	120		
Zn octamethyltetraphenylporphodimethene	519	31.7	—	—	7
Uroporphomethene					
Acid	500	5.6	43		
	340	0.5	—	9.2	5
Alkaline	460	2.8	75		
Uroporphyrinogen	sh220	3.5	—	—	2

[a] Solvent: URO water, ionic strength $\sim 0.1\ M$, 23°C; Zn OMTPP-ethanol.

also autocatalytic in the dark, since it has in common with other autooxidations the kinetics of free radical initiation, propagation, and termination. It was during a study of these kinetics that I was led to the photoreduction of porphyrins[5] near neutral pH. At neutral pH, the autooxidation has side reactions, but in acid and light, the yield of porphyrin is nearly quantitiative.[2] The oxidation of uro- or coproporphyrinogen is best achieved by titration with iodine; exactly six equivalents are required.

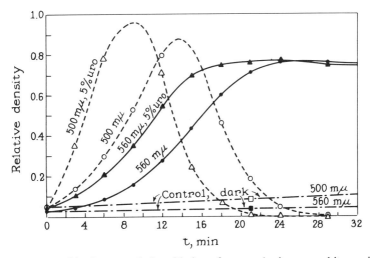

Fig. 6. Photosensitized autocatalytic oxidation of uroporphyrinogen and its sensitization by added uroporphyrin.[2] The dashed line represents the concentration of the intermediate uroporphomethene and the solid line that of the uroporphyrin.

C. Isomerization

Observations of the isomerization of porphyrinogens in acid solution have led to the demonstration of the remarkable thermodynamic stability of uroporphyrinogen. This has important implications with regard to the evolution of these propigments in prebiotic times.

Verification that the thermodynamic stability of a macrocycle exceeds that of all the many possible linear polymers depends upon the demonstration that equilibrium has been established.[19] Four observations provide proof that this condition is obtained. First, since the four similar pyrrole units of uroporphyrinogen contain two different β substitutes, there are four possible isomers of the macrocycle (Fig. 3). Thus, one argument for equilibrium is that the ratio of isomers should be random, i.e., 1/8 I, 1/8 II, 1/4 IV, and 1/2 III. After heating at 98°C for 30 minutes in deoxygenated 1 M HCl, this was found to be so and the yield of intact, cyclized porphyrinogen was about 85%. At the present time, isomers III and IV cannot be separated. The isomer analysis[19] was repeated with a photomultiplier scanning the fluorescent chromatogram (Table 1). Second, the same isomer mixture should be formed beginning with any given isomer. This was shown to be true for isomers I and III. Third, the ratio of isomers should be independent of time, once equilibrium is established. This was also true, limited only by a slow chemical decomposition of the uroporphyrinogen (1% during a 10-minute heating time sufficient to isomerize to equilibrium). Finally, the same random mixture of isomers, and yield of cyclized product should be obtained, beginning with the monomer. As mentioned above, the condensation of PBG in acid produces exactly this result. It was proved that, in all cases, over 95% of the product was in the fully cyclized, porphyrinogen form. Further, the condensation of opsopyrrole dicarboxylic acid with formaldehyde[19] gave the same results as PBG, as does condensation of *iso*-PBG (Table 1). Thus, we can confidently state that the thermodynamic stability of the macrocycle exceeds that of linear polymers.

Von Dobeneck et al.[20] have carried out a kinetic analysis of the randomization from isomer I, assuming that the isomerization proceeds through a single pyrrole inversion at a time. This kinetic constraint is equivalent to the thermodynamic statement that the cyclized porphyrinogen is the more stable species. The ratio of isomers at equilibrium is the same as that arrived at by statistical considerations. The formation of the same random mixture, starting with PBG,[8] the incorporation of added labeled formaldehyde,[19] and the decrease in yield on trapping formaldehyde with dimedone[19] all indicate that the macrocycle truly opened, and that no concerted strictly intramolecular isomerization occurred. The isomerization of coproporphyrinogen I also formed the random mixture of isomers on heating in acid, but the yield was lower. This may be an indication of the ability of the lactone (Fig. 2), specific to the acetic acid side chain, to tie up the reactive formaldehyde. Similar effects of lowered yield and much increased preoxidation of the macrocycle are seen on isomerizing uroporphyrinogen in the presence of small amounts of formaldehyde.[19] In beautifully detailed work on the role of uroporphyrinogen III in the biosynthesis of vitamin B_{12}, Scott et al.[21] have been able to resolve differences in labeling of the four meso positions of uroporphyrinogen in vitro reactions. The difference occurs at the meso-carbon between the two acetic acid groups of isomer III. They may be caused by the increased formation of the lactone (Fig. 2 and Mauzerall[19]). These and similar studies open the possibility of a detailed understanding of both the chemical and biological formation of these intriguing isomers.

IV. THERMODYNAMIC STABILITY AND EVOLUTION

The thermodynamic stability of the porphyrinogens is a striking example of the fact that cyclic systems prevail in nature. It is interesting that Carother's work, which led to the discovery of the nylon family of copolymers, arose during his extended studies of the factors determining cyclic versus linear polymerization.[22] In the systems we are discussing, the condensation reaction is favored thermodynamically, mainly because of the gain in bond energies. The system is assumed to be at equilibrium. In either cyclic or linear polymerization, the number and kinds of bonds are the same. For a given number of bonds formed, therefore, the ratio of cyclic to linear polymer is determined by secondary effects: namely, factors of symmetry, and the restricted rotation in the cyclic versus the linear chains. In addition to these variable effects, a more persistent effect is also present. In either addition or elimination condensations, the number of free molecules is greater by one for an N-mer cyclization than for an $(N + 1)$-mer linear polymerization. Thus, cyclizations are entropically favored by the greater number of independent particles in a cyclization. This effect is enhanced boundlessly in the presence of solvent.[23] In other words, since the equilibrium constant for cyclization has dimensions of one greater unit of concentration than does that of the linear polymerization, the ratio of cyclic to linear polymer will be inversely proportional to monomer concentration at equilibrium.

It is important to note that this "dilution effect" favoring cycles, which is usually explained in kinetic terms, is perfectly valid for equilibrium polymerization. An estimate of the effects of concentration can be made by calculating the equivalent concentration of the reactive ends of the polymer. For the tetrapyrrylmethane this end-to-end distance is maximally 18 Å, and the mean distance[8] is 9 Å; these distances correspond to concentrations of 0.5 M and

4 M. These concentrations are to be compared with a monomer concentration of $< 10^{-3}$ M. Thus, cyclization is favored by a factor of $> 10^3$. In fact when opsopyrroledicarboxylic acid was condensed with formaldehyde, the yield of uroporphyrinogen fell from 80% in dilute solution to 50% at 0.5 M pyrrole. I believe these considerations are generally valid for the Rothemund and similar condensations. A careful study of these equilibria as a function of temperature would be most valuable.

A corollary of this view is that the smallest energetically allowed macrocycle will be favored. For the pyrrylmethanes, the trimer forms a strained cycle, while the tetramer is free of any strain and is, thus, favored. Macrocycles with more than four pyrrole nuclei, which are capable of oxidation to aromatic planar heterocycles, are now known.[24,25]

A consideration of the stability of macrocycles is highly relevant to the elucidation of the evolutionary origin of the porphyrins. In the older literature, the occurrence of porphyrins in biological systems was assumed to be associated with their great aromatic stability. However, since both the enzymic and simplest chemical synthesis of porphyrins proceed via the porphyrinogen, it is apparent that this argument is inadequate. Proof that the porphyrinogens are the thermodynamically favored condensation of pyrroles and aldehydes is, thus, critical to a theory of the chemical evolution of the porphyrin pigment. The occurrence of isomer III in the biosynthesis of porphyrins is readily explained as the isomer most probable to form in the random condensation under prebiotic conditions, even near neutral pH. The porphyrinogens are strong reducing agents, and they could easily transfer hydrogen to other organic molecules. However, of more use for preoxygen photosynthesis[26,27] would be their possible elimination of hydrogen. Ultraviolet light present in the preoxygen atmosphere could catalyze this reaction. Experiments carried out (1959) to test this point were inconclusive, and should be repeated using modern methods. It has been argued[26] that uroporphyrin may be the prebiotic photosynthetic pigment, since it photooxidizes many organic compounds, itself becoming reduced to phlorin or porphomethene. These, or the final disproportionation products, porphyrinogens, could eliminate hydrogen and reform the porphyrin. This photosynthetic cycle could be the useful thermodynamic analogue of modern photosynthesis. It would store chemical energy in the products, oxidized organic compounds and hydrogen, thus forming the thermodynamic gradient favorable to life processes of that time. The oxidized organic compounds would be invaluable synthetic reagents in a reducing environment. The escape of hydrogen could have allowed a net production of these needed oxidized organic intermediates. The evolution of the photosynthetic path to oxygen superseded these reactions and led to the modern era of photosynthesis. With the large energy gap between molecular oxygen and most organic compounds now available, utilization of this energy by respiration was possible, and, thus, the path to more complex organisms was opened.

ACKNOWLEDGMENTS

I am very grateful to Dr. Sam Granick, who triggered many of these problems and whose constant probing led to much of their resolution. Dr. S. F. MacDonald kindly supplied the isoporphobilinogen and known isomers of uroporphyrin prepared by laborious synthesis.

REFERENCES

1. H. Fisher and H. Orth, "Die Chemie des Pyrrols," Vol. II, Part 2, pp. 420–423. Akad. Verlagsges., Leipzig, 1940.
2. D. Mauzerall and S. Granick, *J. Biol. Chem.* **232**, 1141–1162 (1958).
3. S. Sano and S. Granick, *J. Biol. Chem.* **236**, 1173–1180 (1961).
4. S. Sano and K. Tanaka, *J. Biol. Chem.* **239**, 3109–3110 (1964).
5. D. Mauzerall, *J. Am. Chem. Soc.* **84**, 2437–2445 (1962).
6. P. Rothemund, *J. Am. Chem. Soc.* **58**, 625–627 (1936).
7. D. Dolphin, *J. Heterocycl. Chem.* **7**, 275–283 (1970).
8. D. Mauzerall, *J. Am. Chem. Soc.* **82**, 2605–2609 (1960).
9. I. T. Kay, *Proc. Natl. Acad. Sci. U.S.A.* **48**, 901–905 (1962).
10. A. Triebs, E. Herrmann, E. Meissner, and A. Kuhn, *Justus Liebigs Ann. Chem.* **602**, 170–176 (1957).
11. P. Rothemund and C. L. Gage, *J. Am. Chem. Soc.* **77**, 3340–3342 (1955).
12. A. H. Corwin, A. B. Chivvis, and C. B. Storm, *J. Org. Chem.* **29**, 3702 (1964).
13. R. G. Ackerman, W. H. Brown, and G. F. Wright, *J. Org. Chem.* **20**, 1147–1158 (1955).
14. R. Ott and A. Zinke, *Oesterr. Chem.-Ztg.* **55**, 156 (1954).
15. N. L. Allinger and V. Zalkow, *J. Org. Chem.* **25**, 701–704 (1959).
16. J. Sharp and D. Mauzerall, unpublished observations at University of California, San Diego, 1966.
17. A. M. P. Goncalves and R. P. Burgner, *J. Chem. Phys.* **61**, 2975–2976 (1974).
18. G. V. Ponomorev, R. P. Evstigneeva, and N. A. Preobrazhenskii, *Zh. Org. Khim.* **7**, 169–173 (1971); *Chem. Abstr.* **74**, 112032w (1971).
19. D. Mauzerall, *J. Am. Chem. Soc.* **82**, 2601–2605 (1960).
20. H. von Dobeneck, B. Hansen, and E. Vollmann, *Z. Naturforsch., Teil B* **27**, 922–924 (1972).
21. A. I. Scott, N. Georgopapadakou, K. S. Ho, S. Klioze, S. Lee, S. L. Lee, G. H. Temme, III, and C. A. Townsend, *J. Am. Chem. Soc.* **97**, 2548–2550 (1975).
22. H. Mark and G. S. Wibey, *High Polym.* **1**, 1–265 (1940).
23. G. Gee, *Chem. Soc., Spec. Publ.* **15**, 67 (1961).
24. M. J. Broadhurst, R. Grigg, and A. W. Johnson, *J. Chem. Soc., Perkin Trans.* **1**, 2111–2116 (1972).
25. D. Dolphin and R. Woodward, personal communication.
26. D. Mauzerall, *Ann. N.Y. Acad. Sci.* **206**, 483–494 (1973).
27. D. Mauzerall, *Phil. Trans. Roy. Soc. London* **B273**, 287–294 (1976).

4

Oxophlorins (Oxyporphyrins)

P. S. CLEZY

I. INTRODUCTION

That the bile pigments of higher animals are derived from the blood pigment, hemoglobin, appears to have been known to scientists of the last century, although the chemistry of this transformation has remained obscure until relatively recently. A number of investigators in the early part of this century reported the production of green compounds when hemochromes were treated with oxygen in the presence of reducing agents. This work has been summarized by Fischer and Lindner[1] and by Lemberg.[2] The first identification of authentic bile pigment from one of these reactions was made by Lemberg.[2]

In studying the coupled oxidation of heme in pyridine as a model for bile pigment formation, Lemberg and his colleagues[3] isolated an intermediate, which they concluded was the iron complex of a porphyrin substituted by oxygen at a methene carbon. The properties of the compound suggested that it was a tautomeric species in which the enol form was favored as a metal chelate while the keto isomer predominated in the free porphyrin. The Munich group[4-6] obtained a similiar series of compounds by the action of hydrogen peroxide on several hemochromes. Their results included analytical figures for the oxidation product of coproporphyrin and the preparation of a benzoate derivative of the hydroxyl function. These data supported the conclusions of Lemberg's group regarding the structure of this type of compound —for which the name oxyporphyrin was suggested—although, for reasons that will become evident later, the term "oxophlorin" is now preferred.

For over 20 years there was little activity in this field, and then a decade ago several laboratories, stimulated by two factors, showed a renewed interest in this class of tetrapyrroles. First, there was continuing interest in the *in vivo* breakdown of hemoglobin to bile pigment, and, secondly, the oxophlorins became important intermediates in a synthetic pathway to porphyrins proper (see Chapter 6, Volume I). In the following sections, the chemistry of the oxophlorins is summarized.

II. PREPARATION OF THE OXOPHLORINS

A. From *b*-Oxobilanes

The most versatile synthesis of oxophlorins is achieved by the cyclization of *b*-oxobilane-1', 8'-dicarboxylic acids in the presence of trimethyl orthoformate/trichloroacetic acid. No symmetry restraints apply to this synthetic procedure, and a wide variety of oxophlorins have been prepared by this method, which has been discussed in detail in Chapter 6, Volume I.

B. From Pyrroketones and Pyrromethanes

As an extension of the Macdonald porphyrin synthesis,[7] the condensation of 5,5'-diformylpyrroketones and pyrromethane-5,5'-dicarboxylic acids provides an approach to the oxophlorins. In the development of this sequence, the main problem to be solved was the preparation of 5,5'-diformylpyrroketones, as general methods for the synthesis of this class of compound were not available when the work commenced.

5,5'-Diformyl-3,4,3',4'-tetramethylpyrroketone (1) (see Table 1)[8-13] was prepared from the hexamethylpyrroketone (2), using sulfuryl chloride under carefully controlled conditions,[8] but attempts to extend the procedure were unsuccessful, and other oxidizing agents were studied. Lead tetraacetate with some 5,5'-dimethylpyrroketones yielded crystalline 5,5'-diacetoxymethyl derivatives (3, 4), while on some occasions, when lead dioxide was included with the tetraacetate, further oxidation to the 5-acetoxymethyl-5'-formyl-pyrroketone (5) or to the 5,5'-diformylpyrroketone (6) occurred.[9] Bonnett et al.,[10] using tert-butyl hypochlorite, converted the 5,5'-dimethylpyrroketone (7) into the 5-formyl-5'-hydroxymethyl derivative (8). The precise oxidation

TABLE 1

Pyrroketones[a]

$$R^2 \overset{R^3 \quad O \quad R^4}{\underset{R^1 \quad NH \quad HN \quad R^6}{\diagup}} R^5$$

Structure No.	R^1	R^2	R^3	R^4	R^5	R^6
1[8]	CHO	Me	Me	Me	Me	CHO
2	Me	Me	Me	Me	Me	Me
3[9]	CH$_2$OAc	Me	Br	Me	Br	CH$_2$OAc
4[9]	CH$_2$OAc	Br	Me	Me	Br	CH$_2$OAc
5[9]	CH$_2$OAc	Me	H	H	Me	CHO
6[9]	CHO	Me	H	H	Me	CHO
7	Me	Et	Et	Et	Et	Me
8[10]	CHO	Et	Et	Et	Et	CH$_2$OH
9[9]	CHO	Me	P^{Me}	P^{Me}	Me	CHO
10[9]	CHO	Me	P^{Et}	P^{Et}	Me	CHO
11[9]	CHO	Me	Et	Et	Me	CHO
12[11]	CHO	Me	A^{Et}	A^{Et}	Me	CHO
13[12]	CHO	Me	Me	Me	CO$_2$Et	CHO
14[12]	CHO	Me	Et	Me	CO$_2$Et	CHO
15[12]	CHO	Me	Me	Me	Ac	CHO
16[12]	CHO	Me	Br	Me	CO$_2$Et	CHO
17[12]	CHO	Me	H	Me	CO$_2$Et	CHO
18[12]	CHO	Et	Me	Me	Et	CHO
19[13]	CHO	Me	Br	Me	Ac	CHO
20[9]	H	Me	Me	Me	Me	H
21[9]	H	Me	P^{Me}	P^{Me}	Me	H

[a] A^{Et}, CH$_2$CO$_2$Et; P^{Me}, CH$_2$CH$_2$CO$_2$Me; P^{Et}, CH$_2$CH$_2$CO$_2$Et.

level at the α-positions of the pyrroketone is not critical as far as oxophlorin formation is concerned, but the procedure is generally more reliable when 5,5′-diformylpyrroketones are employed, and, hence, other methods for the production of this type of compound were sought.

Osgerby and Macdonald[14] have reported that diethyl pyrromethane-5,5′-dicarboxylates can be oxidized to the corresponding pyrroketones by lead tetraacetate followed by lead dioxide. Attempts to use the same procedure with the readily available 5,5′-diformylpyrromethanes[15] were disappointing, but the transformation was achieved by using bromine followed by sulfuryl chloride, or in some cases by sulfuryl chloride alone.[9] The method is particularly efficient when the Vilsmeier–Haack imine salt precursor of the formyl group is used as substrate. The mechanistic details of the reaction have not been closely studied, but the spectroscopic changes observed suggest that pyrromethene intermediates are involved. This has been substantiated in some measure by showing that pyrroketones can be generated from pyrromethenes by the same procedure.[12]

The use of bromine and sulfuryl chloride has been extended to produce 5,5′-diformylpyrroketones by the oxidation of 5-formyl-5′-methylpyrromethanes.[12] The Vilsmeier–Haack imine salt intermediates of these compounds, when treated with bromine (1 equivalent) followed by sulfuryl chloride (3 equivalents), are oxidized at both the terminal methyl and at the linking methylene group to yield carbonyl functions.

The bromine/sulfuryl chloride procedure has made available a wide variety of 5,5′-diformylpyrroketones (9–19). Both symmetrical and unsymmetrical members of the series have been obtained, and the presence of negative groups does not interfere with the process. It is necessary, however, to protect, by bromination, unsubstituted β-positions prior to the use of sulfuryl chloride, as hydrogenolysis of the carbon–chlorine bond is much more difficult than a similar cleavage of a carbon–bromine bond.[12,16]

Ceric ammonium nitrate in buffered aqueous acetic acid has recently been reported[16a] as an effective oxidant for converting pyrromethanes into pyrroketones. With the inclusion of suitably protected aldehydic functions the method has been employed as an alternative route to the 5,5′-diformylpyrroketones.

Initially, condensations between 5,5′-diformylpyrroketones and pyrromethane-5,5′-dicarboxylic acids to yield oxophlorins were effected in ethanol–hydrogen bromide.[8,10,17] Later work has shown that trifluoroacetic acid is a better solvent for this reaction, which then requires no additional catalytic acid.[18] In this solvent, alkyl-substituted diformylpyrroketones furnished oxophlorins in 70% yield. The presence of electronegative groups in the pyrroketone reduced the yield of oxophlorin to about half that value, although this effect was minimized by the addition to trifluoroacetic acid of the two reactants as a finely ground mixture over an extended period of time.[12]

While condensation of a 5,5'-diformylpyrroketone with a pyrromethane-5,5'-dicarboxylic acid, or its α,α'-free derivative, is mechanistically more favored than the reaction of a 5,5'-diformylpyrromethane with an α,α'-unsubstituted pyrroketone,[8] the latter procedure has been used to produce oxophlorins.[18] For example, the pyrroketones (**20, 21**) have given oxophlorins

TABLE 2

Oxophlorins

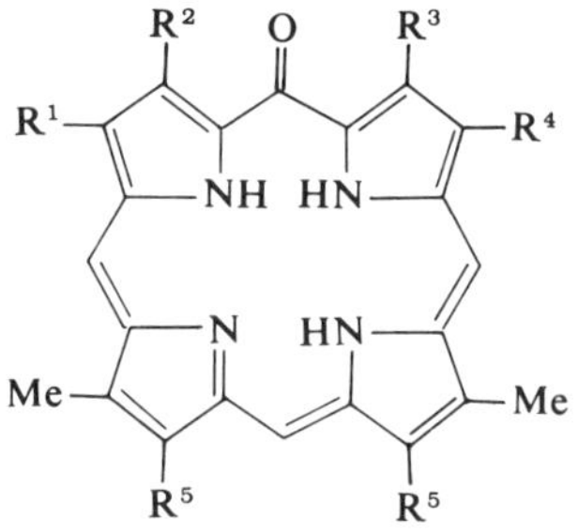

Structure No.	R^1	R^2	R^3	R^4	R^5
22[17]	Me	Me	Me	Me	P^{Et}
23[18]	Me	H	H	Me	P^{Et}
24[18]	Br	Me	Me	Br	P^{Et}
25[18]	Me	Et	Et	Me	P^{Me}
26[18]	Me	Me	Me	Me	P^{Me}
27[18]	Me	P^{Me}	P^{Me}	Me	P^{Me}
28[18]	Me	P^{Me}	P^{Me}	Me	Me
29[18]	Me	P^{Et}	P^{Et}	Me	Me
30[11]	Me	A^{Et}	A^{Et}	Me	Me
31[18]	Me	Br	Me	Br	P^{Et}
32[13]	Me	Br	Me	Ac	P^{Me}
33[12]	Me	Me	Me	CO$_2$Et	P^{Me}
34[12]	Me	Et	Me	CO$_2$Et	P^{Me}
35[12]	Me	H	Me	CO$_2$Et	P^{Me}
36[12]	Me	Br	Me	CO$_2$Et	P^{Me}
37[12]	Me	Me	Me	Ac	P^{Me}
38[20]	Me	Et	Me	Et	P^{Me}
39	Me	CH:CH$_2$	Me	CH:CH$_2$	P^{Me}
40[13]	Me	H	Me	Ac	P^{Me}
41[13]	Me	Ac	Me	Ac	P^{Me}
42	Me	CHOHCH$_3$	Me	CHOHCH$_3$	P^{Me}

in low yield with 5,5'-diformylpyrromethanes in the presence of acetic anhydride. The role played by the anhydride in this reaction has been discussed.[18]

The oxophlorins prepared by the pyrroketone plus pyrromethane procedure include the octamethyl[8] and octaethyl[10,19] derivatives, together with the symmetrically substituted representatives (22–30) (Table 2).[11–13,17,18,20] The general approach is limited by the same symmetry restraints as the Macdonald porphyrin synthesis; one of the dipyrrolic units has to be symmetrical to avoid the formation of porphyrin mixtures. Nevertheless, oxophlorins (31–42) related to the biologically important protoporphyrin series of isomers can be made by this approach.

C. By Oxidation of Metalloporphyrins

It has been mentioned (Section I) that iron derivatives of porphyrins undergo a coupled oxidation at a *meso*-position of the porphyrin nucleus. When the porphyrin involved is substituted in such a way that the four *meso*-carbon atoms are equivalent, this procedure becomes a useful preparative route to the oxophlorins.

Bonnett and Dimsdale[21] have reported a comparative study of hydrogen peroxide oxidation of a range of metal derivatives of octaethylporphyrin. Pyridine solutions of the Ni(II), Cu(II), Zn(II), Fe(III), and Co(III) complexes of this porphyrin were not appreciably affected by treatment with hydrogen peroxide at 50–60°C for a short period. On the other hand, Co(II), Fe(II), and Mn(II) complexes reacted readily with this reagent to give *meso*-oxygenated derivatives in high yield. The Mn(III) complex reacted less readily. Reaction seemed to occur in cases where the chelating metal had available a readily accessible higher oxidation state. In this context, it is surprising that the Mn(III) complex reacted at all, although there is evidence[22] that the porphyrin ligand stabilizes a Mn(IV) oxidation level.

In this reaction sequence, both the substrate and the metal suffer oxidation, and this has led Bonnett and Dimsdale[21] to suggest a mechanism for this reaction, which parallels that proposed for the hydroxylation of aromatic systems by Fenton's reagent ($FeSO_4$–H_2O_2). This involves the formation of an intermediate metalloporphyrin–peroxide complex,[23] which breaks down with oxidation of the metal to the Fe(III) state to give a hydroxyl radical that attacks the porphyrin ring system at a *meso*-position.

This mechanism has recently been criticized,[24,25] principally on the grounds that Fe(III) octaethylporphyrin is not attacked by hydroxyl (or benzoyloxy) radicals. On the other hand, zinc and magnesium complexes of octaethyl-porphyrin react readily with benzoyl peroxide to give metal derivatives of *meso*-benzoyloxyoctaethylporphyrin[24] by way of a π-cation radical and an isoporphyrin intermediate (see Scheme 1; M, Mg or Zn; R, PhCO). It has, therefore, been suggested[24,25] that porphyrin Fe(II) chelates react with hydroxyl radicals in much the same way (Scheme 1; M, Fe; R, H) to give a

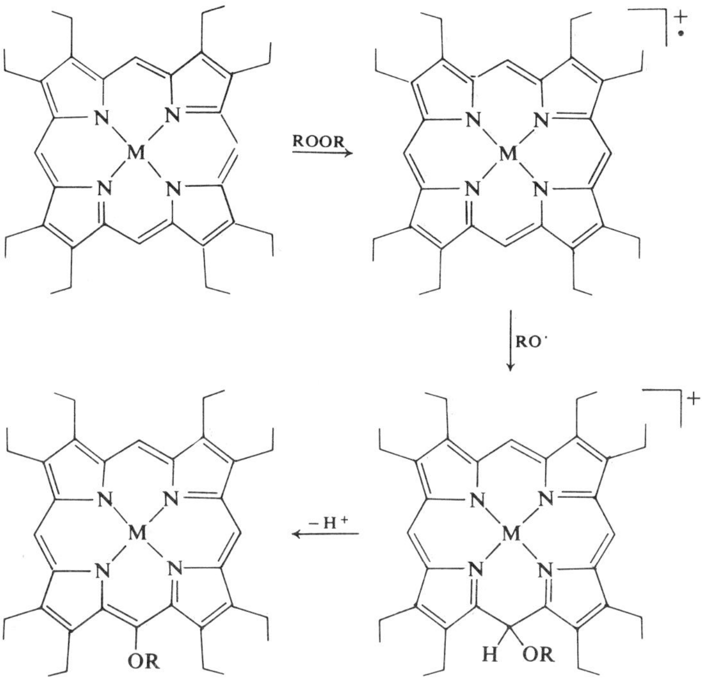

Scheme 1

metallooxophlorin derivative, which may then undergo further oxidation of the metal.

Barnett *et al.*[26] have reported that while Tl(III), Fe(III), Ni(II), and Cu(II) porphyrin complexes are not attacked by Tl(III) trifluoroacetate (TTFA) in the absence of trifluoroacetic acid, the magnesium and zinc chelates of octaethylporphyrin react readily with this reagent (1 equivalent) or with Tl(III) nitrate (1 equivalent). A green species is produced initially ($\lambda_{max} =$ 652–683 nm; no Soret) and this rapidly gives way to a red compound possessing the typical two-banded absorption pattern of a metalloporphyrin.

Demetalation furnishes octaethyloxophlorin. In a similar way, copro-oxophlorin-I tetramethyl ester has been prepared. The product derived from the reaction of TTFA with a porphyrin–magnesium complex is usually contaminated by small amounts of starting porphyrin. This complication does not occur with the zinc derivatives, which makes them more attractive starting materials. In similar reactions, Pb(IV) and Hg(II) trifluoroacetate produced oxophlorins, although in lower yields than obtained with TTFA.

Spectroscopic analysis of the red intermediate showed it was a metal chelate of *meso*-trifluoroacetoxyoctaethylporphyrin. The transient green species revealed esr and visible absorption characteristics similar to those recorded for π-cation radicals. Barnett et al.[26] have suggested that loss of a further electron from this intermediate gives a π-dication[27] which, as a strong electrophile,[28] reacts with a trifluoroacetate ion to yield a metalloisoporphyrin (**43**, M, Zn; R, CF$_3$CO). Deprotonation results in the formation of the *meso*-trifluoroacetoxymetalloporphyrin (**44**, M, Zn; R, CF$_3$CO). However, there is no spectroscopic evidence to support the formation of an intermediate π-dication. As an alternative it seems possible that the reaction proceeds by the pathway illustrated in Scheme 1, utilizing a trifluoroacetoxy radical, although trifluoromethyl substituted by-products were not isolated.

D. By Oxidation of Porphyrins

Benzoyl peroxide has been shown to give *meso*-benzoyloxyoctaethylporphyrin in 30% yield when allowed to react at 90–100°C with octaethylporphyrin.[10,29,29a,29b] Presumably free radical intermediates are involved. Hydrolysis of the benzoate affords the oxophlorin.

Some years ago, the Munich group[30] recorded the formation of tetrapyrrolic derivatives called "xanthoporphyrinogens" when lead dioxide in acetic acid was allowed to react with a range of porphyrins. These workers considered the products to be tetraketones (**45**), and this structure has been

45

R
46 CH:CH$_2$
47 Et

confirmed recently by Inhoffen et al.[19] Treatment of octaethylxanthoporphyrinogen with acetic anhydride and hydrobromic acid at 140–160°C gives octaethyloxophlorin.[19] Ketotetrapyrroles with oxidation levels between those of oxophlorins and xanthoporphyrinogens have also been reported.[19,31,32,32a]

III. PROPERTIES OF THE OXOPHLORINS

A. Tautomeric Properties

Recent work has established beyond doubt the tautomeric nature of the oxophlorins,[33] first reported by Lemberg and Fischer. The free bases have the ketonic structure (48), while the enolic tautomer (49; R, H) can be recognized in many derivatives. In particular, enolic esters are readily prepared and a number of acetate (49; R, Ac), benzoate (49; R, PhCO), and carbonate (49; R, CO₂Me) derivatives have been reported.[8,10,12,13,17,18,33-35] The keto form of the oxophlorins (48) is a vinylogous amide and, therefore, it is not surprising that these compounds react with triethyloxonium tetrafluoroborate to give the ethyl ether (49; R, Et). However, common methylating agents (diazomethane, methyl iodide, dimethyl sulfate) will not convert an oxophlorin into its methyl ether.

The formation of enolic derivatives of the oxophlorins clearly points to the nucleophilic character of the *meso*-oxygen atom in this series. With a favorable substitution pattern, this aspect of their chemistry is further illustrated in intramolecular reactions. Various attempts by Carr et al.[35] to prepare β-oxyprotoporphyrin dimethyl ester (46) were thwarted by a reaction of this type when a neighboring 2-chloroethyl substituent, destined to provide a vinyl group, yielded a cyclic ether (50) in spite of a number of attempts to protect the *meso*-oxygen function. The formation of the lactone (51), which arose during the methanolysis of the copper derivative of 75, further illustrates this type of reaction.[11]

B. Electronic Spectra

The oxophlorins are typically blue in color, which points to a changed absorption pattern from the one found for the red porphyrins. Nevertheless, the electronic spectrum of an oxophlorin is still characterized by a strong Soret band, which indicates that, in spite of the *meso*-carbonyl group, a high degree of conjugation is maintained, presumably through mesomeric contributions from dipolar forms such as 52. Apart from the Soret maximum, the spectrum of an alkyl-substituted oxophlorin is typically two-banded with one maximum between 580 and 590 nm and the major absorption near 635 nm (Fig. 1). Jackson *et al.*[33] have drawn attention to the similarity of this spectral pattern to that of a phlorin,[36] which has led to the use of the term "oxophlorin." Some oxophlorins, substituted by electronegative groups, exhibit a more complicated electronic spectrum (Fig. 2).[12,13,18]

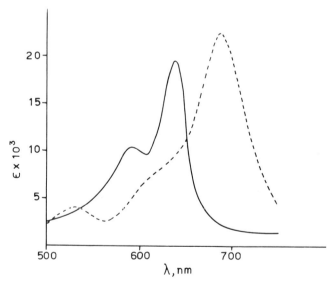

Fig. 1. The electronic spectra of the oxophlorin 25 in $CHCl_3$ (——) and glacial acetic acid (----).

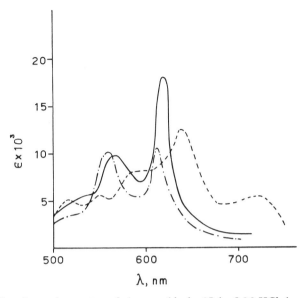

Fig. 2. The electronic spectra of the oxophlorin **25** in 5 M HCl (–·––·–) and the oxophlorin **36** in CHCl₃ (– – – –) and 10 M HCl (——).

C. Infrared Spectra

The infrared spectrum (irs) of an oxophlorin is characterized by a strong absorption maximum in the region 1560–1580 cm⁻¹ (cf. pyrroketones, Chapter 6, Volume I). This is a very low frequency for a carbonyl group absorption and indicates a high degree of polarization of the C–O bond, which is consistent with dipolar forms like **52** contributing to the oxophlorin structure.

D. Basicity

A striking feature of the chemistry of the oxophlorins is the increased basicity of the system relative to porphyrins. Their pK value has been judged[33] to be near 10, which is similar to the basicity of the phlorins assessed by Woodward[36] to have a pK of 9. Acetic acid is a convenient solvent in which to observe the monocationic derivative of an oxophlorin which has been formulated as compound **53**. In agreement with this structure, these ions possess an infrared maximum at 1560 cm^{-1}, and their electronic spectrum closely resembles a phlorin salt (**54**).[36] The major absorption in the electronic spectrum of an oxophlorin monocation is found near 700 nm (Fig. 1).

Strong acids produce dicationic derivatives of the oxophlorin system for which the structure **55** seems likely. The low frequency infrared carbonyl maximum has now disappeared, and the two-banded visible spectrum is similar to that of a normal porphyrin dication, although the longer wavelength maximum is stronger in an oxophlorin than in most other porphyrins. This latter feature is particularly marked in electronegatively substituted oxophlorins (Fig. 2).

Jackson et al.[33] have observed that a C-protonated species must be in equilibrium with the above oxophlorin cations. From a study of the nmr spectrum of an oxophlorin in CDCl$_3$, these workers noticed that the addition of a small quantity of CF$_3$CO$_2$D caused the highest field methene proton (assigned to the *meso*-position opposite the carbonyl group) to exchange with deuterium. This presumably occurs by way of the intermediate dication **56**, or perhaps the monocation **57**, although such species can only be present in very low concentration, as there is no absorption data to support their existence. No further exchange occurred at room temperature over several days,[33] but at 65°C all three methene protons in octaethyloxophlorin have been reported to deuterate rapidly.[10] The fact that one *meso*-proton exchanges preferentially in an oxophlorin is an important observation, as it points to a convenient procedure for introducing, specifically, tritium or deuterium labels into the porphyrin nucleus. As mentioned earlier (Chapter 6, Volume I), this has obvious applications in biosynthetic studies.[35]

Bonnett *et al.*[10] have demonstrated that the oxophlorins are not only stronger bases than porphyrins but are stronger acids as well. For example, treatment of octaethyloxophlorin with ethanolic potassium hydroxide produces the monoanion **58** which spectroscopically resembles the monocation. This reagent has no appreciable effect upon the spectrum of octaethylporphyrin.

E. NMR Spectra

The nmr spectrum of an oxophlorin dication[8,10,12,19,33] is consistent with this species being formulated as a *meso*-hydroxyporphyrin derivative. The three methene protons are at low field (about τ 0.0 to -1.0), which indicates the presence of the usual ring-current effect.

Nmr data for the monocation[33] are also in accord with the structure (**53**) proposed for this species. The *meso*-carbonyl function reduces conjugation in the macrocycle, and this is reflected in a diminished ring-current effect. In consequence, the resonances arising from substituents attached to the nucleus are found at higher field than normal. Particular note should be made of the position of the methene signals, which in the monocation occur between τ 0.5 and 2.0, the region where the bridge proton of a pyrromethene salt occurs.

Unexpectedly, the oxophlorin-free bases gave poorly resolved nmr spectra.[17,33,37] When the spectra of compounds **22** or **26** were determined at room temperature, in CDCl$_3$, the only signals discernible were those due to resonances arising from the esterifying alcohol (CH$_3$ or CH$_2$CH$_3$). However, as the temperature was lowered, the resonances due to the ring methyls sharpened, and at -30°C were observed between τ 7.2 and 7.4. This temperature effect upon the nmr spectrum of an oxophlorin was first reported by Jackson *et al.*,[37] who observed the methene protons (τ 3.0–3.5) as well as the ring methyls. Other workers have not been able to resolve the *meso*-protons in their spectra, but a temperature-dependent signal has been noted at τ 9.5, and this possibly arises from the NH protons.[18]

In connection with these results, it is interesting to note some details of the nmr spectra of the furan (**59**) and thiophen (**60**) analogues of the oxophlorins.[38]

These compounds, unlike their tetrapyrrolic counterparts, gave well-resolved nmr spectra in $CDCl_3$ under normal operating conditions. The positions of the methene protons (τ 2.0–3.0), ring methyls (τ 7.3–7.4), and the NH protons (τ 8.3, exchanges with D_2O) help confirm the assignments made on poorly resolved spectra obtained with the tetrapyrrolic oxophlorins. It is obvious from all three sets of spectra that greatly reduced ring current effects are operating in these molecules compared with the normal porphyrin macrocycle. This is consistent with the *meso*-carbonyl structures proposed for these nuclei.

F. ESR Spectra

Esr studies with the oxophlorins have revealed the existence of a paramagnetic species,[17,33,37,39] which is undoubtedly responsible for the broadening effect upon the nmr resonances outlined above. In agreement with this observation, the furan and the thiophen analogues of the oxophlorins, which gave normal nmr spectra, are not paramagnetic.[38] For some time, the paramagnetism of the oxophlorins was cautiously accepted as an inherent property of this nucleus, and was thought to be associated with a low-energy thermally attainable triplet state. However, doubt has been cast on this recently by the isolation of a stable green radical (61) obtained by oxidation of octaethyl-oxophlorin.[40,40a]

	R	X
59	Me	O
60	Et	S

The esr data for this radical are identical with those reported by Bonnett *et al.*[39] for the same oxophlorin. Thus, it would seem unnecessary to ascribe special properties to the oxophlorin system. Samples, prepared in various laboratories, for which paramagnetic properties have been reported were presumably contaminated by small quantities of a similar radical formed by autooxidation. The stable radical (61) has been oxidized further to give the dioxo derivative (62), which had been reported earlier.[31]

G. Metallo Derivatives

One of the best known reactions of the porphyrin macrocycle is its chelation with transition metal ions, and the oxophlorins provide no exception to this. Available evidence points strongly to the chelation of divalent metal ions by the hydroxy tautomer to give complexes formulated as compound **63**. This is not unexpected, as the keto form is obviously unsuited structurally to complex with this type of ion. Thus, the copper and zinc chelates are red, have a two-banded visible spectrum similar to the metallo derivatives of porphyrins, and show no evidence of the carbonyl frequency in the irs, which characterizes the oxophlorins.[33]

In agreement with their *meso*-hydroxy structure, the copper and zinc chelates show a certain resemblance to phenols. For instance, with potassium carbonate in acetone they yield green anions, which are best described as a hybrid of the mesomeric forms (**64** and **65**), as they exhibit a strong irs absorption at 1580 cm^{-1}. These anions react readily with methyl iodide or

63   64   65

p-toluenesulfonyl chloride.[33] Removal of the metal from the methoxy derivative provides a method of synthesis for the *meso*-methoxyporphyrins (**49**; R = Me) which cannot be obtained directly from the oxophlorin.

On the other hand, Bonnett and Dimsdale[21] have suggested that trivalent metal ions form complexes with the keto isomer. These workers have prepared Fe(III) (**66**), Co(III) (**67**), and Mn(III) (**68**) derivatives of octaethyl-oxophlorin, and all three chelates show a strong irs maximum near 1550 cm^{-1} due to the polarized *meso*-carbonyl group. The absence of a counter ion (analytical data), and the fact that the electronic spectra of **67** and **68** resemble those of normal Co(III)[41] and Mn(III)[22,42–45] complexes only after acidification, also support the *meso*-keto formulation of these chelates. The spectroscopic change that occurs upon acidification presumably arises from *O*-protonation to yield **69** which has the conjugation pattern of a normal porphyrin.

The formulation of these trivalent metal derivatives as complexes of the *meso*-carbonyl tautomer is in contrast with the structure assigned[33] to the iron complex (**70**). However, **70** was isolated by crystallization from methanolic

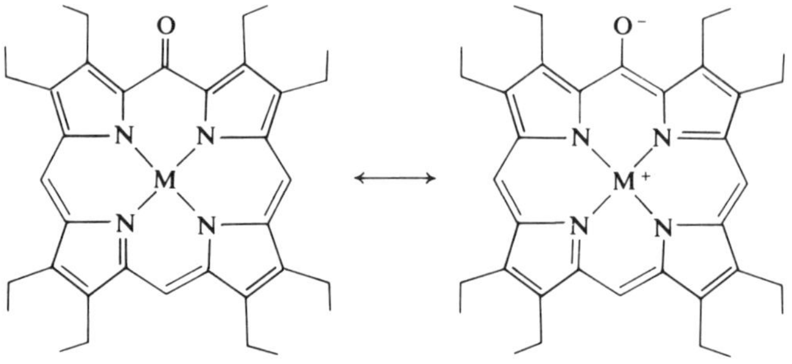

M

66 Fe(III)H$_2$O
67 CO(III)(pyridine)$_2$
68 Mn(III)

hydrogen chloride so it is probable that addition of hydrogen chloride occurred to give the *meso*-hydroxy derivative.

The Fe(II) chelates of the oxophlorins give pyridine hemochromes with maxima at positions similar to those obtained for the same derivative of the parent porphyrin, although, characteristically, the relative intensities of the two bands are reversed ($\epsilon_\beta > \epsilon_\alpha$ for oxophlorin hemochromes).[3,8,21] A similar change in absorption pattern has been reported for the copper chelates of the oxophlorins.[18]

Ni(II) oxyoctaethylporphyrin has been formylated by the Vilsmeier–Haack procedure to give the γ-formyl derivative.[40a]

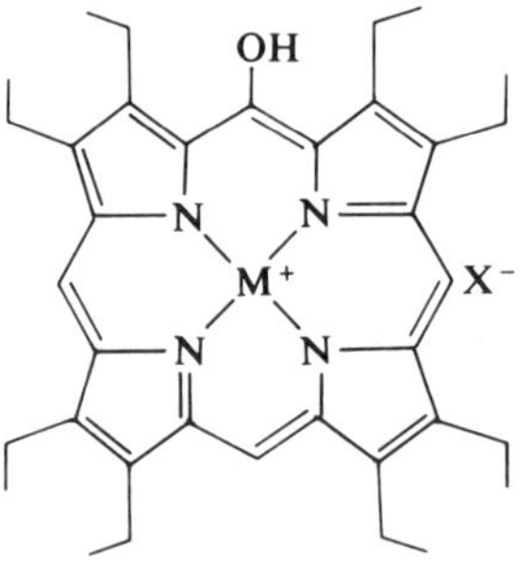

69 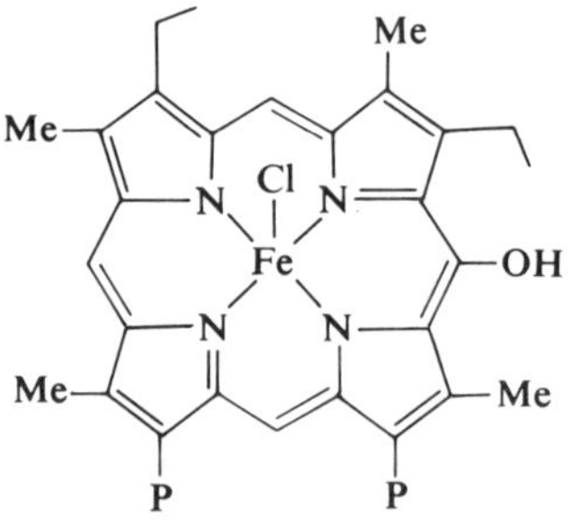   70
                                                     P = CH$_2$CH$_2$CO$_2$H

H. Mass Spectra

The mass spectra of the oxophlorins[19,33] and their metal chelates resemble those given by porphyrins. The molecular ion is usually the base peak, which points to the high stability of this macrocyclic system. Their enolic derivatives

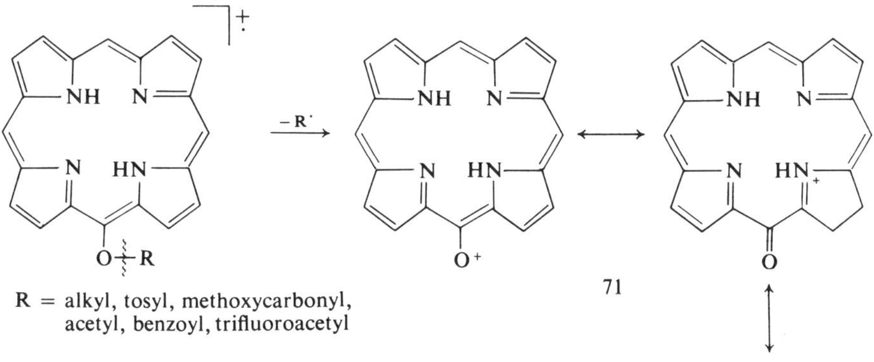

R = alkyl, tosyl, methoxycarbonyl,
acetyl, benzoyl, trifluoroacetyl

71

etc.

Scheme 2

usually cleave in the way illustrated in Scheme 2. In this they differ from most *meso*-substituted porphyrins, which normally fragment by loss of the full substituent.[46] More prominent in the mass spectrum of a *meso*-acetoxy-porphyrin is the loss of ketene to give the corresponding oxophlorin molecular ion, which provides the major peak in the spectrum.[33] Fragmentation of a *meso*-substituent in a porphyrin is facilitated by the steric interaction of this group with the neighboring β-substituents,[47] but the added stability of the oxygen-containing ion (71) in the oxophlorin case presumably leads to retention of this atom.[26]

Some interesting fragmentations have been noted when propionic ester groups flank the *meso*-oxygen function.[11] For example, the oxophlorins (29 and 28) lose ethanol and methanol, respectively; in both cases, this is followed by ejection of the elements of ketene. Their acetate derivatives (after initial loss of ketene) and their metal chelates behave in a similar fashion upon electron impact. In an analogous manner, the acetic ester derivative (30) suffers successive loss of ethanol and carbon monoxide.

I. Reduction

Reduction of the oxophlorin system is an important step in the synthesis of the porphyrin nucleus by way of *b*-oxobilane intermediates. In this connection, the use of sodium amalgam and catalytic hydrogenation followed by diborane has already been discussed (Chapter 6, Volume I). It has also been reported[18] that treatment of an oxophlorin with phosphorus pentasulfide yields the *meso*-unsubstituted porphyrin in low yield. However, the method of choice for removal of the *meso*-oxygen function involves reduction of the *meso*-acetoxy derivative.[34] This procedure has been described earlier (Chapter 6, Volume I) and has been employed in the preparation of a range of electronegatively substituted porphyrins, synthesized by way of oxophlorins, although not involving *b*-oxobilane intermediates.[12] Hydrogen in the presence of palladium on charcoal reduces the *meso*-acetoxyporphyrin to a colorless porphyrinogen at which stage the acetoxy group is either eliminated or removed by hydrogenolysis. Oxidation[34,48] furnishes the *meso*-unsubstituted macrocycle.

While hydrogenation of fully substituted *meso*-acetoxyporphyrins yields leuco compounds, less substituted members of the class have been reported to give *meso*-acetoxychlorins.[49] For example, reduction of **72** affords the chlorin **78**, while hydrogenation of **73** produces the chlorin **79** (Table 3). At no stage in these reactions were leuco compounds formed. In an intermediate case (**74**) where two neighboring positions were unsubstituted, a mixture of a *meso*-acetoxychlorin and a *meso*-unsubstituted porphyrin were produced upon hydrogenation. Even in examples where only one β-free position occurs, the production of small amounts of the *meso*-acetoxychlorin has been noted.[16]

From these results it would seem that two competing reactions are possible when *meso*-acetoxyporphyrins are hydrogenated in the presence of palladium on charcoal. In one case, the conjugated system is reduced, and this leads to a porphyrinogen from which the acetate is lost. In less substituted systems, the

TABLE 3

Oxophlorin Enol Esters

Structure No.	R^1	R^2	R^3	R^4	R^5	R^6	R^7	R^8
72[49]	P^{Me}	P^{Me}	Me	H	H	H	H	Ac
73[49]	Me	Me	Me	H	H	H	H	Ac
74[49]	P^{Et}	P^{Et}	Me	Me	P^{Et}	H	H	Ac
75[11]	A^{Et}	A^{Et}	Me	Me	Me	Me	Me	Ac
76[13]	$CH:CH_2$	Me	$CH:CH_2$	Me	P^{Me}	P^{Me}	Me	Ac
77[13]	$CH:CH_2$	Me	$CH:CH_2$	Me	P^{Me}	P^{Me}	Me	Bz
78[49]	P^{Me}	P^{Me}	Me	H	H	H_2	H_2	Ac
79[49]	Me	Me	Me	H	H	H_2	H_2	Ac

molecule is reduced at its periphery, and chlorins are the products. The two sequences seem to be mutually exclusive because in no case have *meso*-unsubstituted chlorins been reported. Steric factors probably dictate how the tetrapyrrolic system is orientated on the catalyst surface, and this decides the reaction path.

IV. OXIDATION OF HEMES

A. Verdoheme

Coupled oxidation of a pyridine solution of heme by oxygen or by hydrogen peroxide in the presence of either ascorbic acid or hydrazine produces a species called verdohemochrome.[2,3,50-52] This compound is characterized by a three-banded absorption spectrum with the main maximum at 657 nm. It is a true hemochrome with iron present in an octahedral complex. If a chloroform solution of this hemochrome is washed with dilute hydrochloric acid, the pure green color of the pyridine derivative gives way to an olive green species, which has a single broad maximum at 640 nm. From this solution, the pyridine-free verdoheme has been isolated in a crystalline condition.[51] Addition of pyridine reforms the verdohemochrome.

Lemberg[53] has shown that verdohemochromes can be converted by ammonia into monoazohemochromes (**80a**) in which a tertiary nitrogen atom occupies the site of one of the methene carbon atoms of the original heme. From this, it can be concluded that the verdohemes, although tetrapyrrolic, lack one *meso*-carbon atom. Foulkes *et al.*[51] have formulated them as the oxygen-bridged derivatives **80b** or **81**. Unlike iron chelates of porphyrins, the metal in verdoheme is labile and, under mild hydrolytic procedures, bile pigments are formed in high yield. Alkali converts the verdoheme into a bile pigment–iron complex from which the bile pigment itself can be derived with

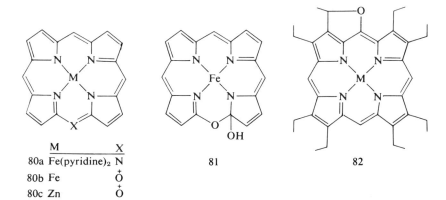

M	X
80a	Fe(pyridine)$_2$ N
80b	Fe $\overset{+}{O}$
80c	Zn $\overset{+}{O}$

81

82

weak acids.[23] The conversion of verdoheme into bile pigment is purely hydrolytic and proceeds in the absence of oxygen. These observations are consistent with the hemiacetal (81) or oxonium salt (oxaporphyrin) (80b) formulation of verdoheme, but it is difficult to see how structures proposed by the Munich group[4-6] or by Rackow[54,55] in which the porphyrin macrocycle is intact could yield bile pigments by such mild hydrolytic procedures. In addition, it has been demonstrated by Drabkin's group[56,57] that liberation of carbon monoxide occurs during verdohemochrome formation and not during its transformation into biliverdin.

B. Oxidation of Oxophlorins

The nature of the further oxidation of the oxophlorin system depends on the experimental conditions of the reaction.[32a,40a] The formation of the stable radical (61) and its oxidation to the dioxo derivative (62) has already been mentioned (Section III, F). The latter system (62) has also been obtained by chemical oxidation of magnesium, zinc, and nickel (II) chelates of oxophlorins or by irradiation of zinc octaethyloxophlorin in a polar solvent such as pyridine. Photooxidation of the same complex in nonpolar solvents yielded the zinc chelate of an oxaporphyrin with elimination of carbon monoxide.[58] Irradiation of zinc octaethyloxophlorin under anaerobic conditions or irradiation of the nickel complex in the presence or absence of oxygen gave the cyclic ether (82, M = Ni or Zn). It is likely that all oxidations involve the initial formation of the π-radical (61).

As mentioned in Section I, an oxophlorin (oxyporphyrin) was first recognized as an intermediate of heme oxidation. The availability of oxophlorins and their iron chelates has now enabled the oxidation of heme derivatives to be studied in greater detail.[20,21,33] For example, a pyridine solution of the Fe(III) chelate of β-oxymesoporphyrin dimethyl ester (47), when treated with oxygen in the dark, furnished a verdohemochrome from which the verdoheme derivative (83) was obtained with methanolic hydrogen chloride.[33] The electronic spectra of these compounds agreed well with existing data of this class, and

83   84

analytical results provided good support for Lemberg's formulation of verdoheme. Brief treatment with alkali, followed by the addition of methanolic hydrogen chloride, gave mesobiliverdin-IXβ dimethyl ester (**84**). In a similar manner mesobiliverdin-IXα dimethyl ester (**87**) was derived from α-oxy-mesoporphyrin dimethyl ester (**38**)[20] and the reaction has been studied with octaethylporphyrin as well. Thus, the pioneering work of Lemberg relating to the pathway of heme oxidation *in vitro* has been confirmed.

However, it has not yet been possible to examine this oxidation in the special and important case of α-oxyprotoporphyrin (**39**), as this compound has resisted all attempts at synthesis. As described above (Section III, A), extensive efforts to obtain β-oxyprotoporphyrin dimethyl ester (**46**) resulted in the formation of the cyclic ether (**50**), and similar results were to be expected in the α-series. However, the acetate (**76**) and benzoate (**77**) derivatives of α-oxyprotoporphyrin have been obtained in the following manner.[13] Hydro-genolysis of the bromooxophlorin (**32**) furnished the β-unsubstituted compound (**40**), which as the iron complex of its enol acetate derivative reacted with acetyl chloride in the presence of stannic bromide to give the diacetyl-oxophlorin (**41**), after removal of iron and hydrolysis of the enol acetate. Reduction of **41** with sodium borohydride produced the hydroxyethyl analogue (**42**) which, as its enol acetate or benzoate derivative, was dehydrated to give **76** or **77**, respectively.

Hydrolysis of these enolic esters afforded an oxophlorin, but it was not stable enough to be isolated and purified, which has prevented a detailed study of this important intermediate. However, it was found[13] that the electronic spectrum of a freshly prepared sample of the oxophlorin closely resembled the absorption data obtained by Lemberg *et al.*[3] for the compound isolated as an early intermediate in the coupled oxidation of heme. In addition, the oxidation of the iron chelate of the oxophlorin gave a product which was identical with biliverdin-IXα (**86**) in respect to melting point (m.p.), mixed m.p., electronic spectrum, and tlc behavior.[13] However, the yield of this bile pigment was very low, presumably due to the instability of the original oxophlorin.

C. Heme Catabolism

It is likely that an oxophlorin iron chelate is also involved in the *in vivo* oxidative breakdown of hemoglobin to give biliverdin. In the absence of α-oxyprotoporphyrin, Jackson and Kenner have prepared tritiated derivatives of both α- and β-oxymesoporphyrins and in collaboration with Kondo and Nicholson[59] have studied, in rats, the conversion of these compounds and their iron chelates into bile pigments. The free oxophlorins were poor precursors of bile pigment, as was the Fe(III) chelate of β-oxymesoporphyrin.

In the latter case, the radioactive bile pigment that was produced did not involve the formation of mesobiliverdin-IXβ. On the other hand, α-oxymeso-ferriheme was converted efficiently by the rat into bile pigment and specifically into the expected mesobiliverdin-IXα.

These results suggest that an α-oxyhemin derivative is an early intermediate formed in the oxidative breakdown of hemoglobin to bile pigment. The manner in which the oxophlorin is generated *in vivo* is uncertain, as are the steps between oxophlorin and biliverdin. It has been suggested that, in the later stages of this tranformation, hydroxylation or peroxidation occurs at an α-position of a pyrrole ring adjacent to the oxo group of the oxophlorin.[23,60] Another hypothesis[59,61] involves direct addition of oxygen to yield the peroxide **85**. Loss of carbon monoxide in a cheletropic fragmentation would give biliverdin-IXα (**86**) after rupture of the oxygen–oxygen bond and loss of the chelating iron (Scheme 3).

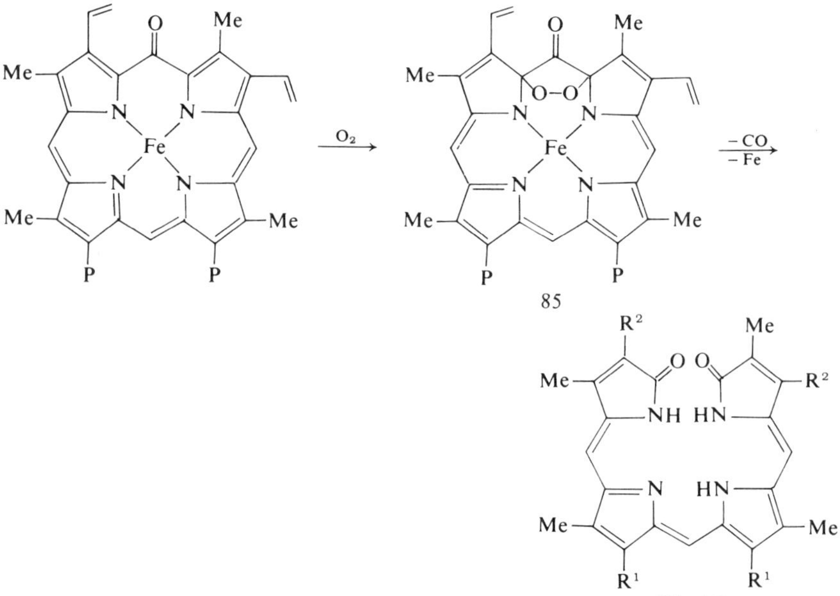

85

	R¹	R²
86	P	CH:CH₂
87	Pᴹᵉ	Et

Scheme 3

However, if hemoglobin catabolism proceeds via an oxophlorin derivative, it is difficult to reconcile the monoformyl bilatriene **91** of Nakajima[62] as part of that pathway, although the attractions of such an intermediate have recently been discussed again.[63,63a] Jackson and Kenner[61] have pointed out that a reductive step would be necessary to open the oxophlorin ring system to give this formyl derivative. In addition, it has been established that the α-methene carbon of heme is lost as carbon monoxide during bile pigment formation *in vivo*,[64,65] and it is not easy to see how this unit could be removed from **91** to give biliverdin (Scheme 4).

Fuhrhop and Mauzerall[66] have recently reported the formation of a bilatriene, similar to Nakajima's compound, by photooxidation of the magnesium chelate of octaethylporphyrin (**88**). There has been some discussion whether the product should be formulated in a cyclic (**92**) or open-chain form (**90**).[63,66] It has been suggested that the photooxidation proceeds via the peroxide **89**, and it seems likely that if Nakajima's compound has the structure **91** then it is formed from a heme derivative through a similar peroxide intermediate.[63]

A study[67] of the uptake of ^{18}O molecular oxygen during bile pigment formation has shown that the resultant bilirubin has incorporated two atoms of ^{18}O/molecule, while no labeled oxygen came from $H_2{}^{18}O$. While this result is in keeping with the production of biliverdin via the peroxide (**85**; Scheme 3), it seemed to exclude verdoheme from the *in vivo* pathway because in this case

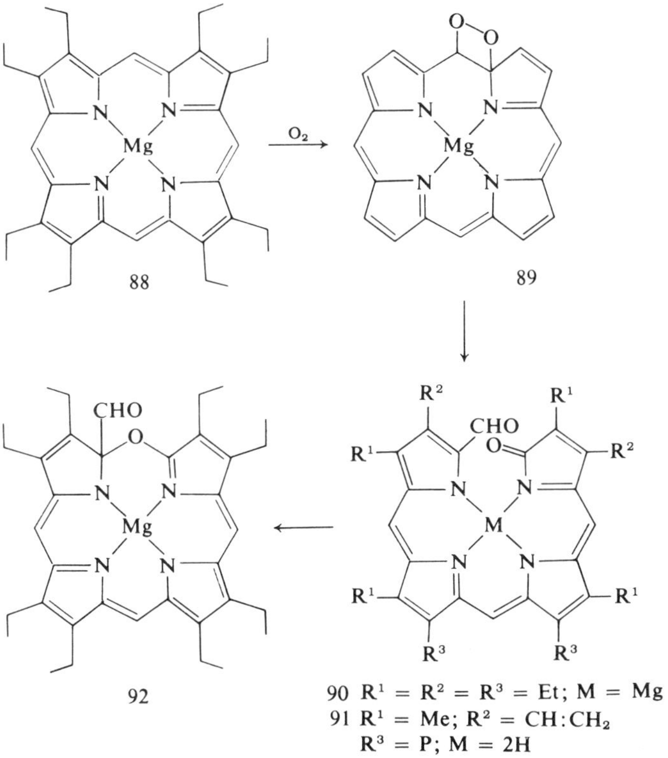

90 $R^1 = R^2 = R^3 = Et$; $M = Mg$
91 $R^1 = Me$; $R^2 = CH:CH_2$
     $R^3 = P$; $M = 2H$

Scheme 4

the bilirubin obtained would contain only one labeled oxygen atom/molecule. Results by Anan and Mason[60] support this expectation.

On the other hand a report[58] that compounds of the biliverdin type are interconvertible with verdoheme-like products allows that the formation of verdoheme *in vitro* could proceed by the pathway proposed for bile pigment production *in vivo* (Scheme 3). However, it is more likely that the two oxidative processes are different and that the formation of verdoheme *in vitro* follows a mechanism similar to that suggested for the decarbonylation of the zinc derivative of octaethyloxophlorin.[58]

Recent results have cast doubt on the simple addition of oxygen to the oxyheme system to yield the peroxide **85**. A one step reaction of this type requires the involvement of singlet oxygen and evidence for the participation of this species has not been forthcoming.[58a,67a] Moreover, bile pigment formation studies in living rats using $^{16}O_2$ enriched with $^{18}O_2$ has revealed that the terminal oxygen atoms of bilirubin are derived from separate oxygen molecules.[63a] Perhaps an intermediate of the type **93** is involved; this could collapse to the iron chelate of biliverdin by loss of hydrogen peroxide and carbon monoxide.

93

Apart from the verdoheme question, the main difference between the *in vivo* and *in vitro* oxidation of heme derivatives centers on the selectivity of the processes. In general, naturally occurring bile pigments have the IXα-type structure, although the tegument pigment of the cabbage butterfly *Pieris brassicae* has been shown to be biliverdin-IXγ (**96**).[68] Perhaps the announcement of this exception will herald the discovery of others![68a] It has been noted[69] that enzymatic reduction of biliverdin-IXβ to bilirubin is slow, compared with reduction of the α-isomer, and, thus, the absence of the β-compound from natural bile[70] could be due to its elimination in another fashion. Nevertheless, the natural oxidative breakdown of heme is believed to be highly selective, while the *in vitro* process seems not to be so. The degree of selectivity of the oxidation of heme outside the cellular environment has been the subject of extensive investigation.[71-76] However, the isolation of the four possible biliverdins, as their crystalline dimethyl esters, from the oxidation of heme seems to have finally resolved this question.[77]

The way in which the living system directs oxidation to the α-methene is not entirely clear. Woodward[36] has suggested the formation of an intermediate phlorin (94), formed by cyclization of a propionic acid, to explain this selectivity. The methene carbon opposite the saturated bridge of a phlorin is particularly susceptible to electrophilic attack; reaction of 94 with $^+$OH or its equivalent or perhaps with the electrophilic hydroxyl radical[78,79] would produce 95 with reduction of iron. Opening of the lactone ring affords an oxophlorin.

On the other hand, there is evidence that the oxidation of the porphyrin ring system is enzymically controlled. The authenticity of the oxygenase of Nakajima and his colleagues[62,80,81] has been questioned[74,75,82,83] but microsomal preparations of several tissues have been shown to have heme-cleaving activity.[84,85,86] The same property has been observed with chicken macrophages.[87]

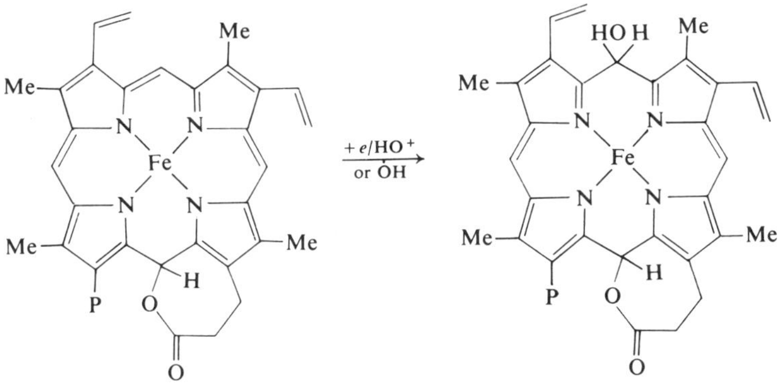

94   95

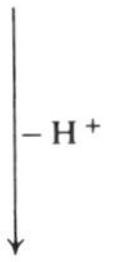

$+ e/\text{HO}^+$ or $\overline{\text{OH}}$

$-\text{H}^+$

Oxophlorin

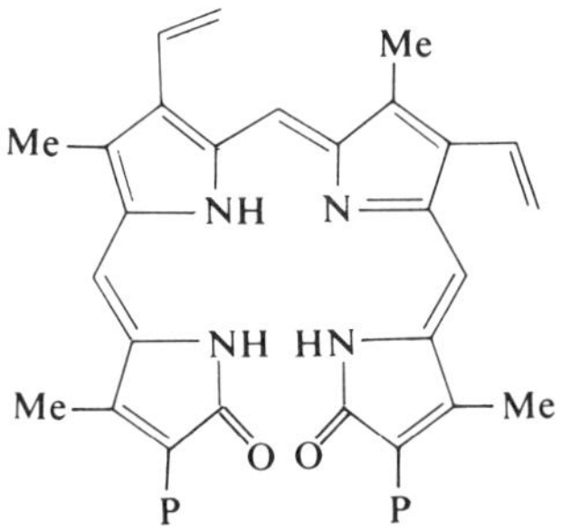

96

O'Carra and Colleran have taken a somewhat different view of the way in which this oxidation at the α-methene carbon is governed.[88] These workers have provided evidence that certain hemoproteins exercise a clear specificity as to the cleavage site in coupled oxidations.[69,89] In particular, myoglobin and a microsomal hemoprotein yield exclusively biliverdin-IXα, possibly by providing a hydrophobic environment about the α-methene carbon of the hemoprotein. Furthermore, it is known that the heme group of the microsomal cytochrome P-450 undergoes a more rapid turnover *in vivo* than its apoprotein.[90] The Irish workers have suggested that this occurrence is due to "accidental" cleavage of the prosthetic group by activated oxygen which normally would be transferred to an external substrate. O'Carra and Colleran conclude that heme released from hemoglobin or elsewhere is scavenged by this microsomal apoprotein (or a derivative) with eventual selective cleavage to yield biliverdin-IXα.

REFERENCES

1. H. Fischer and F. Lindner, *Hoppe-Seyler's Z. Physiol. Chem.* **153**, 54 (1926).
2. R. Lemberg, *Biochem. J.* **29**, 1322 (1935).
3. R. Lemberg, B. Cortis-Jones, and M. Norrie, *Biochem. J.* **32**, 171 (1938).
4. H. Libowitsky and H. Fischer, *Hoppe-Seyler's Z. Physiol. Chem.* **255**, 209 (1938).
5. H. Libowitsky, *Hoppe-Seyler's Z. Physiol. Chem.* **265**, 191 (1940).
6. E. Stier, *Hoppe-Seyler's Z. Physiol. Chem.* **273**, 47 (1942).
7. G. P. Arsenault, E. Bullock, and S. F. Macdonald, *J. Am. Chem. Soc.* **82**, 4384 (1960).
8. P. S. Clezy and A. W. Nichol, *Aust. J. Chem.* **18**, 1835 (1965).
9. P. S. Clezy, A. J. Liepa, A. W. Nichol, and G. A. Smythe, *Aust. J. Chem.* **23**, 589 (1970).
10. R. Bonnett, M. J. Dimsdale, and G. F. Stephenson, *J. Chem. Soc. C* p. 564 (1969).
11. P. S. Clezy, C. L. Lim, and J. S. Shannon, *Aust. J. Chem.* **27**, 1103 (1974).
12. P. S. Clezy and A. J. Liepa, *Aust. J. Chem.* **23**, 2461 (1970).
13. P. S. Clezy and A. J. Liepa, *Aust. J. Chem.* **23**, 2477 (1970).
14. J. M. Osgerby and S. F. MacDonald, *Can. J. Chem.* **40**, 1585 (1962).
15. R. Chong, P. S. Clezy, A. J. Liepa, and A. W. Nichol, *Aust. J. Chem.* **22**, 229 (1969).
16. A. H. Jackson, G. W. Kenner, and J. Wass, *J. Chem. Soc., Perkin Trans. 1* p. 480 (1974).
16a. J. B. Paine III and D. Dolphin, *Can. J. Chem.* **54**, 411 (1976).
17. P. S. Clezy, F. D. Looney, A. W. Nichol, and G. A. Smythe, *Aust. J. Chem.* **19**, 1481 (1966).
18. P. S. Clezy, A. J. Liepa, and G. A. Smythe, *Aust. J. Chem.* **23**, 603 (1970).
19. H. H. Inhoffen, J.-H. Fuhrhop, and F. von der Haar, *Justus Liebigs Ann. Chem.* **700**, 92 (1966.)
20. P. J. Crook, A. H. Jackson, and G. W. Kenner, *Justus Liebigs Ann. Chem.* **748**, 26 (1971).
21. R. Bonnett and M. J. Dimsdale, *J. Chem. Soc., Perkin Trans. 1* p. 2540 (1972).
22. P. A. Loach and M. Calvin, *Biochemistry* **2**, 361 (1963).
23. R. Lemberg, *Rev. Pure Appl. Chem.* **6**, 1 (1956).
24. D. Dolphin, Z. Muljiani, K. Rousseau, D. C. Borg, J. Fajer, and R. H. Felton, *Ann. N.Y. Acad. Sci.* **206**, 177 (1973).

25. D. Dolphin and R. H. Felton, *Acc. Chem. Res.* **7**, 26 (1974).
26. G. H. Barnett, M. F. Hudson, S. W. McCombie, and K. M. Smith, *J. Chem. Soc., Perkin Trans. 1* p. 691 (1973).
27. J. Fajer, D. C. Borg, A. Forman, D. Dolphin, and R. H. Felton, *J. Am. Chem. Soc.* **92**, 3451 (1970).
28. D. Dolphin, R. H. Felton, D. C. Borg, and J. Fajer, *J. Am. Chem. Soc.* **92**, 743 (1970)
29. R. Bonnett and A. F. McDonagh, *Chem. Commun.* p. 337 (1970).
29a. R. Bonnett, P. Cornell, and A. F. McDonagh, *J. Chem. Soc., Perkin Trans. 1* p. 794 (1976).
29b. C. E. Castro, C. Robertson, and H. Davis, *Bio-org. Chem.* **3**, 343 (1974).
30. H. Fischer and A. Stern, "Die Chemie des Pyrrols," Vol II, Part 2, pp. 423–429. Akad. Verlagsges., Leipzig, 1940 (reprinted by Johnson Reprint Corporation, New York, 1968).
31. J.-H. Fuhrhop, *Chem. Commun.* p. 781 (1970).
32. K. M. Smith, *Chem. Commun.* p. 540 (1971).
32a. G. H. Barnett, B. Evans, and K. M. Smith, *Tetrahedron* **31**, 2711 (1975).
33. A. H. Jackson, G. W. Kenner, and K. M. Smith, *J. Chem. Soc., C* p. 302 (1968).
34. A. H. Jackson, G. W. Kenner, G. McGillivray, and K. M. Smith, *J. Chem. Soc. C* p. 294 (1968).
35. R. P. Carr, A. H. Jackson, G. W. Kenner, and G. S. Sach, *J. Chem. Soc. C* p. 487 (1971).
36. R. B. Woodward, *Ind. Chim. Belge* p. 1293 (1962).
37. A. H. Jackson, G. W. Kenner, and K. M. Smith, *J. Am. Chem. Soc.* **88**, 4539 (1966).
38. P. S. Clezy and V. Diakiw, *Aust. J. Chem.* **24**, 2665 (1971).
39. R. Bonnett, M. J. Dimsdale, and K. D. Sales, *Chem. Commun.* p. 962 (1970).
40. J.-H. Fuhrhop, S. Besecke, and J. Subramanian, *J. Chem. Soc., Chem. Commun.* p. 1 (1973).
40a. J.-H. Fuhrhop, S. Besecke, J. Subramanian, Chr. Mengersen, and D. Riesner, *J. Am. Chem. Soc.* **97**, 7141 (1975).
41. A. W. Johnson and I. T. Kay, *J. Chem. Soc.* p. 2979 (1960).
42. G. Engelsma, A. Yamamoto, E. Markham, and M. Calvin, *J. Phys. Chem.* **66**, 2517 (1962).
43. L. J. Boucher, *J. Am. Chem. Soc.* **90**, 6640 (1968).
44. L. J. Boucher, *J. Am. Chem. Soc.* **92**, 2725 (1970).
45. L. J. Boucher, *Ann. N.Y. Acad. Sci.* **206**, 409 (1973).
46. D. R. Hoffman, *J. Org. Chem.* **30**, 3512 (1965).
47. R. B. Woodward, *Angew. Chem.* **72**, 651 (1960).
48. T. T. Howarth, A. H. Jackson, and G. W. Kenner, *J. Chem. Soc., Perkin Trans, 1* p. 502 (1974).
49. P. S. Clezy, V. Diakiw, and A. J. Liepa, *Aust. J. Chem.* **25**, 201 (1972).
50. R. Lemberg, B. Cortis-Jones, and M. Norrie, *Biochem J.* **32**, 149 (1938).
51. E. C. Foulkes, R. Lemberg, and P. Purdom, *Proc. R. Soc. London, Ser. B* **138**, 386 (1951).
52. E. Y. Levin, *Biochemistry* **5**, 2845 (1966).
53. R. Lemberg, *Aust. J. Exp. Biol. Med. Sci.* **21**, 239 (1943).
54. B. Rackow, *Hoppe-Seyler's Z. Physiol. Chem.* **308**, 66 (1957).
55. R. Lemberg, *Nature (London)* **181**, 1131 (1958).
56. G. D. Ludwig, W. S. Blakemore, and D. L. Drabkin, *Biochem. J.* **66**, 38P (1957).
57. W. S. Blakemore, G. D. Ludwig, R. E. Forster, and D. L. Drabkin, *Fed. Proc., Fed. Am. Soc. Exp. Biol.* **16**, 12 (1957).
58. S. Besecke and J.-H. Fuhrhop, *Angew. Chem., Int. Ed. Engl.* **13**, 150, (1974).

58a. A. H. Jackson, *In* "Iron in Biochemistry and Medicine" (A. Jacobs and M. Worwood, eds.), pp. 145–182. Academic Press, London, 1974.

59. T. Kondo, D. C. Nicholson, A. H. Jackson, and G. W. Kenner, *Biochem. J.* **121**, 601 (1971).

60. F. K. Anan and H. S. Mason, *J. Biochem. (Tokyo)* **49**, 765 (1961).

61. A. H. Jackson and G. W. Kenner, *Biochem. Soc. Symp.* **28**, 3–18 (1968).

62. H. Nakajima, *J. Biol. Chem.* **238**, 3797 (1963).

63. P. K. W. Wasser and J.-H. Fuhrhop, *Ann. N.Y. Acad. Sci.* **206**, 533 (1973).

63a. S. B. Brown and R. F. G. J. King, *Biochem. Soc. Trans.* **4**, 197 (1976).

64. T. Sjöstrand, *Acta Physiol. Scand.* **26**, 338 (1952).

65. S. A. Landaw, E. W. Callahan, and R. Schmid, *J. Clin. Invest.* **49**, 914 (1970).

66. J.-H. Fuhrhop and D. Mauzerall, *Photochem. Photobiol.* **13**, 453 (1971).

67. R. Tenhunen, H. Marver, N. R. Pimstone, W. F. Trager, D. Y. Cooper, and R. Schmid, *Biochemistry* **11**, 1716 (1972).

67a. A. H. Jackson, Proceedings of Bilirubin Meeting, Hemsedal, Norway (1974); quoted by P. O'Carra *in* "Porphyrins and Metalloporphyrins" (K. M. Smith, ed.), pp. 123–153. Elsevier, Amsterdam, 1975.

68. W. Rüdiger, W. Klose, M. Vuillaume, and M. Barbier, *Experientia* **24**, 1000 (1968).

68a. D. B. Morell and P. O'Carra, *Ir Med. Sci.* **143**, 181 (1974).

69. P. O'Carra and E. Colleran, *FEBS Lett.* **5**, 295 (1969).

70. C. H. Gray, D. C. Nicholson, and R. Nicholaus, *Nature (London)* **181**, 183 (1958).

71. R. Lemberg and J. W. Legge, "Hematin Compounds and Bile Pigments," pp. 459–460. Wiley (Interscience), New York, 1949.

72. Z. Petryka, D. C. Nicholson, and C. H. Gray, *Nature (London)* **194**, 1047 (1962).

73. C. O'hEocha, *Biochem. Symp.* **28**, 91–105 (1968).

74. E. Colleran and P. O'Carra, *Biochem. J.* **115**, 13P (1969).

75. A. W. Nichol and D. B. Morell, *Biochim. Biophys. Acta* **184**, 173 (1969).

76. W. Rüdiger, *Hoppe-Seyler's Z. Physiol. Chem.* **350**, 1291 (1969).

77. R. Bonnett and A. F. McDonagh, *J. Chem. Soc., Perkin Trans.* **1**, p. 881 (1973).

78. R. O. C. Norman and G. K. Radda, *Proc. Chem. Soc., London* p. 138 (1962).

79. C. R. E. Jefcoate, J. R. L. Smith, and R. O. C. Norman, *J. Chem. Soc. B* p. 1013 (1969).

80. H. Nakajima, T. Takemura, O. Nakajima, and K. Yamaoka, *J. Biol. Chem.* **238**, 3784 (1963).

81. O. Nakajima and C. H. Gray, *Biochem. J.* **104**, 20 (1967).

82. E. Y. Levin, *Biochim. Biophys. Acta* **136**, 155 (1967).

83. R. F. Murphy, C. O'hEocha, and P. O'Carra, *Biochem. J.* **104**, 6C (1967).

84. R. Tenhunen, H. S. Marver, and R. Schmid, *J. Biol. Chem.* **244**, 6388 (1969).

85. T. Yoshida, S. Takahashi, and G. Kikuchi, *J. Biochem. (Tokyo)* **75**, 1187 (1974).

86. M. D. Maines and A. Kappas, *Proc. Nat. Acad. Sci. USA* **71**, 4293 (1974).

87. A. W. Nichol, *Biochim. Biophys. Acta* **222**, 28 (1970).

88. P. O'Carra, "Porphyrins and Metalloporphyrins" (K. M. Smith, ed.), p. 123. Elsevier, Amsterdam, 1975.

89. P. O'Carra and E. Colleran, *Biochem. Soc. Trans.* **4**, 209 (1976).

90. W. Levin, M. Jacobson, E. Sernatinger, and R. Kuntzman, *Drug Metab. Dispos.* **1**, 275 (1973).

5

Irreversible Reactions on the Porphyrin Periphery (Excluding Oxidations, Reductions, and Photochemical Reactions)

J.-H. FUHRHOP

I. REACTIVITY PATTERNS OF THE PORPHYRIN PERIPHERY

This article deals with the main body of known electrophilic substitution reactions on the porphyrin periphery and the few nucleophilic reactions that have been demonstrated so far. Three types of carbon atoms are present in the porphyrin periphery (1) which may be attacked: the methine bridge carbon atoms α–δ, the β-pyrrolic carbons 1–8, and the α-pyrrolic carbons 1′–8′ (1).

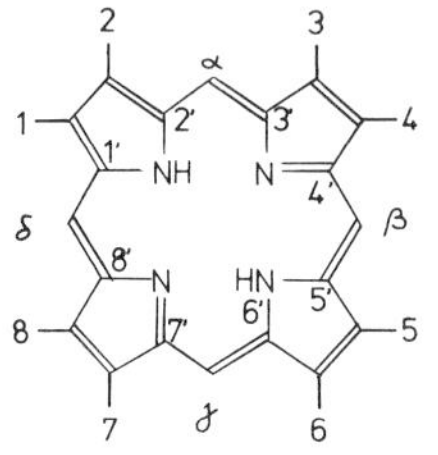

1

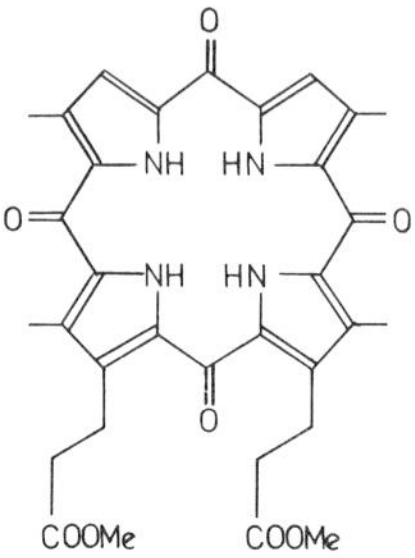

2

The first objective will now be to give an overall picture of the reactivity of the porphyrin in its irreversible substitution reactions. This will be derived from simple models, as well as from a summary of experimental results of porphyrin chemistry.

A. Affinity Criteria from Frontier Orbital Theory Calculations and Experimental Findings

The postulates of the frontier orbital theory of the reactivity of conjugated molecules are the following:

(a) The two electrons occupying the highest occupied π-orbital (HOMO) or the frontier electrons determine the reactivity of the molecule toward an electrophilic reagent. Molecular centers with the largest numerical value of the atomic orbital coefficient should be the most reactive.[1–4]

(b) For reactions of a conjugated molecule with nucleophiles the lowest unoccupied orbital (LUMO) is the frontier orbital. Again, the center of highest electron density in this particular orbital should be the most reactive site.[1–4]

(c) For predictions of the regioselectivity of radical reactions an average of the coefficients of both frontier orbitals is chosen.

In these postulates the important factors such as the influence of other π-electrons, σ-electrons, and solvation effects are neglected. The usefulness of their application can be clarified by comparison with experimental findings. A good, straightforward correlation of the predictions from this model with proven experimental reactivity patterns would then put some confidence into predictions on the reactivity of related, unexplored conjugated systems.

Reactivity indices for a metalloporphyrin (D_{4h} symmetry) are given in Table 1.[5,6] The first prediction which follows from these data is that porphyrins

TABLE 1

π-Electron Densities, Electrophilic and Nucleophilic Reactivity Indices of a Metalloporphyrin[a]

Atomic position	Charge density	Electrophilic[b]	Nucleophilic
α	0.97	0.29 (0.00)	0.16
1	1.01	0.05 (0.06)	0.07
1'	0.93	0.14 (0.19)	0.06
N	1.64	0.07 (0.00)	0.03

[a] For parameters used see ref. 6.
[b] Both values are given for the two almost degenerate highest occupied molecular orbitals.

should be most reactive on methine bridges in *all* types of reactions. This is in gratifying agreement with the bulk of experimental results, which will be described in the later sections. Two illustrative examples that will not be treated in this chapter should be mentioned here: most porphyrins, e.g., deuteroporphyrin dimethyl ester, are oxidized by lead dioxide in high yield to the xanthoporphinogens, e.g., **2**[7,8], and are reduced by a variety of reagents

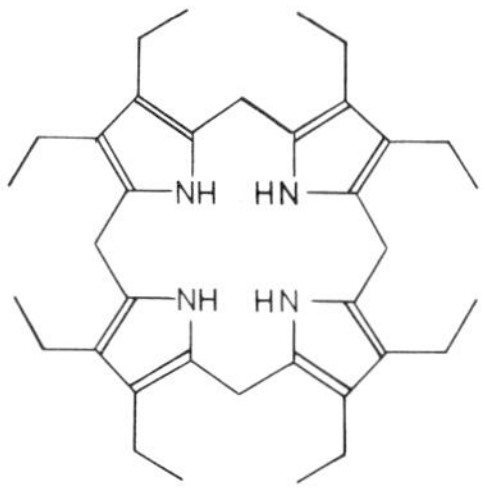

3

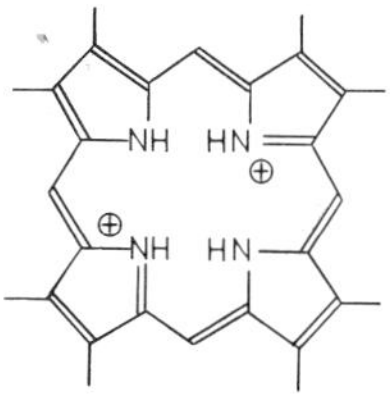

4

to porphinogens,[9] e.g., **3**, without any attack of α- or β-pyrrolic carbon atoms. More examples for selective attack of the *meso*-carbon atoms are given in Sections II, C; III, A; IV, B; and V. Exceptional preferences for β-pyrrolic substitution can be mostly rationalized on grounds of steric hindrance of the *meso*-carbons. Another point suggested by the values in Table 1 is the vulnerability of α-pyrrolic carbon atoms to electrophilic attack or oxidation. This is borne out by many reactions involving metalloporphyrins and oxygen: one usually first observes the oxidation of a methine bridge. This is often coupled with attack of an α-pyrrolic carbon atom by the second oxygen atom or the oxygen or hydrogen peroxide molecule involved (bile pigment formation, see Chapter 5, Volume VI). The β-pyrrolic positions usually remain unaffected.

Prominent exceptions are oxidations with bulky reagents (e.g., Os,O_4) or in strongly acid solution (e.g., H_2O_2 in H_2SO_4). (These reactions are discussed in Chapter 2 of this volume).

B. The Influence of Central Metal Ions

The free porphyrin base (1) may be diprotonated, to form a dication (4), or deprotonated to a dianion (5).[10] Compound (4) bears two positive charges,

and the porphyrin is, therefore, activitated toward nucleophilic attack or hydrogenation, e.g., the Krasnovskii photoreduction,[11] which usually occurs in acid solution. The dianion (5), on the other hand, is rapidly photooxygenated,[12] and this reaction is not observed with neutral base porphyrins (1). Electrophilic substitutions, however, cannot be performed on the dianions (5) because of the highly basic conditions under which these species are formed ($pk_1 + pk_2 \sim 30$).[10] No ordinary electrophile would be active in such a medium.

Introduction of different metal ions into the porphyrin cavity changes the reactivity of the porphyrin π-system in much the same way as the deprotonations and protonations do.[13,14] Substitution of the two NH protons in 1 by metal ions of low electronegativity, e.g., Mg^{2+}, lead to porphyrin ligands of relatively high[14,15] nucleophility and low oxidation potential of the porphyrin ring. A net ring charge of -0.6 has been calculated for magnesium porphyrins and of -0.3 for copper complexes.[13] Although magnesium porphyrins are ideally suited for the formation of cation radicals at low potentials and have been used in photooxygenation reactions[16] and radical substitutions,[16a] their acid lability prevents their general use in electrophilic reactions. Copper[17] or nickel[10] porphyrins represent a good compromise for nucleophilic and acid-stable porphyrins, which can be employed in Friedel–Crafts types reactions. Examples of activation of the porphyrin periphery by metal complex formation are given in Sections II, B and II, C.

On the other extreme of the scale of metalloporphyrins are the tin(IV) complexes that contain a highly electrophilic porphyrin ligand. Here the porphyrin π-system is inert toward the attack of bromine but is readily photohydrogenated by mild reductants, such as tetramethyl ethylene-diamine.[19] Another example is mentioned in Section II, B.

More intricate influences of the central metal ion on the reactivity of the porphyrin ligand than the electrostatic interactions considered above are sometimes found, when reagents with free electrons, e.g., oxygen or carbenes, are considered. In the dark, reactions of molecular oxygen or hydrogen peroxide with iron(II) porphyrins are extraordinarily favored, whereas copper(II) complexes are sometimes best suited for reactions with carbenes[20] (see Section II, B). The important reason for these special activations of the porphyrin ligands is presumably, that some inherently spin-forbidden processes in the courses of the reactions become spin-allowed in para-magnetic metalloporphyrins. A detailed discussion of this aspect of porphyrin reactivity is, however, beyond the scope of this article. The aspects of oxygen activation by iron porphyrins and their relation to carbene chemistry have been summarized in a recent review by Hamilton.[21]

C. Electronic Effects of Substituents and Partial Hydrogenation in the Porphyrin Periphery

Metal complexes of octaethylporphyrin and porphin differ in oxidation potential by about 200 mV, the unsubstituted porphin being more difficult to oxidize or being less electronegative.[22] However, the influence of β-alkyl substituents on any methine bridge substitution is not known. On the other hand, electronegative β-pyrrolic substituents, e.g., a formyl group, largely deactivate the porphyrin toward further electrophilic substitutions, e.g., introduction of a second formyl group by the Vilsmeier procedure.[23]

The same pattern is found for the influence of methine-bridge substituents on porphyrin reactivity. Electron-donating groups, such as alkyls or halogens, cause very little changes, whereas electron-withdrawing substituents such as formyl or cyano groups largely deactivate the porphyrin toward the attack of moderately strong electrophiles[17] (see Sections II, B and III, A for examples).

Strong regioselectivity in substition reactions is found in chlorins (6)[24] and phlorins (7).[24,25] Chlorins are preferentially attacked at the electron-rich methine bridges adjacent to the reduced pyrrole ring, whereas phlorins are substituted on the *meso*-position opposite the methylene bridge. (Examples in Sections II, B; II, C; III, A; and V.)

D. Steric Considerations

Very often reactions that should be favored by strong electronic inter-actions, e.g., the Friedel–Crafts type alkylations and acylations on methine bridges of electronegative metalloporphyrins, do not take place. The reason for this quite general problem of porphyrin chemistry is the enormous steric crowding of substituents in the porphyrin plane, when all β-pyrrolic positions and a methine carbon bear large substituents. The steric strain is largely relieved, when one of the three interacting porphyrin carbon atoms becomes sp^3 hybridized, so that its substituents are kept out of the macrocyclic plane. Therefore, the formation of chlorins (6) or phlorins (7) is thermodynamically

favorable in such highly substituted porphyrins (for examples see Sections II, A and IV, B). This insight into one of the fundamental properties of the porphyrin ring and its most imaginative preparative use is one of main merits of Woodward's chlorophyll synthesis.[24,26]

Another important empirical aspect of porphyrin chemistry is the fact that additions to α-pyrrolic carbon atoms lead immediately to a rupture of the macrocycle (bilatriene formation; see Chapter 6, this volume). No reason for this general observation is obvious to the author. In closely related systems, e.g., tetradehydrocorrins[27] and phthalocyanines,[28] tetrapyrrolic macrocycles with α-pyrrolic substituents are quite stable compounds.

E. Porphyrin–Phlorin Equilibria

Addition of a reagent to a methine bridge, equivalent to the formation of a substituted phlorin, is the most important single reaction in the porphyrin series. Some electron-rich *meso*-substituents, e.g., phenolic oxygen in *meso*-oxyporphyrins, amino nitrogen in *meso*-aminoporphyrins and possibly halogen atoms, may donate their unshared electrons to the porphyrin and form exocyclic double bonds (e.g., oxophlorins, iminophlorins). In such cases, it is sometimes found that the neutral compound is in the form of the porphyrin, whereas a positively charged π-system, either obtained by protonation or oxidation, occurs as a phlorin. These conversions are often fully reversible and demonstrate that the difference between the resonance stabilization energies of porphyrins and phlorins is not large. Three typical examples of such reactions are illustrated by reproduction of the corresponding electronic spectra, which were obtained in acid–base titrations (see Figs. 1, 2, and 3 in Section III, B).

II. INTRODUCTION OF CARBON SUBSTITUENTS INTO THE PORPHYRIN PERIPHERY

A. Intramolecular Cyclizations

Mesoporphyrin dimethyl ester (**8**) is converted by a 1:1 mixture of concentrated sulfuric acid and 65% oleum and subsequent re-esterification with diazomethane into a mixture of the isomeric mesorhodins[29-32] (**9**, m.p.

9   10

262–266°C)[32] and (**10**, m.p. 232–233°C).[32] The yield is almost quantitative. The isomers have been separated by fractionated crystallization.[32] The name rhodin was selected by Fischer for these cyclohexanone derivatives because their electronic spectrum possessing a strong α-band [λ_{max} (ϵ) for **9**:411 nm (133,000) 514 (10,400) 552 (8000) 585 (5300) 638 (9100)[32]] resembled that of rhodin-g_7, a degradation product of chlorophyll b. The stretching vibration of the conjugated ketone carbonyl occurs at 1675 cm^{-1}.[32] Reversibility of rhodin formation has been shown; the copper complexes of the isomer mixture (**9, 10**) yield pure mesoporphyrin in a melt with succinic acid.[33]

If the mesorhodin isomer mixture (**9, 10**) is heated in acid solutions, e.g., refluxed in acetic acid, a color change from violet to green caused by an almost quantitative dehydrogenation to the cyclohexenone derivatives, or "mesoverdins" (**11**, m.p. 265–268°C, **12**, m.p. 233–235°C) is observed. These "verdins" have a chlorin type electronic spectrum [**11**:336 nm (41,400) 380 (40,500) 435 (67,500) 454 (66,000) 704 (11,000)][32] and may again be separated by fractionated crystallization. The ketone carbonyl absorption is shifted to 1612 cm^{-1}.[32] The *exo*-macrocyclic double bonds in the mesoverdins (**11, 12**) have been characterized by catalytic hydrogenation to the corresponding rhodins (**9, 10**)[32] and Azzarello-type pyrazoline ring formation

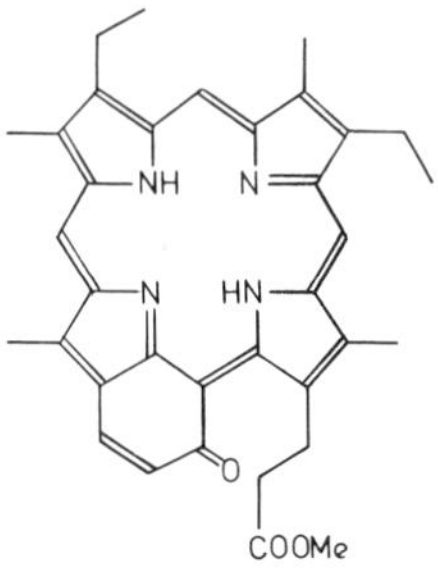

11

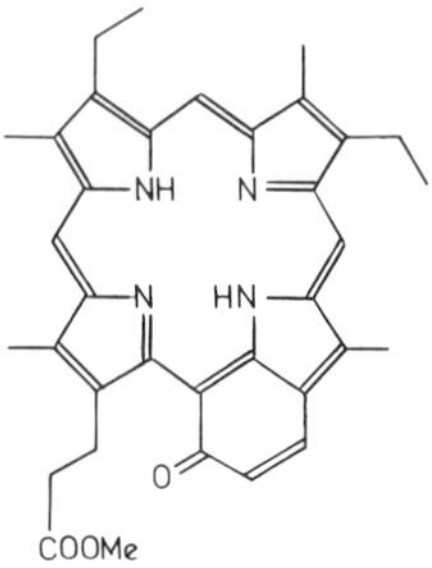

12

with diazomethane[34–36] to form **(13**, m.p. 282–285°C) from **(11**),[32] for example.

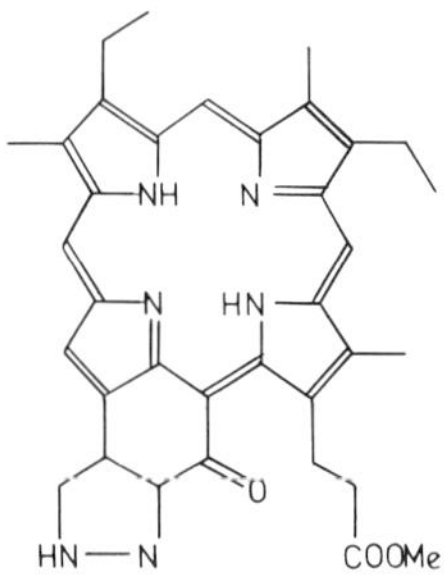

13

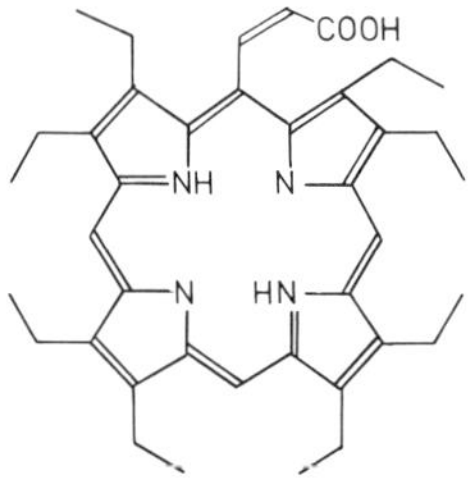

14

A reaction similar to rhodin formation takes place when *meso*-acrylic acid porphyrin derivatives such as **14** are dissolved in concentrated sulfuric acid.

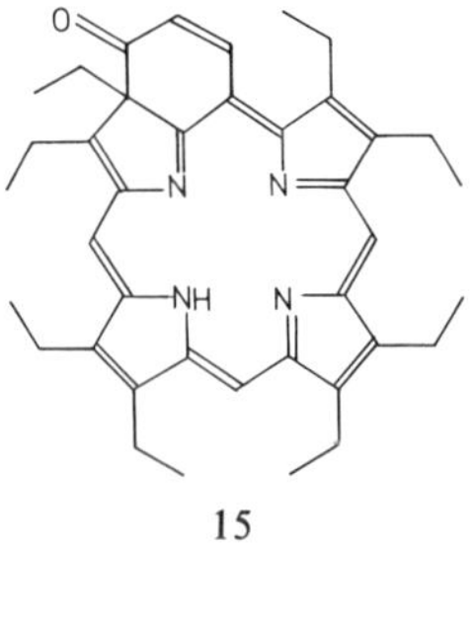

15

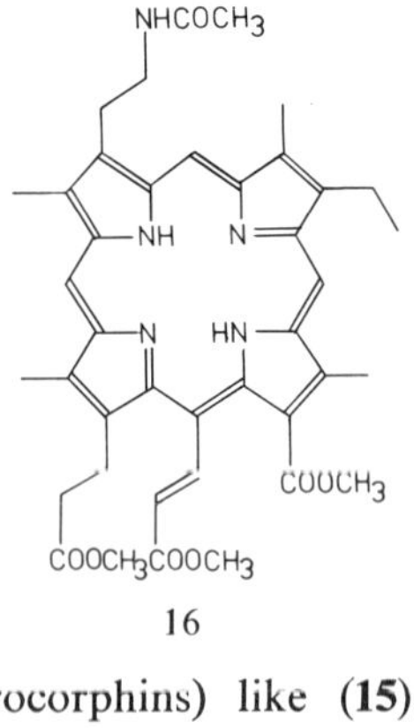

16

Cross-conjugated porphyrins (or tridehydrocorphins) like **(15)**, with a phlorinoid electronic spectrum [335 nm (27,500) 379 (24,000) 471 (42,000) 680,sh (8500) 732 (13,000], are the main products of such condensation reactions.[37] The most characteristic reaction of these tridehydrocorphins is ready formation of rather stable π-radicals with weak reductants, e.g., triethylamine.[6,37]

If a meso-acrylic acid porphyrin, e.g., 16, is heated in acetic acid or toluene, the product (yield $\sim 70\%$) is a purpurin, e.g., 17.[26,37] This reaction was the

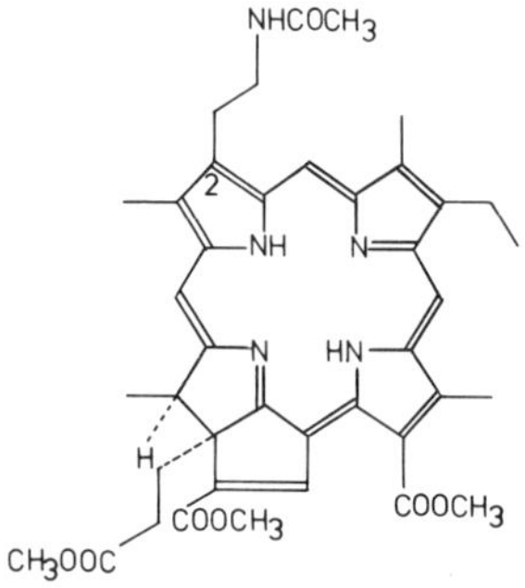

17

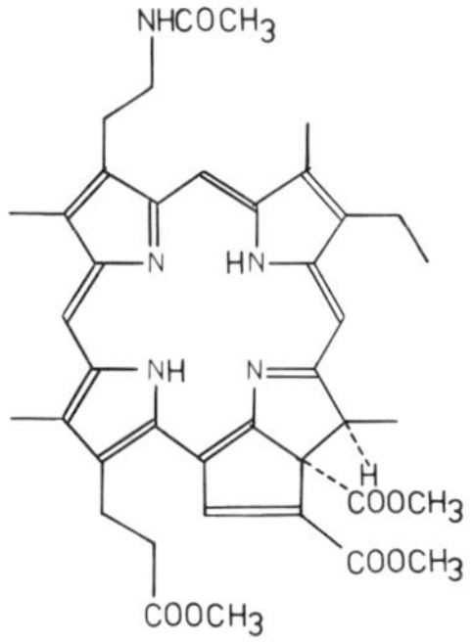

18

first example of a reversible conversion of a porphyrin into a chlorin and an important step on the way to a total synthesis of chlorophyll a[36] (see Chapter 1, this volume). The isomeric purpurin (18) was also present as a minor component of the equilibrium mixture. In the further course of the synthesis, the substituent in position 2 of 17 was converted into a vinyl group by the Hoffmann degradation procedure, and the double bond in the cyclopentenone ring was selectively photooxygenized to yield the keto aldehyde (19).[26] Treatment of 19 with methanolic KOH converted it into the racemic methoxylactone

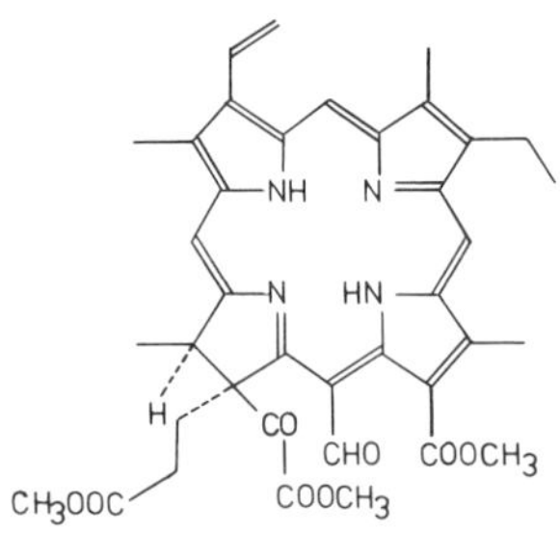

19

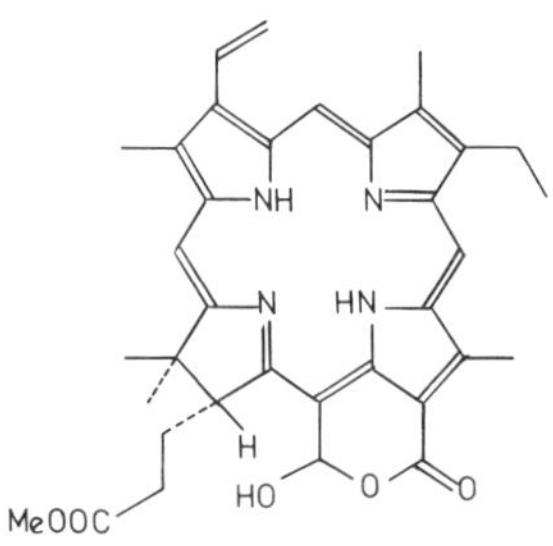

20

(20).[26] This compound had been obtained earlier, although formulated wrongly, as a degradation product from chlorophyll by Fischer, and had been named isopurpurin-5-methyl ester.[38]

When the synthetic β-ketoester (21), after chelation with magnesium, was oxidized to the π-radical, the compound (22) could be isolated in 7% yield

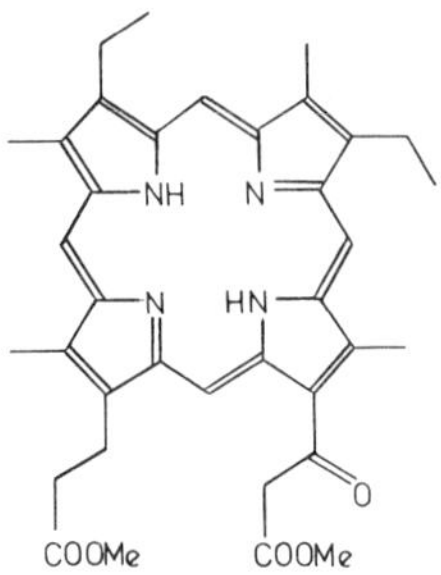

21

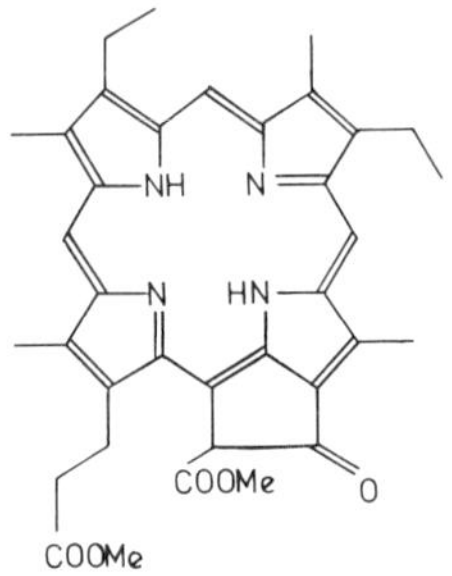

22

by Kenner's group.[39] The yield was raised to 70% when the thallium complex of (21) was irradiated with visible light.[40,41] These radical cyclizations are good models for isocyclic ring formation in the biosynthesis of the chlorophylls.

Vinylic and cross-conjugated double bonds in both rings I and II of protoporphyrin dimethyl ester constitute diene systems capable of undergoing Diels–Alder reactions with activated dienophiles. Thus free-base protoporphyrin dimethyl ester, but not the iron(III) or nickel(II) complexes, gave the adducts **23**, **24**, respectively, with acetylene dicarboxylate (tetracyanoethylene) in refluxing chloroform in 40% (56%) yield.[42,43]

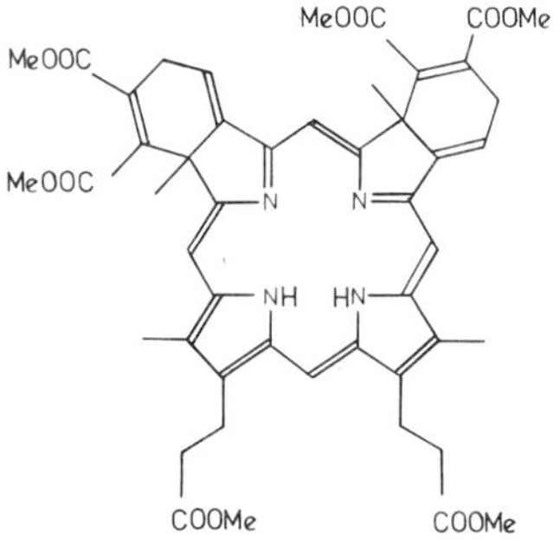

23

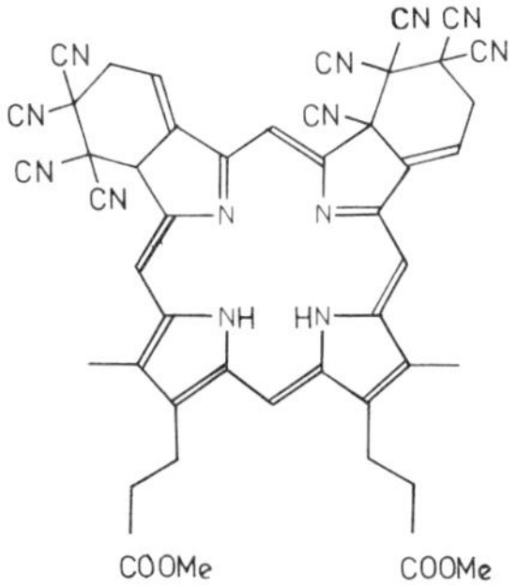

24

More complicated and unusual cyclization reactions between two porphyrin substituents have been described by Fischer,[44-46] and a few examples are given schematically by partial structures (**25–29**).

A final example of cyclization of a porphyrin substituent by C–C bond formation is again from Kenner. He found that the acetamidoporphyrin (**30**) undergoes an intramolecular Vilsmeier reaction followed by a rearrangement to end up in the spirocyclic chlorin (**31**). The conditions were treatment of **30** with phosphoryl chloride and dimethylformamide in pyridine.[47]

Another means of forming new peripheral C–C bonds by intramolecular

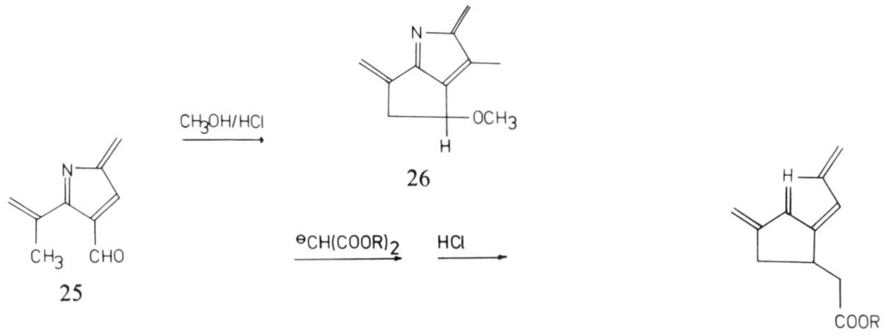

25

26

27

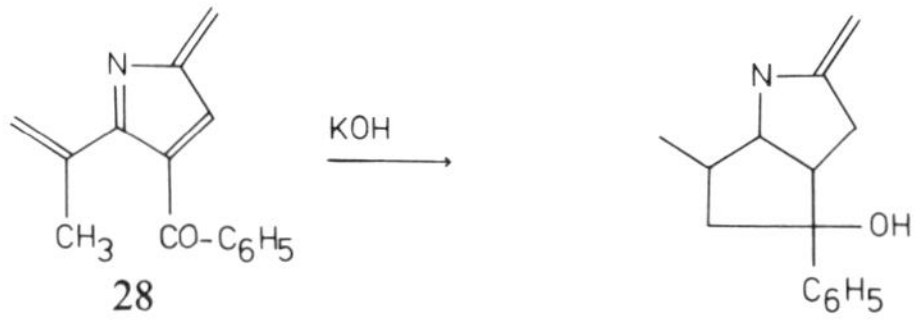

28

29

reaction is given by the pinacol rearrangement, e.g., **32–33**. This reaction was originally found—and misinterpreted—by Fischer[48] and later investigated by Johnson,[49] Bonnett,[50] Klotmann,[51] and Inhoffen.[52,53] The resulting carbonyl group in **33** or similar ketochlorins may be used to introduce further carbon substituents, which will be discussed in the following section.

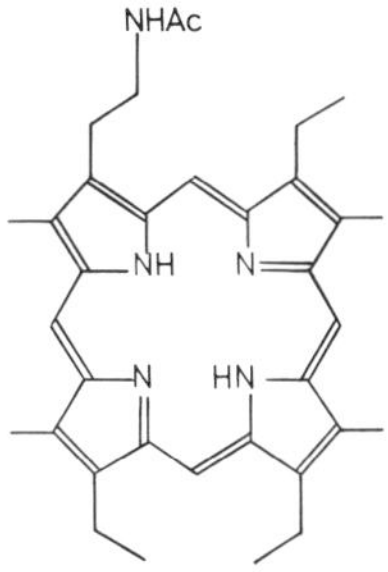

30

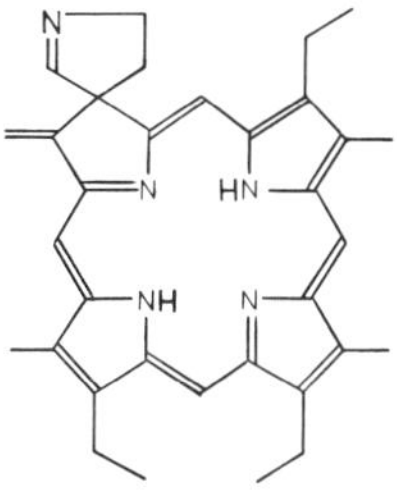

31

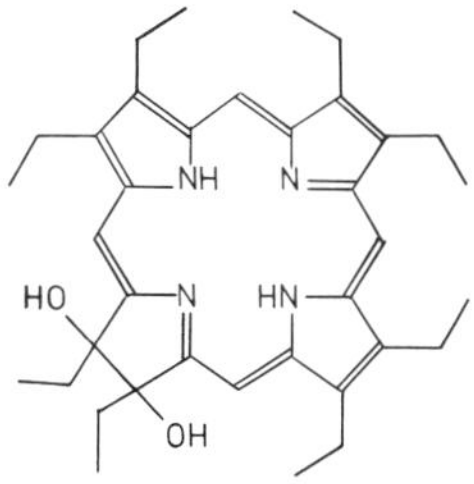

32

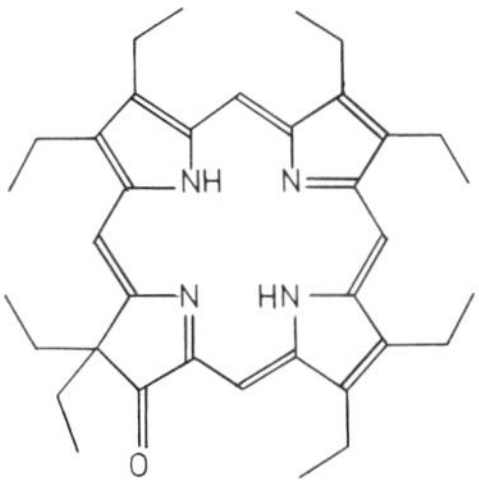

33

B. Introduction of Alkyl, Substituted Alkyl, and Aryl Substituents

Meso-methylation of palladium octaethylporphyrin, to give **34** in 36% yield, has been reported by Johnson.[54] The reagent was methyl fluorosulfonate. The formal addition of a methane molecule to a β-pyrrolic double bond was found, when tin(IV) octaethylporphyrin was first treated with a chloroform–aluminum bromide mixture, reduced with sodium borohydride, and demetalated to give the methyl chlorin (**35**).[55] Hydroxymethyl groups may

be introduced into unsubstituted β-pyrrolic positions with methyl chloromethyl ether and tin(IV) chloride as catalyst. This reaction converted, for example, deuteroporphyrin into **36**.[56] The reaction does not occur on the *meso*-positions.

Recently, a number of reactions involving metalloporphyrins and carbenes have been discovered. The most promising reaction on a preparative scale is the addition of ethyl diazoacetate in refluxing benzene to copper porphyrins in the presence of copper(I) iodide. The overall yield of chlorins (**37a**) and (**37b**) from copper octaethylporphyrin is 60%. Demetalation with sulfuric

(a) R$_1$ = H R$_2$ = COOEt
(b) R$_1$ = COOEt R$_2$ = H

acid was possible, but the precise structures of the four resulting products could not be clarified. One isomer, however, gave, after irradiation, the ethylidene compound (**38**) in 50% yield.[43] A similar addition of various carbenes to zinc *meso*-tetraphenylporphyrin produced, after demetalation, compounds **39a–d** and **40**.[57] In the reaction of copper octaethylporphyrin with

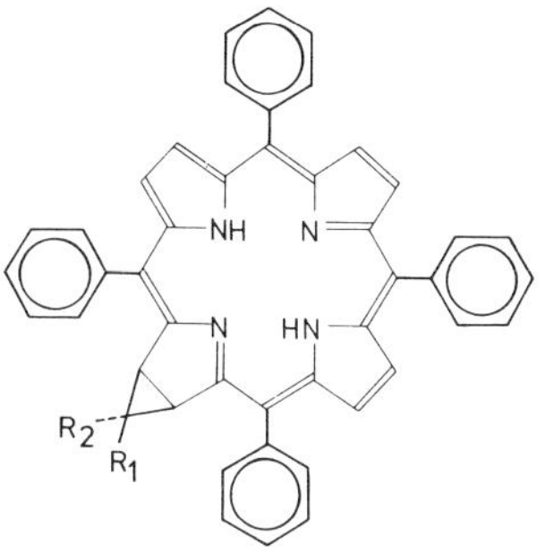

39

(a) $R_1 = R_2 = H$
(b) $R_1 = H$ $R_2 = COOCH_3$
(c) $R_1 = COOCH_3$ $R_2 = H$
(d) $R_1 = R_2 = COOCH_3$

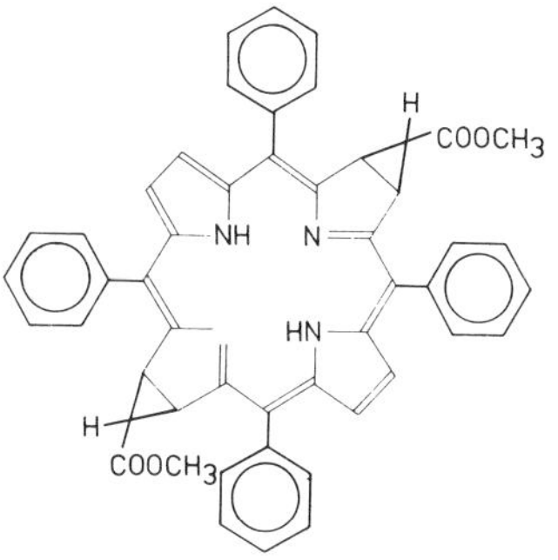

40

ethyl diazoacetate a few percent of the *meso*-ethoxycarbonylethyl derivative are also formed and may be demetalated to (**41**).[43]

The N-substituted tetraphenylporphyrin (**42**)[58,59] rearranged to the homoporphyrin mixture of (**43a,b**) when heated with nickel acetyl acetonate.[60,61]

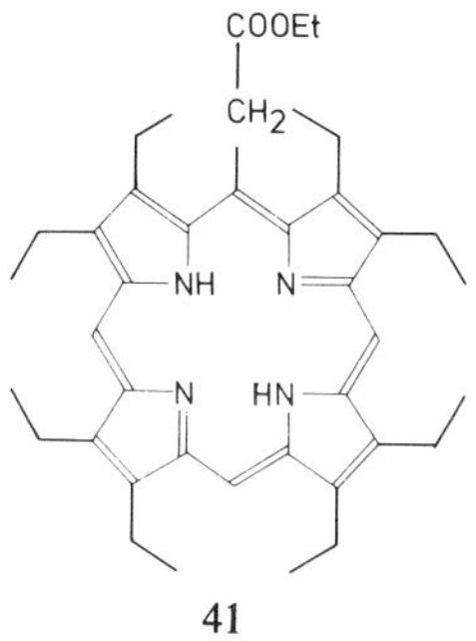

41

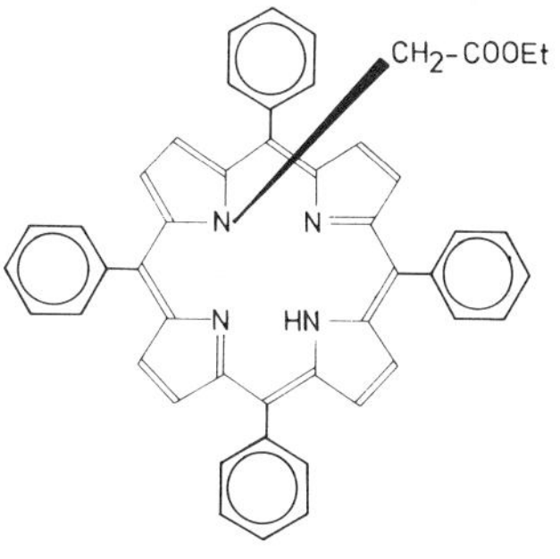

42

The macrocycle of **43a,b** is not planar.[62] The electronic spectra of homoporphyrins (**43a**) [H of new bridge endo with respect to macrocycle: λ_{max} 453 nm (87,000) 584 (6500) 688 (17,000)] and (**43b**) [exo: 449 (87,000) 579 (6800) 675 (15,000)] resemble those of phlorins.[60] At temperatures above 160°C, an equlibrium is reached between the two pairs of epimers, in which the hydrogen of the two carbon bridges is either on the carbethoxy or the phenyl substituted carbon atom.[61]

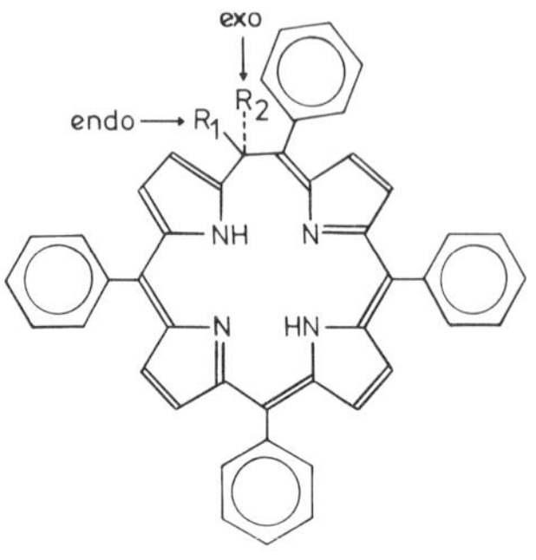

43

(a) R$_1$ = H   R$_2$ = COOEt
(b) R$_1$ = COOEt   R$_2$ = H

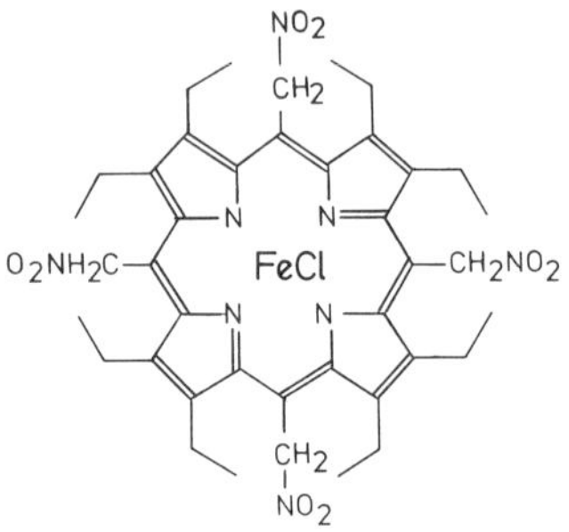

44

The carbene reactions discussed so far constitute mainly addition reactions that may be followed by intramolecular rearrangements of the carbon skeleton. Recently, a surprising high-yield *meso*-substitution reaction with related agents has been achieved. Octaethylhemin was treated with an excess of a 10% methylene chloride solution of dinitrogen tetroxide at room temperature and gave the *meso*-tetranitromethyl porphyrin (44) in 50% yield.[63]

Porphyrins with β-pyrrolic keto functions are readily available by oxidation of porphyrins with hydrogen peroxide or osmium tetroxide and subsequent pinacole rearrangement (see preceding section). Keto functions may also be introduced in all four *meso*-positions of porphyrins (xanthoporphinogens) by oxidation with lead dioxide.[8,64] Both types of carbonyl groupings react readily with Wittig reagents, magnesium or lithium alkyls, and give as products (45 and 46) in moderate to good yields.[65,66] Another methylation method

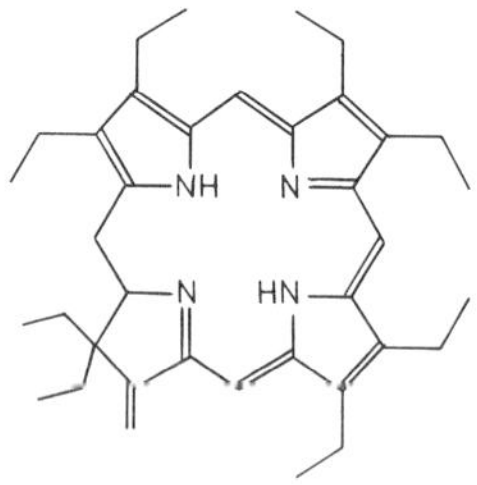

45

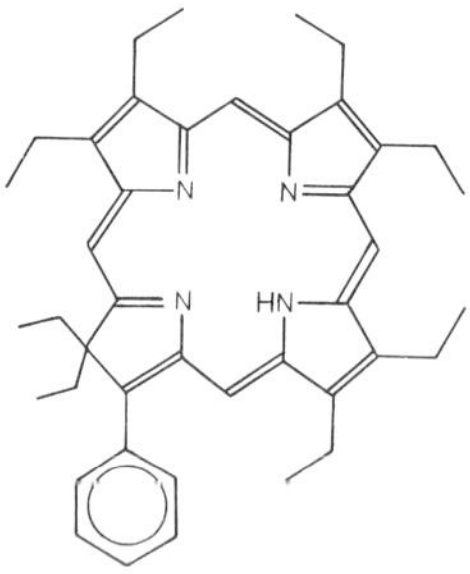

46

is the reduction of porphyrin to the dianion by sodium anthracenide and subsequent addition of methyl iodide. This leads to dimethylporphodimethenes, e.g., 47.[25]

47

48

C. Acylations

Fischer's favorite formylation reagent for unsubstituted β-pyrrolic positons was dichloromethyl methyl ether. This was used with a degradation product of chlorophyll, namely pyrroporphyrin, to obtain **48**. He generally used tin(IV) bromide as the Lewis acid. Iron(III) porphyrins were chosen as substrates, because metal-free porphyrins did not react. *Meso*-substitution was never observed in this reaction, although it has been applied to several different porphyrins.[45,46,67-69]

Inhoffen introduced the Vilsmeier formylation into porphyrin chemistry, which is the only procedure known today by which functional carbon groups may be introduced into the *meso*-position of porphyrins in high yield. Copper[17] or nickel[18] complexes are usually selected as substrates because they activate the porphyrin ligand toward electrophilic attack (see Section I, B), are stable against the phosphoryl chloride used in the reaction, and are easily demetalated, after the reaction has taken place, by concentrated sulfuric acid. α-Formyloctaethylporphyrin (**49**), for example, was obtained from copper octaethylporphyrin in 80% yield.[17] The electronic spectrum of the former is of the phyllo-type [402 nm (82,000) 502 (7500) 532 (4500) 575 (3500) 634 (2000)]; the carbonyl infrared band occurs at 1695 cm^{-1}.

No rationalization has been offered so far for the fact that the usual Friedel–Crafts reactions do not lead to any *meso*-substitution, whereas the closely related Vilsmeier reaction does so in almost quantitative yield. The reaction does, however, only yield the monoformylated product. Although many derivatives (e.g., oxime, semicarbazide, nitrile) are easily formed from **49**, experiments to extend the carbon chain have been largely unsuccessful. Grignard reagents, lithium alkyls, ylides, malonic ester anions, and many other nucleophilic agents were tried without success.[70] Finally, a Knoevenagel condensation in pyridine, catalyzed by titanium tetrachloride,[71] gave the acrylic acid derivative (**50**) in 70% yield.[37] Wittig reagents have also been applied successfully.[71a] Compound (**50**) could be formylated again at the methine bridges adjacent or opposite to the already substituted bridge.[37]

49   50

Copper deuteroporphyrin dimethyl ester yields six mono- and diformylated products in the Vilsmeier reaction. The ratios of *meso* to β-pyrrolic substitution and mono- to diformyl products may be widely varied by the choice of appropriate reaction condition.[23] Only products with the formyl group either in 2,4- or α,β-positions have been isolated in appreciable yields, which shows that the Vilsmeier reaction is also favored in positions with the least steric hindrance.[23] Vilsmeier formylation of hemin dimethyl ester showed that vinyl groups are also attacked, producing acrolein derivatives. The total yield of mono- and diacrolein derivatives was about 20%.[72] The iron(III) complex of chlorin-e_6 trimethyl ester was substituted at the *meso*-positions adjacent to the reduced pyrrole ring and at the vinyl group (51) was isolated in 25% yield from the reaction mixture.[72]

51   52

Acetylation of copper and iron(III) porphyrins has only been reported for β-pyrrolic positions. The reagent was usually acetic anhydride, with tin(IV) chloride[69,73,74] or boron trifluoride[73] as catalysts. The acetyl group deactivates the deuteroporphyrin ligand to a lesser extent than the formyl group. Therefore, usually the 2,4-diacetyl derivative (52) is the only product. However, careful control of reaction condition and the choice of benzene as solvent led to the formation of a 1:1 mixture of 2- and 4-acetyl deuteroporphyrin dimethyl esters.[75]

The best sequence of reactions to introduce nitrile groups or carboxylic functions into porphyrins is Vilsmeier formylation of copper porphyrins, demetalation with sulfuric acid, oxime formation in pyridine, dehydration to the nitrile, and hydrolysis with sulfuric acid. All of these reactions give yields of 80% or better when octaethylporphyrin is the starting material.[17,18]

III. INTRODUCTION OF NITROGEN SUBSTITUENTS INTO THE PORPHYRIN PERIPHERY

A. Nitroporphyrins

Nitration of an unsubstituted porphin dication yields α-mononitroporphin (**53**). This may be nitrated further and was reported to yield α,β-dinitroporphin (**54**) selectively.[76] This selectivity was not found in other porphyrins where

53

54

α,β-dinitro isomers were always found.[77] No β-pyrrolic substition was detected.[76] A similar selectivity was observed in the nitration of deuteroporphyrin dimethyl ester: only α,β-*meso*-position were substituted but not the carbon atoms 2 and 4 of the pyrrole rings.[78] Fully β-alkylated porphyrins are also nitrated in high yields at the methine bridge positions. From octaethylporphyrin, an α,β,γ-trinitro product was obtained.[77] Nitrations are usually carried out on unmetalated porphyrin dications with concentrated nitric acid either in acetic or sulfuric acids. The degree of nitration may be controlled by changing temperature and reaction times.[76,77]

Oxidation-sensitive compounds, e.g., chlorins, are often degraded by concentrated nitric acid.[77] Here nitronium fluoroborate in sulfolane is the reagent of choice. Both mononitrochlorin (**55**) and dinitrochlorin (**56**) were isolated, but further nitration was not possible with this mildly electrophilic reagent.[77] Another possibility of nitration is the formation of porphyrin chlorin π-radicals by iodine oxidation of their magnesium complexes and subsequent treatment with sodium or silver(I) nitrite.[16a] This procedure possesses three advantages over the other methods: (a) it is carried out in solutions of relatively low oxidation potential, which allows nitration of

55   56

chlorins, (b) at neutral pH, so that highly acid labile metal complexes with electronegative π-systems may be used, and (c) the degree of nitration may be controlled by the choice of the nitrite. For example, an isomer mixture of α,β- and α,γ-dinitroporphyrins was obtained in high yield when magnesium octaethylporphyrin was stirred for 6 hours with excess of silver(I) nitrite in chloroform–methanol, followed by demetalation.[16a]

The electronic spectra of *meso*-nitroporphyrins are usually of etio-type, with a small bathochromic shift of the visible bands as compared to the parent compounds [e.g., α-nitrooctaethylporphyrin: 400 nm (128,000), 504 (12,500) 538 (8400) 571 (6300) 623 (5300)]. The irs of nitroporphyrins contain bands close to 1532 and 1378 cm^{-1}. ^{1}H nmr spectra show downfield shifts of the broad NH band by 1–2 ppm, indicating a decrease of ring-current effects in nitrated porphyrins; but effects on methine proton signals are of small magnitude.

If the mixture of α-(β-) mononitrodeuteroporphyrins is heated with succinic acid, the parent porphyrin is formed.[79] It was found that the *meso*-nitro group deactivated the methine bridges of deuteroporphyrin to such an extent that treatment with oleum does not lead to any rhodin formation[79] (see Section II, A). A *meso*-nitrated porphyrin can, however, be brominated in free β-pyrrolic positions.[79] Reduction of nitroporphyrins will be treated in the next section.

B. Aminoporphyrins

Meso-aminoporphyrins are obtained by hydrogenation of the corresponding nitro compounds. Reductions with tin(II) chloride[77] or sodium borohydride and a palladium catalyst[80] have been used. Almost quantitative yields were reported for the latter reaction. To the author's knowledge, porphyrins with β-pyrrolic amino substituents have not yet been prepared from unsubstituted porphyrins.

The electron-donating *meso*-amino substituent changes the physical and chemical properties of the porphyrin macrocycle drastically. The influence of the *meso*-amino group is comparable to that of the *meso*-hydroxy group (see

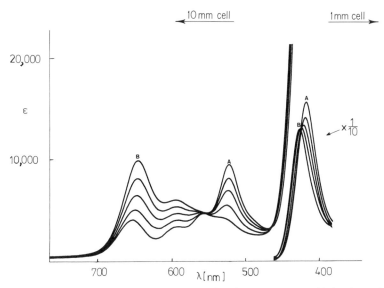

Fig. 1. Aminooctaethylporphyrin (**57**) in chloroform (spectrum A) is titrated with methanolic HCl to form a hydrochloride (spectrum B), for which structure **58** has been proposed.

Chapter 4, this volume). The etio-type electronic spectrum of α-amino-octaethylporphyrin (**57**) (Fig. 1, spectrum A) is shifted by about 20 nm toward longer wavelengths, as compared with that of octaethylporphyrin. Proton magnetic resonance chemical shifts of methine ($+1.0$ and 1.5 ppm) and NH (-3 ppm) proton signals indicate a significant decrease of induced ring current, as compared with octaethylporphyrin. The most dramatic changes are, however, encountered in acid medium. In chloroform–methanolic hydrochloric acid mixture, a green dication with a phlorinoid electronic spectrum (Fig. 1, spectrum B) is formed. In the irs a strong new band at 1602 cm^{-1} is observed on protonation, which may be correlated to a $C{=}N$ stretching vibration.[81] The proton magnetic resonance (pmr) signals of the methine protons occur at $\tau = 3.53$ and 3.19 ppm and disappear within half an hour, when D_2O is added.[82] Another signal is found at 0.27 ppm and is immediately extinguished by addition of D_2O.[82] An obvious rationalization for these properties of α-aminooctaethylporphyrin hydrochloride is the assumption of the phlorin structure (58).[81]

The fact that amino porphyrins often behave like phlorins is also evidenced by their behavior in chemical oxidation: the zinc complex of aminoporphyrin (57) does not yield stable π-radicals on oxidation with iron(III) salts, unlike the parent porphyrin, but is immediately oxygenated to 59.[81] Nitrosation of

59   60

aminoporphyrins with sodium nitrite in tetrafluoroboric acid at $-5°C$ produces the corresponding diazonium fluoroborate, e.g., 60 in 90% yield.

Amino nitrogen leads to phlorin-type structures on protonation, even when it is bound in the form of a Schiff base to a carbon substituent at a *meso*-position.[24,37] This is true for free porphyrin cations, e.g., 61 ($\lambda_{max} = 660$ nm), as well as for metal complexes, e.g., 62 Fig. (2).[37]

61   62

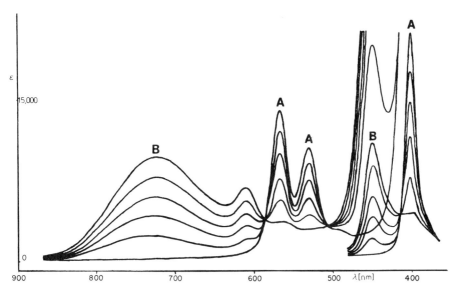

Fig. 2. The copper complex of the Schiff base of (**49**) with β-alanine in chloroform (spectrum A) is titrated with methanolic HCl and a phlorin type absorption spectrum B is observed. This behavior has been described with the structure **62**.

Nitrenes react with porphyrins in a manner similar to carbenes (see Section II, B). Ethoxycarbonylnitrene, for example, was inserted into octaethyl-porphyrin to form homoazaporphyrin (**63**). When copper octaethylporphyrin was used as a substrate, **64** was the only isolable product. Insertion of copper into **63** caused its rearrangement into **64**.[83] The same behavior was observed when Zn(II) was the metal ion.

63 64

Recently, the nucleophilic addition of imidazole and pyridine bases to the methine bridges of porphyrin radicals has been described.[83a,b]

IV. INTRODUCTION OF SULFUR SUBSTITUENTS INTO THE PORPHYRIN PERIPHERY

A. Porphyrin Sulfonates

Unoccupied β-pyrrolic positions of porphyrins may be substituted by sulfonium groups according to Treibs.[84] Some of the isolated sulfonates were dark green, perhaps indicating some *meso*-substitution, but none of them seem to have been identified, since they were first described in 1933. Attempts to sulfonate octaethylporphyrin were unsuccessful.[7] Sulfonic acid groups were, however, often introduced into side chains, e.g., *meso*-phenyl[85,86] or β-pyrrolicpropionic acid,[87] to produce water-soluble porphyrins.

B. Mercaptoporphyrins

Electrophilic thiocyanation of copper porphyrins with thiocyanogen in chloroform and subsequent treatment with 90% (w/w) sulfuric acid at 100°C yields *meso*-mercaptoporphyrins,[88] e.g., **65** [λ_{max} 402 (140,000) 507 (9100)

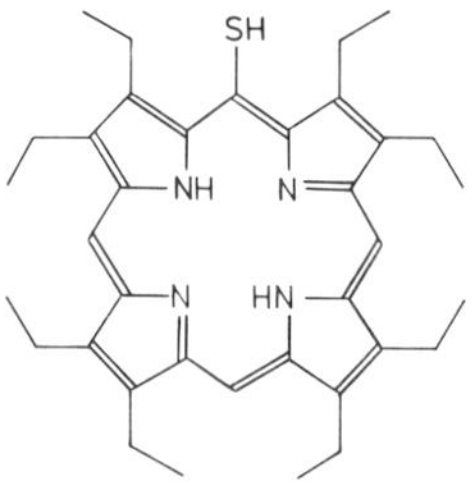

65

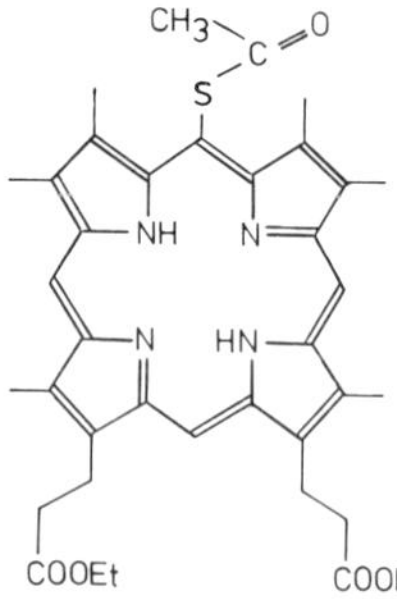

66

550 (8200) 635 (5700)].[88,89] The sulfide group is easily removed thermally, and mass spectrometric molecular ion peaks are of low intensity.[88,89] The sulfur atom does not form double bonds to the methine bridge carbon, as

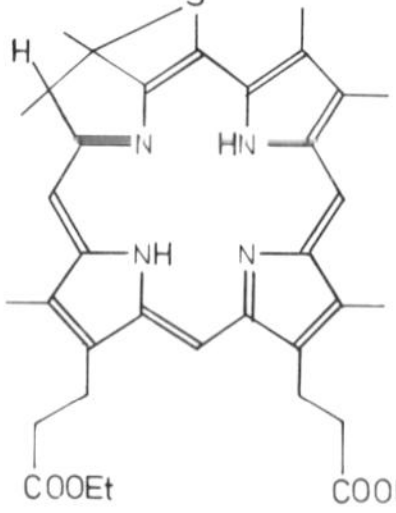

67

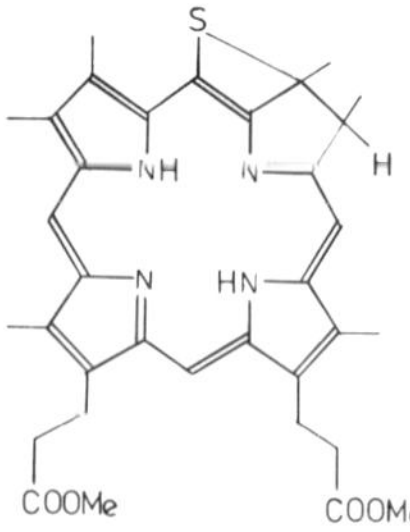

68

amino group nitrogen or hydroxy group oxygen atoms do[90] (see also the preceding section and Chapter 4, this volume). Clezy treated the synthetic mercaptide (66) with ethanolic sulfuric acid (10% v/v) under reflux and isolated a green product with chlorin-type electronic and pmr spectra. He assigned structures 67 and 68 to the presumed mixture of isomers and proposed that an equilibrium between 67 and 68 is rapidly reached at room temperature.[91]

Apart from reductions addition of sulhydryl groups to the methine bridges of porphyrin dications with electron withdrawing β-pyrrolic substituents, e.g., 69 to form 70, constitutes the only example of a nucleophilic reaction of

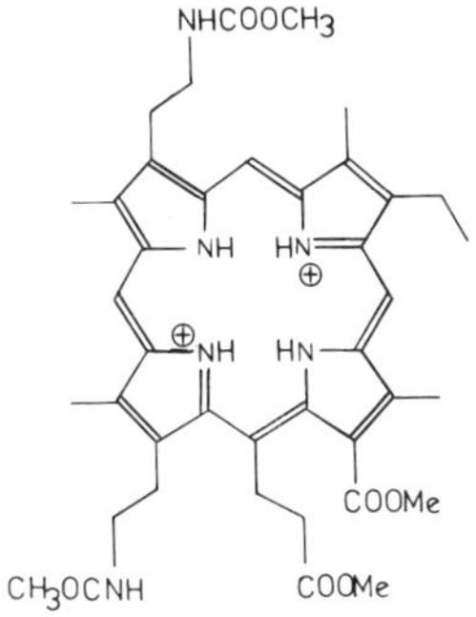

69

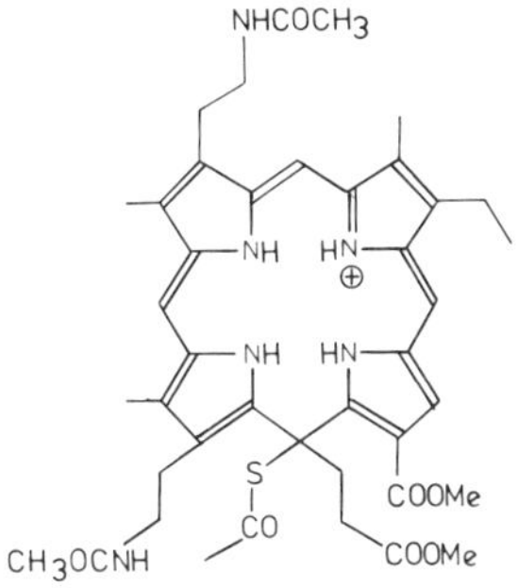

70

porphyrins. Porphyrin dications which do contain a large *meso*-substituent like 69, but no β-pyrrolic carboxylic acid group, do not undergo this reaction.[37]

V. HALOGENATION OF THE PORPHYRIN PERIPHERY

Heating the solid diazonium tetrafluoroborate (60) dispersed in sand *in vacuo* to 160°C (Balz–Schiemann reaction[92,93]) leads to the expected *meso*-fluoroporphyrin (71) in poor yield ($\leq 5\%$). The electronic spectrum is of the

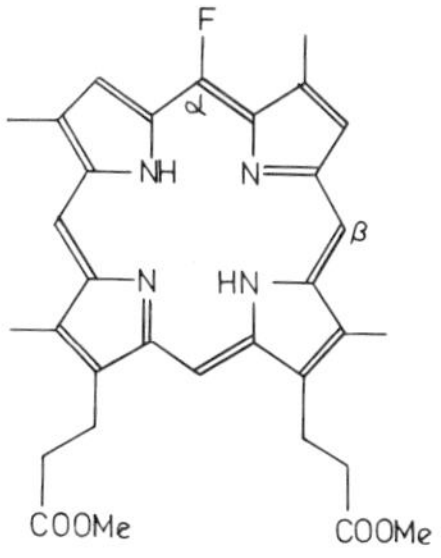

71

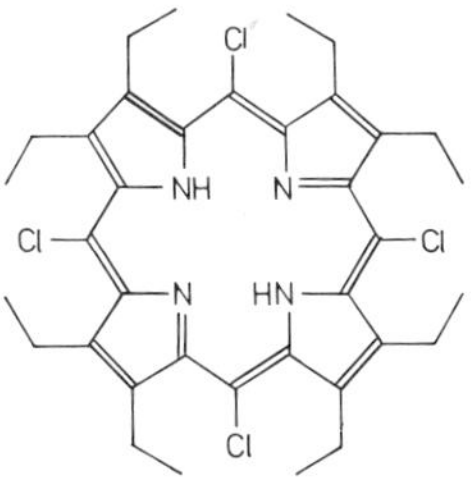

72

phyllo type [399 nm (210,000) 496 (15,700) 528 (4800) 567 (5600) and 621 (1600)]. The methine proton magnetic resonance signals were found at

$\tau = 0.5$ ppm, and those of the β-protons at 0.87 and 1.29 ppm. The fluoroporphyrin obtained was actually a 1:1 mixture of α- and β-fluoro compounds, which could be best seen in the ^{19}F nmr spectrum. This showed only two absorptions at 42.57 and 43.77 ppm upfield from the trifluoroacetic acid signal.[80]

Free-base porphyrins form *meso*-tetrachlorinated products with free chlorine. The reaction is usually carried out with hydrogen peroxide–hydrochloric acid mixtures in tetrahydrofuran or acetic acid.[94–97] Another useful chlorination agent for copper porphyrins is sulfuryl chloride in chloroform.[96] From octaethylporphyrin, 43% of tetrachloro derivative (72), or 20% of α,γ-dichloro- (73) and 24% of α-monochloroporphyrin (74), could be

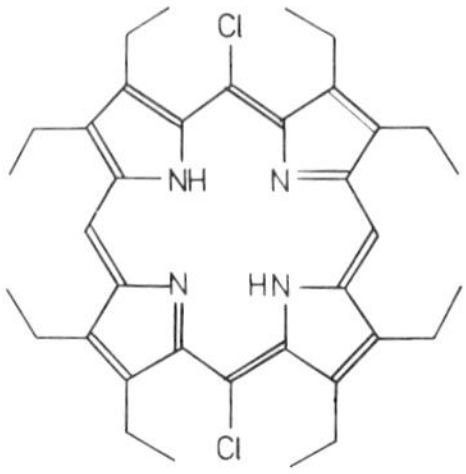

73 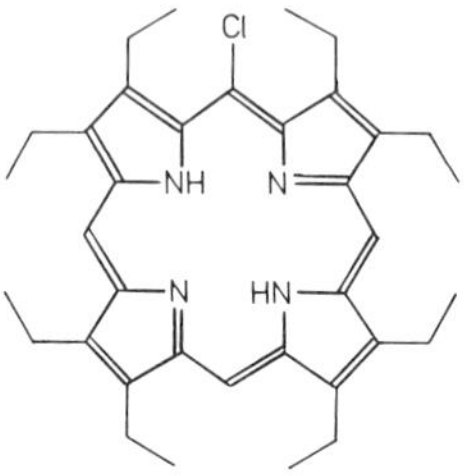   74

obtained by Bonnett by the carefully controlled hydrogen peroxide–hydrochloric acid procedure.[97] α-Chlorooctaethylchlorin was the main product from octaethylchlorin.[97]

Whereas the monochloroporphyrin (74) produces the expected phyllo-type spectrum [406 nm (161,000) 507 (14,800) 540 (5300) 578 (5600) 628 (1600)], the electronic spectrum of the green tetrachloro product (72) is quite unusual. It consists of a long wavelength Soret band [446 nm (188,000)] and only three visible bands, which are considerably shifted toward longer wavelength (Fig. 3), as compared to the parent porphyrin. The most important change, however, is observed in the spectrum of (70) in acid solution, which is of the phlorin-type (Fig. 3).[97,98] The ^{1}H nmr spectrum of 72 in chloroform–trifluoroacetic acid shows a sharp methyl triplet and a broadened methylene

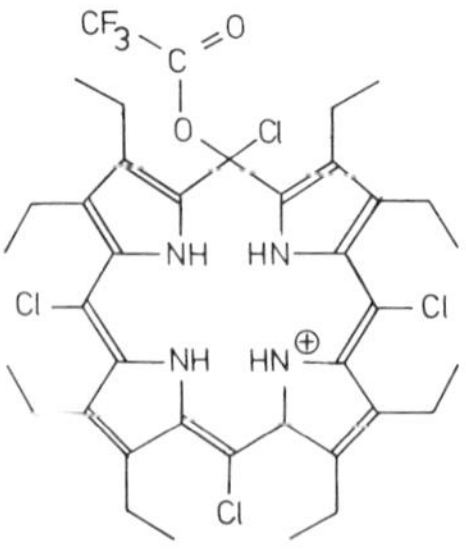

75 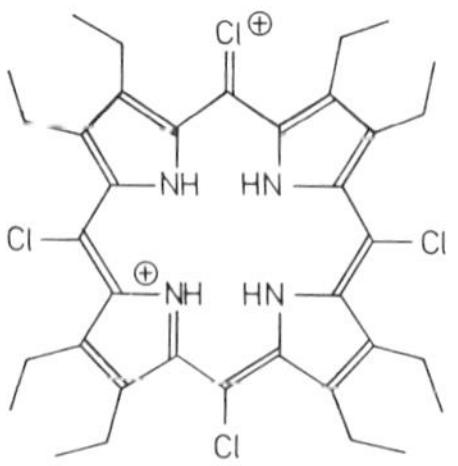   76

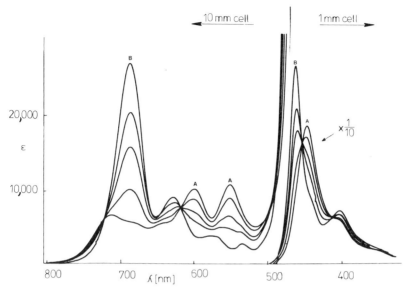

Fig. 3. Protonation of *meso*-tetrachlorooctaethylporphyrin (**72**) in chloroform (spectrum A) by titration with methanolic HCl yields a phlorin-type spectrum. Structures **75** or **76** would offer explanations for this observation, but are thought to be unlikely (see text).

proton signal. Therefore the presence of a phlorin such as **75** seems unlikely. Another explanation would be that one of the electron-releasing chloro substituents is pulled into the plane of the porphyrin conjugation system by its positive charge, thereby overcoming steric barriers. The main resonance structure might then be **76**, in analogy to cations of *meso*-amino- (see Section III, B), *meso*-hydroxy (see Chapter 4, this volume) and *meso*-tetraphenylporphyrins.[85,99] The large electronegativity of chlorine, however, renders this alternative assumption somewhat unattractive, as long as there is no positive experimental evidence for it.

Bromine precipitates crystalline tetrabromo porphyrins from acetic acid solutions of a variety of fully β-substituted porphyrins,[100] but, after various extraction procedures, only the starting porphyrin and degradation products could be isolated.[97] Bromination of octaethylchlorin with hydrobromic acid–hydrogen peroxide in tetrahydrofuran, however, yielded 47% of α-bromooctaethylchlorin (**77**), which could be dehydrogenated with dichlorodicyanobenzoquinone to the corresponding bromoporphyrin [$\lambda_{max} = 409$ nm (184,000) 510 (13,900) 543 (6000) 580 (5700) 630 (1800)].[97]

Bromination of free-base porphin gives β-bromoporphin (**78**),[101] whereas with magnesium porphin, *meso*-tetrabromoporphin (**79**) is obtained.[22]

77   78   79

Deuteroporphyrin dimethyl ester is brominated at the β-pyrrolic carbon atoms 2 and 4.

Attempts to iodinate either octaethylporphyrin or chlorin have been unsuccessful.[97]

ACKNOWLEDGMENTS

This work was supported by the grants from the Ministry of Research and Technology of the Federal Republic (BMFT) within its Technology Program, the Deutsche Forschungsgemeinschaft and the Verband der Chemischen Industrie. I am grateful to Dr. E. Lustig, Dr. J. Subramanian and Dr. W. S. Sheldrick for carefully reading the manuscript and for many valuable suggestions.

REFERENCES

1. K. Fukui, T. Yonezawa, and H. Shingu, *J. Chem. Phys.* **20**, 722 (1952).
2. K. Fukui, T. Yonezawa, C. Nagata, and H. Shingu, *J. Chem. Phys.* **22**, 1433 (1954).
3. K. Fukui, T. Yonezawa, and C. Nagata, *J. Chem. Phys.* **26**, 831 (1957).
4. A. D. McLachlan, *Mol. Phys.* **2**, 233 (1960).
5. J. V. Knop and J.-H. Fuhrhop, *Z. Naturforsch., Teil B* **25**, 729 (1970).
6. J.-H. Fuhrhop and J. Subramanian, *Phil. Trans. Roy. Soc. London B273*, p. 335 (1976).
7. H. Fischer and A. Treibs, *Justus Liebigs Ann. Chem.* **457**, 209 (1927).
8. H. H. Inhoffen, J.-H. Fuhrhop, and F. von der Haar, *Justus Liebigs Ann. Chem.* **700**, 92 (1966).
9. H. Fischer and A. Stern, "Die Chemie des Pyrrols," Vol. II, Part 2, p. 420ff. Akad. Verlagsges., Leipzig, 1940.
10. J. N. Phillips, *Rev. Pure Appl. Chem.* **10**, 35 (1960).
11. A. V. Umrikhina, G. A. Yusupova, and A. A. Krasnovskii, *Dokl. Akad. Nauk SSSR* **175**, 1400 (1967), and references therein.
12. H. Fischer and K. Herrle, *Hoppe-Seyler's Z. Physiol. Chem.* **251**, 85 (1938).
13. M. Zerner and M. Gouterman, *Theor. Chem. Acta* **4**, 44 (1966).
14. J.-H. Fuhrhop and D. Mauzerall, *J. Am. Chem. Soc.* **91**, 4174 (1969).

15. J.-H. Fuhrhop, K. Kadish, and D. G. Davis, *J. Am. Chem. Soc.* **95**, 5140 (1973).
16. J.-H. Fuhrhop, P. K. W. Wasser, J. Subramanian, and U. Schrader, *Justus Liebigs Ann. Chem.* **1974**, 1450 (1974).
16a. G. H. Barnett and K. Smith, *Chem. Commun.* p. 772 (1974).
17. H. H. Inhoffen, J.-H. Fuhrhop, H. Voigt, and H. Brockmann, Jr., *Justus Liebigs Ann. Chem.* **695**, 133 (1966).
18. D. Oldfield and A. W. Johnson, *J. Chem. Soc. C* p. 794 (1966).
19. T. Lumbantobing and J.-H. Fuhrhop, *Tetrahedron Lett.* p. 2815 (1970).
20. H. J. Callot and A. W. Johnson, *J. Chem. Soc., Perkin Trans. 1* p. 1424 (1973).
21. G. A. Hamilton, *in* "Molecular Mechanisms of Oxygen Activation" (O. Hayaishi, ed.), p. 405ff. Academic Press, New York, 1974.
22. R. Schlözer and J.-H. Fuhrhop, *Angew. Chem. Int. Ed. Engl.* **14**, 363 (1975).
23. H. Brockmann, Jr. K.-M. Bliesener, and H. H. Inhoffen, *Justus Liebigs Ann. Chem.* **718**, 148 (1968).
24. R. B. Woodward, *Ind. Chim. Belge* p. 1293 (1962).
25. J. W. Buchler and L. Puppe, *Justus Liebigs Ann. Chem.* **740**, 142 (1970).
26. R. B. Woodward, *Angew. Chem.* **72**, 651 (1960).
27. A. W. Johnson and A. Hamilton, *J. Chem. Soc. C* p. 3879 (1971).
28. F. H. Moser and A. L. Thomas, *J. Chem. Educ.* **41**, 245 (1964); G. Bainhart and B. F. Skiles, U.S. Patent 2,772,284 (1956).
29. H. Fischer, A. Treibs, and H. Helberger, *Justus Liebigs Ann. Chem.* **466**, 243 (1962)
30. H. Fischer and J. Ebersberger, *Justus Liebigs Ann. Chem.* **509**, 19 (1934).
31. H. Fischer and A. Rothaas, *Justus Liebigs Ann. Chem.* **484**, 85 (1930).
32. H. Gröschel, Ph.D. Dissertation, Technische Universitat, Braunschweig, West Germany (1964).
33. H. Fischer and C. G. Schröder, *Justus Liebigs Ann. Chem.* **537**, 250 (1939).
34. G. Azarello, *Gazz. Chim. Ital.* **36**, 40 (1906).
35. C. Djerassi and C. R. Scholz, *J. Org. Chem.* **14**, 660 (1949).
36. W. E. Parham, F. D. Blake, and D. R. Theissen, *J. Org. Chem.* **27**, 2415 (1962).
37. L. Witte and J.-H. Fuhrhop, *Angew. Chem. Int. Ed. Engl.* **14**, 361 (1975).
38. H. Fischer and M. Strell, *Justus Liebigs Ann. Chem.* **540**, 232 (1939).
39. M. T. Cox, T. T. Howarth, A. H. Jackson, and G. W. Kenner, *J. Chem. Soc. Perkin Trans. 1* p. 512 (1974).
40. G. W. Kenner, S. W. McCombie, and K. M. Smith, *Chem. Commun.* p. 844 (1972).
41. G. W. Kenner, S. W. McCombie, and K. M. Smith, *J. Chem. Soc., Perkin Trans. 1* p. 527 (1974).
42. R. Grigg, A. W. Johnson, and A. Sweeney, *Chem. Commun.* p. 697 (1968).
43. H. J. Callot, A. W. Johnson, and A. Sweeney, *J. Chem. Soc., Perkin Trans. 1* p. 1424 (1973).
44. H. Fischer, K. Müller, and O. Leschhorn, *Justus Liebigs Ann. Chem.* **523**, 164 (1936).
45. H. Fischer and H. Kellermann, *Justus Liebigs Ann. Chem.* **524**, 25 (1936).
46. H. Fischer and O. Laubrau, *Justus Liebigs Ann. Chem.* **535**, 17 (1938).
47. G. L. Collier, A. H. Jackson, and G. W. Kenner, *J. Chem. Soc. C* p. 66 (1967).
48. H. Fischer and G. Wacker, *Hoppe-Seyler's Z. Physiol. Chem.* **272**, 1 (1941).
49. A. W. Johnson and D. Oldfield, *J. Chem. Soc.* p. 4303 (1965).
50. R. Bonnett, M. J. Dimsdale, and G. F. Stephenson, *J. Chem. Soc. C* p. 564 (1969).
51. G. Klotmann, Ph.D. Dissertation, Technische Universitat, Braunschweig, West Germany (1964).
52. H. H. Inhoffen and W. Nolte, *Tetrahedron Lett.* p. 2185 (1967).
53. H. H. Inhoffen and N. Müller, *Tetrahedron Lett.* p. 3209 (1969).

54. R. Grigg, A. Sweeney, and A. W. Johnson, *Chem. Commun.* p. 1237 (1970).
55. J.-H. Fuhrhop and T. Lumbantobing, *Tetrahedron Lett.* p. 3771 (1970).
56. H. Fischer and H. J. Riedl, *Justus Liebigs Ann. Chem.* **482**, 214 (1930).
57. H. J. Callot, *Tetrahedron Lett.* p. 1011 (1972).
58. H. J. Callot and T. Tschamber, *Bull. Soc. Chim. Fr.* p. 3192 (1973).
59. H. J. Callot, *Bull. Soc. Chim. Fr.* p. 4387 (1972).
60. H. J. Callot and T. Tschamber, *Tetrahedron Lett.* p. 3155 (1974).
61. H. J. Callot and T. Tschamber, *Tetrahedron Lett.* p. 3159 (1974).
62. B. Chevrier and R. Weiss, *Chem. Commun.* p. 885 (1974).
63. J. C. Fanning, T. L. Gray, and N. Datta-Gupta, *Chem. Commun.* p. 23 (1974).
64. H. Fischer and A. Stern, "Die Chemie des Pyrrols," p. 423ff. Akad. Verlagsges., Leipzig, 1940.
65. H. H. Inhoffen and N. Müller, *Tetrahedron Lett.* p. 3209 (1969).
66. H. H. Inhoffen, J. W. Buchler, and P. Jäger, *Prog. Chem. Org. Nat. Prod.* **26**, 284 (1968).
67. H. Fischer and A. Schwarz, *Justus Liebigs Ann. Chem.* **512**, 239 (1934).
68. H. Fischer and H. Orth, "Die Chemie des Pyrrols," Vol. II, Part 1, p. 290ff. Akad. Verlagsges., Leipzig, 1937.
69. G. V. Ponomarev, R. P. Evstigneeva, V. N. Stromnov, and N. A. Preobrazhenskii, *Khim. Geterotsikl. Soedin.* p. 628 (1966).
70. J.-H. Fuhrhop, unpublished results.
71. W. Lehnert, *Tetrahedron* **28**, 663 (1972).
71a. H. J. Callot, *Bull. Soc. Chim. Fr.* p. 3413 (1973).
72. A. W. Nichol, *J. Chem. Soc. C* p. 903 (1970).
73. H. H. Inhoffen, G. Klotmann, and G. Jeckel, *Justus Liebigs Ann. Chem.* **695**, 112 (1966).
74. H. Fischer and K. O. Deilmann, *Hoppe-Seyler's Z. Physiol. Chem.* **280**, 186 (1944).
75. H. H. Inhoffen, C. Bliesener, and H. Brockmann, Jr., *Tetrahedron Lett.* p. 727 (1967).
76. J. E. Drach and F. R. Longo, *J. Org. Chem.* **39**, 3282 (1974).
77. R. Bonnett and G. F. Stephenson, *J. Org. Chem.* **30**, 2791 (1965).
78. W. S. Caughey, J. O. Alben, W. Y. Fuitimoto, and J. L. York, *J. Org. Chem.* **31**, 2631 (1966).
79. H. Fischer and W. Klendauer, *Justus Liebigs Ann. Chem.* **547**, 123 (1941).
80. M. J. Billig and E. W. Baker, *Chem. Ind. (London)* p. 654 (1969).
81. J.-H. Fuhrhop, *Chem. Commun.* p. 781 (1970).
82. J.-H. Fuhrhop, Habilitationsschrift, Technische Universitat, Braunschweig, West Germany (1972).
83. R. Grigg, *J. Chem. Soc. C* p. 3664 (1971).
83a. S. Besecke, B. Evans, G. H. Barnett, K. M. Smith, and J. H. Fuhrhop, *Angew. Chem. Int. Ed.* **15**, 551 (1976).
83b. G. H. Barnett, B. Evans, K. Smith, S. Besecke, and J. H. Fuhrhop, *Tetrahedron Lett.* p. 4009 (1976).
84. A. Treibs, *Justus Liebigs Ann. Chem.* **506**, 196 (1933).
85. A. Treibs and N. Häberle, *Justus Liebigs Ann. Chem.* **718**, 183 (1968).
86. T. S. Srivastava and M. Tsutsui, *J. Org. Chem.* **38**, 2103 (1973).
87. P. K. Warme and L. P. Hager, *Biochemistry* **9**, 1599, 1606, and 4244 (1970).
88. P. S. Clezy and C. J. R. Fookes, *Chem. Commun.* p. 1268 (1971).
89. S. Besecke, Ph.D. Dissertation, Technische Universitat, Braunschweig, West Germany (1974).

90. J.-H. Fuhrhop, S. Besecke, J. Subramanian, C. Mengersen, and D. Riesner, *J. Am. Chem. Soc.* **97**, 7141 (1975).
91. P. S. Clezy and G. A. Smythe, *Chem. Commun.* p. 127 (1968).
92. G. Balz and G. Schiemann, *Ber. Dtsch. Chem. Ges.* **60**, 1186 (1927).
93. H. Suschitzky, *Adv. Fluorine Chem.* **4**, 1 (1965).
94. H. Fischer and H. Röse, *Ber. Dtsch. Chem. Ges.* **46**, 2460 (1923).
95. H. Voigt, Ph.D. Dissertation, Technische Universitat, Braunschweig, West Germany (1964).
95a. H. Fischer and W. Neumann, *Justus Liebigs Ann. Chem.* **494**, 225 (1932).
96. H. Voigt, Ph.D. Dissertation, Technische Universitat, Braunschweig, West Germany (1964).
97. R. Bonnett, I. A. D. Gale, and G. F. Stephenson, *J. Chem. Soc. C* p. 1600 (1966).
98. H. Hellström, *Z. Phys. Chem. B* **14**, 9 (1931).
99. E. B. Fleischer, *Acc. Chem. Res.* **3**, 105 (1970).
100. H. Fischer and L. Bäumler, *Justus Ann. Chem.* **468**, 58 (1929).
101. E. Samuels and T. S. Stevens, *J. Chem. Soc. C* p. 145 (1968).

6

Chemical Transformations Involving Photoexcited Porphyrins and Metalloporphyrins

FREDERICK R. HOPF and DAVID G. WHITTEN

I. INTRODUCTION

The photochemistry of porphyrins and their metal complexes appears best divided into two categories. First, there are numerous reactions or photoprocesses in which excited states of porphyrins are intermediates, but in which no permanent chemical transformation of the porphyrin occurs. These processes will be discussed in the following section. The second category includes reactions in which the porphyrin system itself undergoes permanent chemical transformation. Examples of the latter include changes in the porphyrin ring and transformations involving the metal, as well as reactions involving extraplanar ligands.

II. PHOTOPROCESSES NOT INVOLVING
PERMANENT TRANSFORMATION OF THE PORPHYRIN

A. Porphyrins as Photosensitizing Agents: Excited States of Porphyrins

Porphyrins and metalloporphyrins, with their sharp and intense long-wavelength electronic transitions and relatively long-lived excited states, are nearly ideal photosensitizing reagents. In general, the excited states associated with free base and most metalloporphyrins are π,π^* states associated with the porphyrin macrocycle.[1-6] These states are not generally perturbed energetically by the substitution of different metals; however, the lifetimes and luminiscence properties are strongly influenced by the metal.[1,2,7-21] Generally, it has been found that reasonably strong prompt fluorescence is observed for free-base and closed-shell metal complexes in solution.[1,8,9] In addition, several open-shell diamagnetic metal complexes show both fluoresence and phosphorescence in solution.[11,13,15,22-24] In contrast, most paramagnetic metal complexes show emission only in rigid glasses at low temperature.[7-10,25] Table 1 summarizes emission properties of several metal complexes in fluid media at room temperature (300°K). A feature of porphyrins that makes them particularly useful as donors of triplet excitation is the relatively small energy gap between the lowest singlet and triplet states and the corresponding high intersystem crossing efficiencies for most compounds. Typical singlet and triplet energies for (free-base) octaalkyl porphyrins are 47 and 40 kcal/mole, respectively,[8] while the values for metal complexes are generally somewhat higher. Table 2 summarizes triplet energies for various metal complexes estimated from phosphorescence data. Although, as mentioned previously, substitution of different metals produces relatively little difference in the excited state energies, the excited state levels are rather

TABLE 1

Luminescence Properties of Porphyrins in Room Temperature (300°K) Solution

Nonluminescent	Fluorescence only	Phosphorescence only	Fluorescence and phosphorescence
Ni(II), VO, Sn(II)	Zn(II), Mg(II)	Co(III)	Pd(II)
Ru(II)L₂, Ru(III)	Sn(IV), Pd(II)	Rh(III)	Pt(II)
Cu(II), Ag,	Al, Cd, *FB*,	Ir	Ru(II)CO
Co(II)	Si(IV), Ge(IV),		
	Ba, Sr, Be,		
	Sc(III), Ti(IV),		
	Zr(IV), Hf(IV),		
	Nb(V), Ta(V)		

TABLE 2

Triplet Energies for Metalloporphyrins Estimated from Onset of Phosphorescence

Complex	E_t (kcal/mole)	Medium	Reference
Fe(III)(*meso*-DME)(OAc)	43	EPA, 77°K	9
Fe(II)(*meso*-DME)	43	EPA, 77°K	9
[Co(II)(*meso*-DME)	43.5	EPA, 77°K	9]^a
[Co(III)(*meso*-DME)(OAc)	43	EPA, 77°K	9]^a
Mn(II)(*meso*-DME)	39.5	EPAF, 77°K	9
Mn(III)(*meso*-DME)(OAc)	44.5	EPAF, 77°K	9
Mg(II)(etio-II)	42.2	EPA, 77°K	8
Mg(II)(etio-II)	40.5	3MeP, 77°K	9
Ba(II)(*meso*-DME)	38.9	EPAF, 77°K	9
Ca(II)(*meso*-DME)	40.9	EPAF, 77°K	9
Zn(II)(*meso*-DME)	41.5	EPA, 77°K	9
Sr(II)(*meso*-DME)	39.7	EPAF, 77°K	9
Cd(II)(*meso*-DME)	40	EPA, 77°K	9
Sn(IV)(*meso*-DME)(OAc)	41	EPA, 77°K	9
Pb(II)(*meso*-DME)	41.5	3MeP, 77°K	9
Hg(II)(*meso*-DME)	38	EPAF, 77°K	9
Cu(II)(*meso*-DME)	44	EPAF, 77°K	8
Pd(II)(*meso*-DME)	44.6	EPAF, 77°K	8
[Ni(II)(*meso*-DME)	43	EPAF, 77°K	8]
Pt(II)(etio-I)	45.3	EPAF, 77°K	11
Pt(II)(etio-I)	45	Decane, 81°K	11
Pd(II)(etio-I)	44	Nonane, 81°K	11
Pd(II)(etio-I)	44.3	EPAF, 77°K	11
Pd(II)(TPP)	42.6	EPAF, 77°K	11
Pt(II)(TPP)	44.6	EPAF, 77°K	11
Pd(II)(OEP)	44.8	Benzene, 300°K	
Ru(II)(OEP)(CO)	46.1	Benzene, 300°K	
Rh(III)(etio-I)(Cl)	44.3	2-MTHF, 77°K	15
Si(IV)(OEP)(Cl)$_2$	41	2-MTHF, 77°K	16
Ph(IV)(OEP)	41	MeOH, 77°K	16
Ge(IV)(OEP)(Cl)$_2$	41.4	2-MTHF, 77°K	16
Sn(IV)(OEP)(Cl)$_2$	41.4	2-MTHF, 77°K	16
Sn(IV)(TPP)(Cl)$_2$	41	2-MTHF, 77°K	16
VO(etio-I)	41	PMA, 74°K	17
VO(TPP)	39.7	PMA, 10°K	17

^{a} Entries in brackets refer to compounds whose luminescence properties have been questioned.

strongly affected by the addition of axial ligands to the tetragonal metalloporphyrin to form square pyramidal or octahedral complexes.[10,24,26,27] Since this process occurs readily in coordinating solvents, it is important to recognize that triplet (as well as singlet) energies will usually be solvent dependent and generally lower in coordinating (i.e., oxygen- or nitrogen-containing) solvents.

Involvement of porphyrin excited states in bimolecular photoprocesses is generally indicated by the observation of quenching or depolarization of luminescence or by direct observation of excited state quenching phenomena in flash photolysis experiments.[28–37] Excited state quenching can be kinetically studied by use of the Stern–Volmer relationship for intensities (Eq. 1) or lifetimes (Eq. 2) where k_q refers to the rate constant for the quenching process (Eq. 3), ϕ_e and ϕ_e^0 refer to the quantum efficiencies (intensities may also be

$$\frac{\phi_e^0}{\phi_e} = 1 + k_q\tau \cdot [Q] \tag{1}$$

$$\frac{\tau^0}{\tau} = 1 + k_q\tau^0[Q] \tag{2}$$

$$P^* + Q \longrightarrow P^0 + Q^* \text{ or products} \tag{3}$$

used) of emission in the presence and absence of quencher, respectively, and τ and τ^0 refer to excited state lifetime in the presence and absence of quencher. If products are formed or a new excited state, Q^*, is observable by emission or direct observation as a consequence of the quenching process, a second type of Stern–Volmer relationship for the product formation (Eq. 4) or its emission (Eq. 5) can be observed where ϕ_{prod} (Eq. 4 and Eq. 5) is the quantum

$$1/\phi_{prod} = 1/\alpha \left(1 + \frac{1}{k_q\tau^0[Q]}\right) \tag{4}$$

$$1/\phi_e = 1/\beta \left(1 + \frac{1}{k_q\tau^0[Q]}\right) \tag{5}$$

efficiency for product formation and α is the efficiency of forming product from quenched P^*, and β is the efficiency of quencher emission per quenching event. The intercept/slope values from plots of Eq. (4) or (5) should equal the slopes of plots of Eq. (1) or (2) to confirm the excited porphyrin as a direct intermediate in a photosensitizing process.

In principle, quenching studies such as those outlined above can be carried out for both excited singlet and triplet states. Frequently, lifetimes of the two states are so different that it is easy to separate the two processes. For porphyrin singlets, intensity quenching of fluoresence is generally the most convenient means of study, while triplets of most porphyrins in solution can be studied by flash spectroscopy or, in the case of those showing substrated solution phosphorescence [Ru(II), Pd(II), Pt(II), Rh, Ir], by intensity quenching. Somewhat more complicated kinetic relationships are encountered where, for example, both states are quenched, whereas only one reacts, but suitable treatments are available for virtually all possible cases.[38–41] Of course, intramolecular photoreactions and reactions proceeding via excitation of ground state complexes cannot be studied by the above techniques, and mechanisms must be developed by other approaches. The very features of the porphyrins and metalloporphyrins that make them excellent photosensitizers (see above) frequently preclude the use of other molecules as photosensitizers for porphyrin photoreactions. Clean sensitization experiments are difficult to perform, and the report[42] that the biacetyl triplet (excited at 436 nm) can sensitize, by triplet energy transfer, the 635-nm emission of protoporphyrin IX, and the conclusion that this emission is phosphorescent rather than fluorescent, is most probably, at least partially, in error.

B. Electronic Energy Transfer

Singlet–singlet energy transfer involving porphyrins, metalloporphyrins, and related compounds has been shown to be an important process in many cases. Especially important is the long-range singlet–singlet energy transfer, which is vital to the light-harvesting process in the photosynthetic apparatus.[43–47] This has been extensively studied by numerous groups using such techniques as fluorescence depolarization to assess the extent and range of this process. It has been shown that singlet energy can migrate extensively until localization in an especially situated low-energy "trap" occurs.[43–50] Singlet energy migration also has been demonstrated between porphyrin units in different chains of hemoglobin where the iron porphyrin of individual heme units has been replaced by porphyrins, such as protoporphyrin IX or zinc protoporphyrins, having longer excited state lifetimes. In the hemoglobin system, it has been found that transfer between unlike chains occurs with that from α to β favored over the reverse.[51] Changes in the rate of such transfer where a heme has been converted to oxyhemoglobin suggest that there are changes in the relative heme-heme orientation in the macromolecule upon oxygenation.[51]

Most fluorescent porphyrins and metalloporphyrins are strongly quenched by (ground state) triplet oxygen. However, the net process of this quenching apparently only enhances rates of S $\rightarrow$ T intersystem crossing in each case. Even though the formation of singlet oxygen is a spin-allowed reaction in such a process (Eq. 6), the S–T energy gap (~ 8 kcal/mole) in the porphyrin is

$$p^{1*} + O_2{}^3 \longrightarrow p^{3*} + O_2{}^{1*} \qquad (6)$$

insufficient to promote oxygen to its lowest ($^1\Delta$) state (23 kcal/mole).[52] Further, it has been found that in other systems, even where the S–T gap is greater than 23 kcal/mole, fluorescence quenching does not produce singlet oxygen.[53] An extensive kinetic investigation of photoperoxidation by Stevens has resulted in an "oxciplex" mechanism in which photoperoxide formation originating from excited singlets of aromatic hydrocarbons proceeds via several noninterconverting complexes.[53,54] Singlet oxygen is formed by reaction of sensitizer triplet with ground-state oxygen (Eq. 7),[42,55-57] and

$$p^{3*} + O_2{}^3 \longrightarrow p^0 + O_2{}^{1*} \tag{7}$$

formation of the peroxide results from collisions of singlet oxygen with the ground state of the sensitizer or other unsaturated molecule.

An important area where excited states of porphyrins and singlet oxygen are frequently, if not universally, linked is the photodynamic inactivation of biologic systems. Several instances of light-mediated damage to biologic systems probably involve the sequence: light absorption by porphyrins or their metal complexes, energy transfer to oxygen (Eq. 7), and subsequent attack of singlet oxygen on unsaturated groups elsewhere in the system. One case where recent investigation has made this sequence fairly clear involves photohemolysis of red blood cell membranes in the genetic disorder erythropoietic protoporphyria.[58-61]

This disorder is found in patients having unusually large concentrations of free-base protoporphyrin.[62,63] The photohemolysis occurs with visible light and has an "action spectrum" corresponding closely to the Soret band of protoporphyrin near 400 nm. Molecular oxygen is required for the process, and the net result is destruction of the red blood cell membrane. Recent experiments[58-61] have shown that protoporphyrin triplet-sensitized singlet oxygen attacks lipid components of the membrane; a major site of reaction appears to be cholesterol, which undergoes the "ene" reaction[64-66] to yield 3β-hydroxy-5α-hydroperoxy-Δ^6-cholestene. The hydroperoxide subsequently breaks down to yield radical products which attack other sites.[61] This molecular mechanism is supported by the observation that massive doses of β-carotene and α-tocopherol, substances which have been shown to quench singlet oxygen, protect patients from photohemolysis.[67] Related disorders have very recently been shown to involve sensitization of singlet oxygen by zinc porphyrin triplets.[68] Several other cases of porphyrin-mediated "photodynamic deactivation," including photooxidation of porphyrins themselves, probably involve similar events.[69-75]

In a series of elaborate investigations, Jori and co-workers[76-82] have found that photooxidative damage occurs to amino acids such as methionine, histidine, tryptophan, and tyrosine, which are located close to heme groups in various systems such as the cytochromes, hemoglobin, and myoglobin.[75,78] It has been found that substitution of the iron porphyrin by free-base hematoporphyrin or closed-shell metal complexes such as the magnesium (II) complex greatly accelerates the photooxidation.[76,77,79] This effect has been used to explore conformations of the macromolecule about the active site in active, but noncrystalline phases. Photooxygenation of the porphyrin-related pigment bilirubin (which may be involved in jaundice phototherapy)[83,84] also probably involves singlet oxygen, since bilirubin efficiently sensitizes its formation,[85] and has been shown to react rapidly with O_2 to yield new products.[86]

As mentioned previously, the porphyrins and those metal complexes having long-lived excited triplets are potentially excellent sensitizers for donation of triplet excitation. Triplet porphyrins have been used to populate triplet excited states of substituted azobenzenes[87] and thioindigo dyes,[88] with concurrent cis-trans isomerization of the acceptor in each case. As with the porphyrins and metalloporphyrins, chlorophyll triplets can efficiently produce singlet oxygen by energy transfer. However, competitive deactivation by energy transfer to β-carotene and related compounds having low-lying short-lived excited states affords at least some measure of protection against destructive photooxidation.

Intramolecular transfer of excitation between porphyrin and metal, porphyrin and extraplanar ligand, and nearby porphyrin moieties have been observed in various investigations. Usually, both absorption and emission spectra of metalloporphyrins originate from π,π^* states; the case of recently prepared ytterbium (III) complexes represents a notable exception.[89]

For the Yb(III) tetrabenzoporphine complexes, absorption spectra are more or less normal; however, emission occurs only from Yb^{3+} F levels in the near infrared (ir) (9800–11000 cm^{-1}).[89] Here it is suggested that efficient porphyrin to metal energy transfer occurs similar to that observed with other rare-earth complexes such as those of europium (III).[90] In systems involving different metalloporphyrins linked by ethylene and p-phenylene bridges, it has been found that efficient intramolecular energy transfer occurs from one porphyrin to the other (Zn → Cu; Zn → Co), even though efficient singlet–singlet energy transfer does not occur.[91] An exchange mechanism has been suggested for the ethylene-bridged Zn–Cu system, while a mechanism involving a charge–transfer complex has been proposed for the Zn–Co system.[91] For Zn(II), Mg(II), and Co(III) porphyrins complexed to the stilbene-like olefinic ligands 4-stilbazole and 1-(1-naphthyl)- 2-(4-pyridyl)ethene (NPE), reversible transfer of triplet excitation from porphyrin to olefinic ligand has been observed.[92–94] In the case of the NPE complexes with Zn(II) and Mg(II) porphyrins, the combination of rapid energy transfer and rapid ligand exchange, coupled with a long triplet lifetime for the metalloporphyrin, allows a quantum chain process of cis → trans isomerization of NPE as outlined in Eq. (8–12), where NPE = cis-NPE and NPE' = $trans$-NPE.[93,94] Several

$$MP-NPE \xrightarrow{h\nu} MP^{1*}-NPE \xrightarrow{isc} MP^{3*}-NPE \qquad (8)$$

$$MP^{3*}-NPE \longrightarrow MP-NPE'^{3*} \qquad (9)$$

$$MP-NPE'^{3*} \longrightarrow MP^{3*}-NPE' \qquad (10)$$

$$MP^{3*}-NPE' + NPE \longrightarrow MP^{3*}-NPE + NPE' \qquad (11)$$

$$MP^{3*}—NPE \text{ (or NPE')} \longrightarrow MP—NPE \text{ (or NPE')} \qquad (12)$$

aspects of the energetically uphill energy transfer step (Eq. 9) are of interest; the low activation energy and large negative entropy of activation suggest that a distorted excited state of NPE is produced in what might be regarded as a "nonvertical" energy transfer step.[94]

C. Electron Transfer Reactions and Exciplex Formation with Excited Porphyrins and Metalloporphyrins

It has long been known that photoexcited porphyrins and related compounds—especially chlorophyll—can function as net electron donors or acceptors or otherwise mediate electron-transfer reactions.[95–107] The dual function of chlorophylls as both electron donors and acceptors is, of course, the cornerstone of photosynthesis.[43–50] Until recently however, precise details of electron-transfer quenching have not been well understood, and results obtained with different porphyrin containing systems indicate that there can be many mechanistic complications. Livingstone and co-workers[108–110] showed several years ago that fluorescence of chlorophyll and related porphyrin derivatives is quenched by quinone and nitroaromatics in solution. Subsequent to Livingstone's work, there have been several investigations indicating that quenching of chlorophyll free-base porphyrin and zinc and magnesium porphyrin fluorescence by electron-deficient aromatic compounds is a rather general phenomenon that occurs in polar solvents and in nonpolar media as well.[26,111–112]

Irradiation of chlorophyll, porphyrins, or metalloporphyrins in polar solvents in the presence of these electon acceptors often leads to the production of detectable transients or permanent products from net electron transfer. For example, it was found that irradiation of zinc porphyrins or chlorophyll in the presence of quinones can lead to steady-state concentrations of the semiquinone radical sufficient for esr detection.[105,113,114] Recent investigations have shown that chlorophyll a, pheophytin, and bacteriochlorophyll can sensitize a "two-electron transfer" in the presence of quinones in ethanol resulting in the formation of the ethanol radical cation and the semiquinone radical anion.[98,114] The authors suggest that formation occurs from the chlorophyll triplet and that the reaction involves decomposition of a ternary exciplex as shown in Eq. (13–20).[98] Insufficient evidence exists to determine

$$Chl \xrightarrow{h\nu} Chl^{1*} \longrightarrow Chl^{3*} \qquad (13)$$

$$Chl^{3*} \longrightarrow (Chl \cdots Q \cdots Et\overset{..}{O}H)^{3*} \qquad (14)$$

$$(Chl \cdots Q \cdots EtOH)^{3*} \longrightarrow (Chl \cdots Q^{-} \cdots Et\overset{+}{O}H) \qquad (15)$$

$$(Chl \cdots Q^- \cdots Et\dot{O}H) \longrightarrow Chl + Q + EtOH \qquad (16)$$

$$(Chl \cdots Q^{\bar{}} \cdots Et\dot{O}H) \longrightarrow Chl + Q^{\bar{}} + Et\dot{O}H \qquad (17)$$

$$2Q^{\bar{}} + 2H^+ \longrightarrow Q + 2H_2Q \qquad (18)$$

$$2Et{-}\dot{O}H^+ \longrightarrow Et{-}O{-}O{-}Et + 2H^+ \qquad (19)$$

$$H_2Q + EtOOEt \longrightarrow Q + 2EtOH \qquad (20)$$

whether a ternary complex exists prior to excitation. Harbour and Tollin[115] have found that irradiation of bacteriochlorophyll and benzoquinone in dry acetone leads to the products of one-electron transfer, the bacteriochlorophyll radical cation and quinone radical anion. Here again, it is suggested that the bacteriochlorophyll triplet state is the electron donor. Chibisov[116] has studied photooxidation of chlorophyll by nitrobenzene and 4-chloro-1,3-nitrobenzene, and they also have concluded, from kinetic evidence, that the chlorophyll triplet is the electron donor.[117]

Flash spectroscopic investigations by Whitten and co-workers have shown that both singlet and triplet states of zinc and magnesium porphyrins are quenched with comparable rates by nitroaromatics and other electron acceptors such as chloro compounds and quinones.[107,113,118–120] These quenching processes occur with rapid rates in both polar and nonpolar solvents; however, a study of the transients produced indicates that different intermediates are produced in media of varying polarity.[107,113,119,120] In polar solvents, both singlet and triplet states of the metalloporphyrin yield free ions that recombine to yield starting materials at rates near the diffusion-controlled limit.[107] In nonpolar solvents, quenching of metalloporphyrin triplets leads to long-lived triplet exciplexes, which do not decay to free ions.[107,119] In certain cases, ternary complexes containing one porphyrin and two molecules of quencher are formed.[118,120] The exciplexes formed between triplet metalloporphyrins and nitroaromatics in nonpolar media are interesting in that charge–transfer interactions appear to play only a limited role.[119] Although correlations between quencher reduction potentials and log k_q exists in several cases, it is clear that "pure" charge–transfer states, $P^{+\cdot}$, $Q^{-\cdot}$, lie substantially higher in energy than the exciplex actually formed.[119]

Through studies of singlet and triplet states of porphyrins, aromatic hydrocarbons, transition metal complexes, and several other systems, a general picture of electron-transfer quenching of excited states is emerging.[120–124] The energy required for the ground-state electron transfer (Eq. 21) can be

$$D + A \longrightarrow D,A \longrightarrow D^+,A^- \qquad (21)$$

$$\Delta G = E_{1/2}[D^+/D] - E_{1/2}[A/D^-] + W_p - W_r \qquad (22)$$

estimated from oxidation–reduction potentials using Eq. (22), where W_p and W_r represent the work required to bring products and reactants together to form the complexes (or ion pairs), D,A and D^+,A^-.[122] For several systems studied to date (where electron self-exchange rates between reactant and product ions are rapid), it has been found that the rate constant for electron-transfer quenching (Eq. 23 or 24) is nearly diffusion-controlled, provided

$$D^* + A \longrightarrow D^+,A^- \tag{23}$$
$$D + A^* \longrightarrow D^+,A^- \tag{24}$$

$E_D^* > \Delta G$ or $E_A^* > \Delta G$, where the excited state is D^* or A^*, respectively. This has been found true for both excited singlets[121,122] and triplets,[123,124] and it appears that electron-transfer quenching can compete even with electronic energy transfer quenching in cases where both processes are exothermic.[124] Where both reactants are neutral, the initial product of electron transfer quenching is an ion pair, D^+,A^-. In such an ion pair, the escape probability for forming the free ions D^+ and A^- is very small,[125,126] especially in nonpolar media and, hence, the yield of ions from such quenching processes in nonpolar media is small. The lifetime of ion pairs found in such a process in solution is very short due to the rapid ($\tau v \ 10^{-11}$ sec)[122] reverse electron transfer to yield the reactant pair, D,A. In very polar media, a significant fraction of the ions can escape from the ion pair[125] and, of course, in the cases where both D and A have charges of the same type, the product ions (or neutral molecules) have higher escape probabilities.[125]

Since porphyrins and their metal complexes are both easily reduced and easily oxidized, it is in retrospect not surprising that electron-transfer quenching should be a prominent pathway. Formation of exciplexes having some charge–transfer contribution can provide an alternate channel for reaction in certain systems, even when direct electron transfer quenching is energetically unfavorable. Although, as mentioned above, such phenomena have long been associated with the chlorophylls and related compounds and with the fluorescent zinc and magnesium complexes, there is no reason such reactions cannot occur with other metalloporphyrins. In fact electron-transfer quenching of excited porphyrins to form detectable products has been observed with such diverse metal complexes as Sn(IV),[106,127] Fe(III),[128,129] Co(III),[130] and Pd(II).[131] In the case of tin (IV) porphyrins, the triplet can be quenched by $SnCl_2$ in pyridine or ethanol to yield ultimately the tin(IV) porphyrin dianion.[127]

Reaction of palladium(II) octaethylporphyrin triplets with the electron-deficient dication paraquat (1,1-bis-N-methyl-4,4'-bipyridine) in butyronitrile yields the π-cation of the porphyrin and the paraquat monocation.[131] In contrast, irradiation of iron(III) tetraphenylporphine in the presence of azobenzene leads to the iron(II) complex[128] and irradiation of cobalt(III) or iron(III) porphyrins in the presence of ligated and/or nonligated amines leads to the corresponding Co(II) and iron(II) complexes.[129,130]

III. PHOTOPROCESSES INVOLVING
PERMANENT TRANSFORMATIONS OF THE PORPHYRIN

A. Alterations of the Porphyrin Ring and/or Side Chains

A brief discussion of electron distribution in the porphyrin macrocycle is useful in reviewing reactions both photooxidations and photoreductions of porphyrins and metalloporphyrins. A useful model is that proposed by Woodward.[132] Each of the pyrrolenine units (1) of the porphyrin macrocycle contains five π-electrons and should be expected to remove electron density from its macrocyclic surrounding to achieve an aromatic sextet. Within the porphyrin ring system, such electron withdrawal can only take place from the methine bridge positions, since the pyrrole rings already contain an aromatic sextet and will resist such withdrawal. Structure 2 represents the extreme resonance contributor (Fig. 1).

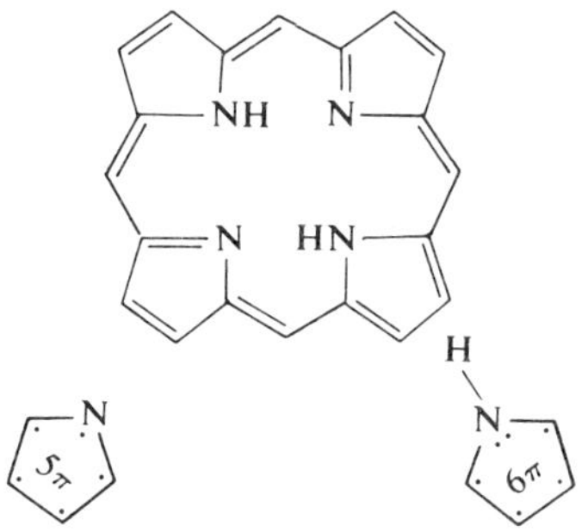

Fig. 1. Model for porphyrin electron distribution as proposed by Woodward.[132]

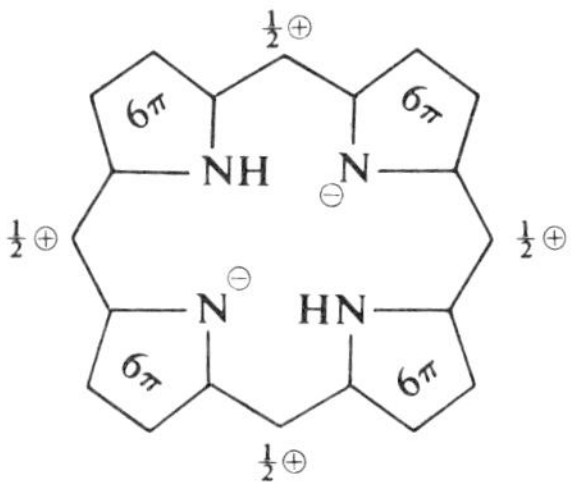

In the formally neutral ligand, i.e., when no electrons are transferred from the metal to the porphyrin and no redox reaction has taken place on the porphyrin, the methine bridges carry a partial positive charge and nucleophilic type additions should occur readily. If, however, electrons are introduced from the central metal ion of metalloporphyrins or by reduction of the ligand, this will have litttle effect on the aromatic system of the pyrrole rings but should have a pronounced effect on the electron density at the methine bridge positions.

If this electron density distribution is considered qualitatively correct, the reactivities of the pyrrole β-carbon atoms should be similar to those in pyrrole, i.e., they should be accessible to electrophilic substitutions. In contrast, the reactivity of the methine bridges in porphyrin complexes with different metals should reflect the great diversity that can be expected from these complexes, corresponding to the variable redox behavior of the central metal ions.[133,134] There is fair qualitative agreement between theory and experiment, inasmuch as most nucleophilic, electrophilic, and free-radical addition reactions of the porphyrin ligand begin at the methine bridges. Extension of this argument would suggest that inductive effects (positive or negative) of functional groups attached either at the periphery of the ring or in extraplanar ligands should influence reactivity, as well as many readily measured physical parameters. Good correlations have been shown.[135,136]

1. PHOTOOXIDATION

a. Free-Base Porphyrins. Mauzerall and Granick[137] and co-workers report the photooxidation of the octomethyl esters of uroporphyrinogen (3) to the corresponding uroporphyrin (6) (Fig. 2) in pH < 8.5 solutions con-

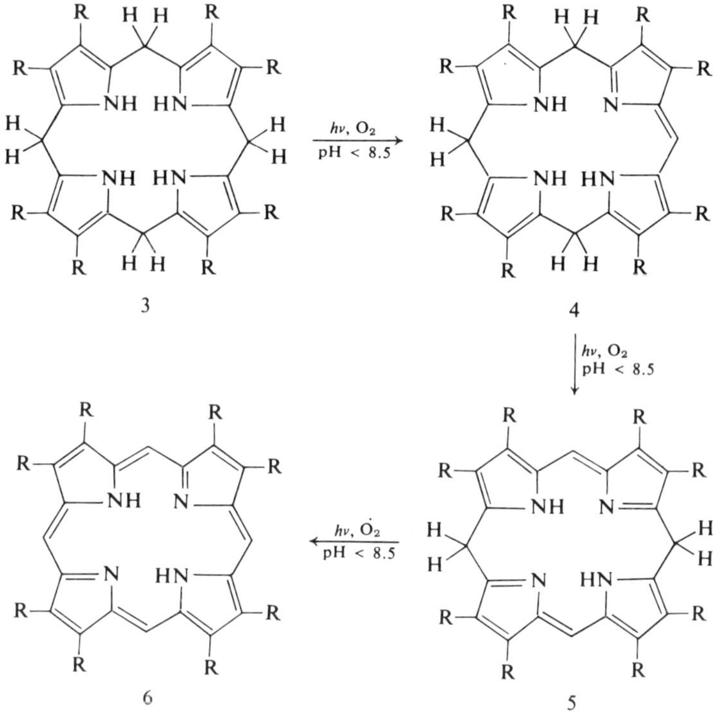

Fig. 2. Photooxidation of uroporphyrinogen to uroporphyrin.

taining oxygen. Visible spectral evidence indicated the oxidation proceeded via the porphomethene (4) and porphodimethene (5) intermediates. White light was used. This reaction exhibited a marked induction period. Control experiments showed that addition of a small percentage of uroporphyrin caused a marked shortening of the induction period. Attempts to obtain an action spectrum by visible spectrometry were unsuccessful. No direct proof of the proposed porphomethene intermediates was obtained. Rigorously degassed solutions of uroporphyrinogen were only slightly oxidized, even when exposed to white light for long periods of time. Various oxidation inhibitors were shown to retard photooxidation. The rate of photooxidation increased with decreasing pH. Yields of porphyrin product also increased as the pH was lowered to an optimum of 1 M HCl. Clearly, both oxygen and porphyrin product are important in this reaction. A plausible explanation may be Eq. (25–27). A

$$\text{Porphyrin} \xrightarrow{h\nu} \text{porphyrin*}^1 \xrightarrow[\text{isc}]{} \text{porphyrin*}^3 \tag{25}$$

$$\text{Porphyrin*}^3 + O_2{}^3 \longrightarrow \text{porphyrin} + O_2{}^{*1} \tag{26}$$

$$O_2{}^{*1} + \text{porphyrinogen} \longrightarrow \longrightarrow \longrightarrow \text{porphyrin} \tag{27}$$

similar photooxidation of a hydroporphyrin involves the use of 1,2-naphthoquinone to convert tetrahydrotetraphenylporphyrin to tetraphenylchlorin.[138]

Recent work[139] on heme-derived porphyrins has shown that unsaturated side chains are photolabile and can be selectively photooxidized by molecular oxygen in organic solvents such as pyridine or dichloromethane. Protoporphyrin, DME, when irradiated into any of the visible bands or the Soret band, is converted to two products (Eq. 28). The reaction occurs via 1,4-

$$\tag{28}$$

addition of oxygen to the vinyl-substituted pyrrolenine rings (Eq. 29), with subsequent decomposition of the peroxide. The absolute quantum yields for air oxidation of protoporphyrin were determined in various solvents, e.g., $\phi = 0.033$ in chloroform and 0.006 in benzene.[69]

Among the earliest porphyrin photooxidation reactions are those reported by Fischer.[140,141] The reactions were run in very basic media on the disodium

$$(29)$$

9   10   11

salts of porphyrins. The products apparently were the nonmetalated analogs of those reported by Fuhrhop[71,133] (formyl biliverdins). As a result of the very basic conditions employed, other nucleophilic additions occurred subsequent to formation of the formyl biliverdin, which complicated structure determination.

b. Metalloporphyrins. Calvin and co-workers[142-144] reported some of the earliest work on photooxidation of metallochlorins to metalloporphyrins. They reported the photochemical conversion of Zn and Mg tetraphenylchlorins by a series of quinones in benzene solution. Kinetic data point to the intermediacy of the metallochlorin triplet state in the reaction. Quantum yields (≈ 0.001 to 0.008) were measured for several series of quinones. Straight-line plots of quantum yields versus oxidation potential of the quinones were obtained. These authors also reported[145] that Zn and Mg tetraphenylchlorins could be photooxidized by molecular oxygen in a manner analogous to that found for ortho- and paraquinones. However, the H_2O_2 initially produced undergoes a secondary reaction with the porphyrin to yield a product similar to that obtained by bleaching chlorophyll in the presence of oxygen.[43,146]

Air photooxidation of a series of porphyrins and their zinc derivatives in organic solvents has been reported by Gurinovich.[147,148] It is suggested that irreversible addition of oxygen occurs with destruction of the conjugated bond system. Partial reduction of the products was effected by photochemical reduction with ascorbic acid/pyridine, but little quantitative data were supplied.

A clear case of photoxidation leading to disruption of the porphyrin macrocycle is reported by Fuhrhop and co-workers.[71,133] Magnesium octaethylporphyrin is photooxygenated with molecular oxygen to form the magnesium formylbiliverdin according to the following sequence shown in Eq. (30). Conclusive pmr, ir, visible, and mass spectral data are given to substantiate the structure of the photooxidation product.

Studies of rate inhibition of this reaction utilizing β-carotene at very low concentration (3×10^{-5} M) indicate the presence of an intermediate of lifetime < 3 μsec. This follows from the maximum rate constant for a diffusion-limited reaction in benzene of $\sim 10^{10}$ M^{-1} sec^{-1} and the failure to observe molecular complexes at these concentrations. However, all triplets

(30)

studied thus far are quenched by oxygen at rates close to the diffusion controlled limit. Considering the solubility of oxygen in benzene, this indicates the lifetime of the triplet state of Mg OEP cannot be more than 0.03 μsec. This clearly suggests that photooxygenation occurs via the excited singlet state of oxygen, probably formed on quenching the triplet state of Mg OEP according to the following sequence (Eq. 31–33).

$$Mg(OEP) \xrightarrow{h\nu} [Mg(OEP)]^{*1} \longrightarrow [Mg(OEP)]^{*3} \qquad (31)$$

$$[Mg(OEP)]^{*3} + O_2{}^3 \longrightarrow Mg(OEP) + (O_2)^{*1} \qquad (32)$$

$$(O_2)^{*1} + Mg(OEP) \longrightarrow \text{oxygen adduct} \qquad (33)$$

Other metalloporphyrins do not photooxygenate in this manner. The authors suggest that Mg OEP uniquely reacts in this way because it has a sufficiently low oxidation potential. By utilizing the fact that metallochlorins have a first oxidation potential ~ 300 mV lower than the corresponding metalloporphyrin, Fuhrhop successfully photooxygenated zinc octaethylchlorin to give products analogous to those obtained from Mg OEP. Now, however, two isomers are formed, as indicated below (structures 15 and 16).

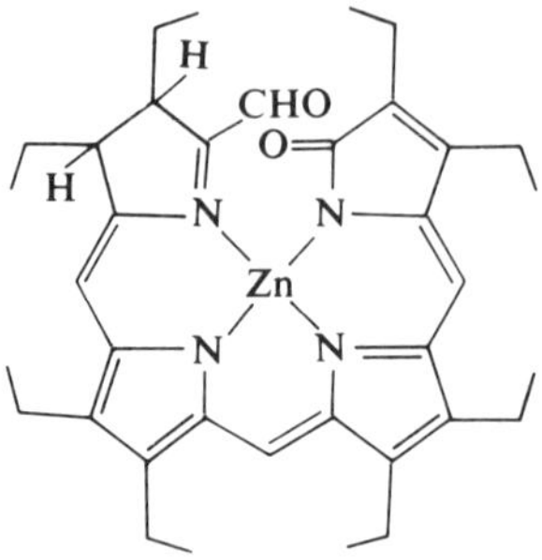

15 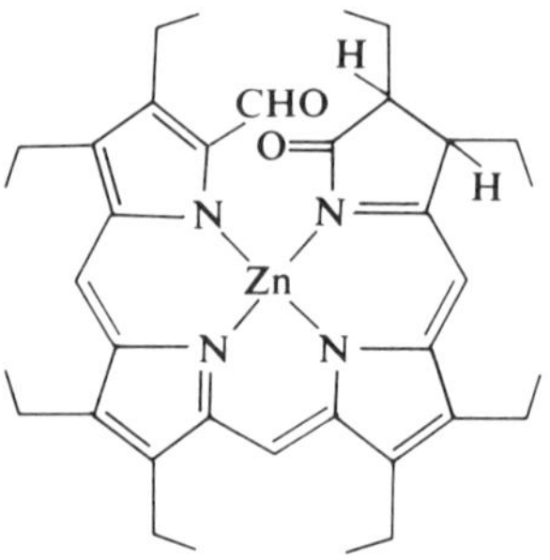   16

Yet a different sequence occurs[149,150] when a bridge carbon is substituted as in octaethyl-α-hydroxyporphinotozinc(II) (17). The reaction (Fig. 3) transforms the compound into a biliverdin residue (23). The corresponding nickel complex undergoes a similar initial reaction to form the nickel analog of 19; however, further irradiation leads to the cyclic ether (24). Further irradiation of 19 in the absence of oxygen also leads to the zinc-containing analog of 24.

Despite speculation by many authors about the mechanistic details of porphyrin photooxidation reactions, little definitive evidence is available. Most of the mechanistic work done in this area has been done with chlorophyll derivatives, rather than with simple porphyrins. Changes in ground state uv-visible spectra have been utilized to obtain information about relatively long-lived intermediate species.[147,148,151,152]

Flash spectroscopic studies have been widely used to investigate reversible photoreactions of chlorophyll at low temperatures.[153]

Coupling of flash photolysis and esr techniques has been a quite useful development.[43] With this technique, light-induced changes in spin properties can be studied. By use of this technique, it has been shown that the light-induced radical in *Rhodospirillum rubrum* arises from a radical of bacteriochlorophyll.[154] Other applications of this technique have been in the study of chlorophyll *a* photooxidation by molecular oxygen[155] and the reversible photooxidation observed in a chlorophyll–benzoquinone system,[156] as well as the study of subchloroplast particles.[43] Recent esr studies have been instructive.[157,158]

Alteration of emission (fluoresence) properties has been widely used to study chlorophyll systems. One such recent case has been the observation of changes in chlorophyll fluorescence upon addition of excess magnesium ions.[159]

Recently chemically induced dynamic electron polarization (CIDEP)[160,161] has been used to study the light-induced reaction between bacteriochlorophyll (BChl) and benzoquinone (BQ) at low temperatures. The results of these

Fig. 3. Photooxidation to biliverdin.$\rightarrow$

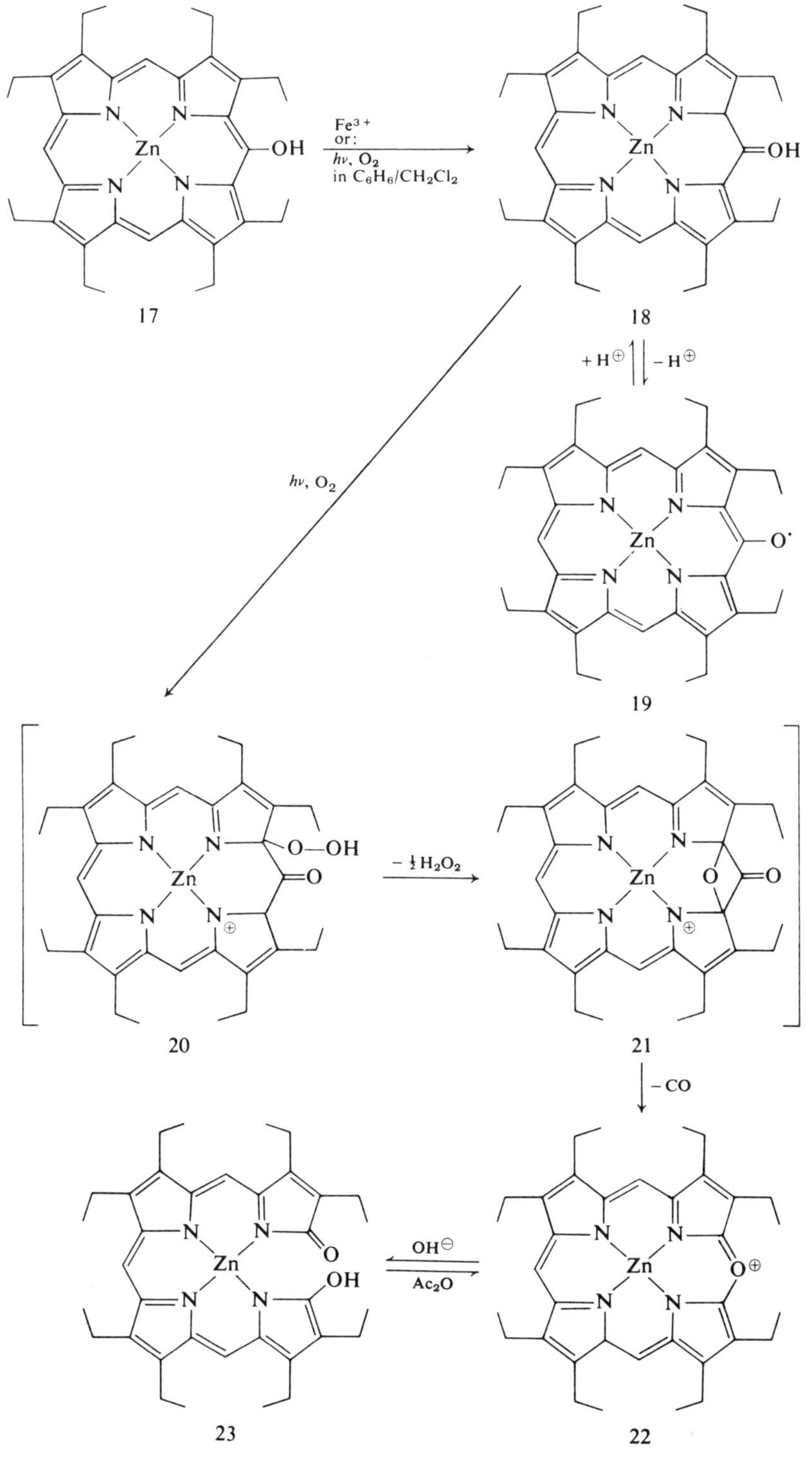

17

18

$+ H^\oplus \updownarrow - H^\oplus$

Fe^{3+}
or:
$h\nu, O_2$
in C_6H_6/CH_2Cl_2

$h\nu, O_2$

19

20

21

$- \frac{1}{2} H_2O_2$

$- CO$

23

22

$\xleftarrow[Ac_2O]{OH^\ominus}$

24

experiments provide good evidence that an initial, direct, light-induced electron transfer occurs between bacteriochlorophyll and benzoquinone, leading to formation of BChl+· and BQ−· Other esr work by the same authors,[162] utilizing continuous illumination of chlorophyll or bacteriochlorophyll in degassed ethanol, suggests that photooxidation occurs via the lowest excited singlet states to produce cation radicals. Chibisov[163] has studied the effect of water on the yield of chlorphyll cation radical during photooxidation. Several recent papers deal with the photooxidation of chlorophyll b.[164,165] Numerous other papers deal with photoinduced effects on chlorophyll,[166,167] as well as the effects of light on various *in vivo* systems related to photosynthesis.[43,168] However, they are of limited value in further understanding the mechanisms of simple photooxidation.

2. PHOTOREDUCTIONS

a. Free-Base Porphyrins. The initial report of reversible photochemical reduction of porphyrins was by Krasnovskii.[169] Since then, a large number of papers by Krasnovskii and others[170–173] have dealt with many aspects of photoreduction under a variety of reaction conditions.[174–179]

Mauzerall[180,181] and co-workers reported a study of the photochemical reduction of free-base uroporphyrin esters under mild conditions with a variety of reducing agents including ascorbic acid, glutathione, ethylenediaminetetraacetic acid, and ethyl acetoacetate. The products of these reactions were shown to be the di-, tetra-, and hexahydroporphyrins (**26–28**) shown in Eq. (33a) (substituents omitted for clarity). Kinetic and spectral evidence, as well as comparison with products formed by nonphotochemical reductants, was utilized in determination of the intermediate photoreduction product structures. Further evidence in support of the phlorin structure (**26**) as the initial dihydrogenation product is provided by thermodynamic studies.[182] The photoreversibility of phlorin formation has been demonstrated.[183] A mixture of phlorin and porphyrin in glycol undergoes conproportionation on

(33a)

irradiation (Fig. 4) to form two porphyrin radical anions, which are converted back to starting materials by a relatively slow dark reaction.

Sidorov[184,185] has investigated the reversible photoreduction of pyridine solutions of tetraphenylporphyrin in the presence of hydrazine or hydrogen sulfide as reductants. From visible, ir, and kinetic evidence, it was proposed that two pathways are operative under these conditions. One leads to phlorin-type products; the other results in disruption of the porphyrin ring.

A subsequent paper by Sidorov[186] on the hydrazine/pyridine photoreduc-tion of tetraphenylchlorin indicates that photoaddition of two atoms of hydrogen occurs only at the central nitrogen atoms and not at the bridge positions. Tetraphenylbacteriochlorin, under the same conditions, was reduced to a colorless product, which reversibly reverted to the starting material upon exposure to air. No evidence was given to indicate whether hydrogenation occurred at the bridge positions or only at the central nitrogen atoms.

Gurinovich and co-workers[187] have shown that photoreduction of pheo-phytin and its derivatives leads to results similar to those for simple porphy-rins when the macrocycle does not contain an integral cyclopentanone ring. The presence of a cyclopentanone ring caused more complex product mixtures.

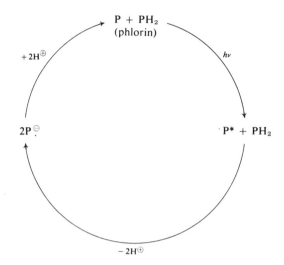

Fig. 4. Conproportionation of porphyrin–phlorin mixtures.

There are numerous reports[169,171–173] of photoreduction in strongly acidic media (e.g., aqueous acid or alcohol/acid). The results obtained are apparently similar to those in basic media. Krasnovskii's group has done extensive work on free-base photoreduction in acidic solutions.[188] A scheme summarizing their results is presented in Fig. 5. Electron-spin resonance evidence implicates the presence of radicals as short-lived intermediates[189] in the overall two-electron processes.

    b. Metalloporphyrins. The first reported demonstration of reversible photochemical reduction of metalloporphyrins was by Seely and Calvin.[190] Zinc tetraphenylporphyrin was reduced, using benzoin or dihydroxyacetone as reductants. Their results are summarized in Fig. 6 (Zn TPP is represented as porphyrin).

    An elegant paper by Seely and Talmadge[191] reports on the photochemical reduction of zinc porphine by ascorbic acid in degassed solutions. Extensive

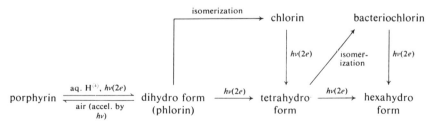

Fig. 5. General scheme for photoreduction in acid media.

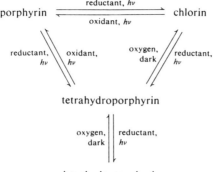

Fig. 6. Photoreduction of metalloporphyrins.

kinetic and spectral evidence is presented which supports the authors' formulation of the initial dihydroderivative as one having hydrogen atoms added to one-bridge carbon and one β-pyrrole carbon. The steps in the proposed mechanism are outlined in Eq. (32–38) (P represents zinc porphine and

$$P \xrightarrow{h\nu} P^{*1} \xrightarrow{isc} P^{*3} \tag{34}$$

$$P^* + QH_2 \longrightarrow PH\cdot + QH\cdot \tag{35}$$

$$PH\cdot + QH\cdot \longrightarrow P + QH_2 \tag{36}$$

$$2PH\cdot \longrightarrow P + PH_2 \tag{37}$$

$$2QH\cdot \longrightarrow Q + QH_2 \tag{38}$$

QH_2 represents reductant). Likewise, the further photoreduction of chlorin was studied, and the steps for its reduction to a tetrahydroporphyrin are given in Eq. (39–44). The back oxidation of PH_2 to P is first order and very large in the presence of partly oxidized ascorbate, probably going via Eq. (45–46), with

$$PH\cdot + PH_2 \longrightarrow P + PH_3\cdot \tag{39}$$

$$PH_3\cdot + QH\cdot \text{ (or Q)} \longrightarrow \text{chlorin} + QH_2 \text{ (or } QH\cdot) \tag{40}$$

$$\text{Chlorin} \xrightarrow{h\nu} \text{chlorin}^{*1} \xrightarrow{isc} \text{chlorin}^{*3} \tag{41}$$

$$\text{Chlorin}^{*3} + QH_2 \longrightarrow \text{chlorin H}\cdot + QH\cdot \tag{42}$$

$$\text{Chlorin H}\cdot + QH\cdot \longrightarrow \text{chlorin} + QH_2 \tag{43}$$

$$2 \text{ Chlorin H}\cdot \longrightarrow \text{tetrahydroporphyrin} + \text{chlorin} \tag{44}$$

$$\text{Chlorin H}\cdot + P \longrightarrow \text{chlorin} + PH\cdot \tag{45}$$

$$\text{Chlorin H}\cdot + PH_2 \longrightarrow \text{chlorin} + PH_3\cdot \tag{46}$$

$$Q + QH_2 \longrightarrow 2QH\cdot \tag{47}$$

$$QH\cdot + PH_2 \longrightarrow QH_2 + PH\cdot \tag{48}$$

Eq. (45) being rate limiting. The variation in quantum yields observed by the authors can be explained if Eq. (47 and 48) are considered as competing reactions for Eq. (34 and 40).

The conversion of PH_2 to chlorin does not occur in the dark, but requires some substance produced only by light. Since excited porphyrins did not react with PH_2 directly (except possibly by quenching), the reduced porphyrin radical PH· was implicated. However, if this is so, and if the dark oxidation of PH_2 proceeds via radicals, it is difficult to explain why no chlorin is formed in the dark. A possible explanation of this discrepancy is proposed: the radical formed by photoreduction of porphyrin differs from and is a stronger reducing agent than the radical formed by oxidation of PH_2. If the former is designated PH·$_{p.r.}$, then Eq. (10) becomes Eq. (49), and Eq. (37) becomes Eq. (50), while all other reactions remain as outlined.

$$P^* + P \longrightarrow 2P \qquad (49)$$

$$\text{Chlorin}^* + \text{chlorin} \longrightarrow 2 \text{ chlorin} \qquad (50)$$

The structures proposed for the intermediates are outlined in Fig. 7. Similar reactions have subsequently been observed by Sidorov[192] for magnesium and calcium porphyrins. However, the authors propose a different set of steps for the photoreduction sequence, i.e., metallochlorins are formed directly from metalloporphyrins while metal PH_2 is a by-product.[193,194]

Shul'ga et al.[195] have attempted to show interconnections between photoreductions of free-base porphyrins and their metallo complexes. Some of their results can tentatively be interpreted in terms of a common photoreduction mechanism for free-base porphyrins and metalloporphyrins.

Suboch and co-workers[187,196] report the photoreduction (rigorously degassed ascorbic acid–propanol–pyridine) of protochlorophyll and a series of its derivatives. Conditions were found for markedly improving the yield of chlorin derivatives.[197,198] These authors have isolated and identified the products of photoreduction of a series of metalloporphyrins. Small but significant differences (electronic, ir, and pmr) were discovered between photochemically prepared chlorins and natural chlorins. These result from the fact that photochemically prepared chlorins have a cis arrangement of hydrogen atoms on the 7,8-linkage (35), whereas natural chlorins have the trans arrangement (36). Cis hydrogenation also occurred upon photoreduction of simple, synthetic metalloporphyrins such as zinc etioporphyrin (37).

To explain this selective cis-photohydrogenation, Suboch proposes direct formation of metallochlorins from metalloporphyrins. At present, the best explanation for the diverse range of experimental results requires dual reaction pathways. Depending upon the reaction conditions, photoreduction follows two parallel paths: (a) reversible metalloporphyrin photoreduction and

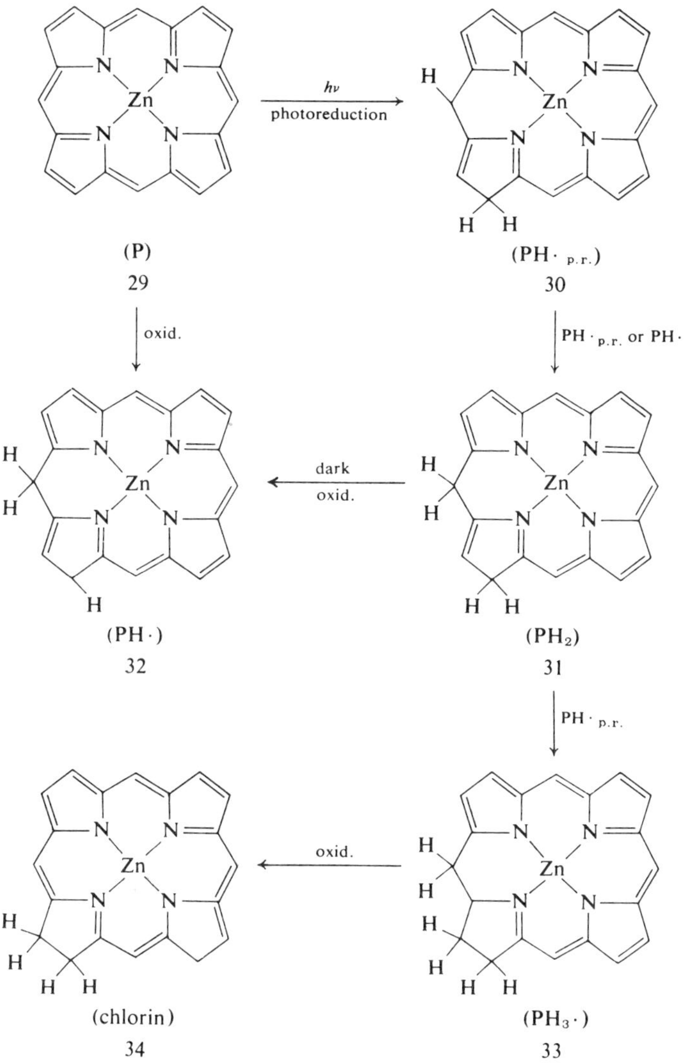

Fig. 7. Porphyrin photoreduction with intermediates.

(b) metalloporphyrin photoreduction to the corresponding metallochlorin (Fig. 8). According to the first pathway, metalloporphyrin (M–P) photoreduction proceeds via a number of intermediates to the dihydrometalloporphyrin (M—PH$_2$), which can either oxidize to starting metalloporphyrin or, by subsequent irradiation, can be converted via a number of intermediates into

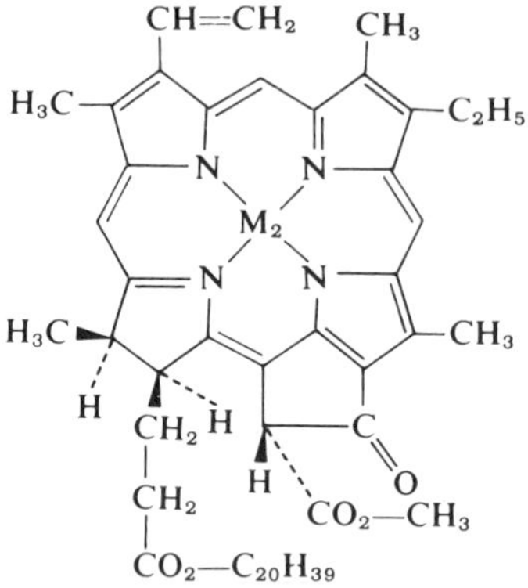

35

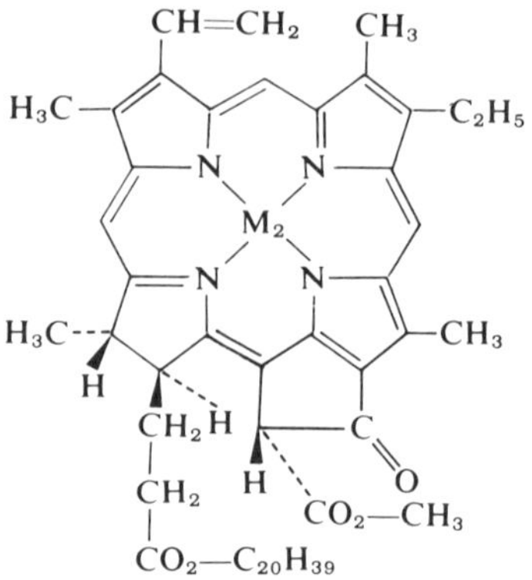

36

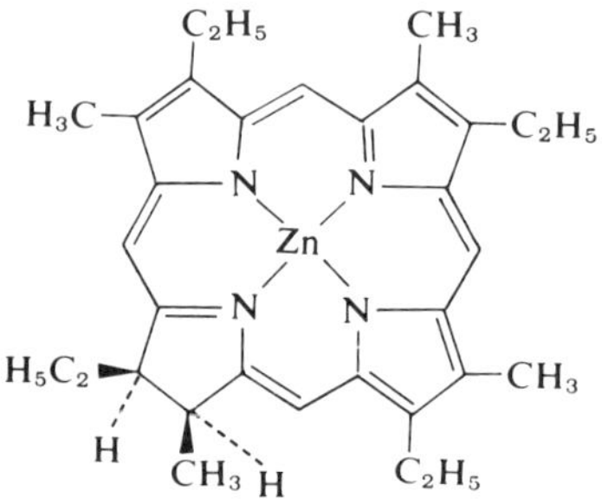

37

tetrahydrometalloporphyrin (M—PH_4) or, in some cases, demetalate to dihydroporphyrin (PH_2).[199]

A parallel pathway effects direct photoreduction to the metallochlorin (M-chlorin). The metallochlorin may react by a different pathway to reversible metallochlorin photoreduction product (M-chlorin H_2), or by further direct photoreduction to a metallo-*cis*-tetrahydroporphyrin (M-*cis*-THP).

Recently, Krasnovskii has reported the reversible H_2S photoreduction of chlorophyll *a* and several other chlorophyll derivatives.[200] Mechanisms of formation were described. Scheer and Katz,[201] utilizing 220 MHz pmr, have characterized the Krasnovskii photoreduction product of chlorophyll *a* as the β,δ-dihydrochlorophyll *a*.

Numerous investigators have made contributions that have led to the present understanding of photoreduction. Visible spectral investigation,[193] specific esr studies,[202-206] and luminescence studies[147,177,207] have all contributed to mechanism elucidation.

Fuhrhop[208] reports a photoreduction unique to Sn(IV) and Ge(IV) octaethylporphyrins. The irradiation of Sn(IV)P or Ge(IV)P leads first to

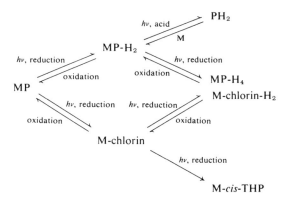

Fig. 8. Proposed mechanism for *cis*-photohydrogenation.

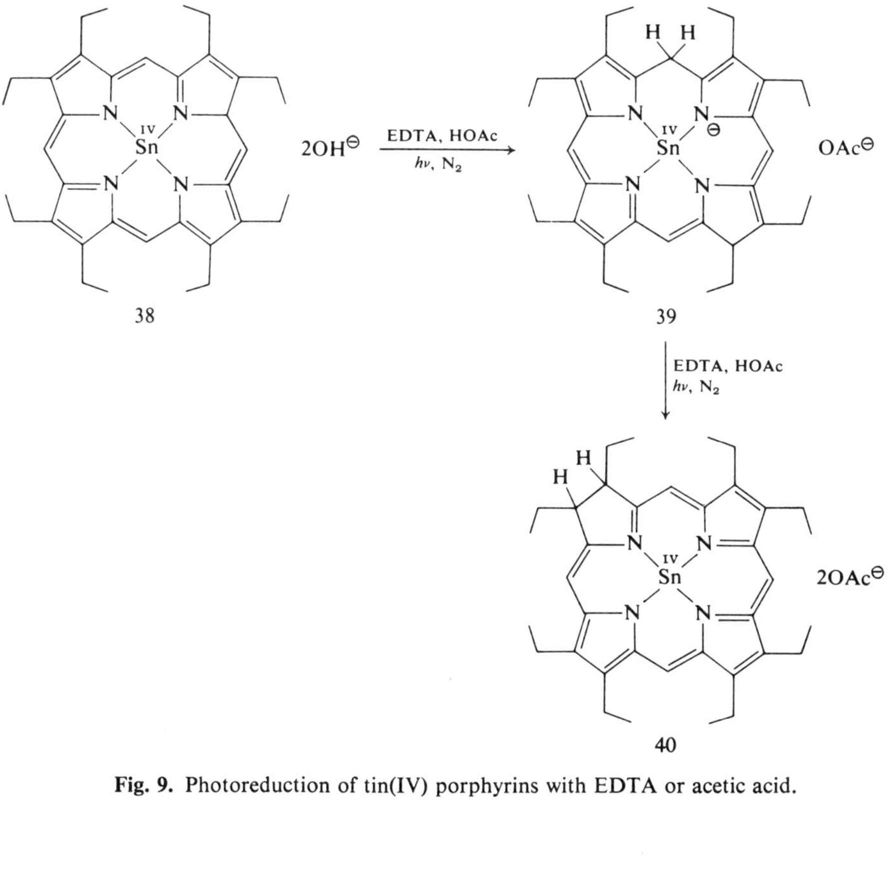

Fig. 9. Photoreduction of tin(IV) porphyrins with EDTA or acetic acid.

metallophlorins, which upon further irradiation rapidly and nearly quantitatively rearrange to metallochlorins without intermediate formation of metalloporphodimethenes (Fig. 9).

The strongly electron-attracting Sn(IV) and Ge(IV) evidently polarize the porphyrin ring so much that even the β-pyrrole carbon atoms acquire a partial positive charge and become more readily reducible. This argument is supported by recent X-ray investigations, which reveal a lengthening of the β,β'-pyrrole bond in Sn(IV) porphyrins, compared to other metalloporphyrins.[209]

$$P^* + QH_2 \longrightarrow PH\cdot_{p.r.} + QH\cdot \qquad (51)$$

Similar mono- and dianions[210] of a series of metalloporphyrins and metallophthalocyanines have been reported as intermediates in hydrazine photoreductions.[211,212] Whitten et al.[92,127] have reported the photochemical reduction of Sn(IV) porphyrins to the Sn(IV) chlorins, with further photoreduction leading to the vic-tetrahydroporphyrintin(IV) (Eq. 51). Under the reaction conditions ($SnCl_2 \cdot 2H_2O$/pyridine), electron transfer from the $SnCl_2$ to the porphyrin excited triplet evidently occurs first, followed by protonation of the resulting Sn(IV) porphyrin dianion (44).[213,214] Preparation of the highly reactive Sn(II) porphyrins is also described by Whitten et al.[94] Activation of the Sn(II) porphyrins by heat or light leads to ring-reduced Sn(IV) porphyrin species, evidently via intramolecular electron transfer.

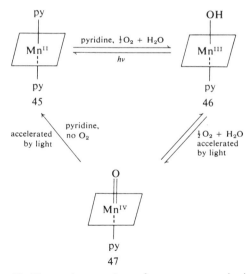

44

B. Changes in Central Metal Oxidation State

1. PHOTOOXIDATION

Although well-characterized photoredox reactions are not published for metalloporphyrins, an example is provided by the related manganous phthalocyanines (Fig. 10). Calvin and co-workers[215,216] report the reversible photoredox or "breathing" of manganous phthalocyanines in pyridine solutions. The complex (45) is assumed to be the species present when manganous phthalocyanine is dissolved in oxygen-free pyridine; it cannot be isolated, however. When air is introduced, oxidation to the trivalent species

Fig. 10. Photoredox reactions of manganese porphyrins.

$$\text{Co(III)(NPE)}_2 \text{ etio I} \xrightarrow{h\nu,\ isc} [\text{Co(III)(NPE)}_2 \text{ etio I}]^{*3}$$

intramolecular
electron
transfer

degradation products ⟵——— NPE⊕ + Co(II)(NPE) etio I

Fig. 11. Photoreduction of cobalt(III) porphyrins.

(46) occurs. This complex also could not be isolated. The complex (46) is photosensitive, intense white light causing reduction to 45; whereas in the dark, disproportionation to 45 and 47 occurs. In the presence of oxygen, 46 slowly oxidizes to 47 in the dark, a process which is accelerated by light. In the absence of oxygen, 47 is reduced to 45, apparently without the intermediate formation of 46. This process is rapid in sunlight and slow in the dark.

2. PHOTOREDUCTION

Few substantiated cases of metalloporphyrin photoreduction are known. Whitten[127] has reported the photoreduction of Co(III) porphyrins to the corresponding Co(II) porphyrins in degassed $SnCl_2 \cdot 2H_2O$/pyridine. Electron transfer occurs between the porphyrin excited triplet and $SnCl_2 \cdot 2H_2O$ to effect reduction. Whitten[94] has also demonstrated that the 1-(1-naphthyl-2-(4-pyridyl) ethene (NPE) complexes of Co(III) etioporphyrin I are photoreduced in polar solvents (Fig. 11). This photoreduction presumably occurs via the intermediate excited triplet state of Co(III) (NPE)$_2$etio I, which undergoes intramolecular electron transfer to form Co(II)(NPE)etio I and NPE+·.

C. Transformations Involving Extraplanar Ligands

All of the examples of permanent chemical change involving extraplanar ligands have been of the type that result from photochemical ejection. At present, few examples of ligand photoejections that lead to permanent alteration of metalloporphyrin structure are known. As more metal–carbonyl–porphyrin complexes are synthesized, they will most surely prove to be a source of a number of interesting photochemical reactions. Photochemical ejection of CO from Ru(II)(CO)(py)TPP to form Ru(II)(py)$_2$TPP has been reported by Chow and Cohen.[217] Sovocool et al.[22] have studied the photochemical ejection of CO (with subsequent ligand substitution) from a number of

$$\text{PH} \cdot_{p.r.} + \text{PH}_2 \longrightarrow \text{P} + \text{PH}_3 \cdot \qquad (52)$$

ruthenium(II) carbonyl porphyrins (Eq. 52). The substituted ligands included pyridine, dimethylsulfoxide, tetrahydrofuran, and several aliphatic and aromatic amines. The porphyrin macrocycles studied included -OEP, -etio I, and -meso IX, dioctadecyl ester.[218] Similar photoejection of CO from degassed solutions of $Os(II)(CO)(py)OEP$ to form $Os(II)(py)_2OEP$ has been observed.[219] Preliminary results[218] indicate that degassed solutions of ruthenium nitrosyl porphyrins [formulated as $Ru(II)(NO)_2P$ by Tsutsui[220]] undergo a slow NO displacement in the dark. For solutions of

$$Ru(II)(CO)(L)P \xrightarrow[\substack{\text{degassed solution} \\ \text{of donor ligand, L}}]{hv,} Ru(II)(L)_2P + CO \qquad (53)$$

$Ru(II)(NO)_2OEP$, this process is accelerated by irradiation into either the visible or Soret bands.

D. Photometalation

One report of photoinduced metalation of porphyrins has appeared in the literature;[221] acceleration of the reaction rate of several $Cu(II)1,3$-diketonates with porphyrins to form $Cu(II)$-porphyrins at spectroscopic concentrations was observed. It appears clear that a relatively long-lived (minutes) $Cu(I)$ species is intermediate in the process. Although the authors suggest a tentative mechanism involving some rather vague electron-transfer steps, it appears more likely that energy transfer from porphyrin to copper complex (or direct excitation of the latter) generates an excited copper(II) species, which undergoes an internal redox reaction to liberate $Cu(I)$ and ligand radical.

Preliminary results[218] of experiments involving irradiation of octaethylporphyrin with several metal salts (K_2PtCl_4, $PtCl_2$, $PdCl_2$, $NiCl_2 \cdot 6H_2O$, $CoCl_2 \cdot 6H_2O$,) in degassed DMF or DMF/NMF solutions [or $Ru_3(CO)_{12}$ in benzene] at room temperature indicate the feasibility of quantitative photometalation in dilute solutions. Selective irradiation at different wavelengths, as well as flash spectroscopic studies, indicate that the relatively inefficient reaction occurs only from the excited state of the metal salts, and not via the porphyrin excited state.

In contrast, Hopf and Whitten[218] have found that degassed solutions of the $PtCl_2 \cdot$ hematoporphyrin complex [formulated as an isolable sitting-atop complex[222]] undergo spectral changes characteristic of formation of $Pt(II)$-hematoporphyrin upon irradiation with visible light. This inefficient reaction evidently involves intramolecular energy transfer as one of the decay modes of the excited state $PtCl_2 \cdot$ hematoporphyrin complex.

ACKNOWLEDGMENT

We wish to thank the U.S. Public Health Service (Grant No. GM 15,238) and the Alfred P. Sloan Foundation for support of this work.

REFERENCES

1. M. Gouterman, *in* "Excited States of Matter" (C. W. Shoppee, ed.), Graduate Studies Texas Tech. U., No. 2, p. 63. Texas Tech. Univ. Press, Lubbach, 1973.
2. C. H. Weiss, H. Kobayashi, and M. Gouterman, *J. Mol. Spectrosc.* **16**, 415 (1965).
3. A. H. Corwin, *Ann. N.Y. Acad. Sci.* **206**, 201 (1973).
4. S. R. Platt, *Radiat. Biol.* **3**, (1956).
5. A. H. Corwin, A. B. Chivvis, R. W. Poor, D. G. Whitten, and E. W. Baker, *J. Am. Chem. Soc.* **90**, 6577 (1968).
6. L. V. Iogansen, *Dokl. Akad. Nauk SSSR* **205**, 390 (1972).
7. R. S. Becker and M. Kasha, *J. Am. Chem. Soc.* **77**, 3669 (1955).
8. R. S. Becker and J. B. Allison, *J. Phys. Chem.* **67**, 2662 (1963).
9. R. S. Becker and J. B. Allison, *J. Phys. Chem.* **67**, 2669 (1963).
10. J. B. Allison and R. S. Becker, *J. Phys. Chem.* **67**, 2675 (1963).
11. D. Eastwood and M. Gouterman, *J. Mol. Spectrosc.* **35**, 359 (1970).
12. G. D. Dorough, J. R. Miller, and F. M. Huennekens, *J. Am. Chem. Soc.* **73**, 4315 (1951).
13. J. B. Callis, M. Gouterman, Y. M. Jones, and B. H. Henderson, *J. Mol. Spectrosc.* **39**, 410 (1971).
14. P. G. Seybold and M. Gouterman, *J. Mol. Spectrosc.* **31**, 1 (1969).
15. L. K. Hanson, M. Gouterman, and J. C. Hanson, *J. Am. Chem. Soc.* **95**, 4822 (1973).
16. M. Gouterman, F. P. Schwarz, P. D. Smith, and D. Dolphin, *J. Chem. Phys.* **59**, 676 (1973).
17. M. P. Tsvirko, K. N. Solov'ev, and V. V. Sapunov, *Opt. Spektrosk.* **36**, 335 (1974).
18. K. N. Solov'ev, *Vestsi Akad. Navuk Bel. SSR, Ser. Fiz.-Tekh. Navuk* No. 3, p. 27 (1962).
19. D. Djuric, *Ark. Farm. (Belgrade)* **11**, 1 (1961).
20. M. P. Tsvirko and V. V. Sapunov, *Opt. Spektrosk.* **34**, 1094 (1973).
21. G. P. Gurinovich, A. P. Losev, and V. P. Suboch, *Photosynth., Two Centuries Its Discovery Joseph Priestley, Proc. Int. Congr. Photosynth. Res., 2nd, 1971* Vol. 1 p. 299 (1972).
22. G. W. Sovocool, F. R. Hopf, and D. G. Whitten, *J. Am. Chem. Soc.* **94**, 4350 (1972).
23. F. R. Hopf, T. P. O'Brien, W. R. Scheidt, and D. G. Whitten, *J. Am. Chem. Soc.* **97**, 277 (1975).
24. S. B. Broyde and S. S. Brody, *J. Chem. Phys.* **46**, 3334 (1967).
25. J. B. Allison and R. S. Becker, *J. Chem. Phys.* **32**, 1410 (1960).
26. D. Mauzerall, *Biochemistry* **4**, 1801 (1965).
27. L. Bajema, M. Gouterman, and C. B. Rose, *J. Mol. Spectrosc.* **39**, 421 (1971).
28. H. Linschitz and K. Sarkanen, *J. Am. Chem. Soc.* **80**, 4826 (1958).
29. R. Livingstone and V. A. Ryan, *J. Am. Chem. Soc.* **75**, 2176 (1953).
30. G. R. Seeley, *in* "The Chlorophylls" (L. P. Vernon and G. R. Seely, eds.), p. 523. Academic Press, New York, 1966.
31. P. A. Shakhverdov, *Flem, Fotoprotessy Mol.* p. 283 (1966).
32. A. K. Chibisov, *Photochem. Photobiol.* **10**, 331, and reference therein (1969).
33. L. Pekkarinen and H. Linschitz, *J. Am. Chem. Soc.* **82**, 2407 (1960).
34. P. J. McCartin, *Trans. Faraday Soc.* **60**, 1694 (1964).
35. R. Livingstone and P. J. McCartin, *J. Phys. Chem.* **67**, 2511 (1963).
36. H. Linschitz, C. Steel, and J. A. Bell, *J. Phys. Chem.* **66**, 2574 (1962).
37. J. S. Connolly, D. S. Gorman, and G. R. Seely, *Ann. N.Y. Avad. Sci.* **206**, 649 (1973).
38. P. J. Wagner, *Mol. Photochem.* **3**, 23 (1971).

39. M. D. Shetlar, *Mol. Photochem.* **6**, 143 (1974).
40. M. D. Shetlar, *Mol. Photochem.* **6**, 167 (1974).
41. M. D. Shetlar, *Mol. Photochem.* **6**, 191 (1974).
42. J. Dalton, C. A. McAuliffe, and D. H. Slater, *Nature (London)* **235**, 388 (1972).
43. J. T. Warden and J. R. Bolton, *Acc. Chem. Res.* **7**, 189 (1974).
44. G. D. Vaughan and K. Sauer, *Biochim. Biophys. Acta* **347**, 383 (1974).
45. V. K. Gorshkov, L. M. Shubin, and E. N. Mukhin, *Biofizika* **19**, 677 (1974).
46. O. Björkman, *Photophysiology* **8**, 1 (1973).
47. D. G. Bishop, *Photochem. Photbiol.* **20**, 281 (1974).
48. J. T. Warden and J. R. Bolton, *Photochem. Photobiol.* **20**, 245 (1974).
49. J. T. Warden and J. R. Bolton, *Photochem. Photobiol.* **20**, 251 (1974).
50. J. T. Warden and F. R. Bolton, *Photochem. Photobiol.* **20**, 263 (1974).
51. J. J. Leonard, T. Yonetani, and J. B. Callis, *Biochemistry* **13**, 1460 (1974).
52. D. R. Kearns, *Chem. Rev.* **71**, 395 (1971).
53. B. Stevens, *Acc. Chem. Res.* **6**, 90 (1973).
54. B. Stevens, *Abstr. Informal. Conf. Photochem., 1974*, p. 53 (1974).
55. C. S. Foote, *Acc. Chem. Res.* **1**, 104, and references therein (1968).
56. C. S. Foote, Y. C. Chang, and R. W. Denny, *J. Am. Chem. Soc.* **92**, 5216 (1970).
57. C. S. Foote, Y. C. Chang, and R. W. Denny, *J. Am. Chem. Soc.* **92**, 5218 (1970).
58. A. A. Lamola, T. Yamane, and A. M. Trozzolo, *Science* **179**, 1131 (1973).
59. F. H. Doleiden, S. R. Fahrenholtz, A. A. Lamola, and A. M. Trozzolo, *Photochem. Photobiol.* **20**, 519 (1974).
60. S. R. Fahrenholtz, F. H. Doleiden, A. M. Trozzolo, and A. A. Lamola, *Photochem. Photobiol.* **20**, 505 (1974).
61. A. M. Trozzolo, F. H. Doleiden, S. R. Fahrenholtz, A. A. Lamola, A. M. Mattuci, and T. Yamane, *Abstr., 2nd Annu. Meet. Soc. Photobiol.* p. 95 (1974).
62. H. Langhof, H. Müller, and L. Rietschel, *Arch. Klin. Exp. Dermatol.* **212**, 506 (1961).
63. I. A. Magnus, A. Jarrett, T. A. Prankerd, and C. Rimington, *Lancet* **2**, 448 (1961).
64. K. Gollnick and G. O. Schenck, *Pure Appl. Chem.* **9**, 507 (1964).
65. G. O. Schenck, *Angew. Chem.* **69**, 579 (1957).
66. A. Nickon and W. L. Mendelson, *J. Am. Chem. Soc.* **87**, 3921 (1965).
67. A. A. Schothorst, J. Van Steveninck, L. N. Went, and D. Suurmond, *Clin. Chim. Acta* **28**, 41 (1970).
68. A. A. Lamola, unpublished results (1974).
69. I. F. Gurinovich, I. M. Byteva, V. S. Chernikov, and O. M. Petsol'd, *Zh. Org. Khim.* **8**, 842 (1972).
70. L. Axelsson, *Plant Physiol.* **31**, 77 (1974).
71. J.-H. Fuhrhop and D. Mauzerall, *Photochem. Photobiol.* **13**, 453 (1971).
72. L. C. Harber, J. Hsu, H. Hsu, and B. D. Goldstein, *J. Invest. Dermatol.* **58**, 373 (1972).
73. A. A. Schothorst, J. Van Steveninck, L. N. Went, and D. Suurmond, *Clin. Chim. Acta* **39**, 161 (1972).
74. J. S. Beilin and G. Oster, *Prog. Photobiol., Proc. Int. Congr., 3rd 1960* p. 254 (1961).
75. M. R. Mauk and A. W. Girotti, *Biochemistry* **13**, 1757 (1974).
76. G. Jori, G. Galiazzo, and E. Scoffone, *Experientia* **27**, 379 (1971).
77. G. Jori, G. Galiazzo, and E. Scoffone, *Biochemistry* **8**, 2868 (1964).
78. G. Jori, G. Gennari, G. Galiazzo, and E. Scoffone, *FEBS Lett.* **6**, 267 (1970).
79. G. Jori, G. Galiazzo, A. M. Tamburro, and E. Scoffone, *J. Biol. Chem.* **245**, 3375 (1970).
80. G. Jori, G. Gennari, and M. Folin, *Photochem. Photobiol.* **19**, 79 (1974).

81. M. Folin, G. Gennari, and G. Jori, *Photochem. Photobiol.* **20**, 357 (1974).
82. G. Galiazzo, A. M. Tamburro, and G. Jori, *Eur. J. Biochem.* **12**, 362 (1970).
83. J. D. Ostrow, *Prog. Liver Dis.* **4**, 447 (1972).
84. J. D. Ostrow, *Semin. Hematol.* **9**, 113 (1972).
85. R. Bonnett and J. C. M. Steward, *Biochem. J.* **130**, 895 (1972).
86. J. J. Lee, B. C. Mathewson, J. E. Wamplen, R. D. Etheridge, and N. U. Curry, *Fed. Proc. Biochemistry II, Abstr.* p. 2522 (1972).
87. P. D. Wildes, J. G. Pacifici, G. Irick, and D. G. Whitten, *J. Am. Chem. Soc.* **93**, 2004 (1971).
88. G. M. Wyman, B. M. Zarnegar, and D. G. Whitten, *J. Phys. Chem.* **77**, 2584 (1973).
89. T. F. Kachura, A. N. Sevchenko, K. N. Solov'ev, and M. P. Tsvirko, *Dokl. Akad. Nauk SSSR* **217**, 1121 (1974).
90. Y. Haas and G. Stein, *J. Chem. Phys.* **79**, 3668 (1971).
91. F. P. Schwarz, M. Gouterman, Z. Muljiani, and D. Dolphin, *Bioinorg. Chem.* **2**, 1 (1972).
92. D. G. Whitten, P. D. Wildes, and I. G. Lopp, *J. Am. Chem. Soc.* **91**, 3393 (1969).
93. P. D. Wildes and D. G. Whitten, *J. Am. Chem. Soc.* **92**, 7609 (1970).
94. D. G. Whitten, P. D. Wildes, and C. A. DeRosier, *J. Am. Chem. Soc.* **94**, 7811 (1972).
95. H. Linschitz and J. Rennert, *Nature (London)* **169**, 193 (1952).
96. I. N. Chernyuk and I. I. Dilung, *Dokl. Akad. Nauk SSSR* **165**, 1350 (1965).
97. G. R. Seely, *J. Phys. Chem.* **73**, 117 (1969).
98. J. R. Harbour and G. Tollin, *Photochem. Photobiol.* **19**, 147 (1974).
99. D. Djuric, *Ark. Farm. (Belgrade)* **12**, 19 and 263 (1962).
100. L. A. Sibel'dina, *Eur. Biophys. Congr., Proc., 1st, 1971* Vol. 4, p. 249 (1971).
101. K. P. Quinlon and E. Fukomori, *J. Phys. Chem.* **71**, 4154 (1967).
102. G. Tollin and G. Green, *Biochim. Biophys. Acta* **60**, 524 (1962).
103. G. Tollin, K. K. Chattergee, and G. Green, *Photochem. Photobiol.* **4**, 592 (1965).
104. G. R. Seely, *J. Phys. Chem.* **69**, 2779 (1965).
105. A. K. Bannerjee and G. Tollin, *Photochem. Photobiol.* **5**, 315 (1966).
106. D. G. Whitten and J. C. N. Yau, *Tetrahedron Lett.* p. 3077 (1969).
107. J. K. Roy and D. G. Whitten, *J. Am. Chem. Soc.* **94**, 7162 (1972).
108. R. L. Livingston, L. Thompson, and M. V. Ramaroao, *J. Am. Chem. Soc.* **74**, 1073 (1952).
109. R. Livingstone and C. L. Ke, *J. Am. Chem. Soc.* **72**, 909 (1950).
110. R. Livingstone, *Q. R. Chem. Soc.* **14**, 174 (1960).
111. S. L. Bondarev, G. P. Burinovich, and V. S. Chernikov, *Izv. Akad. Nauk SSSR, Ser. Fiz.* **34**, 641 (1970).
112. S. Bondarev and G. P. Gurinovich, *Opt. Spektrosk.* **36**, 687 (1974).
113. I. G. Lopp, R. W. Hendren, P. D. Wildes, and D. G. Whitten, *J. Am. Chem. Soc.* **92**, 6440 (1970).
114. R. A. White and G. Tollin, *J. Am. Chem. Soc.* **89**, 1253 (1967).
115. J. R. Harbour and G. Tollin, *Photochem. Photobiol.* **19**, 163 (1974).
116. A. K. Chibisov, *Dokl. Akad. Nauk SSSR* **205**, 142 (1972).
117. B. I. Barashkov and A. K. Chibisov, *Biofizika* **17**, 775 (1972).
118. J. K. Roy and D. G. Whitten, *J. Am. Chem. Soc.* **93**, 7093 (1971).
119. J. K. Roy, F. A Carroll, and D. G. Whitten, *J. Am. Chem. Soc.* **96**, 6349 (1974).
120. D. G. Whitten, J. K. Roy, and F. A. Carroll, *in* "The Exciplex" (M. Gordon and W. R. Ware, eds.), p. 247. Academic Press, New York, 1975.
121. A. Weller, *Pure Appl. Chem.* **16**, 115 (1968).

122. R. Rehm and A. Weller, *Ber. Bunsenges. Phys. Chem.* **73**, 834 (1969).
123. H. D. Ganey and A. W. Adamson, *J. Am. Chem. Soc.* **94**, 8238 (1972).
124. C. R. Bock, T. J. Meyer, and D. G. Whitten, unpublished manuscript (1974).
125. R. C. Jarnigan, *Acc. Chem. Res.* **4**, 420 (1971).
126. C. R. Bock, Ph.D. Dissertation, University of North Carolina, Chapel Hill (1974).
127. D. G. Whitten, J. C. Yau, and F. A. Carroll, *J. Am. Chem. Soc.* **93**, 2291 (1971).
128. D. R. Paulson, R. Ullman, R. B. Sloane, and G. L. Closs, *J. Chem. Soc., Chem. Commun.* p. 186 (1974).
129. M. S. Brookhart and A. H. Corwin, private communication (1974).
130. C. A. DeRosier and D. G. Whitten, unpublished results (1972).
131. C. R. Bock, R. Young, T. J. Meyer, and D. G. Whitten, unpublished results (1974).
132. R. B. Woodward, *Ind. Chim. Belge*, [2] 1293 (1962).
133. J.-H. Fuhrhop, K. Kadish, and D. G. Davis, *J. Am. Chem. Soc.* **95**, 5140 (1973).
134. J.-H. Fuhrhop, *Struct. (Berlin)*, **18**, 1 (1974).
135. J. O. Alben and W. S. Caughey, *Biochemistry* **7**, 175 (1968).
136. B. D. McLees and W. S. Caughey, *Biochemistry* **7**, 642 (1968).
137. D. Mauzerall and S. Granick, *J. Biol. Chem.* **232**, 1141 (1958).
138. G. D. Dorough and J. R. Miller, *J. Am. Chem. Soc.* **74**, 6106 (1952).
139. H. H. Inhoffen, H. Brockman, and K.-M. Bliesener, *Justus Liebigs Ann. Chem.* **730**, 173 (1969).
140. H. Fischer and M. Dürr, *Justus Liebigs Ann. Chem.* **501**, 112 (1933).
141. H. Fischer and K. Herrle, *Hoppe-Seyler's Z. Physiol. Chem.* **251**, 85 (1938).
142. M. Calvin and G. D. Dorough, *J. Am. Chem. Soc.* **70**, 699 (1948).
143. G. D. Dorough and M. Calvin, *Science* **105**, 433 (1947).
144. F. M. Huennekens and M. Calvin, *J. Am. Chem. Soc.* **71**, 4024 (1949).
145. F. M. Huennekens and M. Calvin, *J. Am. Chem. Soc.* **71**, 4031 (1949).
146. A. A. Krasnovskii, *Sovrem. Probl. Fotosint.* p. 64 (1973).
147. G. P. Gurinovich and G. N. Sinyakov, *Biofizika* **10**, 946 (1965).
148. I. F. Gurinovich, G. P. Gurinovich, and A. N. Sevchenko, *Dokl. Akad. Nauk SSSR* **164**, 201 (1965).
149. S. Besecke and J.-H. Fuhrhop, *Angew. Chem.* **86**, 125 (1974).
150. S. Besecke and J.-H. Fuhrhop, *Angew. Chem., Int. Ed. Engl.* **13**, 150 (1974).
151. K. P. Quinlon, *Biochim. Biophys. Acta* **267**, 493 (1972).
152. Axel Madsen, *Prog. Photobiol., Proc. Int. Congr., 3rd, 1960* p. 567 (1961).
153. I. Dilung and I. Chernyuk, *Abh. Dsch. Akad. Wiss. Berlin, Kl. Med.* p. 325 (1966).
154. J. D. McElroy, G. Faher, and D. Mauzerall, *Biochim. Biophys. Acta* **267**, 363 (1972).
155. V. B. Evstigneev, N. A. Sadovnikova, A. P. Kostikov, and L. P. Koyushin, *Dokl. Akad. Nauk SSSR* **203**, 1343 (1972).
156. B. J. Hales and J. R. Bolton, *J. Am. Chem. Soc.* **94**, 3314 (1972).
157. J. Subramanian, J.-H. Fuhrhop, A. Salek, and A. Gassauer, *J. Magn. Reson.* **15**, 19 (1974).
158. J. Fajer, D. C. Forman, R. H. Felton, and L. Vegh, *Proc. Natl. Acad. Sci. U.S.A.* **71**, 994 (1974).
159. R. C. Jennings and G. Forte, *Biochim. Biophys. Acta* **347**, 299 (1974).
160. S. K. Wong and J. K. S. Wan, *J. Am. Chem. Soc.* **94**, 7197 (1972).
161. S. K. Wong, D. A. Hutchinson, and J. K. S. Wan, *J. Am. Chem. Soc.* **95**, 622 (1973).
162. J. R. Harbour and G. Tollin, *Photochem. Photobiol.* **19**, 69 (1974).
163. V. M. Kutyurin, T. D. Slavnova, and A. K. Chibisov, *Biofizika* **18**, 1004 (1973).
164. E. M. Bakuchava, N. N. Drozdova, and A. A. Krasnovskii, *Biokhimiya* **39**, 188 (1974).

165. V. B. Evstigneev and V. S. Chudar, *Biofizka* **19**, 425 (1974).
166. J. P. Chauvet, R. Viovy, and F. Villain, *J. Chim. Phys. Phys-chim. Biol.* **71**, 879 (1974).
167. Y. D. Halsey and W. W. Parson, *Biochim. Biophys. Acta* **347**, 404 (1974).
168. V. P. Oshchepkov and A. A. Krasnovskii, *Fiziol. Rast.* **21**, 462 (1974).
169. A. A. Krasnovskii and K. K. Voinovskaya, *Dokl. Akad. Nauk SSSR* **96**, 1209 (1954).
170. A. A. Krasnovskii and E. V. Pakshina, *Dokl. Akad. Nauk SSSR* **120**, 581 (1958).
171. A. A. Krasnovskii and A. V. Umrikhina, *Dokl. Akad. Nauk SSSR* **122**, 1061 (1958).
172. A. A. Krasnovskii, *J. Chim. Phys.* **55**, 968 (1958).
173. A. A. Krasnovskii, *Annu. Rev. Plant. Physiol.* **11**, 363 (1960).
174. A. N. Sidorov and D. A. Savel'ev, *Biofizika* **13**, 933 (1968).
175. A. M. Shul'ga and G. P. Gurinovich, *Biofizika* **13**, 42 (1968).
176. L. V. Slopolyanskaya, I. M. Byteve, and G. P. Gurinovich, *Vestsi Akad. Navuk Bel. SSR* **16**, 1048 (1972).
177. G. P. Gurinovich, A. I. Patsko, A. M. Shul'ga, and A. N. Sevchenko, *Dokl. Akad. Nauk SSSR* **156**, 125 (1964).
178. A. A. Krasnovskii, *Prog. Photobiol., Proc. Int. Congr., 3rd, 1960* p. 561 (1961).
179. V. P. Suboch and A. M. Shul'ga, *Biofizika* **16**, 603 (1971).
180. D. Mauzerall, *J. Am. Chem. Soc.* **82**, 1832 (1960).
181. D. Mauzerall, *J. Am. Chem. Soc.* **84**, 2437 (1962).
182. D. Mauzerall, *J. Am. Chem. Soc.* **82**, 2601 (1960).
183. D. Mauzerall and G. Feher, *Biochim. Biophys. Acta* **88**, 658 (1964).
184. A. N. Sidorov and A. N. Terenin, *Dokl. Akad. Nauk SSSR* **145**, 1092 (1962).
185. A. N. Sidorov, V. G. Vorobév, and A. N. Terenin, *Dokl. Akad. Nauk SSSR* **152**, 919 (1963)
186. A. N. Sidorov, *Dokl. Akad. Nauk SSSR* **161**, 128 (1965).
187. G. P. Gurinovich, A. P. Losev, and V. P. Suboch, *Photosynth., Two Centuries Its Discovery Joseph Priestley, Proc. Int. Congr. Photosynth. Res., 2nd, 1971* Vol. 1, p. 299 (1972).
188. A. V. Umrikhina, G. A. Yusupova, and A. A. Kranovskii, *Dokl. Akad. Nauk SSSR* **175**, 1400 (1967).
189. G. T. Rikhireva, A. V. Umrikhina, L. P. Kayushin, and A. A. Krasnovskii, *Dokl. Akad. Nauk SSSR* **163**, 491 (1965).
190. G. R. Seely and M. Calvin, *J. Chem. Phys.* **23**, 1068 (1955).
191. G. R. Seely and K. Talmadge, *Photochem. Photobiol.* **3**, 195 (1964).
192. A. N. Sidorov, *in* "Elementary Photoprocesses in Molecules" (B. S. Neporent, ed.), p. 201. Plenum, New York, 1968.
193. A. N. Sidorov, *Dokl. Akad. Nauk SSSR* **158**, 937 (1964).
194. D. A. Soveljev, A. N. Sidorov, R. P. Eustigneeva, and G. V. Ponomarev, *Dokl. Akad. Nauk SSSR* **167**, 135 (1966).
195. A. M. Shul'ga, G. P Gurinovich, and A. N. Sevchenko, *Dokl. Akad. Nauk SSSR* **169**, 1206 (1966).
196. V. P. Suboch, A. P. Losev, and G. P. Gurinovich, *Photochem. Photobiol.* **20**, 183 (1974).
197. V. P. Suboch, A. P. Losev, G. P. Gurinovich, and A. N. Sevchenko, *Dokl. Akad. Nauk SSSR* **194**, 721 (1970).
198. A. A. Krasnoviskii, M. I. Bystrova, and F. Lang, *Dokl. Akad. Nauk SSSR* **194**, 1441 (1970).

199. T. T. Bonnister, *Plant. Physiol.* **34**, 246 (1959).
200. E. V. Pakshina and A. A. Krasnovskii, *Biofizika* **19**, 238 (1974).
201. H. Scheer and J. J. Katz, *Proc. Natl. Acad. Sci. U.S.A.* **71**, 1626 (1974).
202. V. E. Kholmogorov, *Biofizika* **16**, 378 (1971).
203. Z. B. Gribova, V. A. Umrikhin, and L. P. Kayushin, *Biofizika* **11**, 353 (1966).
204. Z. B. Gribova, *Abh. Dsch. Akad. Wiss. Berlin, Kl. Med.* p. 331 (1965).
205. A. V. Umrikhina, N. V. Bublichenko, and A. A. Kranovskii, *Biofizika* **18**, 565 (1973).
206. Z. B. Gribova, *Ul'trafiiolet. Izluch.;* No. 4, *Mater. Vses. Soveshch. Biol. Peistriyu Ul'trafiiolet Izluch., 8th, 1964* p. 41 (1966).
207. G. P. Gurinovich, M. V. Poteeva, and A. M. Shul'ga, *Izv. Akad. Nauk SSSR, Ser. Fiz.* **27**, 777 (1963).
208. J.-H. Fuhrhop and T. Lumbantobing, *Tetrahedron Lett.* p. 2815 (1970).
209. J. L. Hoard, *Science* **174**, 1295 (1971).
210. G. L. Closs and L. E. Closs, *J. Am. Chem. Soc.* **85**, 818 (1963).
211. V. G. Moslov and A. N. Sidorov, *Teor. Eksp. Khim.* **7**, 832 (1971).
212. V. B. Evstigneev, V. G. Moslov, A. F. Mironov, and A. N. Sidorov, *Biofizika* **16**, 999 (1971).
213. H. H. Inhoffen, *Pure Appl. Chem.* **17**, 443 (1968).
214. J. W. Buchler and L. Puppe, *Justus Liebigs Ann. Chem.* **740**, 142 (1970).
215. M. Calvin, P. A. Loach, and A. Yamamoto, *in* "Theory and Structure of Complex Compounds" (B. Jezovska-Tryebistawska, ed.), p. 13. Macmillan, New York, 1964.
216. F. Englsma, A. Yamamoto, E. Markham, and M. Calvin, *J. Phys. Chem.* **66**, 2517 (1962).
217. B. C. Chow and I. A. Cohen, *Bioinorg. Chem.* **1**, 57 (1971).
218. F. R. Hopf and D. G. Whitten, unpublished results (1975).
219. F. R. Hopf, D. G. Whitten, and J. W. Buchler, unpublished results (1975).
220. T. S. Srivastava, L. Hoffman, and M. Tsutsui, *J. Am. Chem. Soc.* **94**, 1385 (1972).
221. A. Bluestein and J. M. Sugihara, *J. Inorg. Nucl. Chem.* **35**, 1048 (1973).
222. I. P. Macquet and T. Theophanides, *Can. J. Chem.* **51**, 219 (1973).

7

Linear Polypyrrolic Compounds

ALBERT GOSSAUER AND JÜRGEN ENGEL

I. INTRODUCTION

This chapter deals with linear polypyrrolic derivatives built from two, three, or four pyrrole rings linked by methylene or methine bridges. Systems of this kind with more than four pyrrole rings are rare and with the exception of some five-,[1] six-,[1-6] and eight-membered[1,6] compounds, are not encountered in the literature. Although the class of compounds mentioned above has been considered occasionally in the scope of several recent reviews on pyrroles,[7-9] as well as porphyrins and related macrocycles,[10-17] no comprehensive treatise on the chemistry and physical properties of linear polypyrrolic compounds has appeared since the classic work of Fischer and Orth.[18,19]

Since the chemistry of pyrroles has been extensively reviewed recently,[20] no comment on the chemistry and properties of pyrrole derivatives used as building blocks for the synthesis of linear polypyrroles has been included in this chapter.

As a matter of convenience, only those members of the latter class of compounds that can be employed as intermediates for the synthesis of porphyrins and related macrocycles will be mentioned here, i.e., those containing pyrrole rings bonded at the α-positions to the —CH_2— or —$CH=$ bridges. Most preparative methods as well as chemical and physical properties described here are, however, to some extent, appropriate to the corresponding derivatives carrying α-β'- or β-β'-linked pyrrole rings.

II. NOMENCLATURE

A rational nomenclature for the compounds that will be dealt with in this chapter is given in Volume I, Chapter 1.

Generally speaking, linear polypyrrolic compounds whose pyrrole rings are linked by CH_2 bridges are denominated polypyrranes. For members with two pyrrole rings (1), the systematic name α,α'-dipyrrylmethanes (2-[pyrrol-2-yl-methyl]-pyrroles) is preferred to the trivial one, i.e., dipyrromethanes, which is incorrect.

Usually the pyrrole rings of dipyrrylmethanes are numbered independently one of another (cf. 2). The methylene bridges are denominated as *meso* positions. On the other hand, compounds of the type (2) are appropriately designated as pyrromethenes (2-[pyrrol-2-yl-methylene]-(2H)-pyrroles), both pyrrole rings and the methine bridge being considered as a whole. The older name, dipyrrylmethenes, is an incorrect one because only one—if any— pyrryl substituent is present in the molecule (see Scheme 1).

Tripyrranes (3) and tetrapyrranes (6), also called bilinogens, as well as their analogues in which one or more CH_2 bridges are replaced by methine bridges, are more conveniently referred to the fully unsaturated compounds, namely, tripyrrins (5) and bilins (12), respectively, and they are numbered consecutively (cf. Scheme 1).

In the tetrapyrrane series, however, the nomenclature based on 1,19-dideoxybilane as the parent, fully reduced system (i.e., 6) is more descriptive and is, therefore, often encountered in the literature.[21,22] A more rational alternative, which is becoming more and more popular in the trivial nomenclature, consists of using the name bilane for compound (6), i.e., without oxygen atoms at the end positions.[17] The bilene, biladiene, and bilatriene nomenclature should be employed to refer to the various oxidation levels that commonly occur (cf., Scheme 1). The name "bilin" must be retained to differentiate the biladiene-ac free bases that occur as bilatrienes (11) (cf.

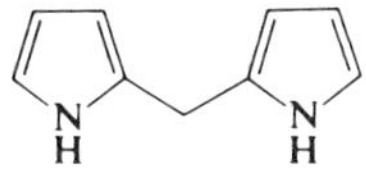

α,α'-Dipyrrylmethanes

1

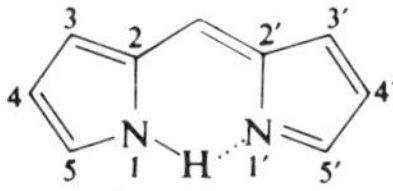

α,α'-Pyrromethenes
[2-(Pyrrol-2-yl-methylene)-(2H)-pyrroles]

2

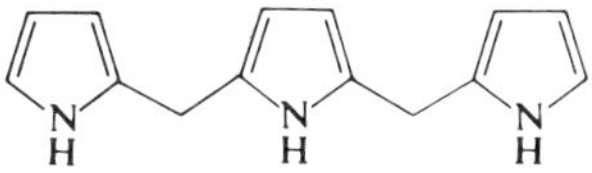

Tripyrranes [5,10,16,17-Tetrahydro-(15H)-tripyrrins]

3

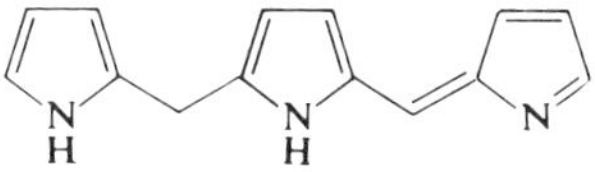

Tripyrrenes [5,16-Dihydro-(15H)-tripyrrins]

4

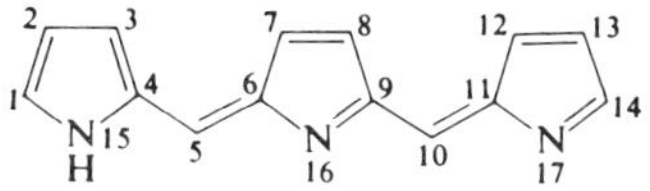

(15H)-Tripyrrins

5

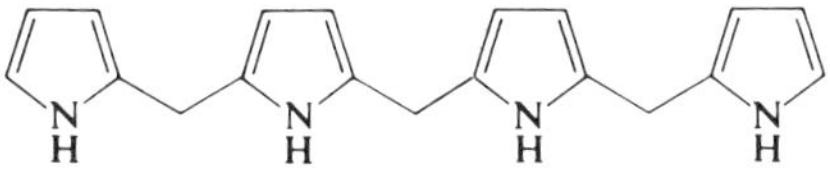

Bilanes [5,10,15,22,23,24-Hexahydro-(21H)-bilins]

6

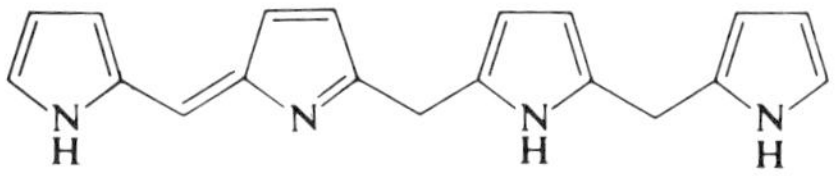

Bilenes-a [5,10,22,24-Tetrahydro-(21H)-bilins]

7

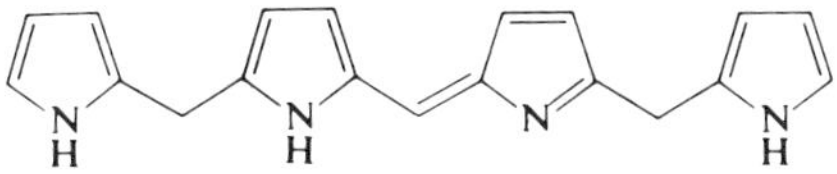

Bilenes-b [5,15,22,24-Tetrahydro-(21H)-bilins]

8

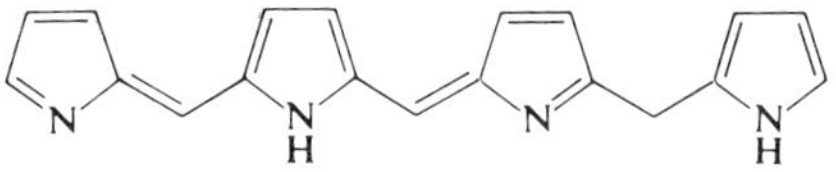

Biladienes-ab [5,23-Dihydro-(21H)-bilins]

9

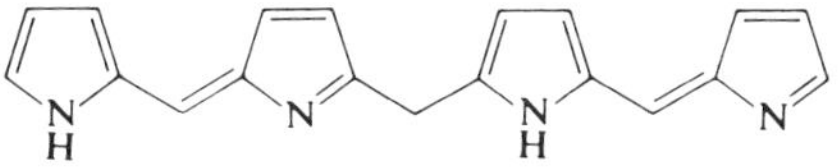

Biladienes-ac [10,23-Dihydro-(21H)-bilins]

10

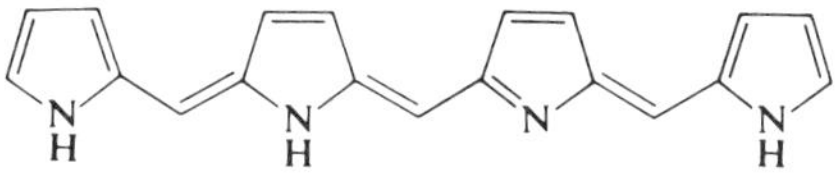

Bilatrienes [22,24-Dihydro-(21H)-bilins]

11

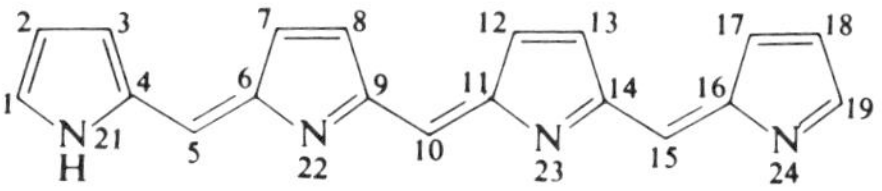

(21H)-Bilins

12

Scheme 1. The different types of linear tetrapyrroles.

Section V, E, 2), with 10 double bonds, from the true bilins (12) which are systems with 11 double bonds. Consequently, the trivial name "tripyrrene" for compounds of type (4) may replace the systematic one, i.e., 5,16-dihydro-tripyrrin (5-[pyrrol-2-yl-methyl]-pyrromethene) in the trivial nomenclature. The fully unsaturated system (5) is denominated tripyrrin (cf. Chapter 1 in Volume I).

For the sake of simplicity, the above mentioned trivial nomenclature has been adopted in this chapter; the corresponding systematic names for the different types of polypyrroles are given in parentheses.

III. BIPYRROLIC COMPOUNDS

A. Dipyrrylmethanes

The parent compound (17) of this series was mentioned in the literature as early as 1907 by Pictet and Rilliet.[23] Its unequivocal synthesis, which was achieved first by the Wolff-Kishner reduction of the α,α'-dipyrrylketone (16),[18] has been improved recently.[24]

Dipyrrylmethanes are obligated intermediates in the synthesis of numerous linear tri- and tetrapyrrolic compounds, as well as in the MacDonald approach to porphyrin synthesis[25] whose merits were demonstrated in the brilliant synthesis of chlorophyll a by Woodward et al.[26,27]

On the basis of incorporation experiments, [13]C- and ^{14}C-labeled dipyrrylmethanes of the types (13 and 14), which may be formed, in principle, by condensation of two porphobilinogen units, have been proposed as intermediates in the biosynthesis of uroporphyrinogens I,[28,29] and III,[30,31] and, therefore, they may act as precursors of heme, chlorophyll, and the vitamin B_{12} chromophore (cf. Chapter 2 in Volume VI).

On the other hand, dipyrrylmethane (13) could be transformed chemically (phosphate buffer, 35°C) into a mixture of uroporphyrinogen I and IV, exclusively.[29,32] Under the same conditions, compound (15) as well as dipyrrylmethane (14) yield uroporphyrinogen II.[29,33]

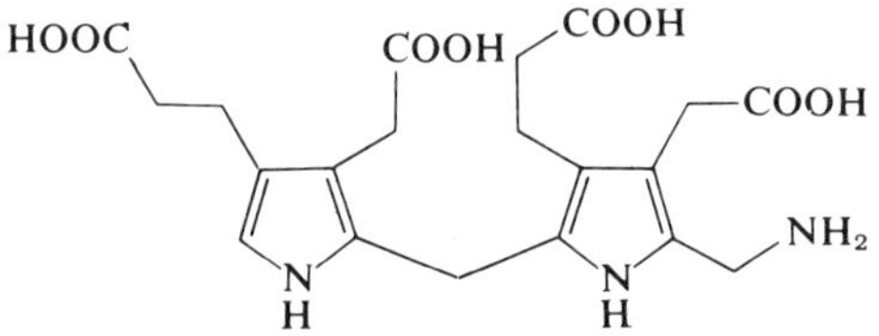

13

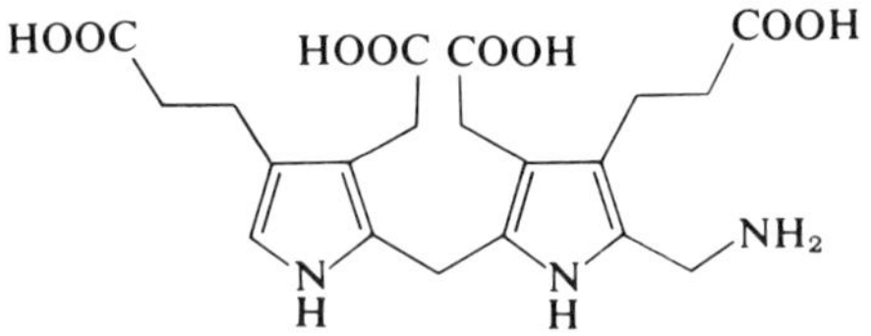

14

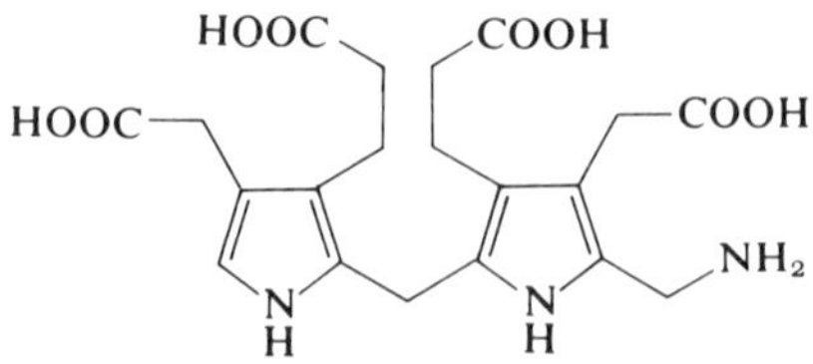

15

1. SYNTHESIS

The following methods for the synthesis of dipyrrylmethanes have been reported:

1. Reduction of pyrromethenes (see Section III, B, 2).
2. Reduction of dipyrrylketones (e.g., **16**) by treatment with zinc hydro-chloric acid,[34] diborane,[35] or metal hydrides, including LiAlH$_4$,[36] NaBH$_4$,[24,35] and KBH$_4$.[35]

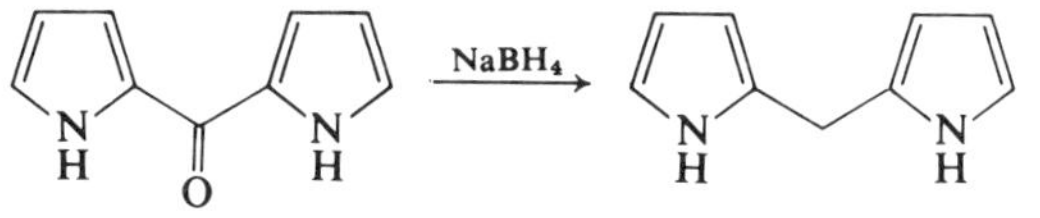

(Ref. 24)

3. "Reductive alkylation" of an α-unsubstituted pyrrole (e.g., **18**) by an α-formylpyrrole (e.g., **19**) in the presence of hydroiodic and hypophos-phorous acid.[37]

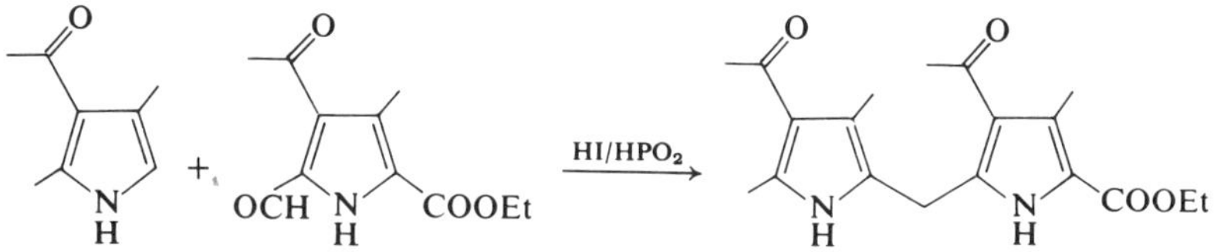

4. Michael-type addition of α-unsubstituted pyrroles at the exocyclic double bond of pyrrol-2-yl-methylene malonic ester derivatives.[38] By this method, dipyrrylmethanes carrying an acetic acid residue at the meso position (e.g., **20**) are readily available.

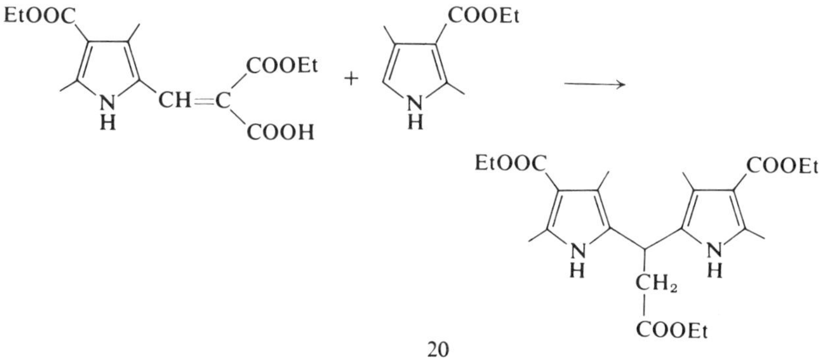

20

5. Acid-catalyzed condensation of two equivalents of an α-unsubstituted pyrrole derivative bearing at least one electron-withdrawing group* with one equivalent of an aldehyde. This method leads to symmetrically substituted dipyrrylmethanes (e.g., **21**):

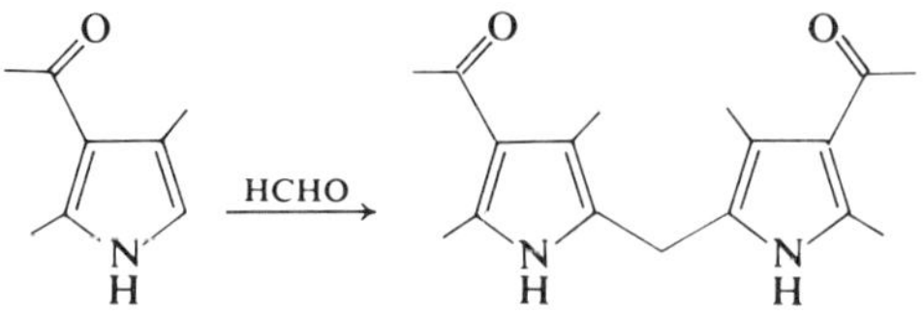

(Ref. 18)

21

By reaction with formaldehyde, dipyrrylmethanes with a methylene bridge joining both pyrrole rings are formed.[39-42] Other aldehydes, such as aliphatic,[18,43,44] aromatic,[18,45,46] or heterocyclic ones,[46,47] afford *meso*-substituted dipyrrylmethanes. By reaction with monoethyl-α-ketosuccinate, dipyrrylmethanes bearing an acetic ester residue at the meso position are obtained.[38]

Frequently, the α-unsubstituted pyrrole may be replaced by the corresponding pyrrole-carboxylic acid or, alternatively, by an α-iodopyrrole.[48]

6. Reaction of an α-halogenomethylpyrrole (e.g., **22**)[26,27,43,49-66] or a quaternized pyrrol-2-yl-Mannich base[01,67] with an α-unsubstituted pyrrole (or, alternatively, a pyrrole-α-carboxylic acid).[55,58,61,68]

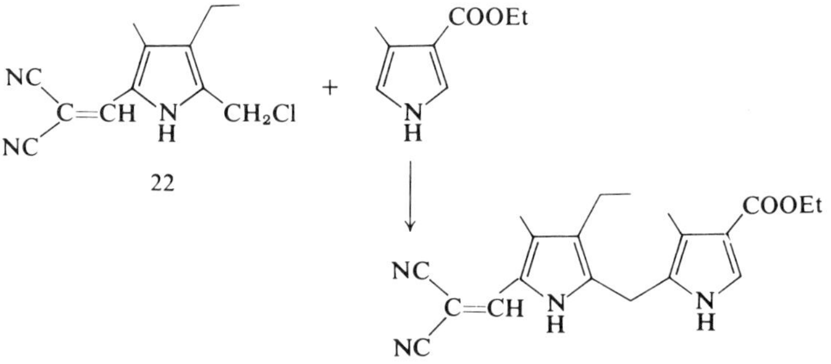

(Ref. 26)

22

An earlier approach to this type of dipyrrylmethane synthesis employed the treatment of a pyrrole Grignard reagent with an α-halogenomethylpyrrole derivative.[69] This method enables the synthesis of dipyrrylmethanes carrying different substituents at both pyrrole rings. Occasionally, however, a sym-

---

* If only alkyl groups are present, dipyrrylmethanes are formed which are very labile towards air.[38a,177] α,α'-Diunsubstituted pyrroles yield, on the other hand, porphyrins (cf. Chapter 3 in Vol. I).

metrically substituted dipyrrylmethane is obtained as a by-product formed by self-condensation of the employed α-halogenomethylpyrrole derivative.[55]

7. Reaction of an α-hydroxymethylpyrrole derivative with an α-unsubstituted pyrrole in acidic methanol.[61] A related method consists in the reaction of the more readily accessible α-acetoxymethylpyrrole derivatives (e.g., **23**) with α-unsubstituted pyrroles (or, alternatively with the corresponding α-carboxylic acids),[61] in the presence of catalytic amounts of toluene-*p*-sulfonic acid in different solvents, including methanol,[70,71] acetic acid,[70–72] or methylene chloride,[73–76] or without catalyst in acetic acid or pyridine.[42,51,77]

A further possibility consists in the employment of tin tetrachloride in methylene chloride at −20°C as condensing agent.[78] The synthesis of dipyrrylmethanes from α-acetoxymethylpyrroles represents by far the most convenient method available until now for the synthesis of derivatives bearing different substituents at both the pyrrole rings.

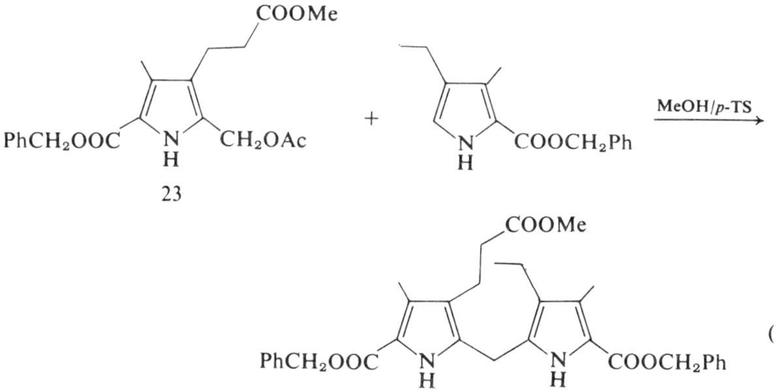

(Ref. 71)

8. Self-condensation of α-acetoxymethyl,[61,72,79–82] α-halogenomethyl,[18,25,55,61,80,81,83,84,85] α-hydroxymethyl,[85] α-methoxymethyl,[32] or α-trimethylammoniummethyl-substituted[177] pyrroles. This method affords symmetrically substituted dipyrrylmethanes.

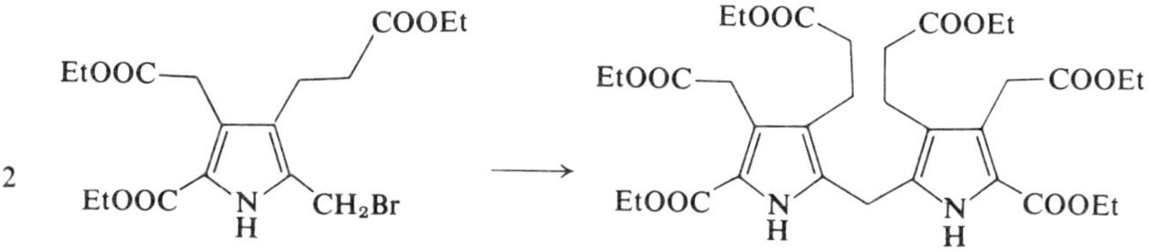

(Ref. 55)

9. Reaction of α-pyridiniummethylpyrroles (e.g., **24**) with the lithium salts of pyrrole-α-carboxylic acids.[58,60,68,86-90]

As the five last methods involve the same intermediates, namely, azafulvenium ions (**26**), they belong, from the mechanistic point of view, to the

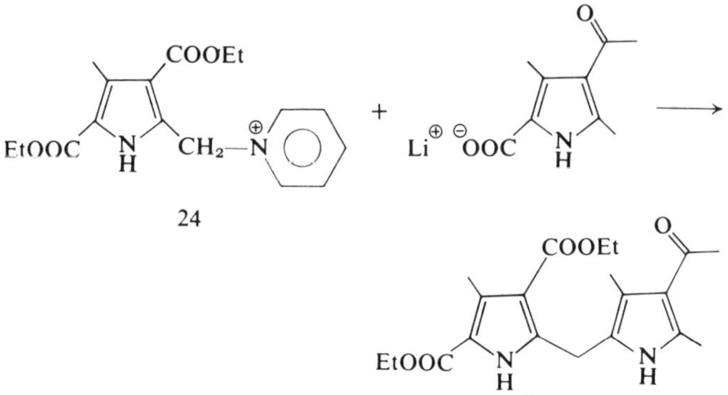

24

(Ref. 86)

same type, and they will be, therefore, considered together. In order to illustrate the reaction mechanism common to the above mentioned methods 5–9, the reaction of an α-halogenomethylpyrrole (**25**, Y = Br) derivative with an α-unsubstituted pyrrole (**27**) will be considered first (method 6). This widely employed method to synthesize dipyrrylmethanes consists essentially in the electrophilic attack of an azafulvenium ion (**26**), generated *in situ* from the corresponding α-halomethylpyrrole by cleavage of a halide ion (Y⁻ = Br⁻), on the unsubstituted α-position of the pyrrole molecule (**27**). The subsequent reaction steps leading to the dipyrrylmethane (**28**) are shown in Scheme 2.

The reaction of α-unsubstituted pyrroles with α-acetoxymethylpyrroles (method 7), as well as with other pyrrole derivatives bearing a Y—CH₂

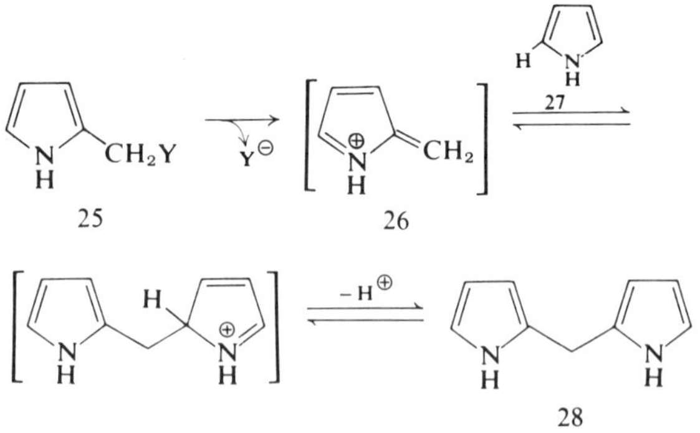

Scheme 2. Mechanism of formation of dipyrrylmethanes via azafulvenium ions. (For the sake of clarity substituents are omitted.)

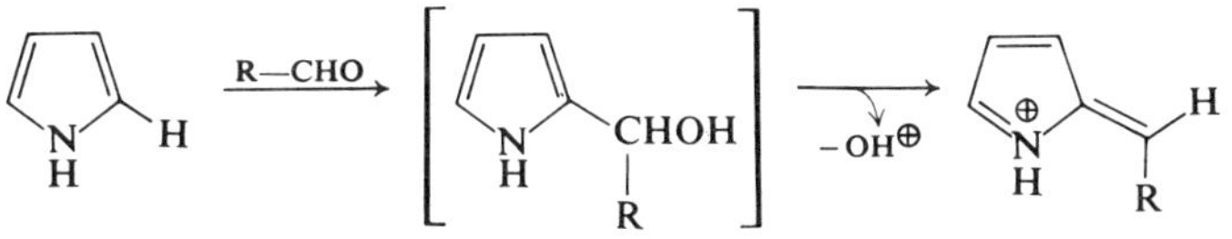

Scheme 3. Pyrrolemethanols as intermediates during the condensation of α-free pyrroles with aldehydes. (For the sake of clarity, substituents are omitted.)

substituent (Y being an anionic leaving group) at the α-position, proceeds in an entirely analogous fashion.

Particular examples of this type of compounds are α-hydroxymethyl-pyrroles (**25**, Y = OH)—so-called "pyrrolemethanols"—which are also capable of reacting with α-unsubstituted pyrroles yielding dipyrrylmethanes.[61] Since one of the methods available for synthesizing α-hydroxymethylpyrroles consists of the acid-catalyzed condensation of α-unsubstituted pyrroles with aldehydes,[91] pyrrolemethanols (**29**) are probably formed as intermediates in the synthesis of dipyrrylmethanes by method 5. The further reactions of (**30**) correspond to those represented in Scheme 2.

Furthermore, the α-position of the pyrrole derivative, where the electrophilic attack of the azafulvenium ion takes place, may be occupied by a labile substituent that can be cleaved as a cation. Labile nuclear substituents in this sense are, for example, hydroxycarbonyl, sulfonic acid,[92] and iodine,[48] or else some substituted methyl groups[93] (see below and cf. method 8).

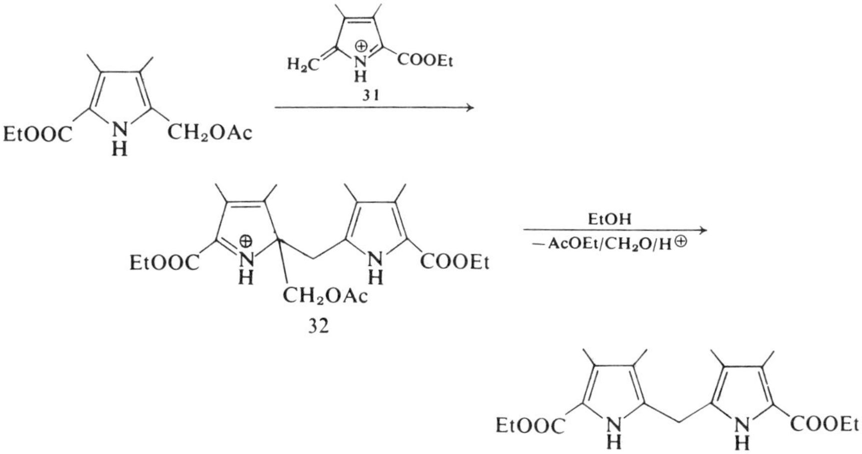

Particularly interesting in this connection is the synthesis of symmetrically substituted dipyrrylmethanes by condensation of two molecules of a pyrrole derivative bearing a halomethyl, hydroxymethyl, methoxymethyl, or an acetoxymethyl substituent (method 8) at the α-position. In these cases, the attack of the primarily formed azafulvenium ion (e.g., **31**) is directed to the occupied α-position of a second molecule of the substrate, and the resulting intermediate (**32**) loses the Y—CH$_2$ group, which probably is finally transformed into formaldehyde,[93,94] yielding the corresponding dipyrrylmethane.

The same kind of reaction probably occurs when α-bromomethylpyrroles (e.g., **35**) react with pyrrole-α-carboxylic acids (e.g., **34**). Although under the reaction conditions, acid-catalyzed decarboxylation of the carboxylic acid may take place, followed by a "normal" condensation of the resulting α-unsubstituted pyrrole with the α-bromomethyl-pyrrole derivative, the possibility of formation of an intermediate such as (**36**), which on decarboxylation affords the dipyrrylmethane (**37**), cannot be ruled out. This possibility is, indeed, substantiated by the higher yield in the reaction of (**34**) with (**35**), compared with that of the condensation of the same α-bromomethylpyrrole derivative (**35**) with cryptopyrrole (**33**).[68]

Likewise, some lithium salts of pyrrolecarboxylic acids react with α-pyridinium methylpyrroles yielding dipyrrylmethanes in excellent yields (method 9). A major limitation to this method, however, is the low nucleophilicity of lithium pyrrole carboxylates with strongly electron-withdrawing substituents.[86] Furthermore, several side reactons have been observed in some special cases.[86,87,89,95]

2. CHEMICAL PROPERTIES OF DIPYRRYLMETHANES

The reactivity of dipyrrylmethanes parallels that of pyrroles. However, as dipyrrylmethanes are more sensitive to acids and oxidants than pyrroles (see below), not all acid-catalyzed electrophilic substitutions, which can be carried out with pyrrole derivatives, are successful in the dipyrrylmethane series.

Furthermore, electrophilic attack on dipyrrylmethanes may take place at the α-positions, which are attached to the methylene bridge whereby cleavage of the molecule occurs most often (see below). For this reason, substituted dipyrrylmethanes react with two equivalents of a diazonium salt yielding aryldiazopyrroles as products.[93,96]

Reaction of trichloroacetyl chloride with α-unsubstituted dipyrrylmethanes proceeds very smoothly affording the corresponding trichloroacetyl derivatives, which can be converted successively in good yields into carboxylic acid esters by treatment with sodium alkoxide.[97]

Stable α-unsubstituted dipyrrylmethanes can be easily acylated by the Friedel–Crafts method[26] and formylated by means of the Gattermann[25,86,98,99] or Vilsmeier reaction,[24,87,100–103] as well as with triethyl orthoformate in the presence of trifluoroacetic acid.[4,104,105] In some cases, methylene ammonium salts have been isolated as intermediates of the Vilsmeier reaction.[87,106]

Instead of the rather sensitive α-unsubstituted derivatives, the corresponding carboxylic acids[101,102] can be employed as substrates in the above mentioned formylation reactions. A convenient reagent to formylate α-unsubstituted dipyrrylmethanes by the Vilsmeier method is benzoyl chloride in dimethylformamide.[24,42,105,107,108]

Most dipyrrylmethanes are stable towards bases. Dipyrrylmethane carboxylic esters can be, therefore, hydrolyzed by treatment with alkali.[55] In one special case, formation of a bipyrrolo [1,2-*a*:2′,3′-*d*]-pyridin-4(9*H*)- one (39) from the dipyrrylmethane β-carboxylate (38) has been reported.[109,110]

38   39

A probably analogous reaction occurs spontaneously during the attempted synthesis of a dipyrrylmethane from the pyrrole derivatives (40) and (41), whereby formation of the pyrrolo[2,3-*f*]indolizine derivative (42) takes place.[57,111,112]

40   41   42

Saponification of nuclear ester groups is, however, not convenient if other substituents, e.g., propionic and/or acetic ester residues, are present in the molecule whose hydrolysis is to be avoided. Sometimes pentachlorophenyl esters can be hydrolyzed selectively under careful basic conditions.[89,90] In any case, *tert*-butyl- or benzyl ester groups can be cleaved without affecting other base-labile groups present in the molecule by treatment with acid or by hydrogenolysis on palladized charcoal, respectively.[60,86,87,107] In the case of dipyrrylmethane-α-*tert*-butyl esters, decarboxylation occurs spontaneously in trifluoroacetic acid solution, the corresponding α-free derivatives being directly obtained (see below).[74,76,82,89,113]

On the other hand, dipyrrylmethanes bearing different ester groups at both α-positions can be transformed stepwise into unsymmetrically substituted derivatives, which are often needed for the synthesis of more complicated molecules.[74,76,88,89,114]

Dipyrrylmethane-α-carboxylic acids are useful intermediates in the synthesis of porphyrins and related macrocycles (cf. Chapter 4 in Volume I). Although they can be employed occasionally as reactants instead of the corresponding α-free compounds, better yields are obtained, in general, from dipyrrylmethanes bearing free α-positions. Decarboxylation of dipyrrylmethane-α-carboxylic acids can be achieved:

1. By substitution of the hydroxycarbonyl group by iodine and successive catalytic hydrogenation.[62,65,113,115]
2. On heating them without solvent in high vacuum[74,76,116] or, alternatively, in dimethylformamide solution[24,105,108] or, else, in alkali in the presence of hydrazine.[25,117]
3. On standing at room temperature in trifluoroacetic acid,[73] or, alternatively, in the presence of catalytic amounts of toluene-*p*-sulfonic acid.[116]

By refluxing the dipyrrylmethane-α-carboxylic acid (**43**) with acetic anhydride a pyrocoll-like compound (**44**) was obtained.[118]

In contrast to pyrocoll derivatives prepared from pyrroles, compound (**44**) binds two molecules of water covalently. The corresponding dipyrrolopyrazindione form could not be obtained.

Owing to the reversibility of the last reaction step in the sequence represented in Scheme 2, dipyrrylmethanes are rather unstable compounds in acidic solutions.[92,93,119–122] In the presence of acids, cleavage of the methylene bridge and recombination of the resulting fragments occur, leading—in the case of unsymmetrically substituted derivatives—to the formation of mixtures. These so-called "scrambling" or "jumbling" reactions are usually accompanied by polymerization and oxidation processes. Accordingly, in the presence of formaldehyde and acids, *meso*-substituted dipyrrylmethanes exchange the *meso*-substituted bridge for a methylene group.[121,122]

Dipyrrylmethanes are only stable to oxygen when they bear electron-withdrawing substituents.[123] Alkyl-substituted derivatives are readily oxidized by air, yielding the corresponding pyrromethenes which are, however, contaminated by many by-products.[102,177] Some alkyl dipyrrylmethane-5,5'-dicarboxylates (e.g., **45**) yield tetrapyrrylethanes (**46**) on oxidation with ferric chloride (cf. ref. 19, p. 6).

Generally, however, stabilized dipyrrylmethanes can be readily converted into the corresponding pyrromethenes by means of bromine,[55,79,124–126] *tert*-butyl hypochlorite,[35] *N*-bromosuccinimide, ferric chloride,[19] lead dioxide, lead tetraacetate,[127] or iodine monochloride in hot methanol. The use of the last mentioned reagent has been recommended.[48] Substitution of α-carboxylic acid groups or β-hydrogens[48] by iodine in some dipyrrylmethanes is also possible, without causing oxidation to the corresponding pyrromethenes.

When dipyrrylmethanes bearing an alkyl group at the methylene bridge are oxidized, cleavage of the meso substituent occurs.[128] *meso*-Aryl substituted dipyrrylmethanes, on the contrary, yield on oxidation the corresponding *meso*-aryl-substituted pyrromethenes.[129]

On the other hand, some *meso*-unsubstituted dipyrrylmethanes have been transformed into α,α'-dipyrrylketones by oxidation with lead tetraacetate–lead dioxide in glacial acetic acid[34,35] or with bromine–sulfuryl chloride.[105,107,130] On oxidation with 30% hydrogen peroxide in alkaline solution, dipyrrylmethane-α,α'-dicarboxylic acids (e.g., **47**) yield colorless products that are derivatives of the so-called propentdyopent (e.g., **48**).[131] By treatment of the latter with reducing agents (e.g., sodium dithionite) a characteristic red coloration ($\lambda_{max} = 525$ nm) appears (cf. Chapter 10 in Volume VI).

Under analogous conditions (30% hydrogen peroxide in pyridine solution), dipyrrylmethanes of the type represented by (**49**) yield 5(2*H*)-dipyrrylmethanones (e.g., **50**), whereas, on oxidation with MnO_2, the corresponding 5(1*H*)-pyrromethenones (**51**) are formed.[132] Dipyrrylmethanes are reductively cleaved by hydroiodic acid in glacial acetic acid (cf. ref. 18, p. 335) or, more appropriately, by "reductive C-alkylation"[37] yielding pyrroles.

3. PHYSICAL PROPERTIES OF THE DIPYRRYLMETHANES

When stabilized by electron-withdrawing substituents, dipyrrylmethanes are readily crystallizable compounds. The molecular structure of (5,5'-diethoxy-carbonyl-3,3'4,4'-tetraethyldipyrrol-2-yl)-methane has been determined by X-ray analysis.[133] Dipyrrylmethanes can be conveniently isolated and purified by chromatography on silica gel. In analytical work, they can be detected by exposing the thin-layer chromatograms to bromine vapor.[61] Dipyrrylmethanes appear then as red spots on the chromatogram.

a. Electronic Spectra. As expected, the uv absorption of dipyrrylmethanes corresponds roughly to that of the pyrrole rings joined together by the methylene bridge.

The characteristic absorption-range for alkyl-substituted pyrroles lies at about 210 nm (so-called "B-band"), whereas pyrrole derivatives bearing electron-withdrawing substituents show, in addition to the B-band bathochromic shifted by 30–60 nm, a long-wavelength absorption in the range 260–300 nm (so-called "K-band").

However, the absorption maxima of dipyrrylmethanes are usually shifted by about 5–6 nm towards longer wavelengths with reference to those of the corresponding substituted pyrroles.[134] This effect may be attributed to homoconjugation[135] between both pyrrole rings, which are joined by the methylene bridge.

b. IR Spectra. As a rule, the ir spectra of dipyrrylmethanes can be interpreted on the basis of the same assignments that have been made for monopyrroles (cf. ref. 20, p. 77ff.). As in the case of pyrroles, dipyrrylmethane derivatives bearing carbonyl substituents at the α-positions show, in addition to a NH-stretching absorption for intermolecular associated species, a concentration-independent absorption band which is shifted to longer wavelengths by 30–40 cm^{-1} with reference to the free NH-stretching mode of pyrrole ($\tilde{v} = 3496$ cm^{-1} in CCl$_4$). Alkyl dipyrrylmethane-α-carboxylates show two concentration-independent absorption maxima owing to the occurrence of two conformers (e.g., **52** and **53**).[136,137] In contrast to pyrrole-β-carboxylic acid esters, which can form only intermolecular associates, dipyrrylmethane-3,3′-dicarboxylates occur in the very stable conformation (**54**) which is twofold hydrogen-bonded. The corresponding N–H absorption lies in the range 3390–3401 cm^{-1}.[138]

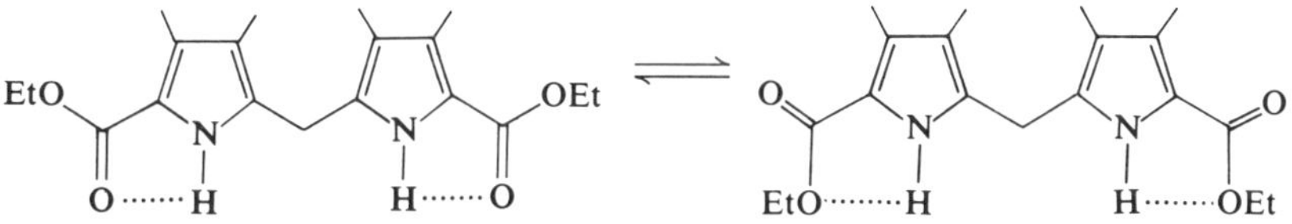

52   53

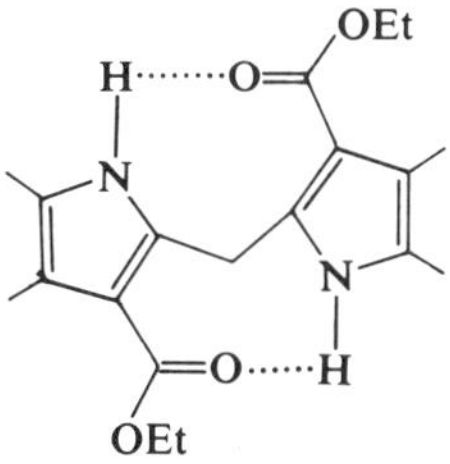

54

c. nmr Spectra. No systematic work has been done on the nmr spectroscopy of dipyrrylmethanes. Most assignments of the absorption maxima can
be made, however, by analogy with the corresponding substituted pyrrole
derivatives (cf. ref. 20, p. 78). A characteristic peak in the nmr spectra of
dipyrrylmethanes is a singlet in the range $\delta = 3.8$–4.0 ppm,[71] which is associated with the hydrogen atoms at the methylene bridge.

In the course of the investigation of the ^1H nmr spectra of a series of pyrrole
derivatives that were prepared as building blocks for the synthesis of several
polypyrrolic compounds related to uroporphyrin III, 12-decarboxyuroporphy-
rin III and phyriaporphyrin III(18-decarboxyuroporphyrin III),[74–76] an
interesting relationship was observed between the chemical shifts of the signals
associated with the methylene protons of β-acetic and propionic ester residues
and the nature of the groups attached to the vicinal α-positions. Thus only
those substituents at the α-positions that exert a conjugative electron-
withdrawing effect on the pyrrole ring bring about a quite constant paramagnetic displacement in the chemical shift of methylene protons at the
neighboring β-positions by about 0.3 ppm, as compared with the corresponding values for derivatives lacking in such substituents (see Table 1).

It is noteworthy that the average values given in Table 1 are rather independent of other substituents present in the molecule. They are, therefore, also

TABLE 1

Average δ^1_H Values (in $CDCl_3$ Relative to TMS) for Methylene Protons of β-Acetic and
Propionic Ester Residues as Dependent on the Substituents at the Vicinal α-Position in Pyrrole
Derivatives

α-Substituent	$-CH_2-COOR$	$-CH_2{}^*-CH_2-COOR$	$-CH_2CH_2{}^*-COOOR$
H J CH_3 CH_2Y (Y = halogen, OAc...) $-CH_2-$⟨pyrrole, N—H⟩	3.5	2.7	2.5
$-COR$ (R = OH, H, alkoxy) $-CH=$⟨pyrrole ⊕, N—H⟩	3.8	3.0	2.6

valid, within a series of related compounds, for many linear polypyrroles such as dipyrrylmethanes, tripyrrenes, biladienes-ac, bilanes, etc., so far as the corresponding peaks are discernible, i.e., not overlapped by other signals in the ^{1}H nmr spectra of the compounds considered.[76]

d. Mass Spectra. The mass spectra of a number of dipyrrylmethanes, including α,β'- and β,β'-derivatives, have been investigated in detail.[139]

With the exception of some dipyrrylmethanecarboxylic acids and benzylic esters that very readily undergo decarboxylation or loss of the benzyl group, respectively, the mass spectra of dipyrrylmethanes all show a fairly intense peak corresponding to the parent ion. The fragmentation patterns are dependent on the nature of the substituents. They can be classified into three main groups characterized by: (a) internuclear cleavage to give monopyrrolic fragments, (b) fragmentation of side chains, and (c) formation of tricyclic fragmentation products. The last type of fragmentation occurs when the methylene bridge is flanked by neighboring carbonyl substituents and accounts for the high intensity of the base peak in the spectrum of dipyrrylmethane (**55**), most likely due to the formation of the very stable fully conjugated tricyclic cation (**56**).

$$M^{\oplus} \; m/e = 462 \; (19\%)$$
55

$$m/e = 389 \; (4\%)$$

$$\xrightarrow[-C_2H_5OH]{m^*}$$

$$m/e = 343 \; (90\%)$$
56

As a rule, cleavage at the methylene bridge, which joins the two pyrrole rings in dipyrrylmethanes, is the predominant process, but fragmentation of the side chains may predominate, particularly if several carbonyl substituents are present in the molecule.

Cleavage at the methylene bridge may either be accompanied by rearrangement of one hydrogen atom from a neighboring substituent or not (e.g., in ion **57**).

The precise nature of the factors controlling the type of internuclear

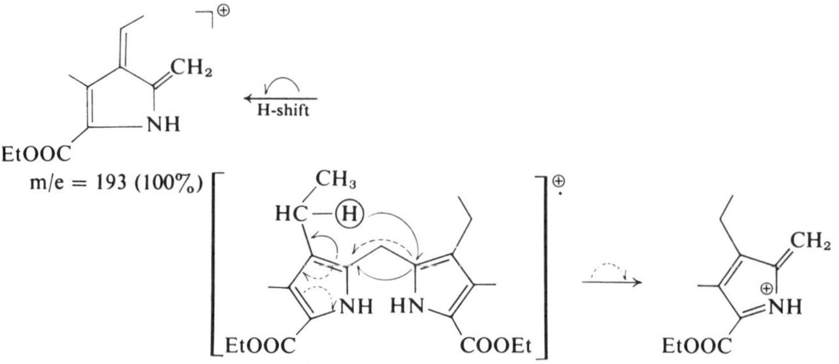

$M \oplus m/e = 374\ (88\%)$

57

cleavage that is likely to occur in any particular case is not yet entirely clear. In the case of unsymmetrically substituted dipyrrylmethanes, it is important to note that the fragment retaining the positive charge in this type of cleavage is the ring bearing the fewer number of electron-withdrawing substituents; i.e., the inductive effect of alkyl groups in stabilizing the positive charge on this fragment is the controlling factor here, whereas in cleavages involving hydrogen rearrangement, hydrogen is transferred from the ring bearing electron-withdrawing substituents, and the positively charged fragment derived from this ring predominates.

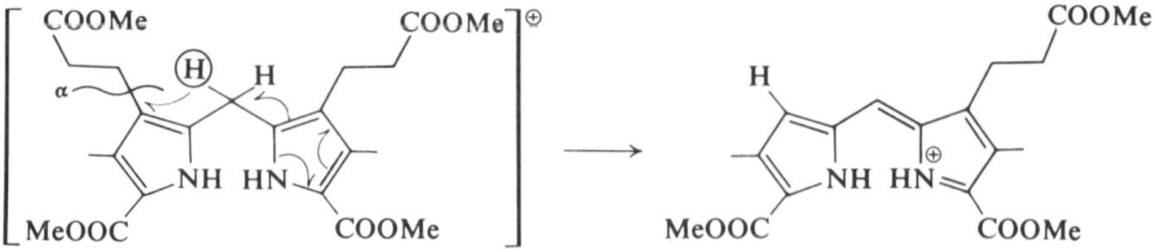

The side chain fragmentations of dipyrrylmethanes are closely analogous in many cases to those observed with monopyrroles.[140] However, in contrast to pyrroles that undergo the more normal cleavage of alkyl substituents at the β-bond, dipyrrylmethanes may lose alkyl side chains by α-cleavage. If one assumes that these "α-type" cleavages are accompanied by rearrangement of one hydrogen atom from a neighboring substituent with concomitant formation of a very stable pyrromethene saltlike cation (e.g., 59), then their relatively more important role in the fragmentations of certain dipyrrylmethanes can be readily explained. On this basis, loss of nuclear propionate groups appears to be a marked feature of the spectra of dipyrrylmethanes bearing such substituents (e.g., 58). It is worth noting that cleavage of the meso-alkyl substituent giving rise to a pyrromethenelike ion (e.g., **59**) is a dominant feature of the mass spectra of *meso*-alkyl-substituted dipyrrylmethanes.

B. Pyrromethenes

Pyrromethenes are obligated intermediates in Fischer's porphyrin synthesis (cf. Chapter 4 in Volume I), which is historically the first discovered method for the preparation of these pyrrole macrocycles.[141] Until now, pyrromethenes have not been obtained from natural sources. A noteworthy physiological property of many pyrromethenes is their sternutatory effect, which is particularly pronounced in the case of derivatives bearing free β-positions.[142,143] Many pyrromethenes are capable of dyeing cotton.

1. Synthesis

The following reactions have been reported to yield pyrromethenes:

1. Oxidation of dipyrrylmethanes (see Section III, A, 2).

2. Treatment of dipyrrylketones with bromine[144,145] or phosgene (cf. ref. 18, p. 365) whereby pyrromethenes and *meso*-chloropyrromethenes, respectively, are obtained.

3. Reaction of α-dichloromethylpyrroles with α-unsubstituted pyrroles or their corresponding α-carboxylic acids (cf. dipyrrylmethane synthesis in Section III, A, 1).[146] α-Trichloromethylpyrroles yield, correspondingly, *meso*-chloropyrromethenes.[146]

4. Halogenation of 5-unsubstituted 2-methylpyrroles[19,147–150] or, alternatively, their corresponding 5-carboxy[148,151] or 5-*tert*-butoxy-carbonyl[58,79] derivatives. This method leads usually to the formation of 5-methyl- (or 5-bromomethyl-)5'-bromopyrromethenes (e.g., **60** and **61**, respectively) by head-to-tail condensation of two molecules of the same pyrrole derivative.

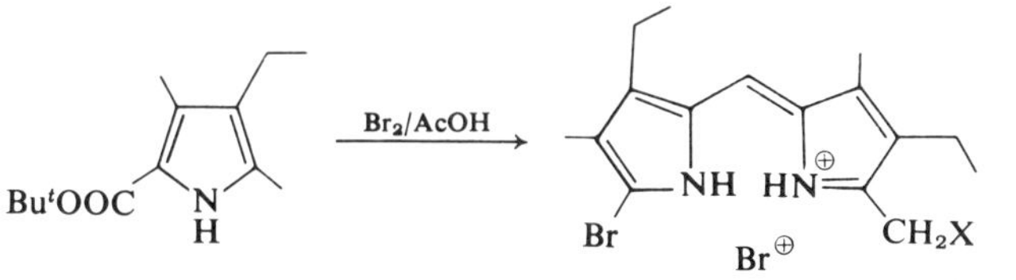

60: X = H
61: X = Br

5. Condensation of a pyrrole aldehyde (e.g., **62**) with an α-unsubstituted pyrrole derivative (e.g., **63**).[19,82,124,152] In certain cases, pyrrole derivatives bearing a halogen atom at the reactive α-position have been employed.[48] This method enables the synthesis of pyrromethenes carrying different substituents on both rings. However, in order to avoid side

reactions (polymerization, self-condensation), this method is usually restricted to pyrrole aldehydes whose 5-position is substituted, and to pyrrole derivatives bearing only one free α-position (see below).

(Ref. 19)

On the other hand, α,α'-diunsubstituted pyrroles may be employed in some cases, if one α-position is deactivated relative to the other by an appropriate neighboring β-substituent.[151] In one case, the pyrrole aldiminium salt, which was obtained as an isolable intermediate of the Vilsmeier–Haak reaction with cryptopyrrole, has been condensed directly with an α-free pyrrole yielding the corresponding pyrromethene.[87]

Condensation of 2-formylpyrroles bearing an acetyl group at the 3-position (e.g., 64) with relatively N–H acidic pyrrole derivatives (65) may lead to the formation of pyrrolo (3,2-*f*)indolizine derivatives (e.g., 67) instead of the expected pyrromethenes (66). The most likely explanation is that intramolecular nucleophilic attack of the pyrrolic nitrogen atom on the acetyl group takes place during the reaction.[153] An analogous reaction was mentioned in the case of dipyrrylmethane-3-carboxylic esters (cf. Section III, A, 2).

6. Condensation of two equivalents of an α-unsubstituted pyrrole (e.g., 68) or an α-pyrrolecarboxylic acid[148,154,155] with formic acid.[19,84,148,154,155] This reaction, as the two subsequent ones, affords symmetrically substituted pyrromethenes (e.g., 69).

68   69   (Ref. 19)

7. Reaction of α-unsubstituted pyrroles (e.g., **70**) with chloromethylmethyl ether in which spontaneous oxidation of the intermediate dipyrrylmethane probably occurs (cf. ref. 19, p. 4).

70   (Ref. 19)

8. Reaction of α-unsubstituted pyrroles (e.g., **71**) with ethyl orthoformate in the presence of boron trifluoride etherate.[156,157]

71   (Ref. 156)

9. Acid-catalyzed self-condensation of α-pyrrole aldehydes (e.g., **72**) under elimination of one formyl group.[19,158,159] Likewise 5,5′-dibromo-pyrromethenes can be obtained from 5-bromo-3-formypyrroles by treatment with acid.[160]

72   (Ref. 159)

The most versatile methods for synthetic purposes are reactions 4–6, and they will be considered, therefore, in detail. As reactions 5–9 are, from the mechanistic point of view, equivalent, they are treated together in Section III, B, 1, b.

a. Synthesis of Pyrromethenes by Halogenation of 5-Unsubstituted-2-Methyl-Pyrroles. The reaction is mostly carried out by treating an acetic acid solution of the pyrrole derivative with an excess of bromine in acetic acid at room temperature. Usually the formed pyrromethene hydrobromide crystallizes out and can be isolated by filtration and purified by crystallization.

The mechanism of this reaction was carefully studied independently by Corwin[147] and Treibs[149] some years ago. By measurements of the reaction rate, it could be demonstrated that bromination of the free α-position of (73) occurs faster than bromination of the α-methyl group. The primarily formed 5-bromo-2-methylpyrrole (e.g., 74) which may be isolated under favorable reaction conditions[147,149] reacts thereupon with the unisolable dibromo compound (75), yielding a pyrromethene hydrobromide (76) that is often obtained as its corresponding perbromide.[161] Successive bromination of the α-methyl group of (76) affords the bromomethylpyrromethene hydrobromide (77).[162] Formation of dibromomethyl derivatives is also possible.[150]

As, in the presence of acids, pyrromethenes can add α-free pyrroles reversibly at the methine bridge (Scheme 4), formation of tripyrrylmethanes (e.g., 78) is possible under the reaction conditions mentioned above.[119,147,]

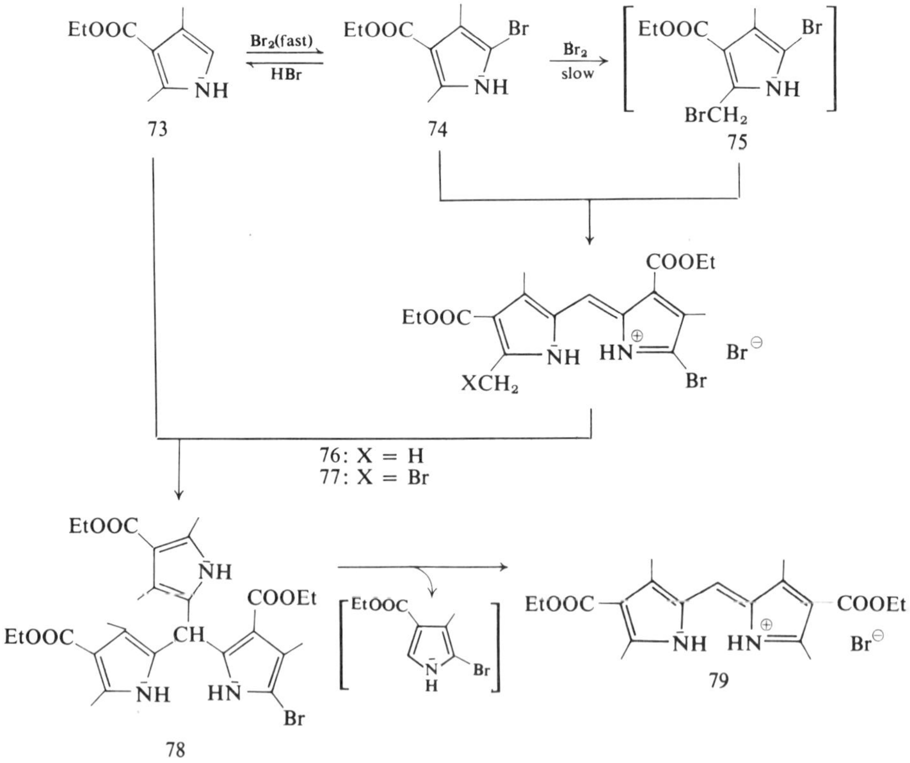

Scheme 4. Synthesis of pyrromethenes by bromination of 5-unsubstituted 2-methylpyrroles.

[157,163-165] Cleavage of one of the three pyrrole rings leads back to a pyrromethene hydrobromide, which in some cases (e.g., **79**) has a *different* structure from the starting one.[166,167]

Closely related to the pyrromethene synthesis represented in Scheme 4 is the reaction of 5 unsubstituted 2-methylpyrroles with an excess (2 moles) of sulfuryl chloride. This reaction, which is usually carried out in ethyl ether under ice cooling, affords 5'-chloropyrromethene hydrochlorides in moderate yields (cf. ref. 19, p. 66ff.).

b. Reaction of α-Pyrrole Aldehydes with α-Unsubstituted Pyrroles. This reaction is usually carried out in a polar solvent (methanol) in the presence of a mineral acid (e.g., toluene-*p*-sulfonic acid–zinc acetate,[82] hydrobromic acid). For the syntheses of pyrromethenes carrying acid-sensitive substituents, the use of phosphorous oxychloride in methylene chloride as catalyst has been recommended.[88] The mechanism of this reaction is, in principle, the same as that of the already described formation of dipyrrylmethanes by condensation of α-unsubstituted pyrroles with aliphatic aldehydes (cf. Section III, A, 1). Substitution of the aliphatic aldehyde for a formylpyrrole leads to an azafulvenium derivative (**30** R = pyrrol-2-yl) which is strongly resonance-stabilized and, therefore, becomes the final product of the reaction. The formed pyrromethene separates usually as a salt of the employed mineral acid. As in the case of method 4 (Section III, B, 1, a), the primarily formed pyrromethene is able to add reversibly a molecule of the α-free pyrrole at the methine bridge, whereby formation of a tripyrrylmethane and, consequently, of a mixture of pyrromethene isomers is possible[119,120,163,166] (cf. Scheme 4).

On the other hand, pyrrole derivatives carrying one free α-position react with formic acid (cf. ref. 19, p. 3) or, alternatively, with methy orthoformate in the presence of boron trifluoride etherate[156,157] as well as with chloromethyl methyl ether (cf. ref. 19, p. 4) yielding symmetrically substituted pyrromethenes. These reactions are probably related to the synthesis mentioned above of pyrromethenes from pyrrole aldehydes and α-unsubstituted pyrroles, owing to the fact that, under the employed reaction conditions, pyrrole aldehydes are formed as intermediates.[104] As the formation of pyrrole aldehydes from α-unsubstituted pyrroles and formic acid is a reversible process,[157,168] two equivalents of a pyrrole aldehyde are able to react in the presence of acid yielding a pyrromethene (method 9).[158,159] In these cases, a formyl group is lost as formic acid.[159]

By condensation of α-acyl pyrroles with pyrrole derivatives bearing a free α-position, *meso*-alkyl-substituted pyrromethenes are obtained.[44,169-173] *meso*-Substituted pyrromethenes are also often formed as by-products when pyrroles are acylated with acetic anhydride in the presence of boron trifluoride etherate.[174] *meso*-Alkyl-substituted pyrromethenes can exist, as free bases, in two tautomeric forms, namely as pyrromethene and as 1,1-dipyrrylethylene derivatives (**80** and **81**, respectively). The predominating form depends on the functional groups attached to the ring as well as on the meso substituent.[173,175,176,177a,178,179]

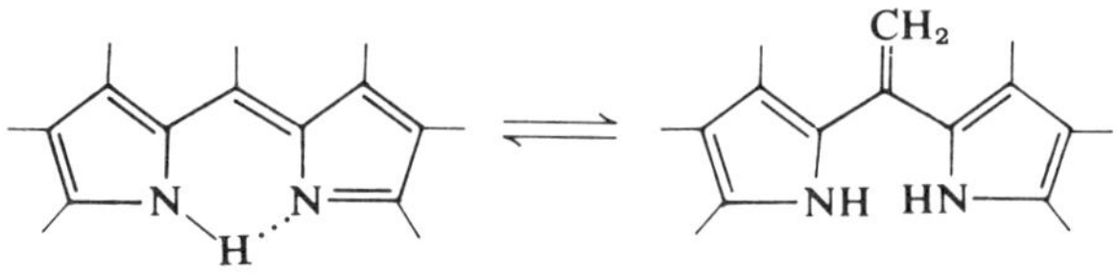

80   81

2. CHEMICAL PROPERTIES OF THE PYRROMETHENES

As already mentioned, pyrromethenes are usually obtained as the corresponding salts of the mineral acid present in the reaction mixture. These salts are mostly stable, red-colored crystalline compounds (λ_{max} in the range 430–470 nm) showing a characteristic green iridescent lustre. The corresponding pyrromethene-free bases, which are usually less stable than the protonated species, particularly when they carry only alkyl substituents (cf. ref. 19, p. 8) may be liberated from the latter by treatment with aqueous ammonia (cf. ref. 19, p. 10). Pyrromethenes are particularly difficult to purify, probably because of their facile protonation and deprotonation; chromatography rarely produces any improvement in purity, and it is, therefore, a prerequisite that any synthesis must be highly efficient to enable convenient isolation of the pyrromethene. Some pyrromethene–free bases can be conveniently purified by sublimation under high vacuum.[180] The most characteristic property of pyrromethenes is their ability to form quite stable metal chelates (see Chapter 8). Some pyrromethenes form stoichiometric-defined compounds with bromine,[161] hexachloroplatinic acid, and several aromatic nitro compounds such as nitrophenol, trinitrotoluene, and picric acid.[181] In contrast to pyrroles, pyrromethenes react only rarely with electrophiles.* This statement is clearly explainable if one takes into account the fact that pyrromethenes behave as valence tautomers whose structures (**82** and **83**, Scheme 5) are interconverted

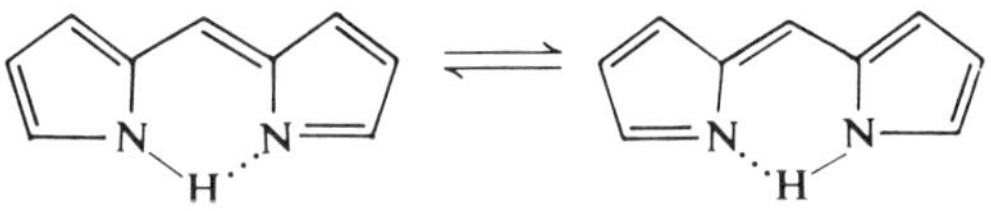

82   83

Scheme 5. Valence-tautomeric structures of pyrromethenes. (For the sake of clarity, substituents are omitted.)

* One exception is the ready condensation of α-unsubstituted pyrromethenes with α-bromomethyl-pyrromethenes which under the conditions of the Friedel–Crafts reaction in the presence of tin tetrachloride yields biladienes-ac[10,23-dihydro-(21*H*)-bilins] (cf. Section V, E, 1).

readily by proton-jump from one nitrogen atom to the other (cf. Section III, B, 3, c).

The predominating tautomer at equilibrium depends mainly on the substituents present at the different ring positions of the pyrromethene molecule.[182,183] Obviously, both tautomers are equally represented in the case of symmetrically substituted derivatives. As the hydrogen bridge in these compounds is still unsymmetric,[184] the tautomeric forms (82 and 83) do not represent, in any case, two resonance structures of a single molecule.

Owing to the ability of the "pyridine-like" nitrogen atom on the pyrrolenine ring to withdraw electrons in contrast to the pyrrolic nitrogen atom, which is a weak electron donor, the reactivity of the pyrromethene molecule as a whole resembles that of a π-electron-deficient heteroaromatic system. Consequently, many nucleophiles such as water,[124] alcohols,[124,125,163,174] bromine,[125] ammonia,[185] sodium bisulfite,[185] hydrogen cyanide,[186] triethyl phosphite,[186] methyl magnesium bromide,[187] cyanacetic acid,[188] azulene,[123] some methyl ketones,* and malonitrile[189] are reversibly added to the exocyclic double bond of pyrromethenes. As mentioned before, pyrroles as well as dipyrrylmethanes[58] and their corresponding α-carboxylic acids[116] add in like manner to pyrromethenes yielding tripyrrylmethane derivatives.

Moreover, nucleophilic substitution of halogen atoms bonded to the α-position of pyrromethenes can be accomplished by treatment with silver or potassium acetate in boiling glacial acetic acid as well as with potassium methylate or with cold methanol,[190] whereby 5(1H)-pyrromethenones and their corresponding imino esters, respectively, are formed.[191] 5,5'-Dibromopyrromethenes (e.g., 84) yield, on treatment with potassium acetate in boiling acetic acid, reduced propentdyopent derivatives (e.g., 85).[192]

A further interesting reaction of 5,5'-dibromopyrromethenes is their conversion into bile pigment derivatives of the glaucobilin- (mesobiliverdin)type (e.g., 86) on treatment with glyoxylic acid in glacial acetic acid.[193]

* Under the employed reaction conditions, the primarily formed meso-acylmethylpyrromethenes rearrange quantitatively to dipyrryltrimethine dyes.[189]

On the other hand, α-bromomethylpyrromethenes can be readily converted into higher alkyl derivatives or into the corresponding methoxymethylpyrromethenes by treatment with Grignard reagents[194] or with boiling methanol,[58,150] respectively. From dibromomethylpyrromethenes, the corresponding dimethoxymethyl derivatives, which are likewise available, can be converted subsequently into pyrromethene aldehydes.[150] These last mentioned compounds are not directly accessible by formulation, owing to the low nucleophilicity of the pyrromethene molecule. Formation of a pyrromethene aldehyde as the main product of the Gattermann formylation of a dipyrrylmethane has been reported.[56] On treatment with sulfuryl choride, a useful reagent for the transformation of α-methylpyrroles into the corresponding aldehydes (cf. ref. 20, p. 290), 5,5'-dimethyl-pyrromethenes yield formylated dipyrrylketones.[107]

Particularly interesting are 5-bromo-5'-methyl- (and 5'-bromomethyl-) pyrromethenes because of their important role as precursors in porphyrin synthesis (cf. Chapter 4 in Volume I, and Section V, E, 1).

Reduction of pyrromethenes can be carried out by treatment with sodium borohydride,[5,6,25,81,145,195,196] sodium amalgam[25,155,197] and catalytically on Pd-black[24,194] or Raney-Nickel.[5,6,118] The obtained dipyrrylmethanes, however, are usually not isolated in a pure state unless they bear electron-withdrawing substituents. On sensitized photooxygenation, pyrromethenes undergo cleavage at the methine bridge.[108]

3. Physical Properties of the Pyrromethenes

a. Electronic Spectra. The characteristic light absorption of pyrromethenes lies in the range of 430–470 nm.[183,199–203] As determined by the liquid crystal-induced circular dichroism technique, the corresponding transition moment is polarized parallel to the "long" axis of the molecule.[202] Protonated species absorb at somewhat (about 20–40 nm) longer wavelengths. This bathochromic shift is particularly pronounced in the case of *meso*-substituted pyrromethenes.[204]

Twisting of the pyrromethene molecule by introducing alkyl groups at the nitrogen atoms brings about a strong bathochromic shift and a decrease of the intensity of the long-wavelength absorption, as compared with protonated pyrromethenes which are planar.[205] This phenomenon, which was observed for the first time by Brunings and Corwin,[124] has been discussed recently in detail by Falk on the basis of advanced MO methods.[206]

The dependence of the planarity of the chromophore on the bulk of the substituent has been investigated for many pyrromethenes bearing alkyl[200] and phenyl groups[207] at the ring positions. Pyrromethenes show a very pronounced fluorescence, not only in solution, but also, as mentioned before, in the solid state.[208]

b. Infrared Spectra. The N–H stretching frequency of pyrromethenes, when measured as free bases in carbon tetrachloride or chloroform, lies in the range 3220–3310 cm^{-1}.[136,138,179,207,209,210] This remarkably low frequency—compared with pyrrole ($\bar{\nu}_{NH}^{CCl_4} = 3496$ cm^{-1})—cannot only be attributed to intramolecular H bonding[211] because in the Z-s-Z structure (see formulas **82** and **83**) whose occurrence in solution is now well established by physical methods, i.e., lanthanide-induced shift technique[182] and dipole moment measurements,[183] the N–H...N angle amounts to only 117°.* Most likely, it can be explained by the lower electron density at the pyrrolic nitrogen atom owing to the presence of the π-deficient pyrrolenine ring in the molecule[212] or else by proton mesomerism between the two valence tautomers (**82** and **83**) (see Section III, B, 2).[213] Latter interpretation agrees with nmr spectroscopic data of symmetrically substituted pyrromethenes (see below).

All pyrromethenes show a characteristic intensive absorption at 1600–1680 cm^{-1} (so-called pyrromethene band) which can also be found in the spectra of chlorins,[214] tripyrrenes (cf. Section IV, B, 3), biladienes-ac (cf. Section V, E, 3), but not of porphyrins.[215] The origin of this band is not yet known.[179]

c. nmr Spectra. Until now the only systematic work which has been done on ^{1}H nmr spectroscopy of pyrromethenes is that of Falk and co-workers.[180] The most characteristic peak of this class of compounds is a singlet in the range $\delta = 6.6$–7.9 ppm associated with the proton at the methine bridge.[150,216,217] Protonation at the pyrrolenine nitrogen atom or formation of diamagnetic metal chelates causes a shift of roughly 0.5 ppm.[217]

Owing to the rather uncomplicated nmr spectra obtained from symmetrically substituted pyrromethenes, both rings become equivalent when observed with the nmr average frequency, i.e., rapid interconversion of both tautomers (**82** and **83**) (see Section III, B, 2) takes place by intra- as well as intermolecular proton transfer from one nitrogen atom to the other.[180] ^{13}C nmr spectroscopic data have only been available for the pyrromethene derivative (**87**), the assigned values (δ in ppm from carbon disulfide) being

87

* It is well known that strong hydrogen bonds require a linear arrangement of the participating atoms.[211a] Such a situation is present, for instance, in the pyridine–pyrrole associate where the frequency of the N–H stretching mode of pyrrole is shifted to lower frequency by only 245 cm^{-1}.[211b]

given in parentheses on the formula. Additional resonances were observed at 35.4, 47.9, and 59.6 ppm.[218]

d. Mass Spectra. The mass spectra of several pyrromethenes have been investigated in detail.[139] With the exception of some α-bromomethyl derivatives, which do not give parent ions, owing to the ease with which cleavage of bromine, hydrogen bromide, or $\dot{C}H_2Br$ occurs, the most striking feature of pyrromethene spectra is the relatively high intensity of the molecular ion; this clearly confirms once again the well-known stability of the fully conjugated pyrromethene system.

Cleavage at the methine bridge does not occur, to a noticeable extent, compared to the fragmentation of the side chains. Alkyl groups are often removed by α-cleavage. The fragmentation patterns of pyrromethenes bearing nuclear carboxylic ester groups are largely dominated by cleavages of the ester groups although nuclear alkyl cleavages are also important in some cases.

IV. TRIPYRROLIC COMPOUNDS

A. Tripyrranes [5,10,16,17-Tetrahydro-(15*H*)-tripyrrins]

Like tetrapyrranes (cf., Section V, A) tripyrranes are probably involved as intermediates in the polycondensation of α,α'-diunsubstituted pyrroles with aldehydes to porphyrinogens whose oxidation leads finally to the formation of porphyrins. Until now, however, they have not been isolated from the reaction mixtures from which porphyrins are obtained.

On the other hand, tripyrranes (**88** and **89**), which might be considered biologic precursors of heme, chlorophyll, and the vitamin B_{12} chromophore (cf. Chapter 2 in Volume VI), have been transformed chemically (phosphate buffer, 35°C) into mixtures of uroporphyrinogens.[33,219] Furthermore,

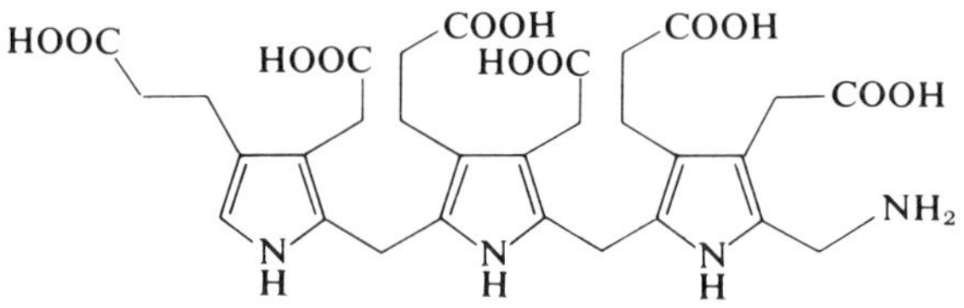

88

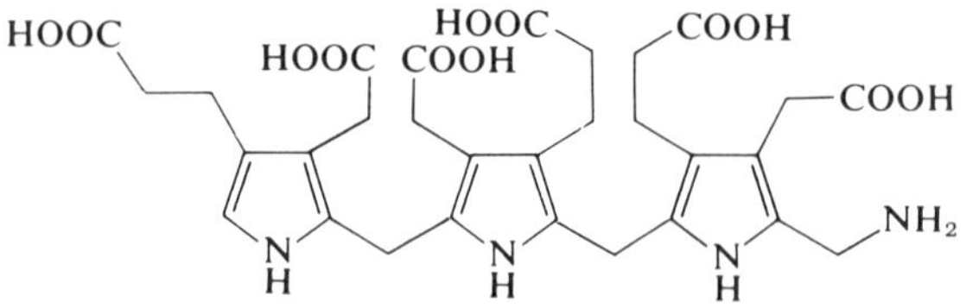

89

compound (88) could be enzymatically incorporated into uroporphyrinogen I.[66]

1. SYNTHESIS

Tripyrranes are available by one of the following methods:

1. Catalytic hydrogenation,[65,219] or, alternatively, sodium borohydride reduction[87,220] of tripyrrenes (e.g., 90). The synthesis of tripyrranes from tripyrrenes represents probably the most convenient procedure to avoid the formation of by-products, which often occurs as a consequence of the low stability of tripyrranes to oxygen and acids (see below). Reduction of a *meso*-oxodihydrotripyrrin with sodium borohydride has also been found successful in preparing the corresponding tripyrrane.[35]

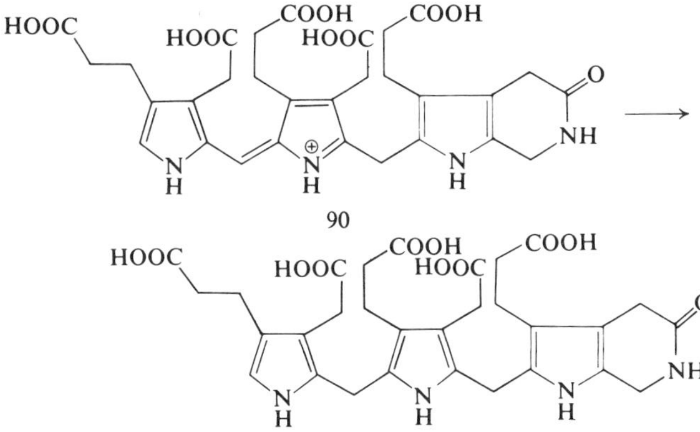

90

(Ref. 65)

2. Condensation of α-halomethylpyrroles[61,220,221] or, alternatively, of α-acetoxymethylpyrroles (e.g., 91)[61,70,71,222] with dipyrrylmethanes bearing an unsubstituted α-position (e.g., 92).

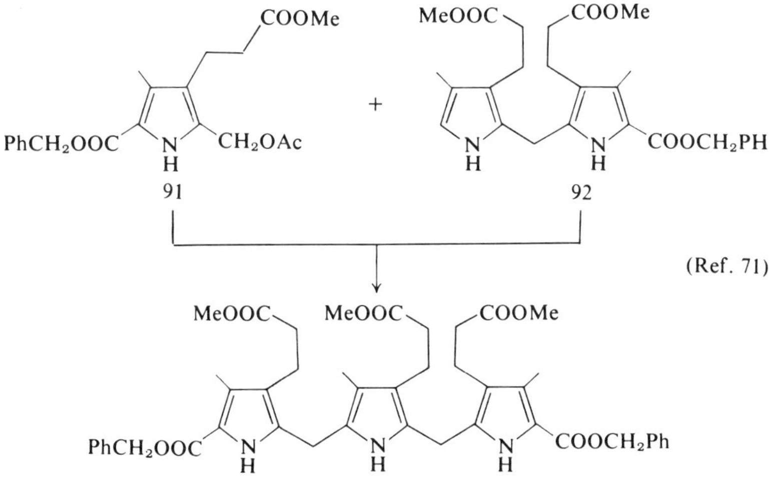

(Ref. 71)

3. Reaction of the lithium salt of a dipyrrylmethane-α-carboxylic acid (e.g., **93**) with a pyridiniummethylpyrrole derivative (e.g., **94**).[68,86]

(Ref. 86)

4. Reaction of two equivalents of a halomethyl (e.g., **95**)[61] or an α-alkoxymethylpyrrole[18,223] with one equivalent of an α,α'-diunsubstituted pyrrole. The latter can be replaced in some cases by the corresponding Grignard derivative (e.g., **96**).[165]

The three last mentioned reactions can be probably interpreted on the basis of the same mechanisms as described for the corresponding synthesis of dipyrrylmethanes from pyrroles (cf. Section III, A), respectively.

2. CHEMICAL PROPERTIES OF THE TRIPYRRANES

As expected, the chemical behavior of tripyrranes parallels that of dipyrryl-methanes (cf. Section III, A, 2). The sensitivity to acids and oxygen is, however, by far more pronounced in the case of tripyrranes, some of which decompose even in the crystalline state on exposure to the air.[71]

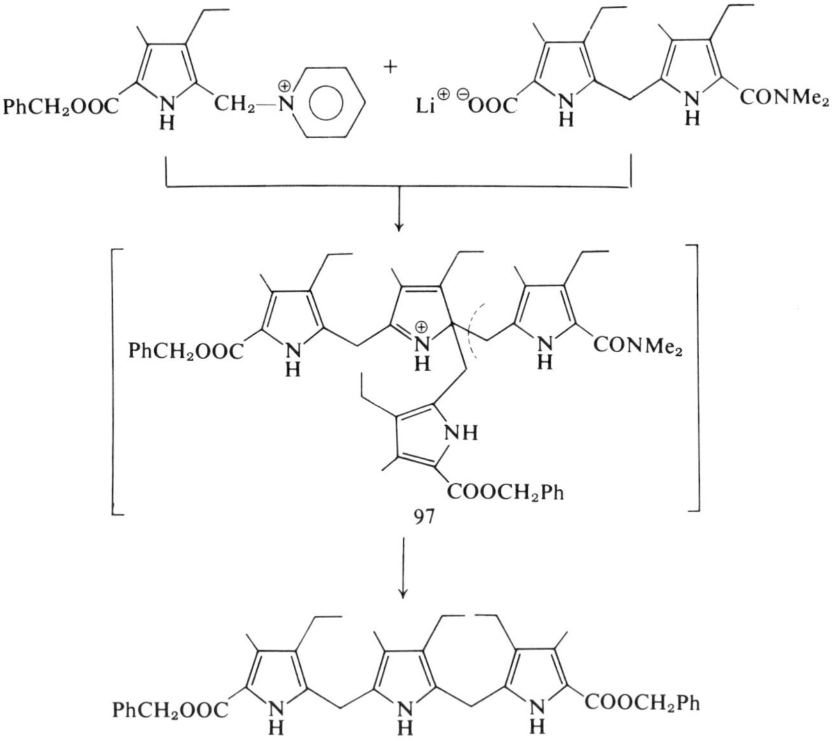

97

98

Tripyrranes, like dipyrrylmethanes, can react with electrophiles at the α-positions, which are linked by the methylene bridges. Such reactions may give rise to the formation of unexpected products (e.g., **98**) during tripyrrane synthesis from pyrrole derivatives. Generally electron-withdrawing groups promote the reaction leading to the desired products and prevent the formation of nonlinear adducts (e.g., **97**).[68]

When stabilized by electron-withdrawing groups, tripyrranes are readily crystallizable compounds. They can be conveniently isolated and purified by chromatography on silica gel. In analytical work, bromine vapor has been found to be useful in detecting and identifying tripyrranes, which give a characteristic violet color with this reagent on thin-layer chromatograms.[61]

3. SPECTROSCOPIC PROPERTIES OF THE TRIPYRRANES

In general, the spectroscopic properties of tripyrranes correspond to those of dipyrrylmethanes. The effects of substituents on the main uv-absorption of the chromophore obey the general rules valid for pyrrole derivatives (cf.

Section III, A, 3). Some examples can be found in the experimental parts of the works published by Ballantine *et al.*[35] and Broadhurst *et al.*[220]. The characteristic ^1H nmr absorption of the protons of the methylene bridges in tripyrranes lies in the 3.7–3.5 ppm range.[61,65,71,220]

Compounds **99** and **100** have been examinated in detail by mass spectrometry.[139] The parent peaks are relatively weak. By far, the most abundant ion in the spectra of both compounds arises by cleavage of the molecular ion at one of the methylene bridges. As in the case of dipyrrylmethanes, this process may occur with or without migration of a hydrogen atom, thus **100** yields the fragments **101** and **102**, respectively (cf. Section III, A, 3, d).

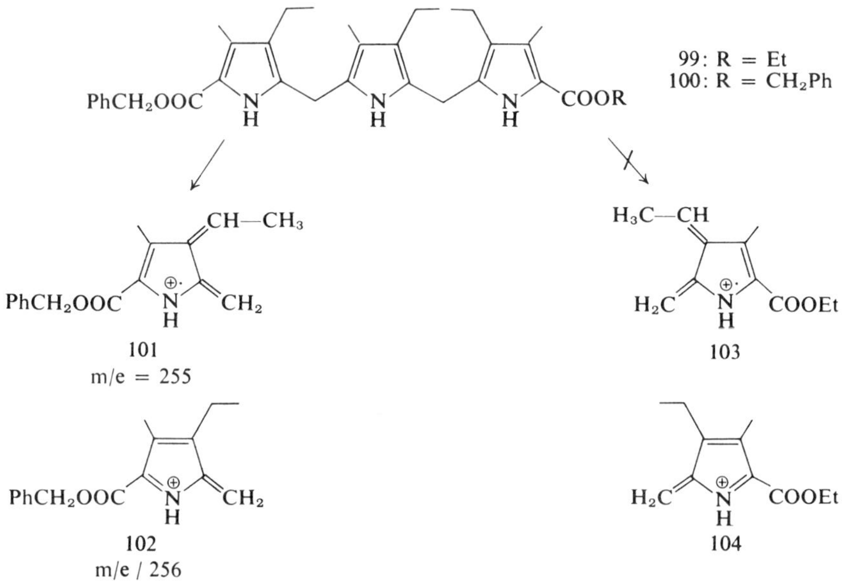
99: R = Et
100: R = CH₂Ph
PhCH₂OOC
COOR
CH—CH₃
CH₂
101
m/e = 255
H₃C—CH
H₂C
COOEt
103
102
m/e / 256
104

It is interesting to note that the benzyl ethyl diester (**99**) undergoes cleavage to give the fragments (**101** and **102**) but not the alternative fragments (**103**) and/or (**104**) bearing the ethoxycarbonyl group. This has been attributed to the greater ability of the benzyl group to assist in the stabilization of the positive charge in the fragment ions.

B. Tripyrrenes [5,16-Dihydro-(15*H*)-tripyrrins]

This class of linear polypyrroles is thus far interesting because they are useful intermediates for the stepwise synthesis of unsymmetrically substituted biladienes-ac [10,23-dihydro-(21H)-bilins], which are precursors of porphyrins and related macrocycles (cf. Scheme 7, p. 243).

1. Synthesis

Tripyrrenes can be synthesized by the following methods:

1. Condensation of an α-pyrrole aldehyde (e.g., **105**) with a dipyrryl-methane bearing an unsubstituted α-position (e.g., **106**) in the presence of mineral acid.[65,74,76,87,219]

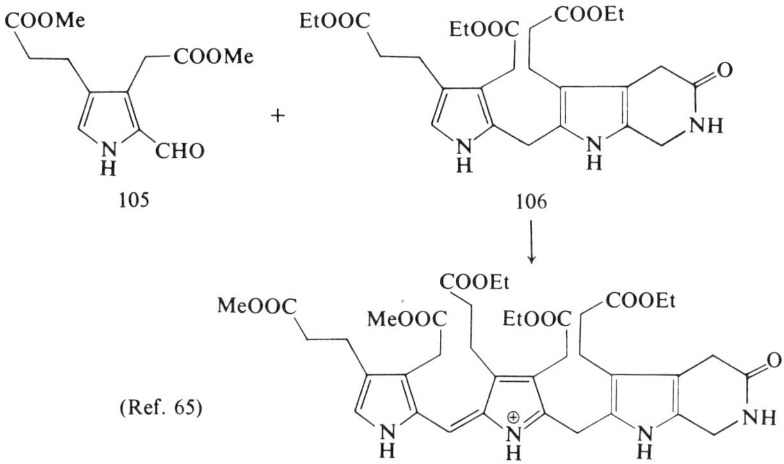

105

+

106

↓

(Ref. 65)

107

As dipyrrylmethane-α-carboxylic acids readily decarboxylate (cf. Section III, A, 2), they can be condensed directly with pyrrole aldehydes to pyrrylmethylpyrromethenes using toluene-p-sulfonic acid as catalyst.[116]

Owing to the deactivation towards electrophilic attack brought about by the pyrromethene moiety on the pyrrole ring of the pyrrylmethylpyrromethene molecule, α,α'-diunsubstituted dipyrrylmethanes react with pyrrole aldehydes, giving pyrrylmethylpyrromethenes in excellent yields (93–96%) without detectable formation of biladienes-ac as by-products.[111,112,224]

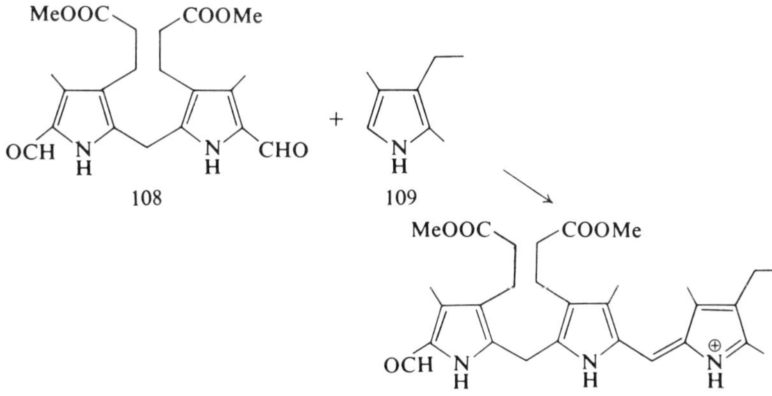

108 + 109

110

2. Condensation of α-formyldipyrrylmethanes (e.g., **108**) with α-unsubstituted pyrrole derivatives (e.g., **109**).[6] Dipyrrylmethane aldiminium salts, which are occasionally isolated as intermediates of the Vilsmeier–Haack reaction of dipyrrylmethanes, can be employed instead of the corresponding formyl compounds.[87]

3. Condensation of the α-methoxymethylpyrromethene (**111**) with cryptopyrrolecarboxylic acid (**112**) yielded the tripyrrene (**113**).[58]

2. CHEMICAL PROPERTIES OF THE TRIPYRRENES

Some reactions of the tripyrrenes reveal interesting peculiarities of this class of polypyrrole compounds. Thus, when the temperature is raised, tripyrrenes readily undergo autocondensation giving porphyrins.[224] This reaction, which is specially characteristic of tripyrrenes with electron-withdrawing substituents, evidently proceeds through formation of tripyrrylmethanelike intermediates, as pointed out in the case of pyrromethenes and α-unsubstituted pyrroles (cf. Section III, B, 1, a).

An important place in the study of the reactivity of tripyrrenes is occupied by the question of the possibility of a prototropic rearrangement in the series of these compounds. Evidently, the α-position of the pyrrole ring in tripyrrenes reacts readily in electrophilic substitution reactions, while the α-position of the pyrromethene moiety, on the contrary, will be deactivated to electrophilic reagents. In fact, both tripyrrenes (**114** and **115**) have been prepared by independent synthesis, and their ^{1}H nmr spectra showed that they are, indeed, individual compounds.

The characteristic differences can be followed according to the values of the chemical shifts of the proton in the 14-position. For the tripyrrene (**114**) it is

6.47 ppm; for the tripyrrene (115) 7.43–7.46 ppm. The prototropic rearrangement proceeds under relatively vigorous conditions. Thus, the tripyrrene (115), which, in contrast with its isomer (114), does not interact with 2-formyl-3,4-dimethylpyrrole under normal conditions, is converted to a biladiene-ac when it is boiled in methanol with a five-fold excess of HBr.[224] As mentioned before, tripyrrenes can be readily hydrogenated to the corresponding tripyrranes (cf. Section IV, A, 1).

3. SPECTROSCOPIC PROPERTIES OF THE TRIPYRRENES

Owing to the rather limited number of tripyrrenes that have been synthesized until now, few spectroscopic data are available for this class of compounds. The characteristic light absorption of tripyrrenes (as free bases) lies in the range of the excitation energy of the pyrromethene chromophore, i.e., near 455 nm (cf. Section III, B, 3, a). The corresponding hydrobromides absorb around 490 nm.[65,87]

As in the case of pyrromethenes, the ir spectra of tripyrrenes are characterized by a strong absorption band near 1600 cm^{-1}.[76] The ^{1}H nmr spectra of tripyrrene hydrobromides show two characteristic singlets at $\delta = 7.1$–7.4 and $\delta = 4.3$–4.5 ppm (relative to TMS) associated with the protons at the methine and methylene bridges, respectively.[76,225]

C. Tripyrrins

Hitherto only one member of this class of pyrrole pigments is reported in the literature. The tentative structure (116) was suggested for this compound, which was obtained as a by-product of the bromination of ethyl 2,4-dimethylpyrrole-3-carboxylate in acetic acid.[149,165] Under these conditions, however, formation of a prodigiosine-like compound could also be possible.[226]

116

V. TETRAPYRROLIC COMPOUNDS

A. Bilanes [Bilinogens or 5,10,15,22,23,24-Hexahydro-(21H)-bilins]

As already mentioned (Section IV, A) bilanes are probably involved as intermediates in the acid-catalyzed polycondensation of α,α'-diunsubstituted pyrroles with aldehydes to porphyrinogens which, on oxidation, finally yield porphyrins (cf. Chapter 3, Volume I). Presumably, the mechanism of formation of polypyrranes under these conditions is the same as that of dipyrrylmethanes; this has been considered in detail in Section III, A, 1. However, as the reaction does not stop at the tetrapyrrane stage, whose cyclization to porphyrinogens is a thermodynamic-favorable process,[227] tetrapyrranes are not isolated from the reaction mixtures from which porphyrinogens[129,228-230] or porphyrins can be obtained.

Bilanes carrying the substitution pattern of uroporphyrinogen III have been suggested as possible intermediates in the biosynthesis of the vitamin B_{12} chromophore.[231,232] On the other hand, a corresponding bilane is probably involved in the enzymatic synthesis of uroporphyrinogen I.[233,234] Some bilanes of the just mentioned types have been synthesized by means of chemical methods.[74,219,238]

1. SYNTHESIS

The synthesis of bilanes has been achieved by the following methods:

1. Hydrogenation of bilenes. Thus bilene-b derivatives (e.g., **117**) can be reduced to the corresponding bilanes (e.g., **118**) in the presence of palladium as catalyst.[07,110,116,235] Recently, Franck and Rowold[219] reported the synthesis of a bilane related to uroporphyrin I via catalytic hydrogenation of the corresponding bilene-a. The obtained product was cyclized, without isolation, to uroporphyrinogen I.

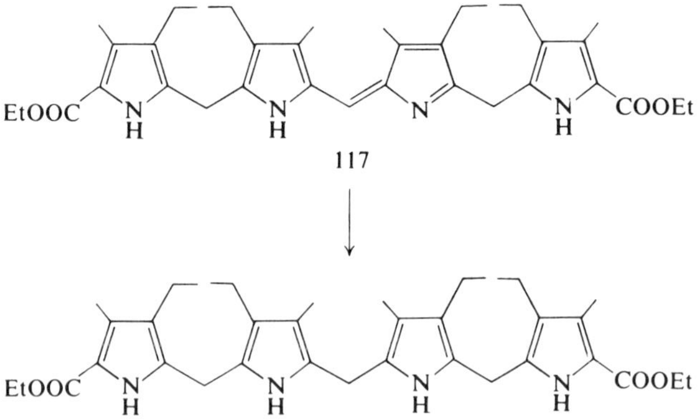

117

(Ref. 235)

118

2. Reduction of a-oxobilanes (e.g., **119**) or alternatively b-oxobilanes with diborane in tetrahydrofuran ethyl acetate.[87,89,90,236,237]

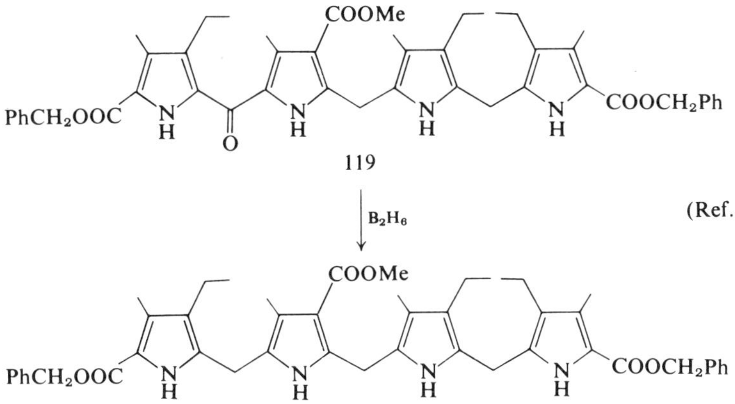

119

B₂H₆

(Ref. 237)

120

3. Catalytic hydrogenation (PtO₂) of biladiene-ac hydrobromides (e.g., **121**) in methanolic solution in the presence of a base (e.g., morpholino-methyl polystyrene). Under these conditions the corresponding bila-trienes (21,24-dihydrobilins) are actually reduced.

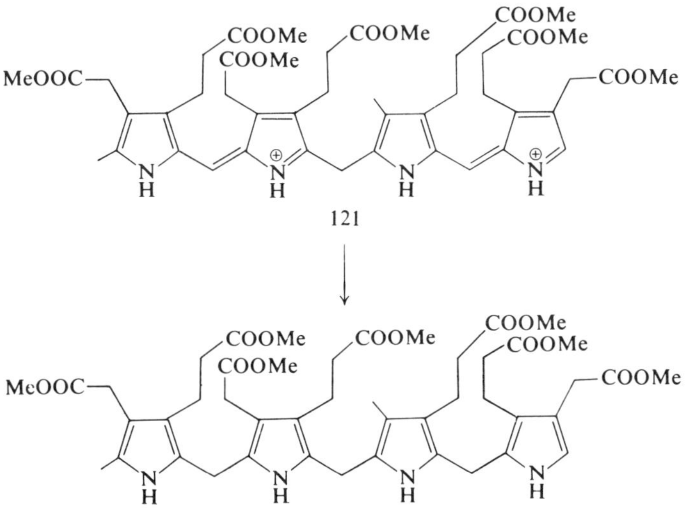

121

122

By this procedure which, until now, represents the most reliable one for synthesizing bilanes, several derivatives related to uroporphyrinogen III and 12-decarboxyuroporphyrinogen III (e.g., **122**) bearing ester or methyl groups as well as hydrogen atoms at the 1,19-positions have been synthesized and fully characterized spectroscopically.[74,238]

2. CHEMICAL PROPERTIES OF THE BILANES

As bilanes are usually very sensitive to oxygen and acids that bring about oxidation and jumbling reactions[87,227,239] (cf. Section III, A, 2), their handling requires exceptionally careful techniques. Consequently, very little is known about their chemical properties, which presumably ought to correspond to those of dipyrrylmethanes and tripyrranes. In fact, only some derivatives bearing electron-withdrawing substituents (preferably alkoxy-carbonyl groups) at the terminal α-positions have been isolated, until now, as crystalline compounds relatively stable in air.[235,238]

In spite of earlier attempts to cyclize bilanes to porphyrins, in which only mixtures of isomers could be isolated, the chemical transformation of a synthetic bilane, having the constitution of a porphobilinogen tetramer, into uroporphyrinogen I has been recently reported. Under the conditions employed (phosphate buffer at 32°C), no isomerization of the substrate occurs.[219] On the other hand, bilanes can be transformed, without isomerization, into porphyrins when they are carefully oxidized to bilenes-b and the crude products readily cyclized.[236]

3. SPECTROSCOPIC PROPERTIES OF THE BILANES

From the scarce spectroscopic data that are avilable for bilanes, the following generalizations can be made. (a) Their electronic spectra resemble those of dipyrrylmethanes and tripyrranes (cf. Section III, A, 3, a). (b) The protons at the methylene bridges absorb near $\delta = 3.8$ ppm (relative to TMS).[238] It is noteworthy that, when measured in $CDCl_3$, four discrete broad signals are observable for the protons bonded to the nitrogen atoms in the ^1H nmr spectra of bilanes bearing slightly different substituents at the pyrrole rings.[238] (c) The main features of the mass spectra of bilanes correspond to those of dipyrrylmethanes and tripyrranes. In most cases, a weak molecular ion is observable. The most favorable process involves cleavage at the methylene bridges giving pyrrole, dipyrrylmethane, and tripyrrane fragments.[238]

B. Bilenes-a [5,10,22,24-Tetrahydro-(21H)-bilins]

1. SYNTHESIS

Bilenes-a are accessible by the following methods.

1. Condensation of the methoxymethylpyrromethene (**123**) with the dipyrrylmethane-α-carboxylic acid (**124**).[58]

2. Acid-catalyzed condensation of a tripyrrane-α-carboxylic acid (e.g., **125**) with an α-formylpyrrole (e.g., **126**).[87,222] By this method, a bilene-a

(Ref. 222)

related to uroporphyrin I has been recently synthesized.[219] The product was characterized by field desorption (FD) mass spectrometry.

2. PROPERTIES OF BILENES-a

Bilenes-a are rather unstable compounds whose properties have not been investigated in detail.

The cyclization of bilenes-a to porphyrins has been found to yield mixtures of isomers,[240] probably because they are unstable under the prevailing acidic conditions and undergo redistribution reactions. This instability limits the usefulness of bilenes-a as intermediates for porphyrin synthesis, although a monomethoxyporphyrin has been prepared by this route.[222]

In acidic solution, bilenes-a bearing electron-withdrawing substituents (e.g., CO_2Et, Ac) at the β-positions are probably formed by a prototropic rearrangement of bilenes-b.[241] The transformation is associated with a change of the light absorption of the solution from $\lambda_{max}^{CHCl_3} = 525$ nm to $\lambda_{max}^{CHCl_3} = 490$ nm (for the bilene-b and bilene-a chromophores, respectively). ^1H nmr data support these assignments, thus the chemical shift of the methine proton changes from $\delta = 8.4$ ppm (for the bilene-b) to $\delta = 7.5$ ppm (for the bilene-a), whereas the two singlets at $\delta = 4.4$ and 4.3 ppm associated with the protons at the methylene bridges of the bilene-b appear at $\delta = 4.7$ and 3.8 ppm in the case of the bilene-a.

C. Bilenes-b [5,15,22,24-Tetrahydro-(21H)-bilins]

1. SYNTHESIS

Bilenes-b can be prepared by the following methods:

1. Dehydrogenation of bilanes (e.g., **127**) (cf. Section V, A, 3) by the action of *tert*-butyl hypochlorite.[87,89,90,236,237,242] The products obtained by this method are often contaminated by small amounts of the corresponding bilenes-a.

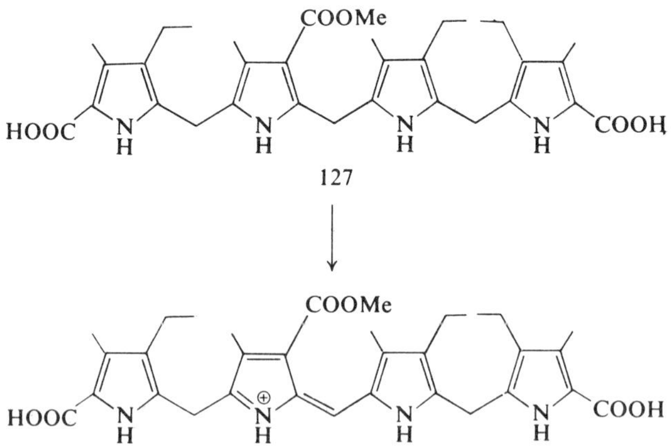

127

(Ref. 237)

2. Reaction of 5,5′-dimethoxymethylpyrromethenes (e.g., **128**) with two equivalents of an α-unsubstituted pyrrole (e.g., **129**).[22,243,244]

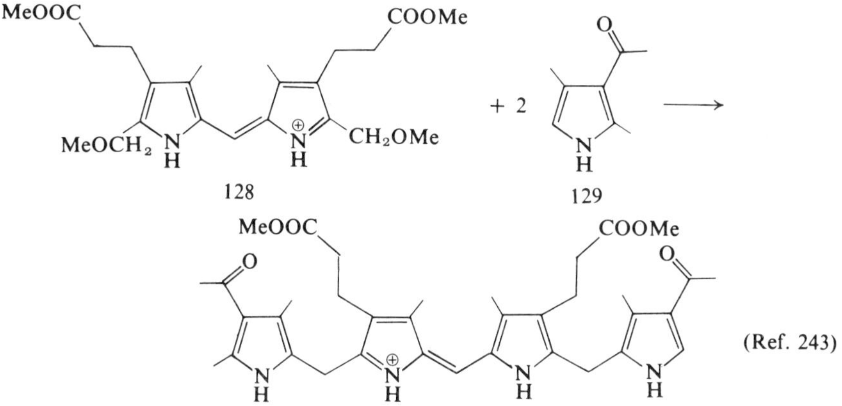

(Ref. 243)

3. Condensation of an α-formyldipyrrylmethane (e.g., **130**) with an α-unsubstituted dipyrrylmethane or, alternatively, the corresponding α-carboxylic acid (e.g., **131**).[2–4,87,116,225,241,245,247,248] As dipyrrylmethane-

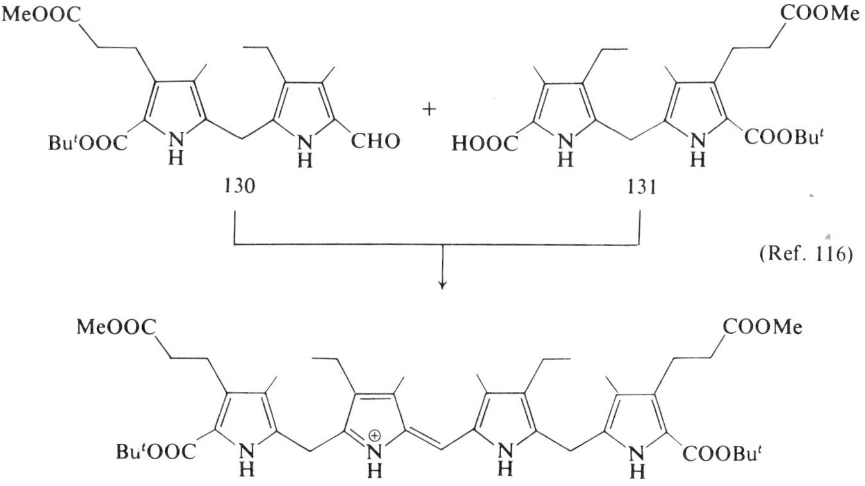

(Ref. 116)

α-carboxylic acids are, in general, less reactive than the corresponding α-unsubstituted derivatives, self-condensation of the employed dipyrryl-methane aldehyde with elimination of one formyl group (cf. Section III, B, 1) may occur in some cases. The obtained compound is, therefore, a symmetrically substituted bilene-b instead of the expected condensation product of both reactants.[225]

On the other hand, both the dipyrrylmethanecarboxylic acid and the dipyrrylmethane aldehyde should bear at least one electronegative group as a

ring substituent in order to avoid side reactions promoted by the acid catalyst. Otherwise tripyrrenes, rather than bilenes-b, can be obtained as reaction
products.[225]

Condensation of α-acyldipyrrylmethanes instead of α-formyldipyrrylmethanes with α-unsubstituted dipyrrylmethanes leads to the formation of
10-substituted bilenes-b. In the particular case of compound (132), which was
prepared as an intermediate for the synthesis of deoxophylloerythroetio-
porphyrin,[88] the more reactive formyl group at C-1 was previously transformed into the corresponding aldimine in order to ensure the formation of
the desired condensation product (133).

132   133

2. PROPERTIES OF THE BILENES-b

As in the case of bilenes-a, very few details about the chemistry and physical
properties of the b-isomers are as yet available. As mentioned before (cf.
Section V, C, 1), bilenes-b, bearing electron-withdrawing substituents at the
β-positions, probably rearrange to bilenes-a under acidic conditions, such isomerizations being associated with changes of the electronic and ^{1}H nmr spectra
of the corresponding compounds. As the methylene bridges of bilenes-b are
in a similar environment to the methylene bridge of biladienes-ac (Section
V, E, 2), it would be expected that the respective hydrogen atoms are acidic
enough to bring about prototropic changes of the molecular structure.

Such rearrangements are probably responsible for the formation of porphyrin mixtures on cyclization of bilenes-b carrying electronegative substituents, for the resulting bilenes-a are less stable and jumbling reactions may
occur.[241]

It is noteworthy that bilenes-b bearing alkyl groups at the β-positions cyclize in high yield to porphyrins, the reaction products being homogeneous. As no change in the light absorption ($\lambda_{max}^{CHCl_3} = 508$ nm) of bilenes-b bearing alkyl groups at the β-positions was observed in acidic media, no rearrangement to bilenes-a should occur, owing to the somewhat lower acidity of the methylene hydrogen atoms in these compounds.[241]

D. Biladienes-ab [5,23-Dihydro-(21*H*)-bilins]

Until now no member of this class of tetrapyrrole pigments has been conclusively characterized. The facile conversion of the formiminobilene-b dihydrochloride (**133**), whose main absorption band lies at $\lambda_{max}^{95\%\,EtOH} = 475$ nm, into a product with a different visible spectrum ($\lambda_{max}^{95\%\,EtOH} = 469$ nm), believed to be the corresponding biladiene-ab, is mentioned once in the literature.[88]

E. Biladienes-ac [10,23-Dihydro-(21*H*)-bilins]

By far the most useful linear tetrapyrrolic intermediates, from the point of view of the synthesis of porphyrins and related macrocycles, are the biladienes-ac (cf. Scheme 7).

1. SYNTHESIS

Many derivatives in this series have been synthesized by means of the following methods:

1. Condensation of two equivalents of an α-formylpyrrole (e.g., **134**) with a 5,5'-diunsubstituted dipyrrylmethane or, alternatively, with the corresponding α,α'-dicarboxylic acid (e.g., **135**).[22,74–76,101,106,240,248,249,251–255]

2. Condensation of a 5,5'-diformyldipyrrylmethane (e.g., **136**) with two equivalents of an α-unsubstituted pyrrole.[105,249–251,253,254,256] 5,5'-

Diacetyldipyrrylmethanes yield, under analogous conditions, 5,15-dimethyl-10,23-dihydrobilins.[106]

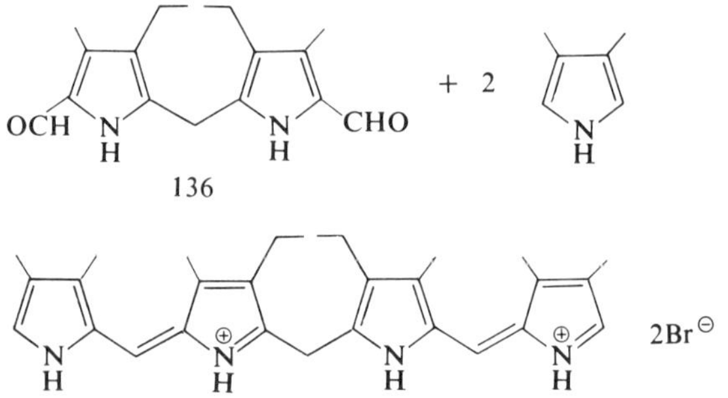

(Ref. 249)

3. Condensation of an α-unsubstituted tripyrrene (e.g., **137**) with an α-formylpyrrole (e.g., **138**).[74–76,106,111,114,224,257,258]

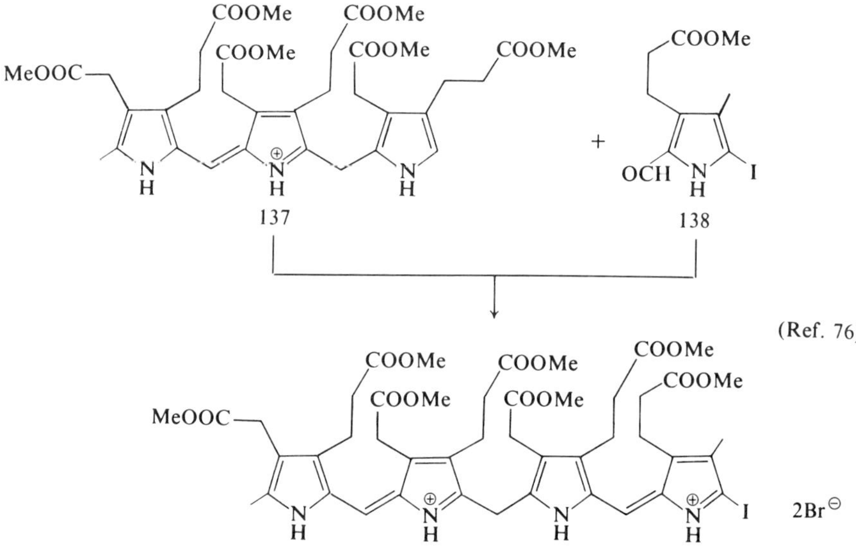

(Ref. 76)

4. Reaction of a 5-bromo-5'-bromomethylpyrromethene (e.g., **139**) with a 5-unsubstituted 5'-alkylpyrromethene (e.g., **140**) in the presence of tin tetrachloride in methylene chloride or nitromethane. The obtained tin biladiene-ac complexes are subsequently demetalated by treatment with hydrobromic acid in methanol.[10,251,260–265] Of course, it must be pointed out that only the two last methods enable the synthesis of biladiene-ac derivatives bearing different substituents at all the pyrrole rings.

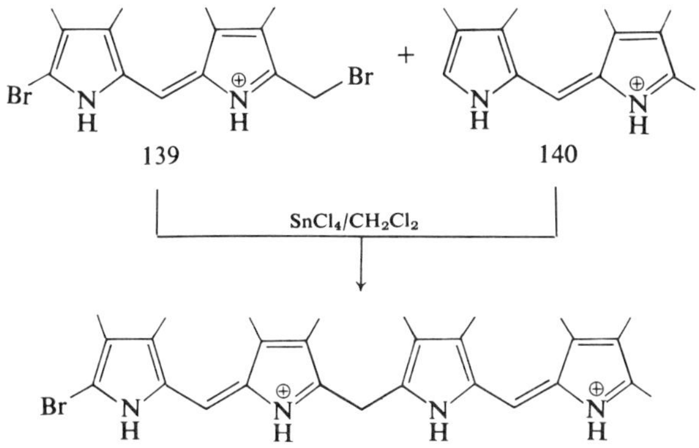

139   +   140

SnCl$_4$/CH$_2$Cl$_2$ (Ref. 261)

Whereas the mechanism of reactions 1–3 can be rationalized on the basis of the acid-catalyzed (preferably by HBr) synthesis of pyrromethenes, which has been considered in detail in Section III, B, 1, the formation of 10,23-dihydrobilins by reaction 4 represents a unique example of an electrophilic substitution at a pyrromethene ring position (cf. Section III, B, 2), which proceeds very smoothly in usually high yields.

2. PROPERTIES OF THE BILADIENES-ac

With the aid of the above mentioned reactions, biladienes-ac are obtained as crystalline red dihydrobromides (e.g., **141**). On treatment with weak bases such as alcohols or dimethylsulfoxide, the corresponding yellow mono-protonated species are obtained; the latter must be formulated as 22,23-dihydro-21H-bilin hydrobromides (e.g., **142**). In the presence of stronger bases (e.g., piperidine,[195] and morpholinomethyl polystyrene[74,76,238]), the corresponding green 22,24-dihydro-21H-bilin free bases (e.g., **143**) are formed.[195] The cleavage of a proton from the C-10 methylene bridge is

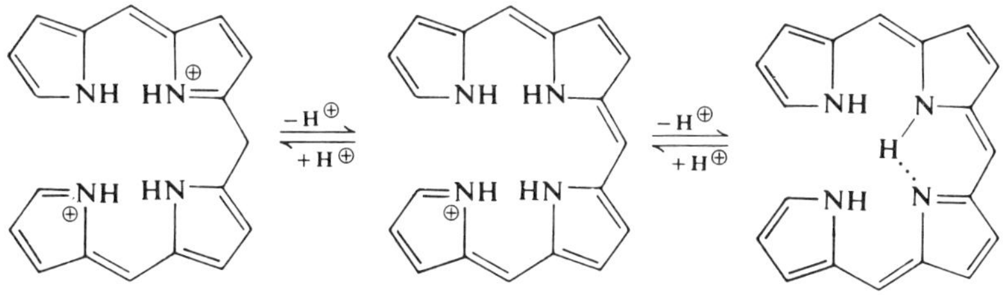

141   142   143

Scheme 6. Deprotonation of biladiene-ac salts by bases. (For the sake of clarity, substituents are omitted.)

facilitated by electron-withdrawing substituents (e.g., nitro or ester groups) in the molecule. Thus, examples of the conversion of biladiene-ac salts into bilatriene free bases are observed in such cases, even in chloroform.[195,240]

Furthermore, the mobility of the methylene hydrogen atoms was demonstrated in the case of a 1,19-diethoxycarbonyl biladiene-ac (**144**) by H–D exchange experiments. Thus, in deuterated trifluoroacetic acid solution, the resonance associated with the C-10 protons becomes 25% weaker after 20 minutes and disappears after 17 hours at room temperature.[195] As Scheme 7

144

F_3COOD

shows, several kinds of biladiene-ac derivatives can be transformed into different macrocylic tetrapyrroles such as porphyrins,[10-13,22,74,76,111,112,114,151,240,248,251,252,260-266] azaporphyrins,[10,75] corroles,[75,195,267,268] 1-alkyl tetradehydrocorrin metal chelates,[74,75,261] and metal tetradehydrocorrin salts bearing alkyl,[74,106,251,255,257] alkoxy,[269] or alkoxycarbonyl groups[106,254] at the 1,19-positions (cf. Chapter 10, Volume II). Despite the significance of these reactions from the mechanistic point of view, very little is known about the structure of the involved intermediates. Particularly, the acidity of the methylene hydrogen atoms of biladiene-ac salts and their corresponding metal complexes seem to play a decisive role on their base-catalyzed oxidative cyclization to 1,19-disubstituted metal tetradehydrocorrin salts.[195] Biladienes-ac bearing two methyl groups at C-10, which are, therefore, unable to loose a proton from the methylene bridge, do not cyclize to metal tetradehydrocorrin salts.[250]

As 1,19-disubstituted biladiene-ac metal complexes can also be transformed into metal tetradehydrocorrin salts by cleavage of a hydride ion from the methylene bridge,[270] it may be speculated that these reactions involve an electrocyclic isomerization of the corresponding bilin (cf. Section V, G), which are systems with 24π-electrons and, therefore, isoelectronic with the 1,19-disubstituted tetradehydrocorrins. In fact, the formation of a trans configurated nickel 1,19-dimethyl tetradehydrocorrin bromide[270] agrees with

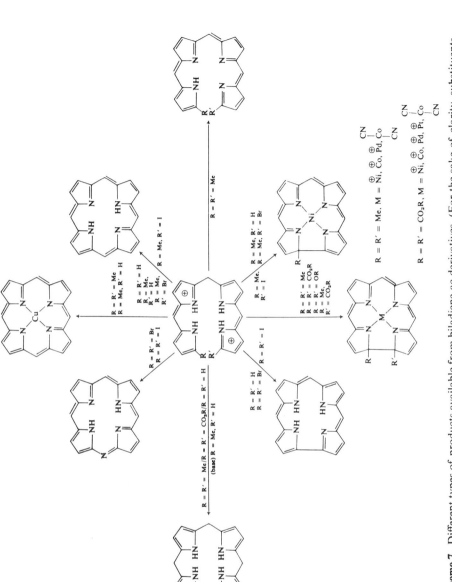

Scheme 7. Different types of products available from biladiene-ac derivatives. (For the sake of clarity, substituents at the β-positions are omitted.)

a thermally allowed conrotatory ring closure. Certainly, the influence of the metal ions in such reactions (cyclizations have been observed only in the presence of bivalent nickel,[250] cobalt,[250] palladium,[271] and platinum ions[259]) is, as yet, not clearly understood (cf. Chapter 8 in Volume II). On the other hand, the 1,19-disubstituted tetradehydrocorrin macrocycle can be reopened to 1,19-disubstituted biladienes-ac by gentle reducing agents.[272] Under the conditions of "reductive methylation," the corresponding 10-methyl derivatives are obtained.[273]

3. SPECTROSCOPIC PROPERTIES OF THE BILADIENES-ac

As mentioned above, the uv–vis spectra of biladienes-ac are strongly dependent on the proton concentration of the medium, the observed changes in

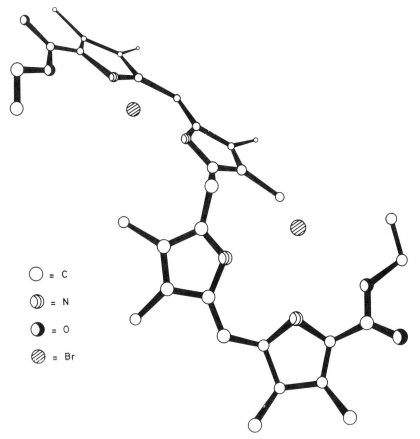

Fig. 1. Perspective drawing of the structure of 1,19-diethoxycarbonyl-2,3,7,8,12,13,17,18-octamethylbiladiene-ac dihydrobromide.

color of the solutions being associated with changes of the structure of the chromophore (cf. Section V, E, 2). Owing to the presence of two pyrromethene moieties, the ir spectra of biladienes-ac show a characteristic absorption band near 1600 cm^{-1}.[254] When measured in CDCl$_3$, symmetrically substituted biladiene-ac dihydrobromides show characteristic ^{1}H nmr absorptions at $\delta = 7.3$ ppm (protons at C-5 and C-15) and 5.3 ppm (protons at C-10). When 1 mole of piperidine is added, the corresponding protons of the resulting bilatriene hydrobromide absorb at $\delta = 6.7$ and $\delta = 5.8$, respectively. In the presence of five equivalents of piperidine, the ^{1}H nmr spectrum of the bilatriene-free base is observed with signals at $\delta = 5.9$ ppm and $\delta = 5.3$ ppm, respectively.

In the case of unsymmetrically substituted bilatriene-abc free bases, three signals are observed at $\delta = 5.4$, 5.8, and 6.1 ppm corresponding to the protons of the methine bridges.[76] As mentioned before, exchange of the protons at C-10 by deuterium occurs in acidic media.

Because of the very low volatility of biladiene-ac salts, no molecular ion is usually observed in the mass spectra of these compounds. In some cases, the corresponding zinc complexes are more suitable for characterization.[274]

Figure 1 shows the structure of 1,19-diethoxycarbonyl octamethylbiladiene-ac dihydrobromide as determined by X-ray diffraction. The dihedral angle between the planes of the two pyrromethene moieties linked together by the methylene bridge equals 107.2°.[275]

F. Bilatrienes-abc [22,24-Dihydro-(21H)-bilins]

As mentioned before, biladienes-ac free bases occur as 22,24-dihydro-21H-bilin tautomers. The electron spectra of bilatrienes and their salts (cf. Section V, E, 2) resemble those of phlorin free bases and their salts, respectively, the

145

hν/MeOH/NH$_4$OH

146

chromophores of both types of compounds being similar.[195,276] In contrast to the corresponding protonated species, the less polar bilatriene free bases can be conveniently purified by chromatography.[76,238] Bilatrienes can be readily hydrogenated catalytically to the corresponding bilanes (cf. Section V, A, 2).

G. Bilins

Until now only one member of this class of linear tetrapyrroles is known. Compound (146) was obtained by ultraviolet irradiation of the 10,23-dihydrobilin (145) in methanolic solution. It crystallizes as blue needles ($\lambda_{max}^{CHCl_3} =$ 305, 385, 705 nm). No further properties of this compound have been reported.[101]

REFERENCES

1. L. I. Fleiderman, A. F. Mironov, and R. P. Evstigneeva, *Zh. Obshch. Khim.* **45**, 197 (1975).
2. P. S. Clezy and A. J. Liepa, *Chem. Commun.* p. 767 (1969).
3. V. D. Rumyantseva, A. F. Mironov, and R. P. Evstigneeva, *Zhr. Org. Khim.* **7**, 828 (1971).
4. P. S. Clezy, A. J. Liepa, and N. W. Webb, *Aust. J. Chem.* **25**, 1991 (1972).
5. L. I. Fleiderman, A. F. Mironov, and R. P. Evstigneeva, *Zh. Obshch. Khim.* **43**, 886 (1973).
6. L. I. Fleiderman, A. F. Mironov, and R. P. Evstigneeva, *Khim. Geterotsikl. Soedin.* p. 1146 (1973).
7. A. H. Corwin, *Heterocycl. Compd.* **1**, 277 (1950).
8. T. S. Stevens, *in* "Chemistry of Carbon Compounds" (E. H. Rodd, ed.), Vol. 4/A, p. 28. Elsevier, Amsterdam, 1957.
9. E. Baltazzi and L. I. Krimen, *Chem. Rev.* **63**, 511 (1963).
10. R. L. N. Harris, A. W. Johnson, and I. T. Kay, *Q. Rev., Chem. Soc.* **20**, 211 (1966).
11. A. W. Johnson, *Chem. Br.* **3**, 253 (1967).
12. A. W. Johnson, *Pure Appl. Chem.* **23**, 375 (1970).
13. A. W. Johnson, *Pedler Lect.* (1974), *Chem. Soc. Rev.* **4**, 1 (1975).
14. H. H. Inhoffen, J. W. Buchler, and P. Jäger, *in* "Progress in the Chemistry of Organic Natural Products" (L. Zechmeister, ed.), Vol. **26**, p. 284. Springer, Vienna and New York, 1968.
15. T. A. Melent'eva, N. D. Pekel, and V. M. Berezovskii, *Russ. Chem. Rev. (Engl. Transl.)* **38**, 926 (1969).
16. K. M. Smith, *Q. Rev., Chem. Soc.* **25**, 31 (1971).
17. A. H. Jackson and K. M. Smith, *in* "The Total Synthesis of Natural Products" (J. Ap Simon, ed.), p. 143. Wiley, New York, 1973.
18. H. Fischer and H. Orth, "Die Chemie des Pyrrols," Vol. I, Akad. Verlagsges., Leipzig, 1934 (reprint by Johnson Reprint Corp., New York, 1968).
19. H. Fischer and H. Orth, "Die Chemie des Pyrrols," Vol. II, Akad. Verlagsges., Leipzig, 1937 (reprint by Johnson Reprint Corp., New York, 1968).
20. A. Gossauer, "Die Chemie der Pyrrole," Springer-Verlag, Berlin and New York, 1974.
21. R. Lemberg and J. W. Legge, "Hematin Compounds and Bile Pigments," p. 105. Wiley (Interscience), New York, 1949.

22. A. W. Johnson and I. T. Kay, *J. Chem. Soc.* p. 2418 (1961).
23. A. Pictet and A. Rilliet, *Ber. Dtsch. Chem. Ges.* **40**, 1170 (1907).
24. R. Chong, P. S. Clezy, A. J. Liepa, and A. W. Nichol, *Aust. J. Chem.* **22**, 229 (1969).
25. G. P. Arsenault, E. Bullock, and S. F. MacDonald, *J. Am. Chem. Soc.* **82**, 4384 (1960).
26. R. B. Woodward, *Angew Chem.* **72**, 651 (1960).
27. R. B. Woodward, W. A. Ayer, J. M. Beaton, J. M. Bickelhaupt, R. Bonnett, P. Buchschacher, G. L. Closs, H. Dutler, J. Hannah, F. P. Hauck, S. Itô, A. Langemann, G. LeGoff, W. Leimgruber, W. Lwowski, J. Sauer, Z. Valenta, and H. Volz, *J. Am. Chem. Soc.* **82**, 3800 (1969).
28. J. Pluscec and L. Bogorad, *Biochemistry* **9**, 4736 (1970).
29. R. B. Frydman, A. Valasinas, and B. Frydman, *Biochemistry* **12**, 80 (1973).
30. R. B. Frydman, A. Valasinas, H. Rapoport, and B. Frydman, *FEBS Lett.* **25**, 309 (1972).
31. A. R. Battersby, K. H. Gibson, E. MacDonald, C. N. Mander, and J. Moron, *J. Chem. Soc., Chem. Commun.* p. 768 (1973).
32. A. Rowold, Dissertation, Münster (1974).
33. B. Frydman and R. B. Frydman, *Acc. Chem. Res.* **8**, 201 (1975).
34. J. M. Osgerby and S. F. MacDonald, *Can. J. Chem.* **40**, 1585 (1962).
35. J. A. Ballantine, A. H. Jackson, G. W. Kenner, and G. MacGillivray, *Tetrahedron, Suppl.* **7**, 241 (1966).
36. A. Treibs and H. Derra-Scherer, *Justus Liebigs Ann. Chem.* **589**, 188 (1954).
37. B. V. Gregorovich, K. S. Y. Liang, D. M. Clugston, and S. F. MacDonald, *Can. J. Chem.* **46**, 3291 (1968).
38. G. G. Kleinspehn and A. H. Corwin, *J. Am. Chem. Soc.* **76**, 5641 (1954).
38a. G. M. Badger, R. L. N. Harris, and R. A. Jones, *Aust. J. Chem.* **17**, 1002 (1964).
39. A. H. Corwin and W. M. Quattlebaum, Jr., *J. Am. Chem. Soc.* **58**, 1081 (1936).
40. A. H. Corwin, W. A. Bailey, Jr., and P. Viohl, *J. Am. Chem. Soc.* **64**, 1267 (1942).
41. A. H. Corwin and J. L. Straughn, *J. Am. Chem. Soc.* **70**, 1416 (1948).
42. P. S. Clezy, C. J. R. Fookes, D. J. K. Lau, A. W. Nichol, and G. A. Smythe, *Aust. J. Chem.* **27**, 357 (1974).
43. G. M. Badger, R. L. N. Harris, and R. A. Jones, *Aust. J. Chem.* **17**, 987 (1964).
44. P. A. Burbidge, G. L. Collier, A. H. Jackson, and G. W. Kenner, *J. Chem. Soc. B* p. 930 (1967).
45. H. Shinohara, K. Honda, and E. Imoto, *Nippon Kagaku Zasshi* **81**, 1163 (1960); *Chem. Abstr.* **56**, 3440 (1962).
46. J. P. Nagarkatti and K. R. Ashley, *Synthesis* p. 186 (1974).
47. M. Strell, A. Zocher, and E. Kopp, *Chem. Ber.* **90**, 1798 (1957).
48. A. Treibs and H. G. Kolm, *Justus Liebigs Ann. Chem.* **614**, 176 (1958).
49. A. H. Corwin and W. M. Quattlebaum, Jr., *J. Am. Chem. Soc.* **64**, 922 (1942).
50. E. I. Filippovich, T. A. Palagina, T. S. Postnikova, R. P. Evstigneeva, and N. A. Preobrazhenskii, *Khim Geterosikl. Soedin.* p. 734 (1954).
51. E. I. Filippovich, V. N. Luzgina, R. P. Evstigneeva, and N. A. Preobrazhenskii, *Zh. Obshch. Khim.* **33**, 2130 (1963).
52. E. I. Filippovich, V. N. Luzgina, L. I. Korsantseva, R. P. Evstigneeva, and N. A. Preobrazhenskii, *Zh. Obshch. Khim.* **36**, 1383 (1966).
53. E. I. Filippovich, T. N. Pilipenko, T. V. Demidkina, and N. A. Preobrazhenskii, *Khim. Geterosikl. Soedin.* p. 1045 (1970).
54. F. Morsingh and S. F. MacDonald, *J. Am. Chem. Soc.* **82**, 4377 (1960).

55. E. J. Tarlton, S. F. MacDonald, and E. Baltazzi, *J. Am. Chem. Soc.* **82**, 4389 (1960).
56. A. F. Mironov, R. P. Evstigneeva, A. V. Chumachenko, and N. A. Preobrazhenskii, *Zh. Obshch. Khim.* **34**, 1488 (1964).
57. A. F. Mironov, V. D. Rumyantseva, and R. P. Evstigneeva, *Zh. Org. Khim.* **9**, 642 (1973).
58. J. Ellis, A. H. Jackson, A. C. Jain, and G. W. Kenner, *J. Chem. Soc.* p. 1935 (1964).
59. Y. Sklyar, E. Sklyar, R. P. Evstigneeva, and N. A. Preobrazhenskii, *Zh. Org. Khim.* **1**, 167 (1965).
60. G. L. Collier, A. H. Jackson, and G. W. Kenner, *J. Chem. Soc. C* p. 66 (1967).
61. P. S. Clezy and A. J. Liepa, *Aust. J. Chem.* **23**, 2443 (1970).
62. V. N. Luzgina, E. I. Filippovich, and R. P. Evstigneeva, *Zh. Obshch. Khim.* **41**, 2294 (1971).
63. T. A. Melent'eva, L. V. Kazanskaya, and V. M. Berezovskii, *Zh. Obshch. Khim.* **41**, 921 (1971).
64. L. V. Kazanskaya, T. A. Melent'eva, and V. M. Berezovskii, *Zh. Obshch. Khim.* **42**, 2523 (1972).
65. J. M. Osgerby, J. Pluscec, J. M. Kim, F. Boyer, N. Stojanac, H. D. Mah, and S. F. MacDonald, *Can. J. Chem.* **50**, 2652 (1972).
66. R. B. Frydman, A. Valasinas, H. Rapoport, and B. Frydman, *FEBS Lett.* **25**, 309 (1974).
67. B. Frydman, S. Reil, A. Valasinas, R. B. Frydman, and H. Rapoport, *J. Am. Chem. Soc.* **93**, 2738 (1971).
68. A. Hayes, G. W. Kenner, and N. R. Williams, *J. Chem. Soc.* p. 3779 (1958).
69. H. Fischer, E. Baumann, and H. J. Riedl, *Justus Liebigs Ann. Chem.* **475**, 237 (1929).
70. A. M. d'A. Rocha Gonsalves, G. W. Kenner, and K. M. Smith, *Tetrahedron Lett.* p. 2203 (1972).
71. J. A. S. Cavaleiro, A. M. d'A. Rocha Gonsalves, G. W. Kenner, and K. M. Smith, *J. Chem. Soc., Perkin Trans.* 1 p. 2471 (1973).
72. G. W. Kenner and K. M. Smith, *Ann. N.Y. Acad. Sci.* **206**, 138 (1973).
73. J. A. S. Cavaleiro, G. W. Kenner, and K. M. Smith, *J. Chem. Soc., Perkin Trans.* 1 p. 1188 (1974).
74. J. Engel and A. Gossauer, *J. Chem. Soc., Chem. Commun.* p. 570 (1975).
75. J. Engel and A. Gossauer, *J. Chem. Soc., Chem. Commun.* p. 713 (1975).
76. J. Engel and A. Gossauer, *Justus Liebigs Ann. Chem.*, p. 1637 (1976).
77. P. S. Clezy and V. Diakiw, *Aust. J. Chem.* **26**, 2697 (1973).
78. A. R. Battersby, E. Hunt, M. Ihara, E. McDonald, J. B. Paine, III, F. Sato, and J. Saunders, *J. Chem. Soc., Chem. Commun.* p. 994 (1974).
79. A. W. Johnson, I. T. Kay, E. Markham, R. Price, and K. B. Shaw, *J. Chem. Soc.* p. 3416 (1959).
80. A. M. Fargali, R. P. Evstigneeva, and N. A. Preobrazhenskii, *Zh. Obshch. Khim.* **34**, 898 (1964).
81. A. F. Mironov, T. R. Ovsepyon, R. P. Evstigneeva, and N. A. Preobrazhenskii, *Zh. Obshch. Khim.* **35**, 324 (1965).
82. A. M. d'A. Rocha Gonsalves, G. W. Kenner, and K. M. Smith, *Chem. Commun.* p. 1304 (1971).
83. S. F. MacDonald, *J. Chem. Soc.* p. 4176 (1952).
84. S. F. MacDonald and D. M. MacDonald, *Can. J. Chem.* **33**, 573 (1955).
85. R. J. Motekaitis, D. H. Heinert, and A. E. Martell, *J. Org. Chem.* **35**, 2504 (1970).
86. A. H. Jackson, G. W. Kenner, and D. Warburton, *J. Chem. Soc.* p. 1328 (1965).
87. A. H. Jackson, G. W. Kenner, and G. S. Sach, *J. Chem. Soc. C* p. 2045 (1967).

88. M. E. Flaugh and H. Rapoport, *J. Am. Chem. Soc.* **90**, 6877 (1968).
89. P. J. Crook, A. H. Jackson, and G. W. Kenner, *J. Chem. Soc. C* p. 474 (1971).
90. R. P. Carr, A. H. Jackson, G. W. Kenner, and G. S. Sach, *J. Chem. Soc. C* p. 487 (1971).
91. A. A. Ponomarev, I. M. Skvortsov, and A. A. Khorkin, *Zh. Obshch. Khim.* **33**, 2687 (1963).
92. A. Treibs and H. Bader, *Chem. Ber.* **91**, 2615 (1958).
93. A. Treibs and G. Fritz, *Justus Liebigs Ann. Chem.* **611**, 162 (1958).
94. A. Treibs and W. Ott, *Justus Liebigs Ann. Chem.* **615**, 137 (1958).
95. A. Hayes, A. H. Jackson, J. M. Judge, and G. W. Kenner, *J. Chem. Soc.* p. 4385 (1965).
96. A. Treibs and H. Derra-Scherer, *Justus Liebigs Ann. Chem.* **589**, 196 (1954).
97. J. W. Harbuck and H. Rapoport, *J. Org. Chem.* **37**, 3618 (1972).
98. H. Fischer and P. Halbig, *Justus Liebigs Ann. Chem.* **447**, 123 (1926).
99. A. F. Mironov, R. P. Evstigneeva, and N. A. Preobrazhenskii, *Zh. Obshch. Khim.* **35**, 1938 (1965).
100. G. G. Kleinspehn and A. E. Briod, *J. Org. Chem.* **26**, 1652 (1961).
101. E. Bullock, R. Grigg, A. W. Johnson, and J. W. F. Wasley, *J. Chem. Soc.* p. 2326 (1963).
102. G. M. Badger, R. A. Jones, and R. L. Laslett, *Aust. J. Chem.* **17**, 1157 (1964).
103. B. Franck, D. Gantz, and F. Hüper, *Angew. Chem.* **84**, 432 (1972).
104. P. S. Clezy, C. J. R. Fookes, and A. J. Liepa, *Aust. J. Chem.* **25**, 1979 (1972).
105. P. S. Clezy, C. L. Lim, and J. S. Shannon, *Aust. J. Chem.* **27**, 1103 (1974).
106. D. Dolphin, R. L. N. Harris, J. L. Huppatz, A. W. Johnson, and I. T. Kay, *J. Chem. Soc. C* p. 30 (1966).
107. P. S. Clezy and A. J. Liepa, *Aust. J. Chem.* **23**, 2461 (1970).
108. P. S. Clezy and A. J. Liepa, *Aust. J. Chem.* **23**, 2477 (1970).
109. A. H. Corwin and R. C. Ellingson, *J. Am. Chem. Soc.* **66**, 1146 (1944).
110. A. H. Corwin and S. R. Buc, *J. Am. Chem. Soc.* **66**, 1151 (1944).
111. A. F. Mironov, M. A. Kulish, V. V. Kobak, B. V. Rozynov, and R. P. Evstigneeva, *Zh. Obshch. Khim.* **44**, 1407 (1974).
112. A. F. Mironov, V. D. Rumyantseva, O. A. Kusukila, B. V. Rozynov, M. I. Struchkova, and R. P. Evstigneeva, *Zh. Obshch. Khim.* **44**, 2022 (1974).
113. M. T. Cox, A. H. Jackson, G. W. Kenner, S. W. MacCombie, and K. M. Smith, *J. Chem. Soc., Perkin Trans.* 1 p. 516 (1974).
114. J. A. P. Baptista de Almeida, G. W. Kenner, K. M. Smith, and J. Sutton, *J. Chem Soc., Chem. Commun.* p. 111 (1975).
115. G. G. Kleinspehn and A. H. Corwin, *J. Am. Chem. Soc.* **82**, 2750 (1960).
116. A. H. Jackson, G. W. Kenner, and K. M. Smith, *J. Chem. Soc. C* p. 502 (1971).
117. R. P. Evstigneeva and N. A. Preobrazhenskii, *Zh. Obshch. Khim.* **36**, 806 (1966).
118. B. Harrel and A. H. Corwin, *J. Am. Chem. Soc.* **78**, 3135 (1956).
119. A. H. Corwin and J. S. Andrews, *J. Am. Chem. Soc.* **58**, 1086 (1936).
120. J. S. Andrews, A. H. Corwin, and A. G. Sharp, *J. Am. Chem. Soc.* **72**, 491 (1950).
121. A. Treibs and H. G. Kolm, *Justus Liebigs Ann. Chem.* **614**, 199 (1958).
122. A. Treibs, F. H. Kreuzer, and N. Häberle, *Justus Liebigs Ann. Chem.* **733**, 37 (1970).
123. A. Treibs and R. Zimmer-Galler, *Chem. Ber.* **93**, 2542 (1960).
124. K. J. Brunings and A. H. Corwin, *J. Am. Chem. Soc.* **64**, 593 (1942).
125. K. J. Brunings and A. H. Corwin, *J. Am. Chem. Soc.* **66**, 337 (1944).
126. A. H. Corwin and K. J. Brunings, *J. Am. Chem. Soc.* **64**, 2106 (1942).
127. H. Shinohara, K. Honda, S. Misaki, and E. Imoto, *Nippon Kagaku Zasshi* **81**, 1740 (1960); *Chem Abstr.* **56**, 3441 (1962).

128. A. Treibs and F. Reitsam, *Chem. Ber.* **90**, 777 (1957).
129. D. Dolphin, *J. Heterocycl. Chem.* **7**, 275 (1970).
130. P. S. Clezy, A. J. Liepa, A. W. Nichol, and G. A. Smythe, *Aust. J. Chem.* **23**, 589 (1970).
131. H. Fischer and H. von Dobeneck, *Hoppe-Seyler's Z. Physiol. Chem.* **263**, 125 (1940).
132. J. H. Atkinson, R. S. Atkinson, and A. W. Johnson, *J. Chem. Soc.* 5999 (1964).
133. R. Bonnett, M. B. Hursthouse, and S. Neidle, *J. Chem. Soc. Perkin Trans. 2* p. 1335 (1972).
134. U. Eisner and P. H. Gore, *J. Chem. Soc.* p. 922 (1958).
135. L. N. Ferguson and J. C. Nandi, *J. Chem. Educ.* **42**, 529 (1965).
136. G. M. Badger, R. L. N. Harris, R. A. Jones, and J. M. Sasse, *J. Chem. Soc.* p. 4329 (1962).
137. A. F. Mironov, L. D. Miroshnichenko, R. P. Evstigneeva, and N. A. Preobrazhenskii, *Khim. Geterotsikl Soedin.* p. 74 (1965).
138. L. P. Kuhn and G. G. Kleinspehn, *J. Org. Chem.* **28**, 721 (1963).
139. A. H. Jackson, G. W. Kenner, H. Budzikiewicz, C. Djerassi, and J. M. Wilson, *Tetrahedron* **23**, 603 (1967).
140. H. Budzikiewicz, C. Djerassi, A. H. Jackson, G. W. Kenner, D. J. Newman, and J. M. Wilson, *J. Chem. Soc.* p. 1949 (1964).
141. H. Fischer and J. Klarer, *Justus Liebigs Ann. Chem.* **448**, 178 (1926).
142. H. Fischer, *Chem. Ber.* **47**, 3273 (1914).
143. H. Fischer and A. Kirstahler, *Justus Liebigs Ann. Chem.* **466**, 179 (1928).
144. H. Fischer and H. Orth, *Justus Liebigs Ann. Chem.* **489**, 62 (1931).
145. P. S. Clezy and A. W. Nichol, *Aust. J. Chem.* **18**, 1835 (1965).
146. R. P. Evstigneeva, L. I. Arkhipova, and N. A. Preobrazhenskii, *Zh. Obshch. Khim.* **31**, 2972 (1961).
147. A. H. Corwin and P. Viohl, *J. Am. Chem. Soc.* **66**, 1137 (1944).
148. S. F. MacDonald and R. J. Stedman, *Can. J. Chem.* **32**, 896 (1954).
149. A. Treibs and H. Bader, *Justus Liebigs Ann. Chem.* **627**, 182 (1959).
150. A. Markovac and S. F. MacDonald, *Can. J. Chem.* **43**, 3364 (1965).
151. G. V. Ponomarev, S. M. Nasralla, A. G. Bubnova, and R. P. Evstigneeva, *Khim. Geterotsikl. Soedin.* p. 202 (1973).
152. G. S. Marks, D. K. Dougall, E. Bullock, and S. F. MacDonald, *J. Am. Chem. Soc.* **82**, 3183 (1960).
153. P. S. Clezy and A. J. Liepa, *Aust. J. Chem.* **24**, 1933 (1971).
154. S. F. MacDonald and R. J. Stedman, *Can. J. Chem.* **33**, 458 (1955).
155. S. F. MacDonald and K. H. Michl, *Can. J. Chem.* **34**, 1768 (1956).
156. A. H. Cook and J. R. Majer, *J. Chem. Soc* p. 482 (1944).
157. A. Treibs, E. Herrmann, E. Meissner, and A. Kuhn, *Justus Liebigs Ann. Chem.* **602**, 153 (1957).
158. H. Fischer and W. Zerweck, *Chem. Ber.* **55**, 1949 (1922).
159. A. J. Castro, J. G. Tertzakian, B. T. Nakata, and D. A. Brose, *Tetrahedron* **23**, 4499 (1967).
160. H. v. Dobeneck and F. Schnierle, *Tetrahedron Lett.* p. 5327 (1966).
161. H. Fischer, E. Baumann, and H. J. Riedl, *Justus Liebigs Ann. Chem.* **475**, 205 (1929).
162. H. Fischer and H. J. Riedl, *Hoppe-Seyler's Z. Physiol. Chem.* **207**, 193 (1932).
163. A. H. Corwin and J. S. Andrews, *J. Am. Chem. Soc.* **59**, 1973 (1937).
164. A. H. Corwin and J. S. Andrews, *J. Am. Chem. Soc.* **62**, 418 (1940).
165. A. Treibs and R. Zimmer-Galler, *Hoppe-Seyler's Z. Physiol. Chem.* **318**, 12 (1960).

166. J. H. Paden, A. H. Corwin, and W. A. Bailey, Jr., *J. Am. Chem. Soc.* **62**, 418 (1940).
167. A. H. Corwin and K. W. Doak, *J. Am Chem. Soc.* **77**, 464 (1955).
168. A. Treibs and H. G. Kolm, *Justus Liebigs Ann. Chem.* **606**, 166 (1957).
169. H. Fischer and H. Höfelmann, *Hoppe-Seyler's Z. Physiol. Chem.* **251**, 218 (1938).
170. A. Treibs and K. Hintermeier, *Justus Liebigs Ann. Chem.* **592**, 11 (1955).
171. E. I. Filippovich, R. P. Evstigneeva, and N. A. Preobrazhenskii, *Zh. Obshch. Khim.* **30**, 3253 (1960).
172. E. I. Filippovich, R. P. Evstigneeva, and N. A. Preobrazhenskii, *Zh. Obshch. Khim.* **31**, 2968 (1961).
173. L. D. Miroshnichenko and R. P. Evstigneeva, *Izv. Akad. Nauk SSSR, Ser. Fiz.* **27**, 50 (1963).
174. A. Treibs and F. H. Kreuzer, *Justus Liebigs Ann. Chem.* **718**, 208 (1968).
175. A. Treibs and F. Reitsam, *Justus Liebigs Ann. Chem.* **611**, 194 (1958).
176. A. Treibs and F. Reitsam, *Justus Liebigs Ann. Chem.* **611**, 205 (1958).
177. A. Treibs and H. Scherer, *Justus Liebigs Ann. Chem.* **577**, 139 (1952).
177a. A. Treibs and W. Seifert, *Justus Liebigs Ann. Chem.* **612**, 242 (1958).
178. L. D. Miroshnichenko, E. I. Filippovich, R. P. Evstigneeva, and N. A. Preobrazhenskii, *Dokl. Akad. Nauk SSSR* **134**, 1100 (1960).
179. L. D. Miroshnichenko, R. P. Evstigneeva, E. I. Filippovich, and N. A. Preobrazhenskii, *Zh. Obshch. Khim.* **31**, 2975 (1961).
180. H. Falk, S. Gergely, and O. Hofer, *Monatsh. Chem.* **105**, 853 (1974).
181. A. Treibs and P. Dieter, *Justus Liebigs Ann. Chem.* **513**, 65 (1934).
182. H. Falk, S. Gergely, and O. Hofer, *Monatsh. Chem.* **105**, 1004 (1974).
183. H. Falk, S. Gergely, and O. Hofer, *Monatsh. Chem.* **105**, 1019 (1974).
184. H. Falk, O. Hofer, and H. Lehner, *Monatsh. Chem.* **105**, 366 (1974).
185. A. Treibs and R. Zimmer-Galler, *Justus Liebigs Ann. Chem.* **664**, 140 (1963).
186. A. Treibs and F. H. Kreuzer, *Justus Liebigs Ann. Chem.* **721**, 116 (1969).
187. H. Booth, A. W. Johnson, F. Johnson, and J. Langdale-Smith, *J. Chem. Soc.* p. 650 (1963).
188. A. C. Jain and G. W. Kenner, *J. Chem. Soc.* p. 185 (1959).
189. P. Bamfield, A. W. Johnson, and J. Leng, *J. Chem. Soc.* p. 7001 (1965).
190. H. Fischer and J. Aschenbrenner, *Hoppe-Seyler's Z. Physiol. Chem.* **229**, 71 (1934).
191. H. Fischer and W. Fröwis, *Hoppe-Seyler's Z. Physiol. Chem.* **195**, 50 (1931).
192. H. v. Dobeneck, *Hoppe-Seyler's Z. Physiol. Chem.* **270**, 223 (1941).
193. H. v. Dobeneck, E. Brunner, and H. Reinhard, *Tetrahedron Lett.* p. 5331 (1966).
194. S. F. MacDonald and A. Markovac, *Can. J. Chem.* **43**, 3247 (1965).
195. D. Dolphin, A. W. Johnson, J. Leng, and P. van den Broek, *J. Chem. Soc. C* p. 880 (1966).
196. H. von Dobeneck, E. Brunner, and U. Deffner, *Z. Naturforschung* **22b**, 1005 (1967).
197. S. F. MacDonald, *J. Am. Chem. Soc.* **82**, 4384 (1957).
198. D. A. Lightner and D. C. Crandall, *Tetrahedron Lett.* p. 1799 (1973).
199. F. Pruckner and A. Stern, *Z. Phys. Chem., Abt. A* **180**, 25 (1937).
200. R. W. Guy and R. A. Jones, *Aust. J. Chem.* **18**, 363 (1965).
201. H. v. Dobeneck and E. Brunner, *Hoppe-Seyler's Z. Physiol. Chem.* **341**, 157 (1965).
202. H. Falk, O. Hofer, and H. Lehner, *Monatsh. Chem.* **105**, 169 (1974).
203. H. Falk and O. Hofer, *Monatsh. Chem.* **106**, 97 (1975).
204. R. A. Jeffreys and E. B. Knott, *J. Chem. Soc.* p. 1028 (1951).
205. H. Falk and O. Hofer, *Monatsh. Chem.* **105**, 995 (1974).
206. H. Falk and O. Hofer, *Monatsh. Chem.* **106**, 115 (1975).

207. R. W. Guy and R. A. Jones, *Aust. J. Chem.* **19**, 1871 (1966).
208. A. Stern and H. Molvig, *Z. Phys. Chem., Abt. A* **175**, 38 (1935).
209. C. S. Vestling and J. R. Downing, *J. Am. Chem. Soc.* **61**, 3511 (1939).
210. R. Breslow, R. Boikess, and M. Battiste, *Tetrahedron Lett.* p. 42 (1960).
211. A. J. Castro, J. P. Marsh, and B. T. Nakata, *J. Org. Chem* **28**, 1943 (1963).
211a. W. A. P. Luck, *Naturwissenschaften* **54**, 601 (1952).
211b. F. Cruege, P. Pineau, and J. Lascombe, *J. Chim. Phys.* **64**, 1161 (1967).
212. I. F. Gurinovich, *Zh. Prikl. Spektrosk.* **6**, 657 (1967).
213. J. R. Sabin, *Int. J. Quantum Chem.* **2**, 31 (1968).
214. H. R. Wetherell, M. J. Hendrickson, and A. R. McIntyre, *J. Am. Chem. Soc.* **81**, 4517 (1959).
215. H. Bürger, *in* "Porphyrins and Metalloporphyrins" (K. M. Smith, ed.), p. 527. Elsevier, Amsterdam, 1975.
216. H. Plieninger, H. Bauer, A. R. Katritzki, and U. Lerch, *Justus Liebigs Ann. Chem.* **654**, 165 (1962).
217. F. C. March, D. A. Couch, K. Emerson, J. Fergusson, and W. T. Robinson, *J. Chem. Soc. A* p. 440 (1971).
218. D. Doddrell and W. S. Caughey, *J. Am. Chem. Soc.* **94**, 2510 (1972).
219. B. Franck and A. Rowold, *Angew. Chem.* **87**, 418 (1975).
220. M. J. Broadhurst, R. Grigg, and A. W. Johnson, *J. Chem. Soc. C* p. 3681 (1971).
221. J. L. Davies, *J. Chem. Soc. C* p. 1392 (1968).
222. P. S. Clezy and N. W. Webb, *Aust. J. Chem.* **25**, 2217 (1972).
223. H. Fischer and E. Adler, *Hoppe-Seyler's Z. Physiol. Chem.* **197**, 268 (1931).
224. R. P. Evstigneeva, A. F. Mironov, and L. I. Fleiderman, *Dokl. Akad. Nauk SSSR* **210**, 1090 (1973).
225. J. M. Conlon, J. A. Elix, G. I. Feutrill, A. W. Johnson, M. W. Roomi, and J. Whelan, *J. Chem. Soc., Perkin Trans. 1* p. 713 (1974).
226. A. Bauer, *Justus Liebigs Ann. Chem.* **736**, 1 (1970).
227. D. Mauzerall, *J. Am. Chem. Soc.* **82**, 2601 (1960).
228. A. A. Ponomarev, R. P. Evstigneeva, and N. A. Preobrazhenskii, *Khim. Geterotsikl. Soedin.* p. 185 (1969).
229. H. W. Whitlock and D. H. Buchanan, *Tetrahedron Lett.* p. 3711 (1969).
230. G. V. Ponomarev, R. P. Evstigneeva, and N. A. Preobrazhenskii, *Zh. Org. Khim.* **7**, 169 (1971).
231. A. I. Scott, *Heterocycles* **2**, 125 (1974).
232. A. I. Scott, E. Lee, and C. A. Townsend, *Bioorganic Chem.* **3**, 1 (1974).
233. R. Radmer and L. Bogorad, *Biochemistry* **11**, 904 (1972).
234. R. C. Davies and A. Neuberger, *Biochem. J.* **133**, 471 (1973).
235. A. H. Corwin and E. C. Coolidge, *J. Am. Chem. Soc.* **74**, 5196 (1952).
236. A. H. Jackson, G. W. Kenner, and J. Wass, *J. Chem. Soc., Perkin Trans. 1* p. 480 (1974).
237. T. T. Howarth, A. H. Jackson, and G. W. Kenner, *J. Chem. Soc., Perkin Trans. 1* 502 (1974).
238. A. Gossauer and J. Engel, *Justus Liebigs Ann. Chem.* p. 225 (1977).
239. D. Mauzerall, *J. Am. Chem. Soc.* **82**, 2605 (1960).
240. P. S. Clezy, A. J. Liepa, and N. W. Webb, *Aust. J. Chem.* **25**, 2687 (1972).
241. P. S. Clezy and C. J. R. Fookes, *Aust. J. Chem.* **27**, 371 (1974).
242. A. H. Jackson, G. W. Kenner, G. MacGillivray, and G. S. Sach, *J. Am. Chem. Soc.* **87**, 676 (1965).
243. H. Fischer and A. Kürzinger, *Hoppe-Seyler's Z. Physiol. Chem.* **196**, 213 (1931).
244. G. M. Badger and A. D. Ward, *Aust. J. Chem.* **17**, 1013 (1964).

245. A. F. Mironov, R. P. Evstigneeva, and N. A. Preobrazhenskii, *Tetrahedron Lett.* p. 183 (1965).
246. A. H. Jackson, G. W. Kenner, and K. M. Smith, *J. Chem. Soc. C* p. 1974 (1971).
247. M. T. Cox, R. Fletcher, A. H. Jackson, G. W. Kenner, and K. M. Smith, *Chem. Commun.* p. 1141 (1967).
248. P. S. Clezy and A. J. Liepa, *Aust. J. Chem.* **24**, 1027 (1971).
249. A. W. Johnson and I. T. Kay, *J. Chem. Soc.* p. 1620 (1965).
250. D. Dolphin, R. L. N. Harris, J. L. Huppatz, A. W. Johnson, I. T. Kay, and J. Leng, *J. Chem. Soc. C* p. 98 (1966).
251. R. Grigg, A. W. Johnson, R. Kenyon, V. B. Math, and K. Richardson, *J. Chem. Soc. C* p. 176 (1969).
252. M. K. Kulish, A. F. Mironov, B. V. Rozynov, and R. P. Evstigneeva, *Zh. Obshch. Khim.* **41**, 2743 (1971).
253. H. H. Inhoffen, N. Schwarz, and K.-P. Heise, *Justus Liebigs Ann. Chem.* p. 146 (1973).
254. H. H. Inhoffen, F. Fattinger, and N. Schwarz, *Justus Liebigs Ann. Chem.* p. 412 (1974).
255. T. A. Melent'eva, N. S. Genokhova, N. D. Pekel, T. A. Stetsko, and V. M. Berezovskii, *Zh. Obshch. Khim.* **44**, 929 (1974).
256. A. W. Johnson and W. R. Overend, *J. Chem. Soc., Perkin Trans. 1* p. 2681 (1972).
257. I. D. Dicker, R. Grigg, A. W. Johnson, H. Pinnock, K. Richardson, and P. van den Broek, *J. Chem. Soc. C* p. 536 (1971).
258. A. F. Mironov, V. D. Rumyantseva, R. P. Fleiderman, and R. P. Evstigneeva, *Zh. Obshch. Khim.* **45**, 1150 (1975).
259. J. Engel and H. H. Inhoffen, *Justus Liebigs Ann. Chem.*, p. 767 (1977).
260. P. Bamfield, R. L. N. Harris, A. W. Johnson, I. T. Kay, and K. W. Shelton, *J. Chem. Soc. C* p. 1436 (1966).
261. D. A. Clarke, R. Grigg, R. L. N. Harris, A. W. Johnson, I. T. Kay, and K. W. Shelton, *J. Chem. Soc. C* p. 1648 (1967).
262. P. Bamfield, R. Grigg, R. W. Kenyon, and A. W. Johnson, *J. Chem. Soc. C* p. 1259 (1968).
263. R. Grigg, A. W. Johnson, and M. Roche, *J. Chem. Soc. C* p. 1928 (1970).
264. R. V. H. Jones, G. W. Kenner, and K. M. Smith, *J. Chem. Soc., Perkin Trans. 1* p. 531 (1974).
265. R. L. N. Harris, A. W. Johnson, and I. T. Kay, *J. Chem. Soc. C* p. 22 (1966).
266. A. F. Mironov, V. D. Rumyantseva, B. V. Rozynov, and R. P. Evstigneeva, *Zh. Org. Khim.* **7**, 165 (1971).
267. R. Grigg, A. W. Johnson, and G. Shelton *J. Chem. Soc. C* p. 2287 (1971).
268. R. Grigg, A. W. Johnson, and G. Shelton, *Justus Liebigs Ann. Chem.* **476**, 32 (1971).
269. H. H. Inhoffen, H. Maschler, and A. Gossauer, *Justus Liebigs Ann. Chem.* p. 141 (1973).
270. R. Grigg, A. P. Johnson, A. W. Johnson, and M. J. Smith, *J. Chem. Soc. C* p. 2457 (1971).
271. A. Gossauer, H. Maschler, and H. H. Inhoffen, *Tetrahedron Lett.* p. 1277 (1974).
272. H. H. Inhoffen and H. Maschler, *Justus Liebigs Ann. Chem.* p. 1269 (1974).
273. H. H. Inhoffen, J. W. Buchler, L. Puppe, and K. Rohbock, *Justus Liebigs Ann. Chem.* **747**, 133 (1971).
274. J. Ullrich, Dissertation, Technische Universitaet, Braunschweig (1967).
275. G. Struckmeier, U. Thewalt, and J. Engel, *J. Chem. Soc., Chem. Commun.* p. 963 (1976).
276. R. B. Woodward, *Ind. Chim. Belge* **11**, 1293 (1962).

8

Metal Complexes of Open-Chain Tetrapyrrole Pigments

J. SUBRAMANIAN AND J.-H. FUHRHOP

I. INTRODUCTION

Studies of metal complexes of biliverdin, bilirubin, and similar open-chain tetrapyrroles are often motivated by problems arising in porphyrin chemistry. Porphyrins produce rigid square–planar ligand fields with strong σ-electron donation from the nitrogens into the metal orbitals. The π-backbonding from the metal-d-orbitals into the antibonding π-orbitals of the porphyrin is not significant because of the high energy of the latter.[1-3] This geometric and electronic situation is one of the fundamental preconditions of most biologic redox chains, and therefore, deserves detailed investigation. An important aspect of the problem of the open-chain tetrapyrrole complexes is concerned with the changes in the metal–ligand interactions when the porphyrin macrocycle is cleaved at a methine bridge. Typical questions related to these problems are: Is the ligand field still square–planar as with the porphyrins or tetrahedral as with bispyrromethene complexes? Do open-chain tetrapyrroles stabilize metal ions in their high oxidation states, and do they form stable π-radicals like the porphyrins? How does the metal–ligand interaction in metallophlorins compare with that in metallobilirubinates? Are the metal complexes of open-chain tetrapyrroles kinetically stable? Answering these questions would, of course, also result in a better understanding of the nature of metal ligand bonds in metalloporphyrins.

The special cases of metallobiliverdinates and -bilirubinates, however, are also of some interest of their own. Many of the biologically active tetrapyrrole pigments occur in the form of metal complexes (e.g., with Fe, Co, Mg), and their primary degradation products are probably tetrapyrroles with an intact inner conjugation pathway (see Chapter 4, this volume). To understand these metabolic processes, the properties of the possible products should be known. Furthermore, the plant hormone, phytochrome, contains perhaps a bilatriene chromophore, which is considered to change its conformation in the course of its regulating action.[4,5] Clear-cut correlations of electronic spectra with the results of conformational studies on metallobilatrienes should help to enlighten the virtues of this latter hypothesis.

II. COMPOSITION AND ELECTRONIC AND VIBRATIONAL SPECTRA OF OPEN-CHAIN TETRAPYRROLIC METAL COMPLEXES

Bilirubin (**1**) and urobilins, e.g., stercobilin (**2**), contain derivatives of dipyrromethene as the chromophores. Before the properties of these more complex tetrapyrrole ligands are discussed, the main findings on well-defined models of the subunits will be described.

Dipyrromethenes,[6,7] e.g., **3a-e**, form 2:1 complexes with most bivalent transition metal ions (e.g., Mn, Co, Ni, Cu, Zn), which may be crystallized. The electronic spectra of copper and cobalt dipyrromethenes consist of strong π-π^* absorption bands abound 500 nm ($\epsilon \sim 10^5$) and very weak d-d transitions ranging from 500 to 1000 nm ($\epsilon \sim 300$). From these latter bands, the geometry of the ligand field has been evaluated.[8] Co(II) and Ni(II) complexes of dipyrromethenes are tetrahedral, while the Cu(II) complexes have tetragonally distorted tetrahedral configuration.[8] The degree of distortion depends on the bulkiness of the substituents in the 1,9 positions. Increase in the size of the substituent leads to smaller tetragonal distortion. The ligand field

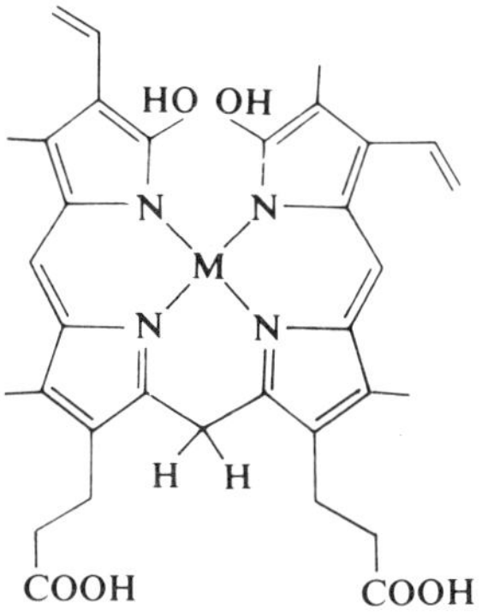

1

(a) M = 2H
(b) M = bivalent metal ions

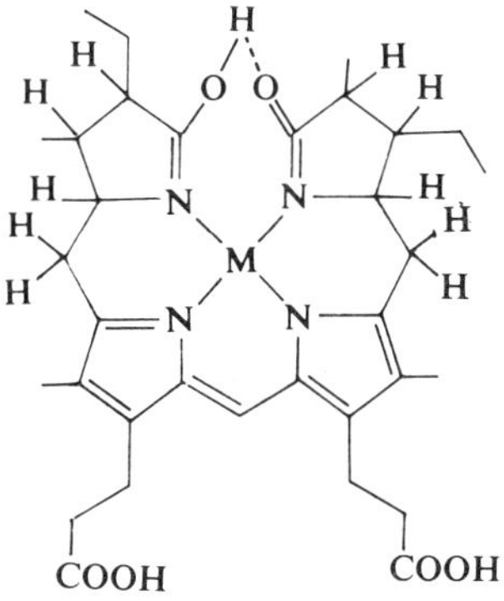

2

(a) M = 2H
(b) M = bivalent metal ions

bands in the optical spectra of the Cu(II) complexes shift to lower energy with increase in the bulkiness of the substituents in the 1,9 positions. A strong-band in the infrared at 1600 cm^{-1} was assigned to the skeletal stretching mode of the pyrrole rings,[8,9] which was slightly shifted in different metal complexes (e.g., free ligand: 1622 cm^{-1}, copper complex: 1606 cm^{-1}, nickel complex: 1605 cm^{-1}, cobalt complex: 1603 cm^{-1}), indicating insignificant changes in the nature of coordinate bonds.

In 1962, O'Carra observed that for bilirubin (BR) (**1a**) the π-π* absorption band at 452 nm shifted to 476 nm when a drop of 2% methanolic zinc acetate

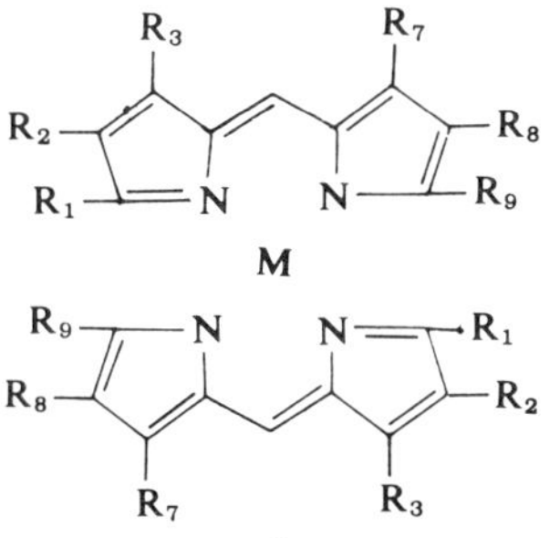

3

(a) M = Ni
   $R_1 = R_3 = R_7 = R_9 = CH_3$
   $R_2 = R_8 = H$

(b) M = Cu
   $R_1 = R_3 = R_7 = R_9 = CH_3$
   $R_2 = R_8 = COOEt$

(c) M = Cu
   $R_1 = R_9 = H$
   $R_2 = R_3 = R_7 = R_8 = CH_3$

(d) M = Cu
   $R_1 = R_3 = R_7 = CH_3$
   $R_2 = R_8 = R_9 = H$

(e) M = Cu
   $R_1 = R_9 = C_6H_5$
   $R_2 = R_3 = R_7 = R_8 = H$

per 5 ml of a neutral 1:1 chloroform–methanol solution of the tetrapyrrole base[10] was added. Ever since this first evidence of the existence of a defined bilirubin metal complex appeared numerous attempts have been made to characterize these rather unstable species. Since bilirubin contains altogether six dissociable protons (four on the pyrrole and pyrrolone units and two on the carboxylic acids, H_6BR), three bivalent metal ions could be chelated. A deep red complex ($\lambda_{max} = 530$ nm), which was formed in N,N-dimethylformamide from bilirubin and zinc acetate and precipitated from the same solvent, did indeed give an elemental analysis which suggested a Zn:BR ratio of 3:1.[11] The base tetramethyl guanidine (TMG) seems to remove all dissociable protons from bilirubin in dimethylformamide.[12] The resulting hexa-anion absorbing at 440 nm has been titrated with zinc salts, and the formation of a complex absorbing at 530 nm was completed at a ratio of Zn:BR of 2:1. Larger excesses of Zn salts lead to decomposition of the chelate.[12] If no base is added, much higher ratios are needed to achieve complete chelation.[13] The overall reaction has been proposed to be[12]

$$H_6BR + 2\,Zn^{2+} + 8TMG \;\rightleftharpoons\; H_2BRZn_2 \cdot 4TMG + 4HTMG$$

The TMG base in the complex should fill the coordination sphere of zinc, and the two remaining hydrogens are presumably on the lactim oxygen. A tentative structure proposed for bis-zinc bilirubinate,[13] which fits all the known data, is given in Fig. 1. In a tris-zinc bilirubinate zinc salt, complexation of the two acid side chains, together with complex formation in each of the pyrromethene units, is the simplest and most plausible explanation.

Many other metal complexes of bilirubin (2) have been characterized spectrophotometrically, e.g., Fe(II) with an absorption peak at $\lambda_{max} =$ 424 nm, Fe(III) (450 nm), Co(II) (423, 470 nm), and Cu(II) (420, 352 nm). All of them are unstable.[14] The only stable complex was obtained with samarium (III) acetate (395 nm, $\epsilon = 46,400$).[14] To our knowledge, no further analysis or an infrared (ir) spectrum of any of these complexes has been reported. Zinc bilirubinate is decomplexed by addition of acetylacetone. The

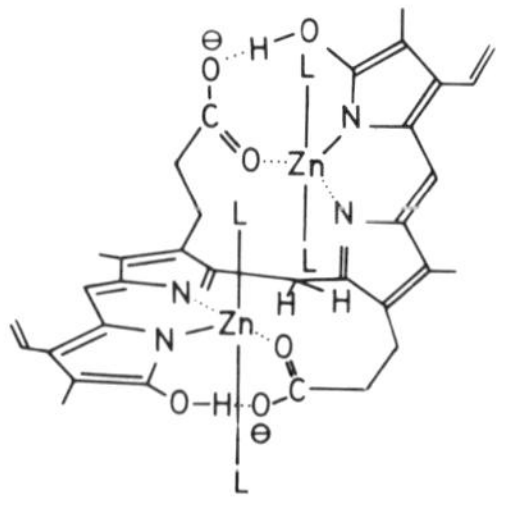

Fig. 1. Hypothetical structure of (Zn)$_2$ bilirubinate (from Kuenzle *et al.*,[13] slightly modified).

general instability of BR metal complexes, as compared to the pyrromethene and biliverdin complexes, has been correlated with the fact that conformations like **4a–c**, which are thought to predominate in solution and which are stabilized by extensive hydrogen bonding, have to be destroyed to form a chelate like the one given in Fig. 1.[15–17]

Optically active urobilins of natural origin also form zinc complexes that show much less specific optical rotation than the parent compounds.[18,19] The metal-to-nitrogen bond in metallobilatrienes is much more stable than in the corresponding biladiene complexes. Zinc(II) and copper(II) biliverdinates (**5b,d**) have been prepared in alcoholic solutions and isolated in pure form as early as 1934 by Lemberg[20] and later by Fischer.[21] Zinc(II), copper(II), as well as manganese(II), cobalt(II), and nickel(II) chelates (**5b,c,d,e**) have recently been prepared from biliverdin and mesobiliverdin dimethyl esters,[22] the metal-to-ligand ratio being always 1:1. The iron complex could only be obtained in solution as a short-lived species. Metal complexes of monomethoxy- (**6**) and diethoxy biliverdins (**7**) have also been described,[22] and will not be considered further in this article.

Synthetic bilatrienes, which have been studied as metal complexes, are octaethyl formylbiliverdinates (**8**)[23–25] and dioxobilatrienes (**9**).[26,27] They also form 1:1 complexes with the metals indicated.

The electronic spectra of these metal complexes are highly dependent on the nature of the metal ion and on the pH of the solution. Manganese biliverdinate dimethyl ester, for example, produces a long-wavelength absorption at 900 nm in chloroform, whereas the corresponding transition in the zinc complex occurs at 700 nm (Fig. 2). If one titrates the manganese complex with hydrochloric acid, its absorption peak shifts to 680 nm (Fig. 3), whereas addition of base to the zinc complex produces a bathochromic shift of about 100 nm. Both these changes are fully reversible and have been tentatively assigned to protonation–deprotonation equilibria, either on an amide oxygen

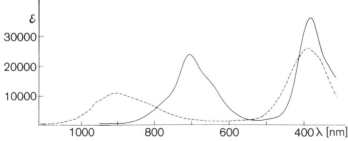

Fig. 2. Electronic spectra of manganese biliverdinate (---, **5c**) and zinc biliverdinate (——— **5b** in CHCl$_3$); from Fuhrhop et al.,[22] with permission.

4a

4b

4c

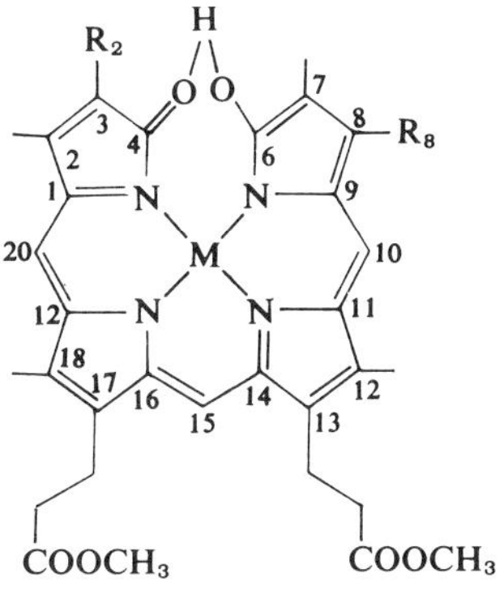

5a $R_2 = R_8 = C_2H_3$ M = 2H
5b $R_2 = R_8 = C_2H_3$ M = Zn
5c $R_2 = R_8 = C_2H_3$ M = Mn
5d $R_2 = R_8 = C_2H_3$ M = Cu
5e $R_2 = R_8 = C_2H_3$ M = Ni
5f $R_2 = R_8 = C_2H_5$ M = 2H
5g M = $R_2 = R_8 = C_2H_5$ M = Zn
5h $R_2 = R_8 = C_2H_5$ M = Mn
5i $R_2 = R_8 = $ M = Cu
5k $R_2 = R_8 = C_2H_5$ M = Ni

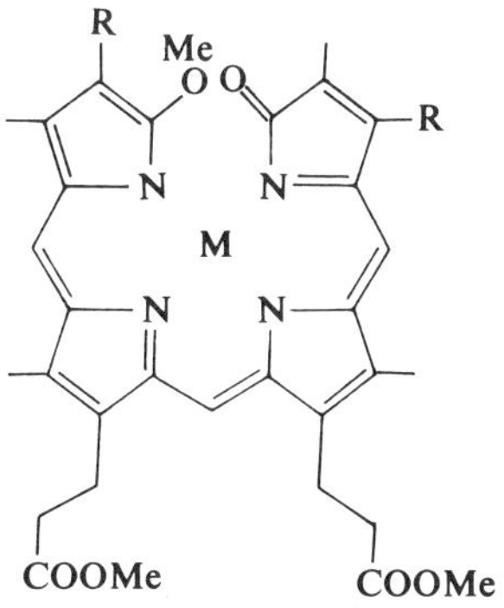

5

6

(a) R = C_2H_3
(b) R = C_2H_5 M = bivalent metal ions

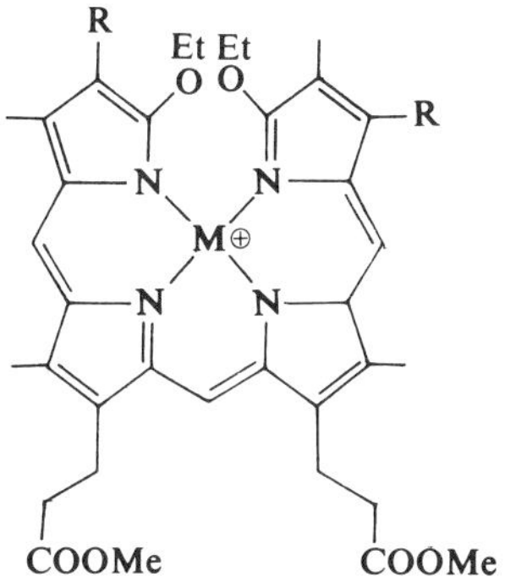

7

(a) R = C_2H_3
(b) R = C_2H_5 M = bivalent metal ions

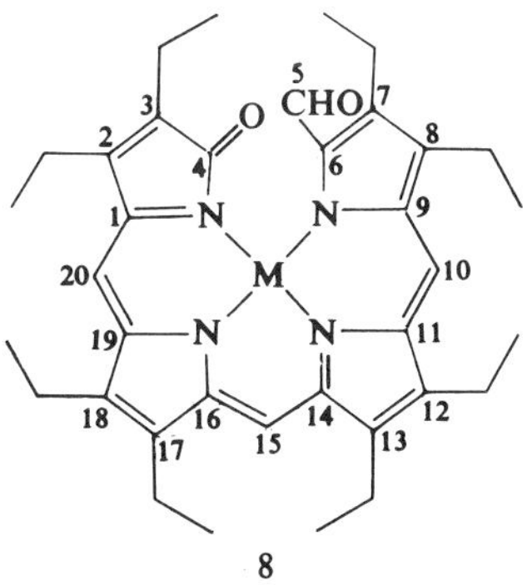

8
(a) M = Cu
(b) M = Ni
(c) M = Zn

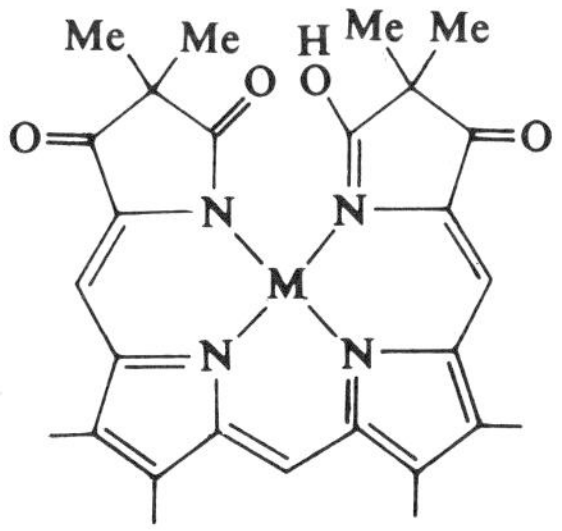

9
(a) M = 2H
(b) M = Ni
(c) M = Pd

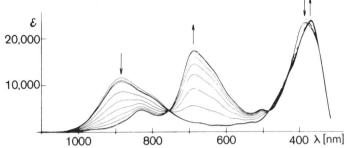

Fig. 3. Spectrophotometric titration of manganese biliverdinate with hydrochloric acid in methanol; from Fuhrhop et al.,[22] with permission.

of **5** or a nitrogen atom, followed by conformational changes of the chromophore.[22] The metal is not replaced in any of these titrations. Similar observations have been made with the metal formylbiliverdinate (**8**). Variations are also observed in the relative intensities of the bands around 700 nm and those around 350 nm. These changes were interpreted on the basis of HMO calculations, assuming various geometries for the bilatriene systems.[4] SCF calculations in the Pariser–Parr framework, including configuration interactions, have also been made, taking the different geometries explicitly into account (see Table 1).[22,27a] It has been observed that, in the all-cis phlorin-type con-

TABLE 1

Electronic Transitions (Singlet–Singlet) in Different Conformations of Bilatrienes[a,b]

Conformation	s–s (in nm)	Oscillator strength (f)
Phlorin type	764	0.54
Conformation I	359	2.44
(Or cis-cis-cis configuration)	323	0.27
	313	0.27
Open	697	1.23
Conformation II	386	2.66
(Or cis-cis-trans configuration)	360	9.16
	290	0.23
Open	672	2.90
Conformation III	383	0.78
(Or trans-cis-trans configuration)	293	0.64

[a] Calculated by Pariser–Parr–Pople SCF method including configuration interactions, using a QCPE program (QCPE catalog X, 1974, Indiana University, program 71, by Bloor and Gilson). See also Fuhrhop and Subramanian.[63]

[b] The SCF parameters were taken from Knop and Fuhrhop.[27a]

figuration, the band in the near infrared occurs at the longest wavelength. The band at 400 nm has a larger extinction coefficient than that of the 700-nm band. In the all-trans open-chain configuration, the band in the near infrared region shifts to shorter wavelength and becomes more intense. Its extinction coefficient is then larger than that of the 350-nm band. This model is capable of explaining qualitatively the observed spectral behavior in the bilatrienes. However, the only X-ray structure known today rather suggests that metallobilatrienes stay in their all-cis configuration on acidification. This case is discussed in Section II. Another interesting effect is that bilatrienes, as well as their metal complexes, produce large induced circular dichroism in cholesteric meso phases.[22] This has also been observed in metal-free dipyrromethenes.[28]

In the infrared, metallomesobiliverdinate dimethyl esters produce skeletal vibration bands [e.g., zinc complex (5g): 1535 cm^{-1}; manganese complex (5h): 1580 and 1545 cm^{-1}] which are somewhat lower in energy than that of the free base (5f: 1595 cm^{-1}).[22]

III. EPR STUDIES

A. Metal Complexes of Dipyrromethenes

EPR studies have been reported for a number of Cu(II) dipyrromethene complexes.[8] The extent of tetragonal distortion has been estimated from the g values and expressed in terms of the dihedral angles between the two dipyrromethene planes. The g values and hyperfine couplings for some Cu–dipyrromethene complexes are listed in Table 2. The shifts in the g values have been correlated with the dihedral angle, 2ω, between the two chelate rings. However, the g values are also dependent on covalent bonding between the metal and the ligand. Hence, caution should be exercised in the calculation of dihedral angles from the g values only.

B. Metal Complexes of Bilatriene

1. Cu(II) COMPLEXES

EPR studies are known only for Cu(II) formyloctaethylbiliverdin (8).[27] The Cu(II), Co(II), and Mn(II) complexes of biliverdin esters 5 did not yield any EPR signals. The long-wavelength form of Cu(II) formylbiliverdin (8a) (see Sections II and IV) yielded EPR signals in fluids and frozen solutions (Fig. 4), similar to those of Cu(II) porphyrins.[27b] The hyperfine couplings of Cu and the superhyperfine couplings from four equivalent ligand nitrogens in Cu(II) formylbiliverdin are very close to the corresponding values for Cu porphyrins and Cu chlorins (see Table 2).

The geometry of the ligand field has been predicted for the Cu-bilatriene

TABLE 2

ESR Parameters for Copper(II) Complexes of Some Tetrapyrrole and Dipyrromethene Systems[a,b]

	Solvent	Temp. (°K)	$\langle g \rangle$	$\langle A \rangle$ Cu	$\langle A \rangle$ 14N	$g_\perp$	g_{11}	$A_\perp$ Cu	$A_\perp$ 14N	A_{11} Cu	A_{11} 14N	Reference
Octaethylformylbiliverdin (8a)	CHCl$_3$	300	2.080	72.16	13.30							
	CHCl$_3$—EBBA	100				2.029	2.174	37.22	16.66	171.7	13.33	27
Octaethylporphyrin	CHCl$_3$	300	2.095	82.55	15.00							
	CHCl$_3$—EBBA	100				2.031	2.173	41.11	18.33	186.66	15.00	27
Octaethylchlorin (OEC)	CHCl$_3$	100	2.093	81.66	15.00							
	CHCl$_3$-EBBA	100				2.015	2.173	38.86	16.66	183.00	13.33	27
Dipyrromethene (3c)	Xylene–benzene 6:4	RT	2.113	58.6								
	Xylene–benzene 6:4	123				2.039	2.224	—	13.1	154.00	11.20	8
Dipyrromethene (3d)	Xylene–benzene 6:4	RT	2.132	56.1								
	Xylene–benzene 6:4	123				2.050	2.248	—	11.2	136.00	9.80	8
Dipyrromethene (3e)	Xylene–benzene 6:4	RT	2.130	—								
	Xylene–benzene 6:4	123				2.070	2.279	—	7.9	103.00	—	8

[a] The hyperfine couplings are expressed in 10^{-4} cm^{-1}.
[b] The estimated angle (2ω) between the two dipyrromethene units are 63° for 3c and 3d, and 73° for 3e.

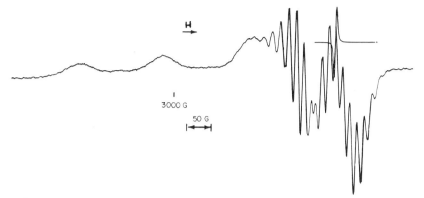

Fig. 4. ESR spectrum of copper octaethylformylbiliverdinate (**8a**) CHCl₃–EBBA mixture at 100°K (from Subramanian *et al.*,[27] with permission).

complex by taking into consideration both the g value and the hyperfine tensors. The Cu–anisotropic hyperfine couplings for square–planar complexes,[29–33] $A_{\parallel}^{Cu}$, lie in the range 0.0180–0.0200 cm⁻¹, whereas for tetrahedral systems[8,34] $A_{\parallel}^{Cu}$ (or A_z^{Cu}) is in the range 0.0100 cm⁻¹. Likewise, A_x^{Cu} and A_y^{Cu} (or $A_{\perp}^{Cu}$) values are considerably smaller (0.0010 cm⁻¹) for the tetrahedral systems than for square–planar complexes (0.0035 cm⁻¹). Thus, from the close similarity of the EPR parameters of Cu formylbiliverdin (**8**) to those of Cu porphyrins, it has been proposed that the ligand in the former system is cyclic, with only moderate deviation from square planarity. This arrangement is only possible when both the rings of the bilatriene ligand are slightly tilted out of plane in order to minimize the steric interactions between the end groups. The moderate distortions from square planarity would lead to a decrease in the hyperfine couplings of Cu, as observed experimentally (see Table 2).

The bonding coefficients of the Cu-N bonds in the bilatriene complex have been computed from the EPR parameters by following the approach of Maki and McGarvey[35] and Kivelson.[36] The bonding coefficients for Cu and N atomic orbitals can be obtained independently[32,36] (a) from the g and Cu hyperfine tensors and (b) from the superhyperfine couplings of the ligand N atoms. For square–planar systems like Cu porphyrins, the coefficient obtained by these two methods agree very well. For Cu formylbiliverdin, the agreement is not as good as that obtained for the porphyrin complex. The discrepancy also indicates small distortions from a square–planar ligand field.

C. Free-Radical Systems from Bilatrienes and Their Ni Complexes

Attempts to generate radical cations by one-electron oxidation of the free–ligand biliverdin and mesobiliverdin have not been successful. The only

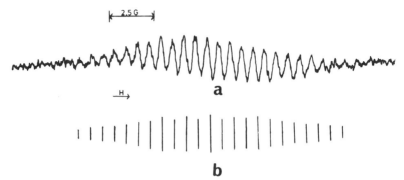

Fig. 5. ESR spectrum of the π-radical of the tetraoxobilatriene (**9a**) CH_2Cl_2 at room temperature (from Subramanian et al.[27]). (a) Experimental; (b) calculated.

bilatriene that has yielded an EPR spectrum so far is the synthetic dioxo bila-triene (**9**). Electrooxidation of this system at 0.30 V versus Ag/AgCl had yielded a free-radical system whose EPR spectrum consists of 23 lines spread over 13 gauss (Fig. 5). The hyperfine couplings and the spin densities calculated by the Hückel–McLachlan method appear in Table 3. From the EPR data and spin density calculations, it appears that considerable spin density is located at the carbon atoms, which are not bonded to protons, and to the oxygen atoms.

The Ni complexes of biliverdins (**5**) have yielded single-line EPR signals (line width 10 G). These systems have low oxidation potentials (-0.100–0.200 V versus SCE), and their EPR behavior is very similar to that of the Ni–dithiolate complexes reported by Maki et al.[37,38] From the g values, from

TABLE 3

Experimental and Calculated Hyperfine Coupling Constants a_i (in Gauss) for the π-Radical of Tetraoxobilatriene (9a) Generated by Electrooxidation in CH_2Cl_2[a][b]

Position	Nucleus	a_i (exptl.)	a_i (calc.)[c,d]
20.21	N	1.30	1.70
22.23	N	0.65	0.35
5.15	H	1.30	0.30
10	H	2.60	1.21

[a] From Subramanian et al.[27]

[b] The isotropic g value at room temperature is 2.0023 ± 0.0005.

[c] Calculated by the Hückel–McLachlan method[19,39] using the relations $a_H = -23\ p^C$ and $a_N = 28.45\ p^N + 2.62\ (p^{C1} + p^{C2})$, where C1 and C2 are carbon atoms directly bonded to the nitrogen.

[d] The Hückel parameters used are $h_N = 1.0$, $\beta_{CN} = 1.0$, $H_{\ddot{N}} = 2.0$, $\beta_{CN} = 0.5$, $h_O = 1.4$, and $\beta_{CO} = 0.9$.

line widths, and also from magnetic susceptibility measurements (1.75 BM as measured by Evans' method [18]) the Ni and Pd complexes of biliverdins have been identified as free-radical systems. At low temperatures, the EPR spectra of the frozen solutions of these compounds exhibit g anisotropy (see Fig. 6

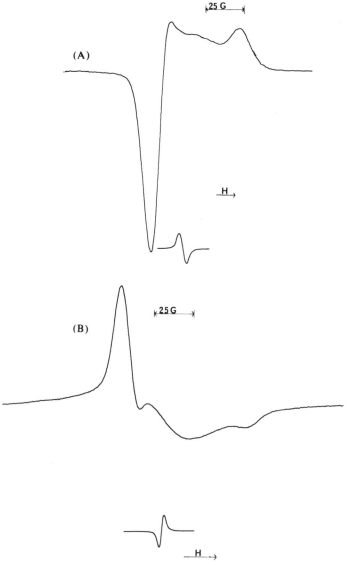

Fig. 6. ESR spectra of (A) nickel biliverdinate (**5e**) and (B) nickel bilatrienate (**9b**) in $CHCl_3$–EBBA at 100°K (from Subramanian *et al.*[27]).

TABLE 4

g Values of Some Nickel and Palladium Bilatrienates in Chloroform–EBBA Mixture at 100°K

System	g_1	g_2	g_3	$\langle g \rangle^a$
Ni octaethylformylbiliverdin (**8b**)	2.012	2.012	1.982	2.002 (2.004)
Ni biliverdin dimethyl ester (**5d**)	2.011	2.011	1.984	2.002 (2.003)
Ni (MBVD)	2.011	2.011	1.984	2.002 (2.003)
Ni tetraoxobilatriene (**9b**)	2.019	2.002	1.970	1.997 (2.001)
Pd tetraoxo bilatriene (**9c**)	2.011	2.011	1.965	1.995 (1.996)

^{a} The isotropic *g* values measured at room temperature are given in parentheses.

and Table 4).[27] Similar to the dithiolate complexes,[35] Ni and Pd bilatrienes also hve $g_\perp < g_{\parallel}$. This trend has been taken as evidence for the presence of Ni(III) character in the Ni bilatrienes. Since the *g* anisotropy is very small ($|g_{\parallel} - g_\perp = 0.049|$), the unpaired electron is considered to be predominantly in the π-orbitals of the ligand.

IV. X-RAY STRUCTURES

To our knowledge, only two types of open-chain tetrapyrrolic metal complexes have been investigated successfully. These are some dipyrromethene chelates of type **3** and two zinc complexes of zinc octaethylformylbiliverdinate (**8**).

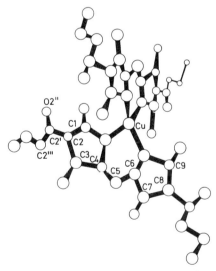

Fig. 7. X-Ray structure of copper dipyrromethanate (**3b**) (from Elder and Penfold,[42] with permission).

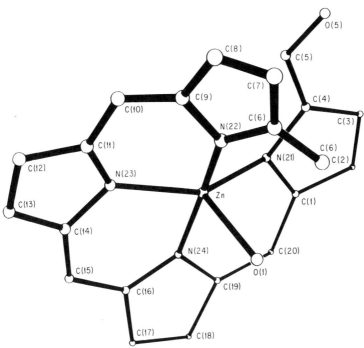

Fig. 8. X-Ray structure of zinc octaethylformylbiliverdinate (**8c**). Ethyl groups are omitted for clarity (from Struckmeier *et al.*,[43] with permission).

The nickel(II) complex of dipyrromethene (**3a**) is a distorted tetrahedron with a dihedral angle of 76.3° between the two ligand planes.[40] Copper(II) chelates are pseudo-tetrahedral.[41,42] The structure of **3b** is reproduced in Fig. 7. These few examples may suffice to establish the fundamental fact about pyrromethene metal complex structures, which is the basically tetrahedral arrangement of the four nitrogen atoms around the central ion. This has also been established in EPR studies of solutions (see Section III).

Recently, the structure of an open-chain tetrapyrrolic ligand, namely Zn(II) octaethylformylbiliverdinate hydrate, has been established for the first time by X-ray method.[43] This complex (**8c**) is almost planar. The slight deviations from planarity are imposed by the acyl substituents at the ends of the cyclic chromophore (Fig. 8). Addition of acid leads to dehydration of the metal complex. The coordination number of zinc changes from five to four. Normally for four-coordinated zinc, a tetrahedral configuration of the ligands is energetically preferred. Such an arrangement of the nitrogen ligands around the zinc is obtained, in the present case, by the formation of a wide bilatriene helix (inner diameter 3.6 Å) and formation of a bis-helical dimer (Fig. 9). In this dimer, each zinc ion is bound to the two nitrogen atoms of both the

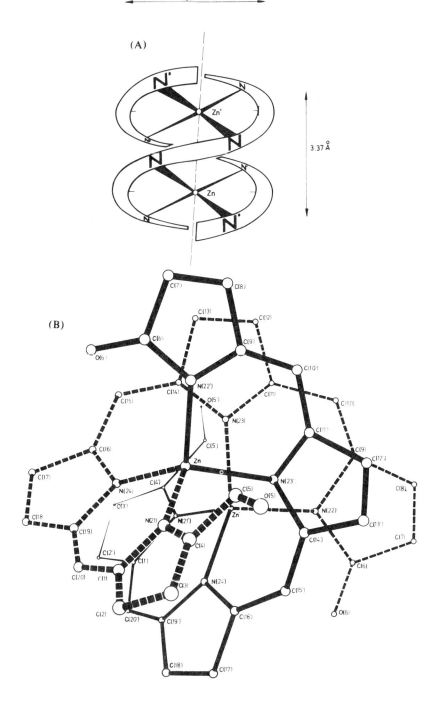

(A)

4.2 Å

3.37 Å

N'   N

Zn'

N

N

N

Zn

N

N'

(B)

C(7')   C(8')

C(13)   C(12)

C(6')   C(9')

O(6')

C(10')

N(22')   C(11)

C(14)   O(5')

C(15)   N(23)   C(10)

C(16)   C(5')   C(11')

Zn   C(9)

C(17)   N(24)   C(4')   N(23')   C(12')

C(18)   C(3')   C(5)   O(5)   C(8)

C(19)   N(21)   N(21')   C(4)   Zn'   N(22)

C(2')   C(1')   C(14')   C(13')

C(20)   C(1)   C(14)   C(7')

C(2)   C(20')   C(3)   N(24')   C(15')   C(16')

C(19')   C(16)   O(6)

C(18')   C(17')

C(15)

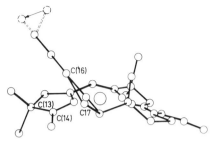

Fig. 10. X-Ray structure of the precorrinoid nickel complex (**10a**) (from Currie and Dunitz,[45] with permission).

chromophore molecules, and the coordination sphere is indeed a distorted tetrahedral one. The distance between the two zinc ions is 3.37 Å. The height of the chromophore helix, as defined by the distance of N-21 and N-23 parallel to the Zn–Zn axis, is 4.2 Å.

A similar structure as for the monomer in Fig. 8 has been determined for the *seco*-corrinoid nickel complex (**10a**). The distortion of planar coordination is stronger due to the large ethoxy group and is in the direction of tetrahedral coordination (Fig. 10).[44] In palladium and platinum complexes (**10b,c**), the tendency toward tetrahedral coordination is less marked than in the nickel complex (**10a**).[45]

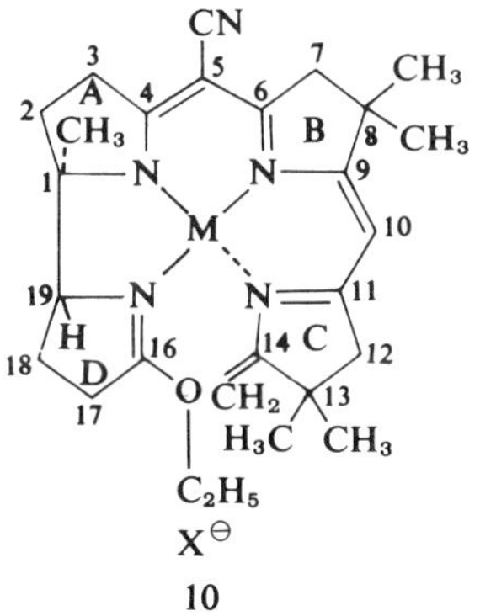

10

(a) M = Ni
(b) M = Pd
(c) M = Pt

←**Fig. 9.** (A) Schematic representation of the dimer of **8c**. (B) X-Ray structure of the dimer of **8c**. The ethyl groups are omitted for clarity (from Struckmeier *et al.*,[43] with permission).

V. CHEMICAL REACTIVITY

Mainly, the redox properties of metallobiliverdinates and ring closure reactions of biliverdinate and precorrinoid metal complexes have been studied. Virtually nothing is known about the influence of bound metal ions on electrophilic and nucleophilic addition or substitution reactions.

A. Oxidation and Reduction

Metal complexes of octaethylformylbiliverdin (8), biliverdin dimethyl ester (5a), and mesobiliverdin dimethyl ester (5f) produce polarographic oxidation and reduction waves at relatively moderate potentials (see Table 5).[22] The chemical reaction underlying these electrochemical observations is very probably the formation of π-radicals. Evaluation by EPR spectroscopy of these radicals, however, proved to be difficult because of the very short lifetime of these transient species. Two exceptions are nickel and palladium complexes of ligands 5a,f and 9, but the reasons for their stability are not yet understood. Their EPR spectra are discussed in Section III.

All other metallobiliverdinate radicals undergo ill-defined, irreversible addition reactions. Therefore, biliverdinate metal complexes are generally unstable toward oxidants. The primary products of oxidation reactions are usually oxoviolins (e.g., 11) from the oxidation of the zinc complex (8c). Surprisingly, formylbiliverdin (8c) in solutions of dry benzene is almost indefinitely stable toward oxygen and light. Only after the addition of small amounts of water or methanol is rapid formation of 11 observed, even in the dark.[23-25] Acetyl biliverdins (e.g., 12) are also quite stable toward light and oxygen and, in addition, are also rather inert against attack of protic solvents.[46,47] Metal-free formybiliverdinates are stabilized by the formation of a

TABLE 5
Midpoint Potentials of Metallobiliverdinates[a]

Metal	meso-Biliverdin dimethyl ester (5f)		Biliverdin dimethyl ester (5a)		Octaethylporphyrin	
	$E_{0.5}^{Ox\,b}$	$E_{0.5}^{Red\,b}$	$E_{0.5}^{Ox\,b}$	$E_{0.5}^{Red\,b}$	$E_{0.5}^{Ox\,b}$	$E_{0.5}^{Red\,b}$
2 H	0.53	−1.00	0.65	−1.14	0.81	−1.46
Mn	0.40	−0.53	0.42	−0.54	1.12	−1.61
Co	0.31	−0.71	0.28	−0.77	1.0	—
Ni	0.21	−0.86	0.30	−0.80	0.73	−1.5
Cu	—	—	—	—	0.79	−1.46
Zn	0.19	−1.0	0.26	−1.10	0.63	−1.61

[a] Versus SCE, from Fuhrhop et al.[22,70]
[b] Oxidations were usually carried out in butyronitrile, reductions in DMSO.

11   12

2,4-dinitrophenylhydrazone derivative.[22] We also found that α-amino-biliverdins reversibly form stable radicals on oxidation.[47a] Catalytic hydrogenation of metallobiliverdinates leads to metal-free bilirubins (BR), in close analogy to the behavior of free-base biliverdins reduction products.[48] A qualitative observation by Velapoldi and Menis indicates that the photo-oxidation of BR to biliverdin is catalyzed by transition metal ions. The most active metal ion was Nd(III), which is thought to produce square–planar complexes with BR.[14] The electrochemical reduction potentials of some bilirubin complexes with divalent metal ions have been determined,[49] but no correlations are apparent in the values observed. The Fe(II) complex is reduced at approximately -0.5 V, the Cd(II) complex at -0.7 V, and the Zn(II) complex at -1.2 V. All these potentials are only slightly lower than those of the uncomplexed metal ions, although complex formation is clearly indicated by the electronic spectra.

To summarize the experimental observations on the redox properties of biliverdinate and bilirubinate complexes, one may state that only in the most stable complexes are the chemical properties of the π-systems or the metal ion significantly changed, as compared with the free components. BR does form such complexes with some trivalent metal ions, especially with those of the rare earth series; biliverdin dimethyl ester and other bilatrienes do so with bivalent transition metal ions. π-Radicals are unstable with both ligands, except for some Ni(II) and Pd(II) complexes of bilatrienes and of α-aminobiliverdins.

B. Ring-Closure Reactions

The formation of tetrapyrrolic macrocycles from the chemically active open-chain tetrapyrroles is kinetically favored over intermolecular addition reactions. This is exemplified by the formation of porphyrinogens in almost quantitative yields from suitable pyrroles, e.g., porphobilinogen,[50] and of "acetone pyrrole" from acetone and pyrrole.[51] No "template effect" of a central metal ion, which would thermodynamically favor intramolecular over intermolecular reactions,[52] is needed. Most of the cyclizations of the open-chain tetrapyrrolic chromophores with extended π-conjugation pathways are little influenced by the presence of metal ions. Some examples of the rather extensive literature on this subject, which is interesting from the synthetic viewpoint, will be discussed to indicate its complexity.

1. BILADIENES TO CORROLES AND TETRADEHYDROCORRINS

The cyclization reactions of biladienes to corroles and tetradehydrocorrins have been observed and thoroughly investigated by Johnson and his co-workers. In the case of 1,19-dideoxybiladienes-a,c (13), where hydrogen atoms are the only substituents at the reaction centers, irradiation or addition of free-radical initiators in the presence of ammonia yielded rapidly up to 80% of the corrole-free bases (15). A radical mechanism involving oxidation of the primary anion (14) has been suggested.[53-55] The 10-dimethyl compound (16) could not be cyclized,[56] as would be predicted from the proposed sequence of reactions from 13 to 15.

If the reacting α-pyrrolic sites are substituted, e.g., with methyl or carbethoxy groups (17), a central transition metal ion with vacant d orbital in the xy-plane must be present if cyclization to a tetradehydrocorrin (e.g., 18) is to be achieved. The metal ion presumably takes up the electrons of the anion

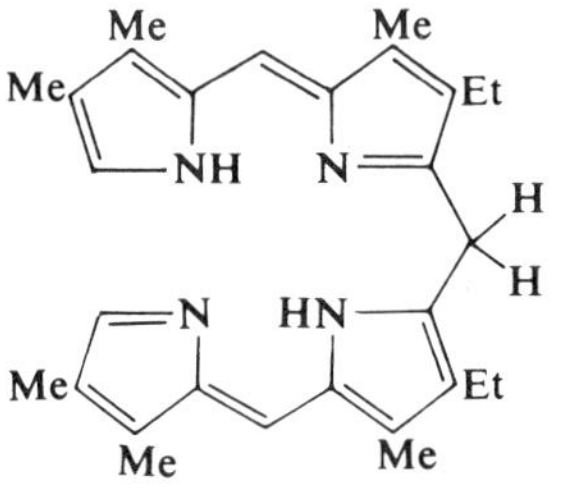

13

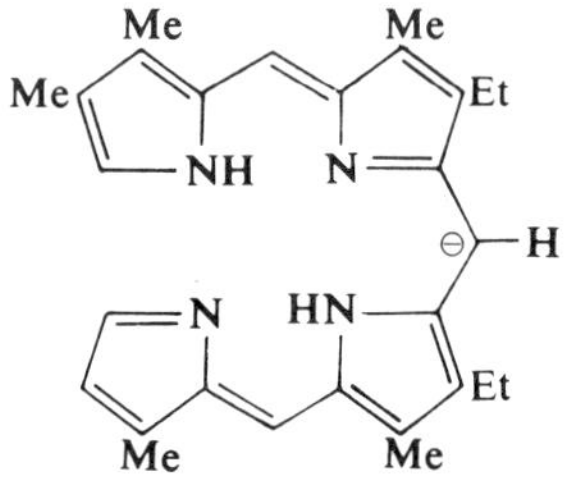

14

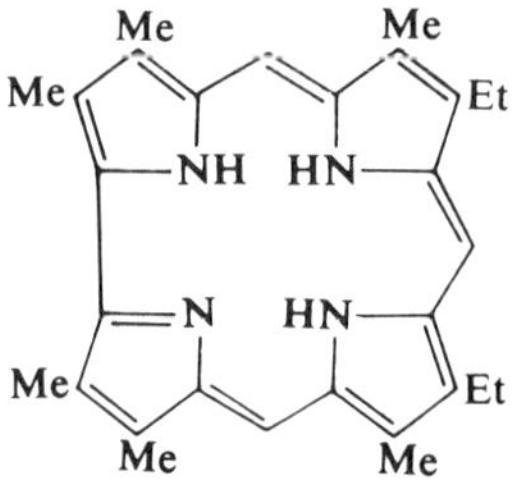

15

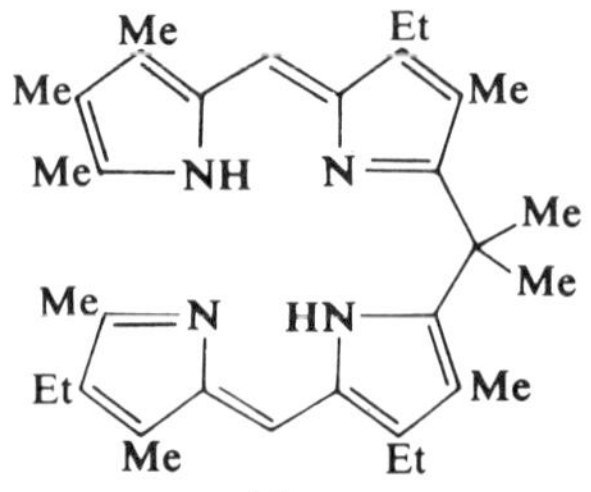

16

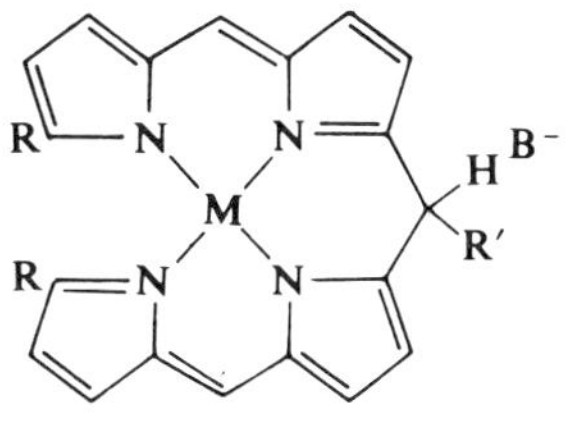

R = CH₃, COOEt

17   18

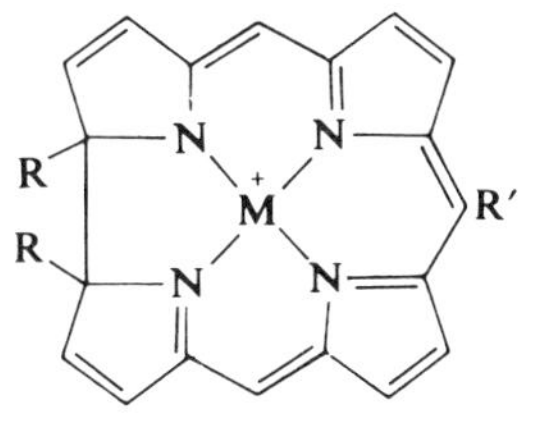

formed by the deprotonation of **17** and is then reoxidized by air.[53–55] Nickel and cobalt ions are generally used. Free-base biladienes of type **17** cannot be cyclized, and weakly bound metal ions (e.g., Zn) and radical initiators do not promote the cyclization. The latter actually inhibits the reaction.[53–55] From 1,19-diethoxybilirubindimethyl ester in the presence of nickel ions, the 1,19-diethoxytetradehydrocorrin nickel complex was obtained by Inhoffen et al.[57] Bilenes-b also cyclize to tetradehydrocorrins in the presence of base and nickel.[58] With copper(II) ions, oxidative cyclization of 1-methyl dideoxybiladienes to porphyrins is generally observed.[58]

2. Seco-CORRINOIDS TO CORRINOIDS

The most thorough study on the role of metal ions on cyclization reactions has been made by Eschenmoser's and Woodward's groups during the course of their total synthesis of vitamin B_{12}. Only a few relevant results will be discussed here. The observations on the stereospecificity in the cyclization, which led to the formulations of the Woodward–Hoffmann rules, have been reviewed elsewhere,[59] and will be omitted here.

Acid-catalyzed reaction with hydrogen sulfide, subsequent metallation with zinc perchlorate, and treatment with base converted the open-chain precorrinoid sodium salt (**19**) into **20**.[60] Treatment of **20** with trifluoroacetic acid presumably leads to an equilibration toward the ring-opened thiolactam (**21**). This was oxidized by benzoylperoxide to the hypothetical intermediate

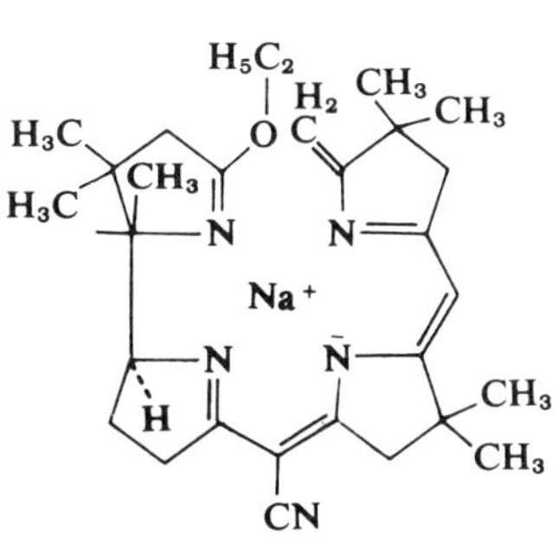

19 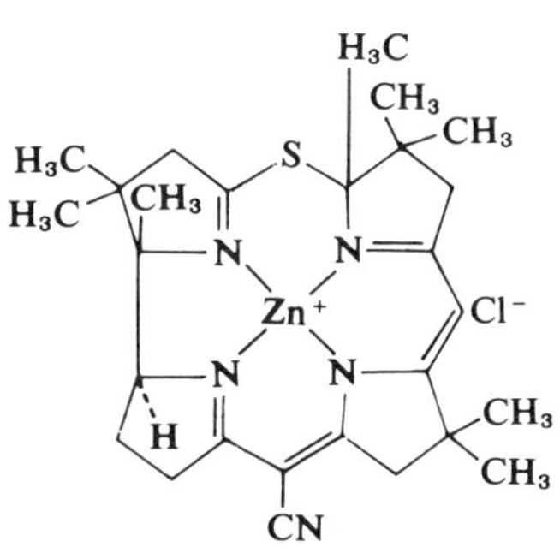   20

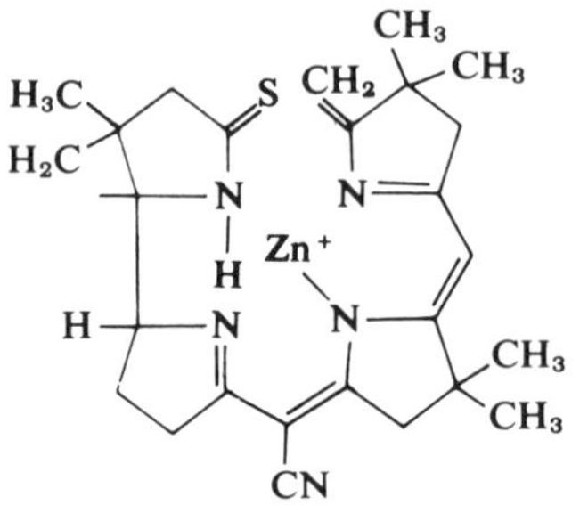

21

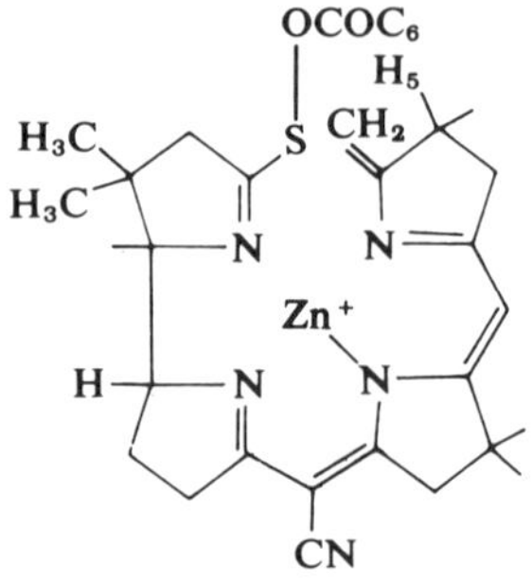

22

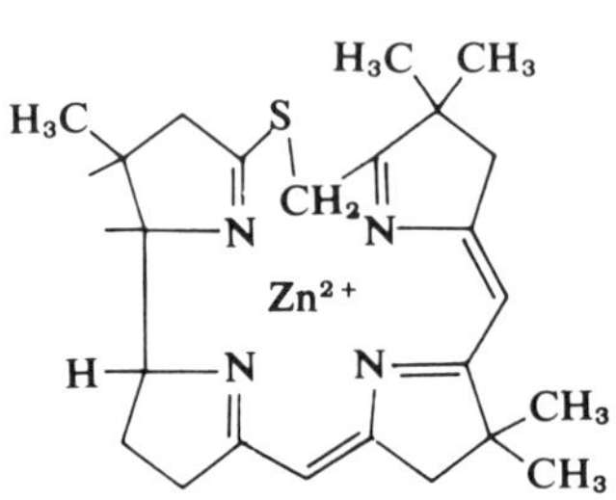

23

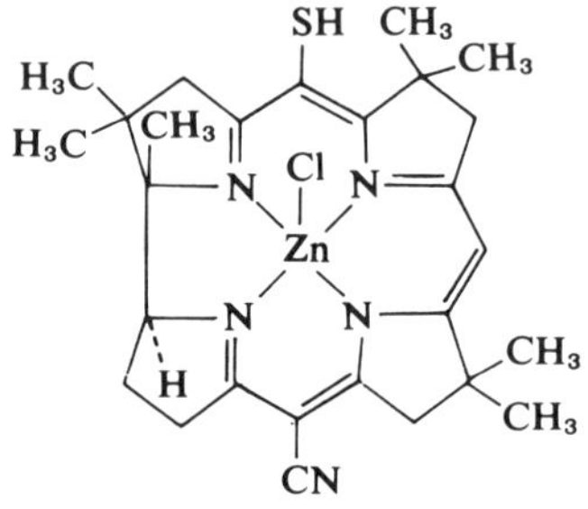

24

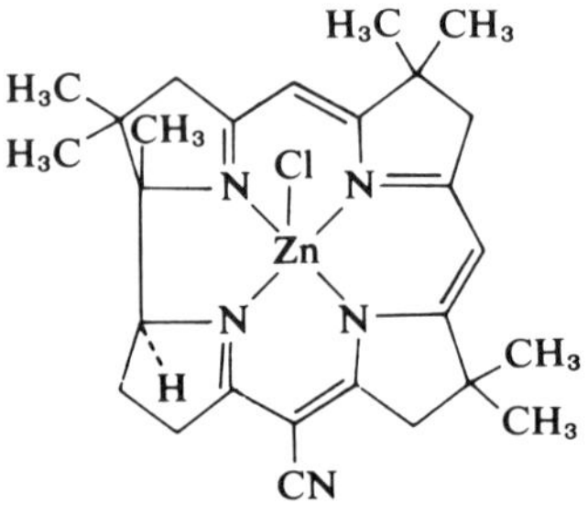

25

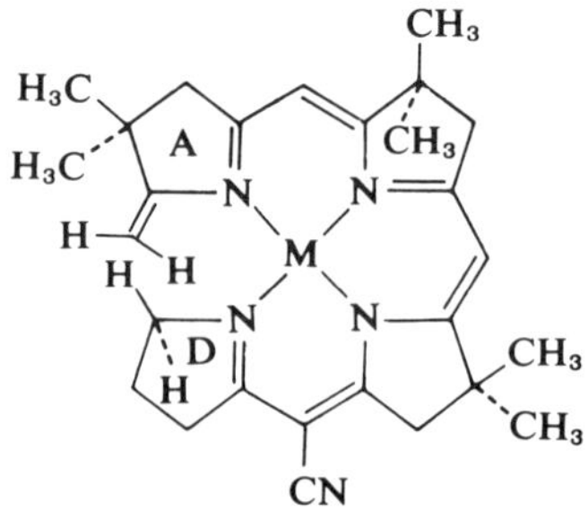

26

22 and cyclized to **23**.[61,62] Treatment of **23** with acid and EDTA removed the zinc and caused the formation of a corrin chromophore. This was again metallated with zinc to form **24** and desulfonated with triphenylphosphine to **25**.[60,61] Both cyclization reactions, **20** → **23** and **19** → **20**, occur with more than 50% yield and do not require central metal ions.

A second important cyclization is the photoreaction exemplified by the transformation **26** → **28**.[62] The rate-limiting step is the migration of a hydrogen atom at ring D to the methylene group in ring A; the central metal ion may be any "electronically inert" metal (e.g., Li^+, Na^+, Mg^{2+}, Zn^{2+}, Cd^{2+}), but not a proton. Whether the reactions **26** → **28** do indeed involve a triplet

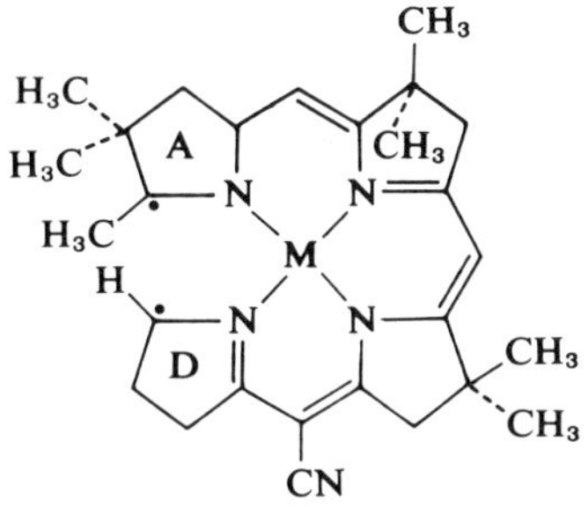

27

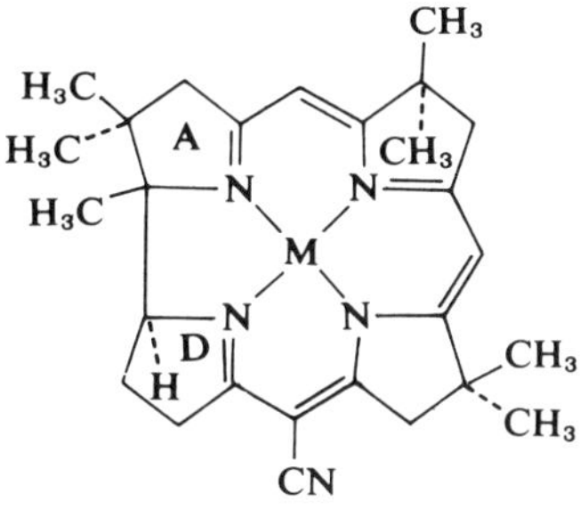

28

state, as suggested by structure **27**, has not been proved experimentally so far. Such an assumption is, however, quite plausible for a photoreaction and is further supported by the experimental finding that triplet-quenching central ions like Cu(II) and Ni(II) inhibit the cyclization. Considering the successful cyclizations of sodium and lithium salts, a "template effect" of the central metal ion on the reaction is probably of minor importance. A third type of cyclization reaction of *seco*-corrinoids occurs under strongly basic conditions and is reminescent of Johnson's tetradehydrocorrin synthesis (see Section V, A). The imine (**29**) is deprotonated to **30** and only cyclizes to **31** if Co(III), Ni(II), or Pd(II) are the central ions.

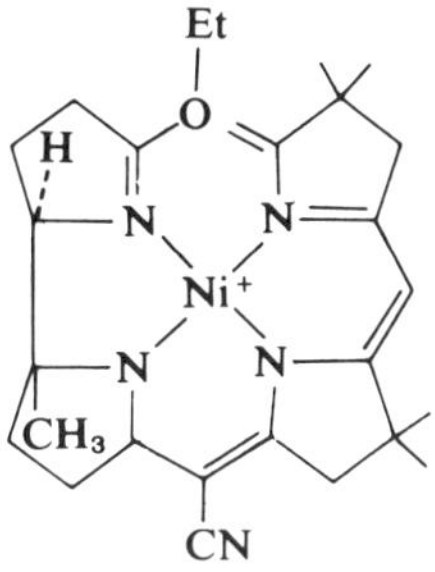

29

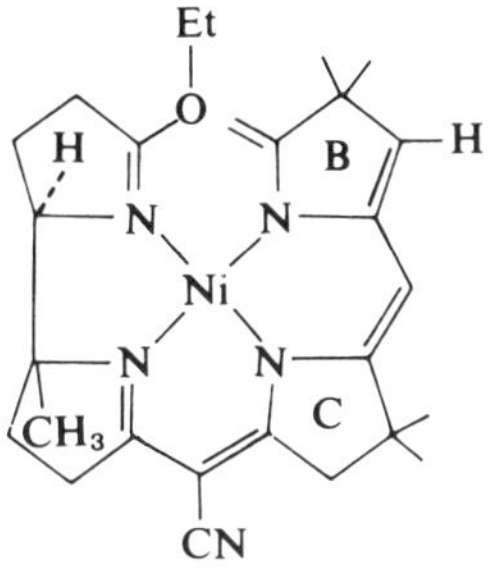

30

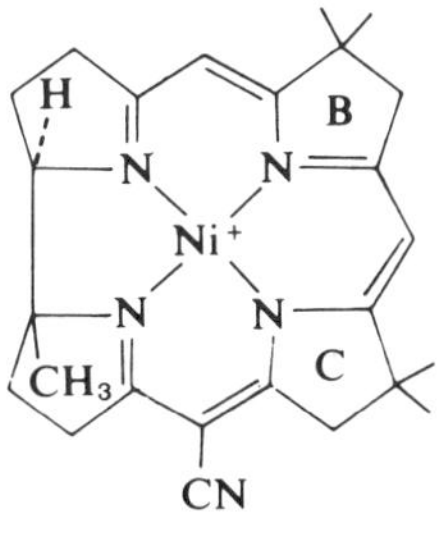

31

3. Metallobiliverdinates to Oxa-, Aza-, and Thiaporphyrins

Zinc complexes of mesobiliverdin dimethyl ester **5g** and related ligands cyclize in high yield to oxaporphyrins (e.g., **32**) in the presence of dehydrating agents such as acetic anhydride (Ac_2O) and dicyclohexylcarbodiimide (DCC). Free-base biliverdins cyclize in very low yield. With sodium hydroxide, the reaction is fully reversible. Ring opening with ammonia leads to zinc aminobiliverdinates, which may be recyclized to zinc azaporphyrins with Ac_2O. Reaction with sodium sulfide yields zinc thiobiliverdinates, which recyclize to zinc thiaporphyrins[47,71]. All these types of porphyrins have been obtained earlier by total synthesis.

COOMe   COOMe
32

4. Summary and Conclusions

All the open-chain tetrapyrrolic chromophores can be cyclized in good to excellent yield to porphyrins or corrinoids under a great variety of reaction conditions. In principle, it does not matter whether the *seco*-macrocycle is held in a more rigid conformation by the three connecting methine bridges (e.g., in biliverdins or *seco*-corrinoids), or in a flexible one, as in the urobilinogens. The reaction sites of an intramolecular cyclization of such chromophores are, on an average, not further apart than 9 Å, and this argument alone suffices to rationalize the formation of macrocycles, when the reaction conditions are favorable. Radical, photochemical, or electrophilic cyclizations, therefore, tend to occur without need for a strongly bound metal ion. Base-catalyzed reactions, however, which involve a chromophore anion as an intermediate, only occur with central ions stable to take up electrons. This is exemplified by Johnson's and Eschenmoser's experiences with *seco*-tetradehydrocorrinoids and -corrinoid metal complexes. Without such "electron sinks," the repulsion between the electrons at the bond-forming site of the conjugated systems is presumably too large and inhibits an approach of the reaction centers. In addition to what has been discussed above, Woodward–Hoffmann rules are applicable to photochemical cyclizations of metal-free systems.

VI. A COMPARISON OF OPEN-CHAIN TETRAPYRROLE CHROMOPHORES WITH RELATED MACROCYCLIC SYSTEMS

Porphyrins (33), phlorins (34), tetradehydrocorrins (35), and biliverdins (36) are selected for this comparison. The major variations in the series of tetrapyrrole pigments, which are of interest in this connection, are observed in these systems. Porphyrins (33) are thought to represent the most aromatic and rigidly planar molecules of the series; in phlorins (34) and tetradehydrocorrins (35) the macrocyclic conjugation is disrupted, and the pyrrole rings can be more easily twisted out of the plane of the macrocycle than in porphyrins. In addition, in biliverdin (36) the rigidity of the planar configuration is absent. Cis-trans isomerizations at the methine bridge double bonds, for which a variety of plausible mechanism can be formulated, would produce a different molecular shape that would range from cyclic (all-cis) to stretched (all-trans). Further variations in the steric appearances of biliverdin molecules are introduced by possible twists around the methine bridge single bonds. This may, for example, convert an almost planar macrocycle into a long helix (see Section V).

A. Electronic Spectra

The outer orbitals (HOMO, LUMO) of metalloporphyrins (33) are degenerate,[1] which leads to "forbidden" and "allowed" electronic transitions.[1,64-66] One finds a strong (Soret) band close to 400 nm, and two bands in the visible region, which are weaker by a factor of about ten (see Fig. 11).[24,67,68] The degeneracy of the outer orbitals is lifted in the three chromophores (34-36), and, therefore, both electronic transitions produce absorption bands of comparable intensity. The energies of the long-wavelength transitions are similar in porphyrins and tetradehydrocorrins and undergo a bathochromic shift of about 250 nm in phlorins and biliverdins. The order of transition energies is well reproduced by SCF calculations (Table 6).[63] Line widths of absorption bands rise with the expected thermal deformability of the ligands (36 > 34 > 35 ≥ 33).

TABLE 6

Experimental and Calculated (singlet–singlet) Electronic Transitions in the Longest Wavelength Region for Some Tetrapyrrole Pigment Zinc Complexes[a]

Chromophore	Phlorin (34)	Biliverdin (36)	Porphyrin (33)	Tetradehydrocorrin (35)
a_{max}(exper.)	810	790	574	570
a_{max}(theor.)	850	760	630	

[a] From Fuhrhop and Subramanian.[63] The assumptions and approximations used are discussed in this paper.

B. Redox Potentials

Metalloporphyrins (33) form stable π-radicals, whereas the radicals of the metal complexes of phlorins,[68] tetradehydrocorrins, and biliverdins[22] usually undergo rapid and irreversible addition reactions. Exceptions are nickel and palladium biliverdinates,[27] aminobiliverdin[47a] and metallooxophlorins[66a] (see Section 4) which form relatively long-lived radicals, and unsubstituted metalloporphinates, which polymerize immediately.[69] We propose that the stability of porphyrin radicals is partly caused by steric inaccessibility of reactive sites to reagents of addition reactions, whereas the flexibility of ligands 34-36 facilitates addition reactions. In general, metalloporphyrins are more difficult to oxidize or reduce than any of the other three ligands. For example, zinc complexes of octaethylporphyrin, octaethyl-α-oxophlorin, and biliverdin are oxidized to the corresponding π-radicals at 0.63, 0.32, and 0.26 V, respectively.

Another interesting fact must be mentioned here. The strong π-electron donating nature of porphyrins, combined with high-energy antibonding π-orbitals that do not permit backbonding, leads to central metal ions of very

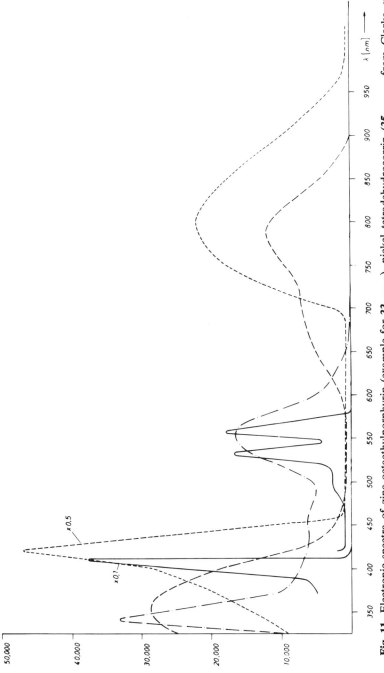

Fig. 11. Electronic spectra of zinc octaethylporphyrin (example for **33**, ———), nickel tetradehydrocorrin (**35**, –·–·–· from Clarke *et al.*[67]), zinc biliverdinate (**36**, –––, from Wasser and Fuhrhop[24b]), and the phlorin of mesoporphyrin (**34**, - - - -, from Peychal-Heyling and Wilson[68]).

high electron density or low oxidation potentials. Manganese (II) ions in the porphyrin cavity, for example, are, therefore, easily oxidized to the rare manganese(III) state. In biliverdinates, however, the antibonding π-orbitals are presumably of much lower energy, which facilitates the backbonding from the metal d(π)-orbital to the ligand. Therefore, stable Mn(II) biliverdinates are formed, and oxidation to Mn(III) could so far not be shown with these complexes. The same is probably true for ligands **34** and **35**.

C. Chemical Reactivity

It is well known that variations of the central metal ions in metalloporphyrins cause strong perturbations of the π-systems.[70] On the other hand, no clear-cut difference between the reactivities of free-base open-chain tetrapyrrole pigments and their various metal complexes so far has been demonstrated. The most active centers in biliverdinates, as with all other chromophores (**33–35**), are the methine-bridge carbon atoms. The regioselectivities of reactions of the biliverdinate have been described briefly in Section V, A and can be summarized here as follows. Reduction occurs at the middle methine bridge (position 10 in **36**), whereas oxidation first takes place on the outer bridges (positions 5 and 15). This behavior is well reproduced by results from SCF calculations and application of the frontier orbital model[63] (see Table 7).

TABLE 7

Electrophilic (Nucleophilic) Reactivity Parameters for Some Tetrapyrrole Pigments[a]

Atomic position	Porphyrin (33)		Phlorin (34)		Tetradehydrocorrins (35)		Biliverdin (36)	
1	0.19	(0.06)	0.07	(0.02)	—	—	0.15	(0.11)
2	0.06	(0.07)	0.11	(0.13)	0.00	(0.05)	0.00	(0.05)
3	—	—	0.02	(0.00)	0.00	(0.00)	0.25	(0.10)
4	—	—	0.13	(0.11)	0.01	(0.08)	0.00	(0.05)
5	*0.30*	*0.16*	—	—	0.30	(0.20)	—	—
6	—	—	—	—	0.00	(0.09)	—	—
7	—	—	—	—	0.02	(0.13)	—	—
8	—	—	—	—	0.02	(0.06)	—	—
9	—	—	—	—	0.01	(0.17)	—	—
10	—	—	*0.15*	*(0.31)*	*0.37*	*(0.00)*	*0.11*	*(0.06)*
11	—	—	0.12	(0.00)	—	—	0.11	(0.12)
12	—	—	0.02	(0.17)	—	—	0.11	(0.00)
13	—	—	0.13	(0.01)	—	—	0.04	(0.13)
14	—	—	0.13	(0.20)	—	—	0.19	(0.00)
15	—	—	0.00	(0.00)	—	—	0.00	*(0.40)*

[a] From Fuhrhop and Subramanian.[63]

ACKNOWLEDGMENT

This work was carried out within the technology program of the Bundesministerium für Forschung und Technologie (BMFT). We also thank the Deutsche Forschungsgemeinschaft and the Fonds der Chemischen Industrie for financial support.

REFERENCES

1. M. Zerner and M. Gouterman, *Theor. Chim. Acta* **4**, 44 (1966).
2. J.-H. Fuhrhop, *Struct. Bonding (Berlin)* **18**, 1 (1974).
3. J.-H. Fuhrhop, *Angew. Chem.* **86**, 363 (1974); *Angew. Chem., Int. Ed. Engl.* **13**, 321 (1974).
4. M. J. Burke, D. C. Pratt, and A. Moscowitz, *Biochemistry* **11**, 4025 (1972).
5. H. W. Siegelman and H. W. Butler, *Annu. Rev. Plant Physiol.* **16**, 383 (1965).
6. H. Fischer and M. Schubert, *Chem. Ber.* **56**, 1202 (1923).
7. A. Gossauer, "Die Chemie der Pyrrole," p. 180ff. Springer-Verlag, Berlin and New York, 1974.
8. Y. Murakami, Y. Matsuda, and K. Sakata, *Inorg. Chem.* **10**, 1728 and 1735 (1971).
9. Y. Murakami and K. Sakata, *Inorg. Chim. Acta* **2**, 273 (1968).
10. P. O'Carra, *Nature (London)* **195**, 899 (1962).
11. D. W. Hutchinson, B. Johnson, and A. J. Knell, *Biochem. J.* **133**, 399 (1973).
12. J. D. van Norman and R. Szentirmay, *Bioinorg. Chem.* **4**, 37 (1974).
13. C. C. Kuenzle, R. R. Pelloni, M. H. Weibel, and P. Hemmerich, *Biochem. J.* **130**, 1147 (1972).
14. R. A. Velapoldi and O. Menis, *Clin. Chem.* **17**, 1165 (1971).
15. A. W. Nichol and D. B. Morrell, *Biochim. Biophys. Acta* **177**, 599 (1969).
16. C. C. Kuenzle, *Biochem. J.* **119**, 395 (1970).
17. D. W. Hutchinson, B. Johnson, and A. J. Knell, *Biochem. J.* **123**, 483 (1971).
18. D. F. Evans, *J. Chem. Soc.* p. 2003 (1969).
19. A. D. McLachlan, *Mol. Phys.* **3**, 233 (1960).
20. R. Lemberg, *Biochem. J.* **28**, 978 (1934).
21. H. Fischer, H. Plieninger, and O. Weissbarth, *Hoppe-Seyler's Z. Physiol. Chem.* **268**, 197 (1941).
22. J.-H. Fuhrhop, A. Salek, J. Subramanian, C. Mengersen, and S. Besecke, *Justus Liebigs Ann. Chem.* **1975**, 1131 (1975).
23. J.-H. Fuhrhop and D. Mauzerall, *Photochem. Photobiol.* **13**, 453 (1971).
24. P. K. W. Wasser and J.-H. Fuhrhop, *Ann. N.Y. Acad. Sci.* **206**, 533 (1973).
25. J.-H. Fuhrhop, P. K. W. Wasser, J. Subramanian, and U. Schrader, *Justus Liebigs Ann. Chem.* **1974**, 1450 (1974).
26. A. Gossauer and H. H. Inhoffen, *Justus Liebigs Ann. Chem.* **738**, 18 (1970).
27. J. Subramanian, J.-H. Fuhrhop, A. Salek, and A. Gossauer, *J. Mag. Res.* **15**, 19 (1974).
27a. J. Knop and J.-H. Fuhrhop, *Z. Naturforsch., Teil B* **25**, 729 (1970).
27b. J. Subramanian, *in* "Porphyrins and Metalloporphyrins" (K. Smith, ed.), p. 555. Elsevier, Amsterdam, 1975.
28. H. Falk, O. Hofer, and H. Lehner, *Monatsh. Chem.* **105**, 169 (1974).
29. E. M. Roberts and W. S. Koski, *J. Am. Chem. Soc.* **82**, 3006 (1960).
30. J. M. Assour, *J. Chem. Phys.* **43**, 2477 (1965).
31. A. McCragh, C. B. Storm, and W. S. Koski, *J. Am. Chem. Soc.* **87**, 1470 (1965).

32. P. T. Manoharan and M. T. Rogers, in "Electron Spin Resonance of Metal Complexes" (Te Fu Yen, ed.), p. 143. Plenum, New York, 1969.
33. W. E. Blumberg, J. Peisach, B. A. Wittenberg, and J. B. Wittenberg, *J. Biol. Chem.* **243**, 1854 (1968).
34. G. F. Kokoszka, C. W. Reimann, and H. C. Allen, Jr., *J. Phys. Chem.* **71**, 121 (1967).
35. A. H. Maki and B. R. McGarvey, *J. Chem. Phys.* **29**, 35 (1958).
36. D. Kivelson and R. Neiman, *J. Chem. Phys.* **35**, 149 (1961).
37. A. H. Maki, N. Edelstein, A. Davidson, and R. H. Holm, *J. Am. Chem. Soc.* **86**, 4580 (1964).
38. A. Davidson, N. Edelstein, R. H. Holm, and A. H. Maki, *J. Am. Chem. Soc.* **2**, 1227 (1963).
39. B. L. Barton and G. K. Fraenkel, *J. Chem. Phys.* **41**, 1455 (1964).
40. F. A. Cotton, B. G. de Boer, and J. R. Pipal, *Inorg. Chem.* **9**, 783 (1970).
41. J. E. Ferguson and C. A. Ramsay, *J. Chem. Soc.* p. 5222 (1965).
42. M. Elder and B. R. Penfold, *J. Chem. Soc.* A p. 2556 (1969).
43. G. Struckmeier, U. Thewalt, and J.-H. Fuhrhop, *J. Am. Chem. Soc.* **98**, 278 (1976).
44. M. Dobler and J. D. Dunitz, *Helv. Chim. Acta* **54**, 90 (1971).
45. M. Currie and J. D. Dunitz, *Helv. Chim. Acta* **54**, 98 (1971).
46. H. Brockmann, Jr., *Phil. Trans. Roy. Soc. London* **B273**, 277 (1976).
47. G. W. Kenner, K. M. Smith, and J. F. Unsworth, *Phil. Trans. Roy. Soc. London* **B273**, 255 (1976).
47a. P. Krüger and J.-H. Fuhrhop, *Justus Liebigs Ann. Chem.*, in press.
48. J.-H. Fuhrhop, unpublished observation.
49. J. D. van Norman and R. Szentiermay, *Anal. Chem.* **46**, 1456 (1974).
50. D. Mauzerall, *J. Am. Chem. Soc.* **82**, 2605 (1960).
51. A. II. Corwin, A. B. Chirris, and C. B. Storm, *J. Org. Chem.* **29**, 3702 (1964).
52. J. P. Candlin, K. A. Taylor, and D. T. Thomson, "Reactions of Transition-Metal Complexes," pp. 87ff. Elsevier, Amsterdam, 1968.
53. A. W. Johnson and I. T. Kay, *J. Chem. Soc.* p. 1620 (1965).
54. D. Dolphin, A. W. Johnson, J. Leng, and P. van den Brock, *J. Chem. Soc.* C p. 880 (1966).
55. D. Dolphin, R. L. N. Harris, J. L. Huppatz, A. W. Johnson, and I. T. Kay, *J.Chem. Soc.*, C p. 30 (1966).
56. D. Dolphin, R. L. N. Harris, J. L. Huppatz, A. W. Johnson, I. T. Kay, and J. Leng, *J. Chem. Soc.*, C p. 98 (1966).
57. H. H. Inhoffen, H. Maschler, and A. Gossauer, *Justus Liebigs Ann. Chem.* **1973**, 141 (1973).
58. I. D. Dicker, R. Grigg, A. W. Johnson, H. Pinnock, K. Richardson, and P. van den Brock, *J. Chem. Soc.*, C p. 536 (1971).
59. R. B. Woodward, *Chem. Soc.*, *Spec. Publ.* **21**, 217 (1967).
60. A. Fischli and A. Eschenmoser, *Angew. Chem.* **79**, 865 (1967).
61. A. Eschenmoser, *Pure Appl. Chem.* **20**, 1 (1969).
62. A. Eschenmoser, *Naturwissenschaften* **61**, 513 (1974).
63. J.-H. Fuhrhop and J. Subramanian, *Phil. Trans. Roy. Soc. London* **B273**, 335 (1976).
64. W. T. Simpson, *J. Chem. Phys.* **17**, 1218 (1949).
65. M. Gouterman, *J. Chem. Phys.* **30**, 1138 (1959).
66. C. Weiss, H. Kobayashi, and M. Gouterman, *J. Mol. Spectrosc.* **16**, 415 (1965).
66a. J.-H. Fuhrhop, S. Besecke, J. Subramanian, C. Mengersen, and D. Riesner, *J. Am. Chem. Soc.* **97**, 7141 (1975).
67. D. A. Clarke, R. Grigg, R. L. N. Harris, A. W. Johnson, I. T. Kay, and K. W. Shelton, *J. Chem. Soc.*, C p. 1648 (1967).

68. G. Peychal-Heyling and G. S. Wilson, *Anal. Chem.* **43**, 545 and 550 (1971).
69. R. Schlözer and J.-H. Fuhrhop, *Angew. Chem.* **87**, 388 (1975).
70. J.-H. Fuhrhop, K. M. Kadish, and D. G. Davis, *J. Am. Chem. Soc.* **95**, 5140 (1973).
71. J.-H. Fuhrhop, P. Koriges, and W. S. Sheldrick, *Justus Liebigs Ann. Chem.* **1977**, in press.

9

Stereochemistry and Absolute Configuration of Chlorophylls and Linear Tetrapyrroles

HANS BROCKMANN, JR.

The field of stereochemistry was almost totally paralyzed throughout the first half of this century, mainly because organic chemists prided themselves erroneously on their ability to visualize the three dimensional structures with which they were dealing.

N. L. Allinger and J. Allinger (1965)

I. INTRODUCTION

To fulfill their various biologic functions as prosthetic groups of enzymes, e.g., in the electron transport system, as blood pigment, or in photosynthesis, tetrapyrroles are nearly always bound to proteins or membranes. Whereas a considerable amount of work has been done to study specific interactions between the chromophores and their chiral complexing ligands, until recently only few investigations about the stereochemistry of porphyrins and related compounds themselves were carried out. With the exception of vitamin B_{12}, the chirality of which was established in the course of its X-ray analysis,[1,2] only for a limited number of tetrapyrroles could the absolute configuration be determined by chemical methods within the last 10 years.

Nothing is known so far about the stereochemistry of porphyrins isolated from cytochromes and related enzymes, although optical activity of the hematoporphyrin derived from cytochrome c[3] and of hematin a[4] could be demonstrated and should be expected for the prosthetic group of cytochrome a_2[5] and for sirohydrochlorin[6,7] on the basis of their constitutional formulas.

The purpose of this chapter is to give a review of the chemical and spectrometrical investigations by which the absolute configuration of chlorophylls and their derivatives, as well as that for a number of bile pigments, has been determined, while the stereochemistry of corrinoids will not be referred to.

II. CHLOROPHYLLS OF HIGHER PLANTS

The designation "chlorophyll" is used as a generic name for magnesium complexes of macrocyclic tetrapyrroles that occur as photosynthetically active pigments in green plants and algae, as well as in phototrophic bacteria. With the exception of protochlorophyll,[8] the biosynthetic precursor of chlorophyll a, of chlorophyll c from brown algae,[9,10] and of chlorophyll e, which could be isolated so far from only one organism,[11,12] chlorophylls are derivatives of di- or tetrahydroporphyrins and contain asymmetric carbon atoms at the periphery of the macrocyclic chromophore.

As early as 1931, on the basis of the constitutional formula he proposed for chlorophyll a, Fischer[13,14] postulated optical activity for phytol-free derivatives of chlorophyll. It was demonstrated 2 years later.[15,16] Stereochemical investigations, however, were scarce in the following 40 years. Early work by Fischer and his group was mainly concerned with configurational correlations between the chlorophylls a and b and bacteriochlorophyll a,[17–20] and it lasted then over 10 years until Linstead et al. published their fundamental studies on the relative configuration of these pigments.[21–24] A decade after this, their absolute configuration could be determined,[25–30] but up to now there seem to exist no investigations regarding the chirality of the algal chlorophylls c, d, and e. For chlorophyll c, (E)-configuration of the acrylic acid side chain has been deduced from ^{1}H nmr coupling constants.[9,10]

A. Relative Configuration between C-17 and C-18 of Chlorophylls a and b

Derivatives of chlorophyll a (1) and chlorophyll b (2) afforded both optically active acids (dihydrohematinimides) with similar values of optical rotation by chromic acid oxidations. From these results it was first deduced that 1 and 2 have the same configuration at C-17 and C-18.[17] This conclusion was then substantiated when the corresponding methyl pheophorbides (3 and 4)

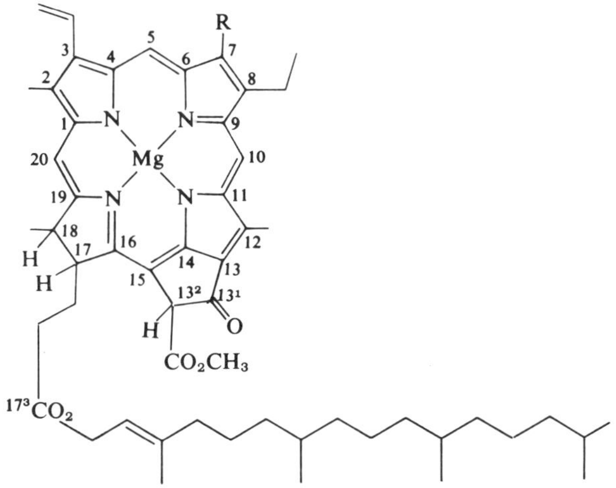

1: R = CH₃
2: R = CHO

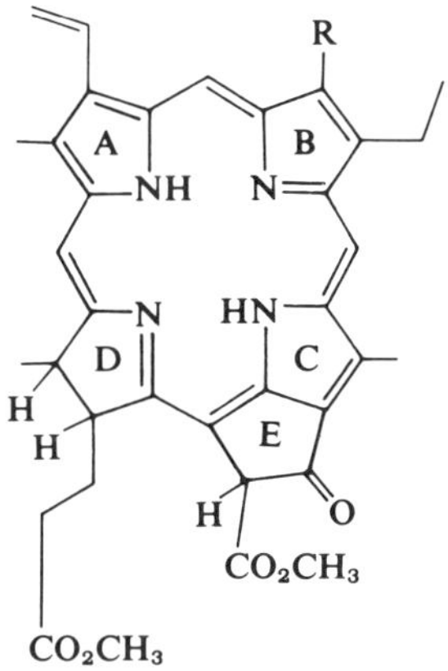

3: R = CH₃
4: R = CHO

could be transformed by decarbomethoxylation and subsequent reduction with palladium in formic acid to 13[1]-deoxo-*meso*-pyromethylpheophorbides (5), the identity of which was proved by mixed melting point, X-ray powder photographs, and optical rotation.[18] More recently, Thomas[31] reduced rhodin-g₇ trimethyl ester (6) with diborane to mesochlorin-e₆ trimethyl ester (7), which was identified by ^{1}H nmr-, mass spectrometry and, in particular, by ORD and CD with a sample of (7) obtained by catalytic hydrogenation of 8.

Without giving any explanation, Fischer and Gibian[31a] in 1942 assumed trans orientation of the 17- and 18-hydrogens of natural chlorins. Extended investigations concerning the stereochemistry of both chlorophyll a and bacteriochlorophyll a by means of oxidative degradation led Linstead *et al.*[21,23] to the same conclusion. Although these authors compared the optically active degradation products only with racemic synthetic compounds, a procedure

5

6: $R^1 = CH{=}CH_2$, $R^2 = CHO$
7: $R^1 = CH_2{-}CH_3$, $R^2 = CH_3$
8: $R^1 = CH{=}CH_2$, $R^2 = CH_3$

which can give rise to wrong deductions,[32] their results are often quoted as a proof for the relative configuration of chlorophyll *a*.

Besides the results of X-ray studies (see Section II, C), the relative stereo-chemistry at C-17 and C-18 is further confirmed by synthesis and nmr measurements. (a) In the course of Woodward's total synthesis of chlorophyll *a*,[33,34] the chiral centers at C-17 and C-18 were produced in a thermodynamically controlled reaction; thus, the side chains must be in the more stable trans configuration. (b) The coupling constant $J_{17,18}$ observed in the ^{1}H nmr spectra of pheophorbides is only about 2 Hz, indicating trans orientation of the corresponding hydrogens.[35] (c) By interaction with a bulky 20-substituent, e.g., a halogen atom, ring D of the macrocycle is turned out of the molecular plane in such a way that 18-H is more strongly influenced by the ring current, while 17-H and 18-CH$_3$ are withdrawn from its paramagnetic deshielding area.[36,37]

B. Relative Configuration between C-13^2 and C-17

Because the β-ketoester system in ring E of chlorophylls can enolize very easily, it was to be expected that in methyl pheophorbides, prepared from chlorophylls by acid treatment, the 13^2-methoxycarbonyl group and the 17-propionic acid side chain are in the more stable trans configuration. This has been proved by a combination of ORD and nmr spectrometric studies of 13^2-alkoxypheophorbides. Wolf et al.[38] succeeded in separating the diastereoisomeric 13^2-methoxymethylpheophorbides (9 and 10), prepared by oxidation of 3 with chloranil in the presence of methanol, and converted them to the corresponding pyro compounds (12 and 13).

Comparison of the ^{1}H nmr spectra of pyromethylpheophorbide *a* (11)

H
H
E
H
CO$_2$CH$_3$ CO$_2$CH$_3$

3

⟶

H
H
H$_3$CO   O
CO$_2$CH$_3$ CO$_2$CH$_3$

9

+

H
H
H$_3$CO   O
CO$_2$CH$_3$ CO$_2$CH$_3$

10

H
H
H   O
CO$_2$CH$_3$ H

11

H
H
H$_3$CO   O
CO$_2$CH$_3$ H

12

H
H
H$_3$CO   O
CO$_2$CH$_3$ H

13

and of **3**, on the one hand, and of **11** with those of **12** and **13**, on the other, revealed that the resonance signals of substituents in the vicinity of the 13^2-methoxycarbonyl group are little affected by its presence or absence. However, a 13^2-methoxy group has a distinct deshielding effect on those neighboring substituents which are located at the same side of the molecular plane. Thus, the relative configuration between C-13^2 and C-17 of **12** and **13** and, in analogy to this, of **9** and **10** could be determined unequivocally by ^{1}H nmr spectroscopy.

The chiroptical spectra of pheophorbides, on the contrary, are more influenced by the presence of a 13^2-methoxycarbonyl substituent than by the presence of a 13^2-methoxy group. Hence, the configurations of **3** and **10** could easily be correlated by ORD and CD spectra.[38]

Exactly the same studies were carried out in the chlorophyll *b* series with the result that **3** and **4** have the same configuration at 13^2.[39]

Because Houssier and Sauer[40,41] have demonstrated that, during the preparation of protopheophytin from protochlorophyll, little or no racemization at C-13^2 occurs, it seems to be desirable to determine by ORD or CD measurements of carefully prepared pheophytins *a* and *b* whether their configuration at C-13^2, and that of the corresponding chlorophylls, is the same as in **3** and **4** or as in the hypothetical 13^2-epimethylpheophorbides. The chiroptical properties of the latter compounds can be estimated by increment calculations.[38]

Opposite configuration at C-13^2 is assumed for the chlorophylls *a'* and *b'* described by Strain and Manning.[42,43] However, no ORD or CD spectra of these compounds have been measured, and the published ^{1}H nmr data are not in agreement with expectations from the work of Wolf *et al.*[38]

C. X-Ray Studies

X-Raycrystallography, the most powerful physical tool in structure determination of natural products, was unsuccessful for a long time in solving stereochemical problems in chlorophyll chemistry. After preliminary two-dimensional analyses,[44,45] Hoppe *et al.* were the first to show trans diaxial orientation of the C-17 and C-18 hydrogen atoms of chlorophyll derivatives by a three-dimensional structure determination of phyllochlorin methyl ester (**14**).[46]

14

The crystal structure of methyl pheophorbide *a* (**3**) was analyzed in 1971 independently by Hoppe's group[47] and by Fischer *et al.*[48] It confirms all the results described in the preceding section. Attempts to solve the problem of the absolute configuration of **3** by application of the Bijvoet technique were not conclusive.[48]

Recently the X-ray analysis of ethyl chlorophyllide *a* has been published.[49] The main purpose of this work, however, was not to solve stereochemical problems but to get information about intermolecular interactions with regard to possible mechanisms of photosynthesis.*

D. Absolute Configuration at C-13² by Indirect Methods

1. ^{1}H NMR OF PHEOPHORBIDES WITH CHIRAL 13²-ALKOXY GROUPS

In addition to compounds **9** and **10**, mentioned in Section II, B, a number of other 13²-alkoxypheophorbides were prepared[50] in order to study their

* In the detailed publication of this work,[49a] the determination of the absolute configuration of ethylchlorophyllide *a* is described as well. A second independent crystal structure analysis of ethyl chlorophyllide *a*[49b] as well as one of ethyl chlorophyllide *b*[49c] were published soon after.

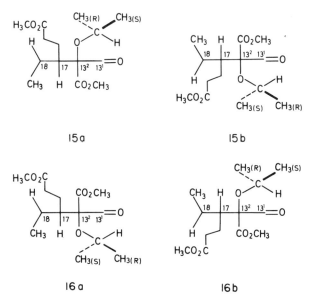

15: R = isopropyl: 16
17: R = (S)-sec-butyl: 18
19: R = (R)-sec-butyl: 20
21: R = (—)-menthyl: 22

spectroscopic properties. In the case of the 13^2-isopropoxymethylpheophorbides a (**15** and **16**), the interesting observation that the diastereotopic methyl groups of the alkoxy substituent give rise to ^{1}H nmr resonances at extremely different frequencies (1.19 and 1.59; 0.96 and 1.26 ppm, respectively) could only be interpreted in terms of preferred conformations of the isopropoxy groups. Inspection of space-filling molecular models revealed that, in these conformations, one of the methyl groups holds a position over the plane of the macrocycle and, thus, is located in the diamagnetic shielding area of the ring current, whereas the other methyl group is directed towards the periphery of the molecule (Fig. 1). Depending on the configuration at C-13^2, either the

Fig. 1. Possible absolute configurations of the 13^2-isopropoxymethylpheophorbides (**15** and **16**) and preferred conformations of their isopropoxy groups. The plane of the macrocycle is depicted as a horizontal line in these projections.

pro-*R* or the pro-*S* methyl group is in the central position and causes the high field resonance signal.

To distinguish chemically between pro-*R* and pro-*S* methyl groups, Richter[51] replaced isopropanol by 2(*S*)-butanol in the alkoxylation reaction. A triplet at 0.58 ppm in the ¹H nmr spectrum of one of the epimers obtained in this reaction clearly demonstrated the central position of C-4′ of the 2(*S*)-butoxy side chain, thus proving $13^2(R)$ configuration for this compound (**18**). The corresponding methyl group in the other epimer absorbed at 1.08 ppm, in agreement with its peripheral position in the $13^2(S)$ compound (**17**). Quite similar results gave the ¹H nmr spectra of 13^2[2(*R*)-butoxy]methylpheophorbides (**19** and **20**) and of 13^2[(−)-menthoxy]methylpheophorbides (**21** and **22**) (Figs. 2 and 3).

In summary, using optically active alcohols in the 13^2-alkoxylation of methyl pheophorbide *a*, ¹H nmr spectroscopy allows one to deduce the relative configuration between the chiral center of the alcohol and of C-13^2 of the alkoxypheophorbides. Because the steric relation between C-13^2 and C-17 in the compounds **15–22** can be derived also from their ¹H nmr spectra or independently from ORD or CD measurements (see Section II, B), these

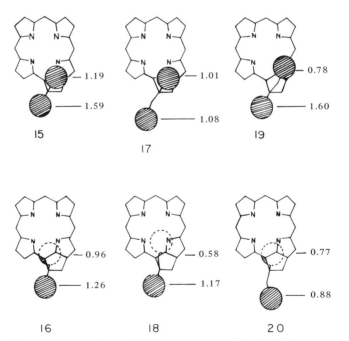

Fig. 2. Preferred conformations of 13^2-*sec*-alkoxy substituents of the pheophorbides (**15–20**). Chemical shifts of the methyl protons are given in ppm relative to tetramethylsilane.

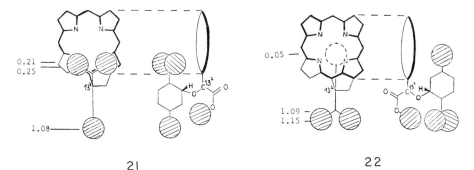

21   22

Fig. 3. Preferred conformations of the 13^2-[(−)-menthoxy] groups of the pheophorbides (**21** and **22**). Chemical shifts of the methyl protons are given in ppm relative to tetramethylsilane.

investigations allowed the determination of the absolute configuration $17(S)$ of chlorophyll a and its derivatives.[51]

2. HOREAU ANALYSES OF [13^1-HYDROXY]PHEOPHORBIDES

Another approach to determine the absolute configuration of chlorophylls at C-13^2 by indirect methods was the application of Horeau's method to [13^1-hydroxy]methylpheophorbides a. Compounds **23** and **24** are obtained by sodium borohydride reduction of methyl pheophorbide a (**3**)[52] and can be separated by repeated thin-layer chromotography (tlc) on silica gel. The four possible absolute configurations of these compounds (**23a,b** and **24a,b**) are schematically shown in Fig. 4.

Although *a priori* neither the relative configuration between C-13^2 and the newly introduced asymmetric center at C-13^1 was known, nor whether C-13, i.e., the macrocycle, or C-13^2 acts as large ligand of the carbinol carbon (C-13^1)

23   24

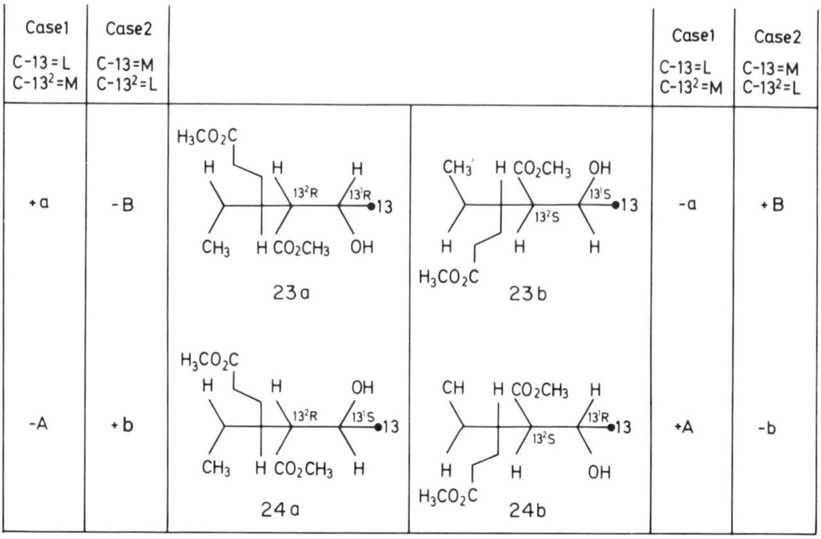

Case1	Case2		Case1	Case2
C-13=L	C-13=M		C-13=L	C-13=M
C-13²=M	C-13²=L		C-13²=M	C-13²=L
+a	-B	H₃CO₂C ... 23a ... 23b	-a	+B
-A	+b	24a ... 24b	+A	-b

Fig. 4. Expected results for the Horeau analyses of [13¹-hydroxy]methylpheophorbides.

in the kinetic resolution of α-phenylbutyric acid anhydride, the following consideration allowed Brockmann and Bode to deduce the absolute configuration at C-13² of 23 and 24.[53] In the case that C-13² acts as the medium-size ligand, according to Horeau's rule,[54,55] the 13¹(R) configurated alcohol (23a) should be preferentially acylated by the levorotatory acid, and the remaining α-phenylbutyric acid should be dextrorotatory. Its value of optical rotation (+a), however, is expected to be lower than that of the unreacted levorotatory α-phenylbutyric acid (−A) in the corresponding experiment with 24a [13¹(S) configuration] because the 13²-methoxycarbonyl substituent cis to the hydroxy group "enlarges" C-13². On the other hand, if C-13² is the large ligand, the Horeau analyses of 23a should afford levorotatory α-phenylbutyric acid with a higher degree of optical purity (−B) than that of the dextrorotatory acid (+b) obtained in the analysis of 24a. For the enantiomeric compounds 23b and 24b, one can predict, in the first case, α-phenylbutyric acids with optical rotations +A and −a; in the second case, with values of −b and +B.

In fact the α-phenylbutyric acids provided by Horeau analyses of 23 and 24 exhibited optical rotations of $[\alpha]_{578}^{22} = -38°$ and $[\alpha]_{578}^{22} = +21.7°$. Hence, the alcohols must have the configurations 23a and 24a (left half of Fig. 4), i.e., 13²(R) configuration is demonstrated for methyl pheophorbide a (3).

Based on the ^1H nmr coupling constants $J_{13^1,13^2}$, Wolf and Scheer[52] assigned structures 23 and 24 to the less and to the more polar diastereoisomer, respectively. Reduction of 23 and 24 with lithium aluminum hydride afforded

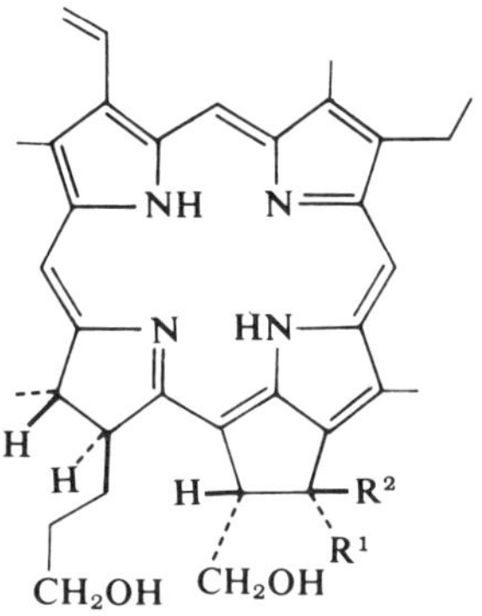

25: $R^1 = OH$, $R^2 = H$
26: $R^1 = H$, $R^2 = OH$

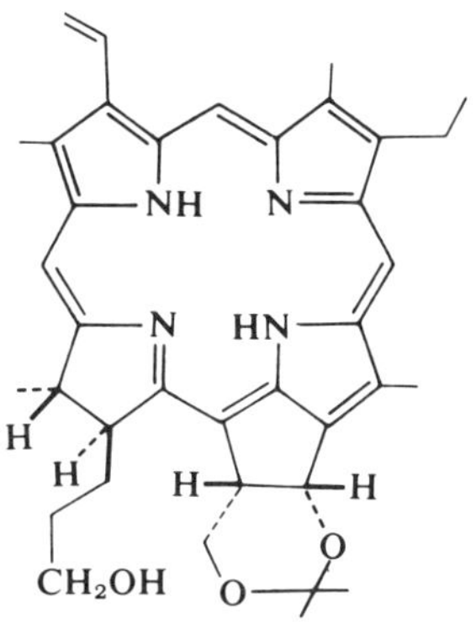

27

the triols (25 and 26). Their configuration between C-13^1 and C-13^2 follows from the fact that only 25 could be converted to the cyclic isopropylidene derivative (27).[53] Knowing the stereochemistry between C-13^1 and C-13^2, it was clear that C-13^2 acts as the large ligand and C-13 as the medium-size ligand in Horeau analyses of 23 and 24.

Interesting results were obtained when the pyro compounds 28–32 were subjected to Horeau analyses.[56] Under normal conditions, only monoacylation of 28 and 29 at the propanol side chain took place. As with compounds 30–32, the primary hydroxy groups, separated by three methylene groups from the next chiral center, caused small but reproducible stereoselectivities (2–5%). The fact that 28 reacted under these conditions preferentially with levorotatory α-phenylbutyric anhydride, while 29 showed the opposite stereoselectivity like the compounds 30, 31, and 32, was interpreted in terms of hydrogen bonding between 13^1- and 17^3-hydroxy groups as depicted in 28a. Because the monoester of 28, in a subsequent Horeau analysis under more vigorous conditions, was shown to be in the 13^1(S) configuration, 17(S) configuration of chlorophyll a has been concluded from these experiments.[56]

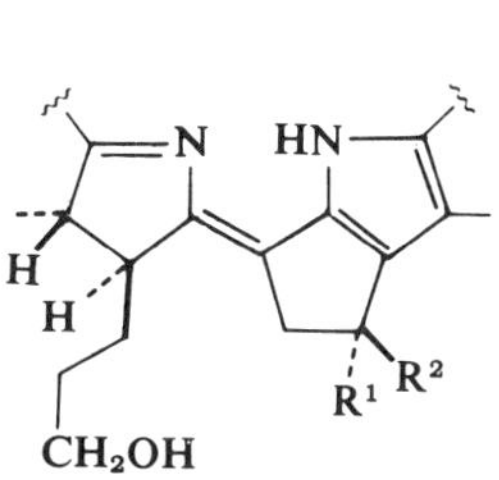

28: $R^1 = H$, $R^2 = OH$
29: $R^1 = OH$, $R^2 = H$
30: $R^1 = H$, $R^2 = OCH_3$
31: $R^1 = OCH_3$, $R^2 = H$
32: $R^1 = R^2 = H$

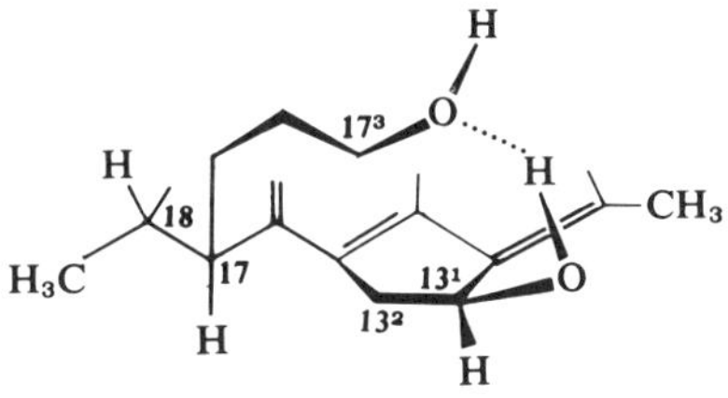

28a

E. Oxidative Degradation

By far, the best chemical proof of the absolute configuration of an optically active compound is to degrade and/or to convert it to another usually smaller compound of known chirality.

Chromic acid oxidation for the degradation of porphyrins was originally developed by Küster[57] in 1899. Using this technique, Fischer,[17,58] Mittenzwei,[20] and Linstead[21,23] have obtained optically active degradation products from chlorophyll and bacteriochlorophyll a derivatives, which were identified as dihydrohematinimide (33).

Compound 33 was first characterized as a silver[17] or a benzylamine salt.[21] For identification on a small scale now, the crystalline p-bromophenacycl ester (34), m.p. 99°C $[\alpha]_D^{20} = -32°$, seems to be the best derivative.[30] Compound 33 can be hydrolyzed to hematinic acid (35), which has been converted to its tri-p-bromophenacyl ester (36). In contrast to 34, 36 exhibits only a low value of optical rotation, $[\alpha]^{20} = +2.6°$.[27,30]

33: R = H
34: R = CH₂COC₆H₄Br

35: R = H
36: R = CH₂COC₆H₄Br

Until quite recently, the absolute configuration of only three 2,3-dialkyl substituted succinic acids was known, i.e., 2,3-dimethyl-,[59,60] 2,3-diethyl-,[61,62] and *threo*-2-ethyl-3-methylsuccinic acid.[62] Therefore, in order to determine the absolute configuration of chlorophylls at C-17 and C-18, Brockmann[30] planned to remove the propionic acid carboxy group so that chromic acid degradation would afford 2-ethyl-3-methylsuccinimide (44) rather than 33.

Because of the experimental difficulties he met trying to decarboxylate pheophorbides, Bode[56] alternatively reduced the propionic acid side chain to a propyl group. Pyromethylpheophorbide a (11), on treatment with lithium aluminum hydride in the presence of zinc chloride, afforded the 13¹ deoxo-alcohol (37), the p-toluenesulfonate (38) of which was again treated with lithium aluminum hydride to give 39. The optically active imide (40) obtained by oxidation of 39 has been identified with an authentic synthetic sample, and its configuration could be correlated to that of 44 by several methods[56] [see Section II, F, 3]. Barbier–Wieland degradation of 13¹-deoxo-*meso*-pyromethylpheophorbide a (5) finally enabled Brockmann and Müller–Enoch[63,64] to reach their goal. Reaction of 5 with phenylmagnesium bromide

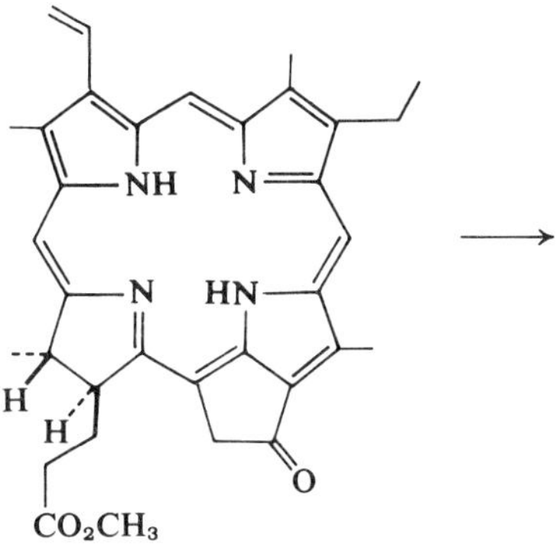

11

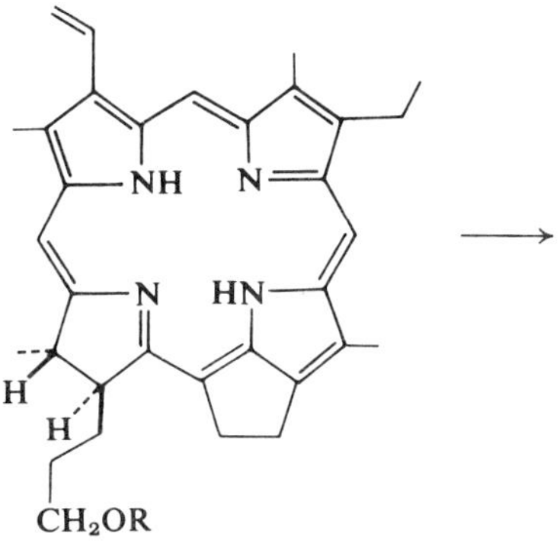

37: R = H
38: R = Tos

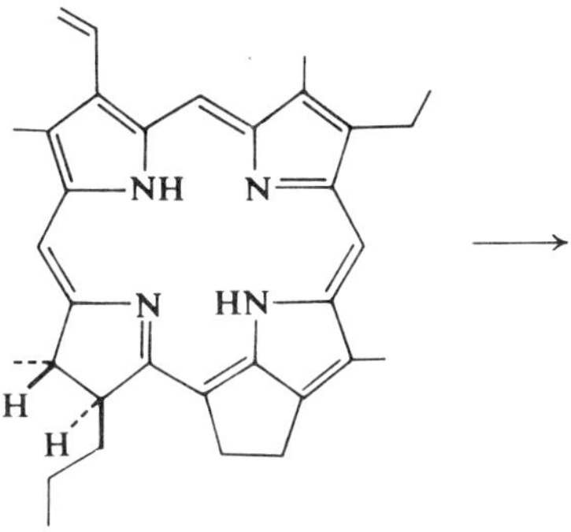

39

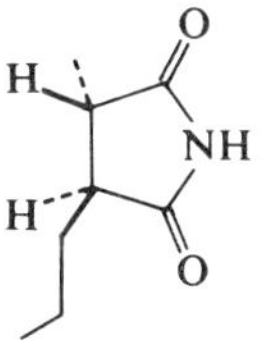

40

led to the tertiary alcohol (41), which in turn by heating with acetic acid gave the olefin (42) in very good yield.

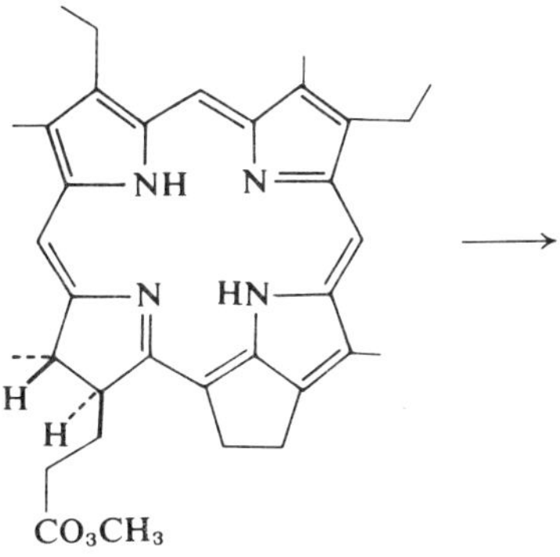

5

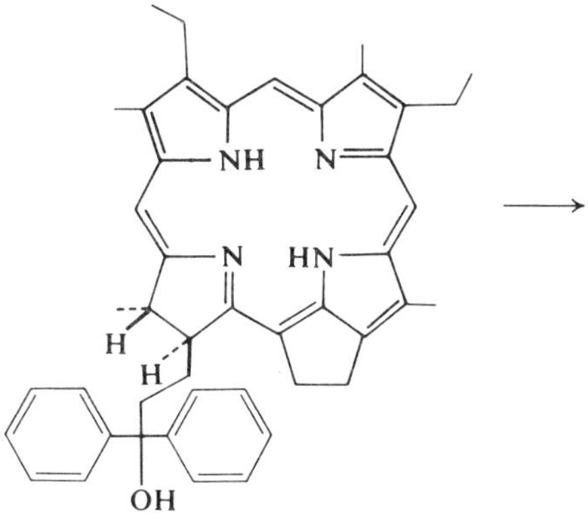

41

42

42

43   −44

Although selective oxidation of the exocyclic double bond of **42** was not achieved because of the sensitivity of the chromophore to oxidants, chromic acid treatment of **42** afforded the imide (**43**) which by anodic crossed coupling with acetic acid could be converted to 2(S)-ethyl-3(S)-methyl succinimide (−**44**).

F. Stereochemistry of Degradation Products

1. DIHYDROHEMATINIMIDES

The stereochemistry of both diastereoisomeric dihydrohematinic acids (**35** and **45**) and of the corresponding imides (**33**) was extensively investigated by Linstead et al.[22] In particular, the behavior of both acids in epimerization reactions was compared with that of the isomeric 2,3-dimethylsuccinic acids, the relative configurations of which were known from the fact that the low melting acid could be resolved into enantiomers.[65] All results indicated that the low melting dihydrohematinic acid is threo configurated (**35**).

By using the (+)-methylbenzylamine salts, Fleming[25,26] succeeded in resolving the *threo*-imide (**33**) prepared from synthetic (**35**). The oily dextrorotatory imide, $[\alpha]_D^{20} = +68°$, could be hydrolyzed to the crystalline tricarboxylic acid (+**35**), which was converted to the tri-*p*-bromophenacyl ester (−**36**), $[\alpha]_D^{20} = -2.8°$. This ester was assigned 2(R), 3(R) configuration because the enantiomer of it (+**36**) could be prepared from (−)-α-santonine of known absolute configuration[66–70], as illustrated in Fig. 5.

Because the degradation imide (**33**) from methylpheophorbide a obtained by Linstead et al.[21] was levorotatory, $[\alpha]^D = -46°$, Fleming[25,26] deduced 17(S),18(S) configuration for chlorophyll a. Final proof of this conclusion came from the direct comparison of samples of +**36** from (−)-α-santonine

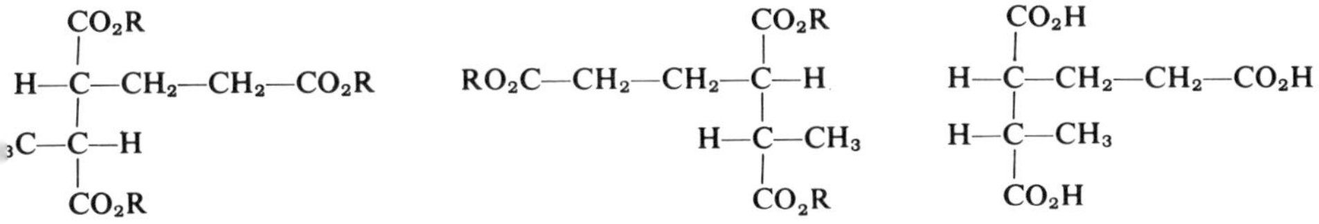

CO₂R
|
H—C—CH₂—CH₂—CO₂R
|
₃C—C—H
|
CO₂R

+35
−36

R = H
R = CH₂COC₆H₄Br

CO₂R
|
RO₂C—CH₂—CH₂—C—H
|
H—C—CH₃
|
CO₂R

−35
+36

CO₂H
|
H—C—CH₂—CH₂—CO₂H
|
H—C—CH₃
|
CO₂H

45

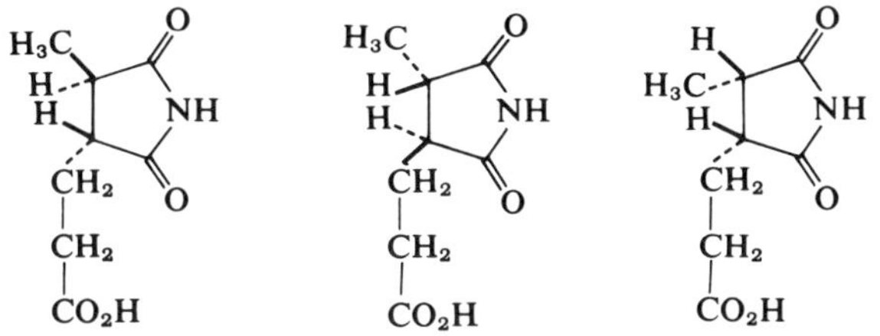

+33   −33   (erythro-33)

and from methyl pheophorbide a,[27,30] the identity of which could be shown unambiguously.

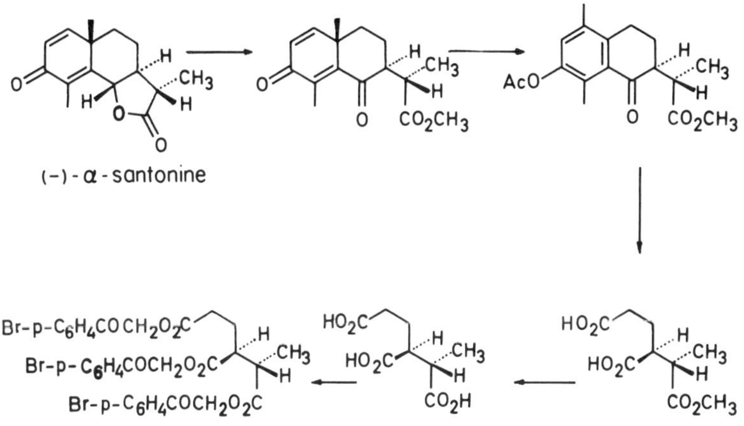

(−)-α-santonine

(+)-36   (−)-35

Fig. 5. Degradation of (−)-α-santonine to (−)-dihydrohemantinic acid.

2. 2-ETHYL-3-METHYLSUCCINIC ACIDS

The fact that the absolute configuration of the low melting 2-ethyl-3-methylsuccinic acid (47) could be correlated with a high degree of certainty[28,62] with the known chiralities of 2,3-dimethylsuccinic acid[59,60] and 2,3-diethyl-succinic acid[61] was one reason for Brockmann[30,71] to choose this acid as the key intermediate in his reaction sequence to determine the absolute configuration of chlorophylls *a* and *b* (see Section II, E). Another reason was that optically active 2-ethyl-3-methylsuccinimides (**44**) were also found to be oxidation products of bacteriochlorophyll a[20,23,29] and (−)stercobilin.[72,73] Moreover, the 2-ethylidene-3-methylsuccinimide obtained by degradation of aplysioviolin,[74] phycocyanin,[75] phycoerythrin,[75] and phytochrome[76] (see Section V, B) can easily be hydrogenated to **44**.[74,77]

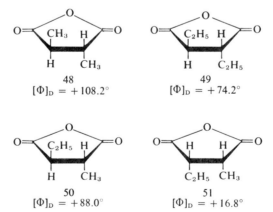

Berner and Leonardsen,[78] who first resolved the racemic acids (**46** and **47**), compared the molar optical rotations of their cyclic anhydrides (**50** and **51**) with those of 2,3-dimethyl- and 2,3-diethylsuccinic anhydride. The finding that the rotation of one of the anhydrides (**50**) is roughly equal to the sum of half of the molar rotations of **48** and **49**, whereas the rotation of the other anhydride (**51**) corresponds to the difference of half to the rotations of **48** and **49**, allowed Berner and Leonardsen to deduce the relative configurations of **50** and **51**. Since the absolute configurations of + **48** and + **49** are known to be 2(R),3(R),[59,61] this type of increment calculation was also used to derive the chiralities of **50** and **51** and the corresponding free acids.[28,79]

The results of quasiracemate studies carried out by Fredga and Terenius[62] indicate also 2(R),3(R) configuration of (+ **47**). In addition to the argumentation of Berner and Leonardsen,[78] Golden and Linstead[24] compared chemical and physical properties of **46** and **47** with those of the diastereoisomeric 2,3-dimethylsuccinic acids. Bode and Brockmann[80] repeated some of these experiments and extended their investigation also on the isomeric 2-methyl-3-

propylsuccinic acids. They substantiated the earlier conclusions regarding the relative stereochemistry, in particular by ^{1}H nmr studies of conformationally fixed cyclic derivatives of **46** and **47**.

Reaction of the levorotatory acids ($-$**46** and $-$**47**) via the diols (**52**) and their ditosylates (**53**) afforded in both cases ($-$)-2,3(S)-dimethylpentane (**54**) of known chirality,[81] thus, 2(S) configuration of both acids is also proved by chemical correlation.[79,82]

$$
\begin{array}{cccc}
\text{CO}_2\text{H} & \text{CH}_2\text{OR} & \text{CH}_3 & \text{CH}_3 \\
\mid & \mid & \mid & \mid \\
\text{H}-\text{C}-\text{C}_2\text{H}_5 & \text{H}-\text{C}-\text{C}_2\text{H}_5 & \text{H}-\text{C}-\text{C}_2\text{H}_5 & \text{H}-\text{C}-\text{CH}_3 \\
\mid & \mid & \mid & \mid \\
\text{CH}-\text{CH}_3 & \text{CH}-\text{CH}_3 & \text{H}-\text{C}-\text{CH}_3 & \text{H}_3\text{C}-\text{C}-\text{H} \\
\mid & \mid & \mid & \mid \\
\text{CO}_2\text{H} & \text{CH}_2\text{OR} & \text{CH}_3 & \text{CH}_2 \\
 & & & \mid \\
 & & & \text{CH}_3
\end{array}
$$

$-$**46**   **52**: R = H   **54**   **55**
$-$**47**   **53**: R = Tos

Although they did not doubt the assignment of the relative stereochemistry of **46** and **47**, Brockmann and Trowitzsch[83,89] tested it unambiguously by chemical methods in the following way. They converted the 1-carboxy group of $-$**46** into a methyl group and the 4-carboxy group into an ethyl group and obtained optically active 3,4-dimethylhexane; (**55**), hence, erythro configuration of the starting acid is proved because the *threo*-acid would have led to the corresponding *meso*-hydrocarbon. In a further experiment, $-$**46** was doubly homologized without attack of the chiral centers via reduction, tosylation, substitution with cyanide, and hydrolysis. The resulting 3-ethyl-4-methyladipic acid (**56**) afforded, on total reduction, 3-ethyl-4(S)-methyl-hexane (**57**), a hydrocarbon that is also accessible from 2(S)-methylbutyric acid (**58**).[84,85]

$$
\begin{array}{ccc}
^1\text{CO}_2\text{H} & \text{CH}_3 & \text{C}_2\text{H}_5 \\
\mid & \mid & \mid \\
\text{H}-\text{C}-\text{C}_2\text{H}_5 & \text{H}-\text{C}-\text{C}_2\text{H}_5 & \text{H}_3\text{C}-\text{C}-\text{H} \\
\mid & \mid & \mid \\
\text{H}-\text{C}-\text{CH}_3 & \text{H}-\text{C}-\text{CH}_3 & \text{H}-\text{C}-\text{CH}_3 \\
\mid & \mid & \mid \\
^4\text{CO}_2\text{H} & \text{C}_2\text{H}_5 & \text{C}_2\text{H}_5 \\
-\textbf{46} & \textbf{55} & \textbf{55}
\end{array}
$$

$$
\begin{array}{ccc}
\text{CO}_2\text{H} & & \\
\mid & & \\
\text{CH}_2 & \text{C}_2\text{H}_5 & \\
\mid & \mid & \\
\text{H}-\text{C}-\text{C}_2\text{H}_5 & \text{H}-\text{C}-\text{C}_2\text{H}_5 & \text{CO}_2\text{H} \\
\mid & \mid & \mid \\
\text{H}-\text{C}-\text{CH}_3 & \text{H}-\text{C}-\text{CH}_3 & \text{H}-\text{C}-\text{CH}_3 \\
\mid & \mid & \mid \\
\text{CH}_2 & \text{CH}_2 & \text{CH}_2 \\
\mid & \mid & \mid \\
\text{CO}_2\text{H} & \text{CH}_3 & \text{CH}_3 \\
\textbf{56} & \textbf{57} & \textbf{58}
\end{array}
$$

3. 2-METHYL-3-PROPYLSUCCINIC ACIDS

The relative configurations of the diastereoisomeric 2-methyl-3-propyl-succinic acids (**59** and **60**) have been established only quite recently by Bode and Brockmann,[80] using the same chemical and physical investigations as in the case of the 2-ethyl-3-methylsuccinic acids. These authors also correlated the absolute configuration of the low melting *threo*-acid (**60**) with those of other threo configurated 2,3-dialkylsuccinic acids.[86] Because (+)-α-naphthyl-ethylamine, as well as (+)-α-methylbenzylamine, is in all cases preferentially acylated by the anhydrides of the levorotatory acids under the condition of kinetic resolution, it was concluded that − **60** is 2(S),3(S) configurated.

By fractional crystallization of the (+)-α-naphthylethylamine salts, Bode[56] was able to resolve racemic **60**. The levorotatory acid, $[\alpha]_D^{20} = -19.1°$, and

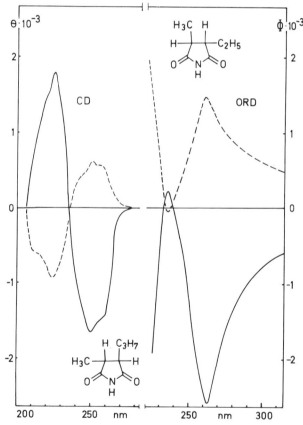

Fig. 6. CD and ORD spectra of 2(S)-methyl-3(S)-propyl succinimide (− **40**) and 2(R)-ethyl-3(R)-methyl succinimide (+ **44**).

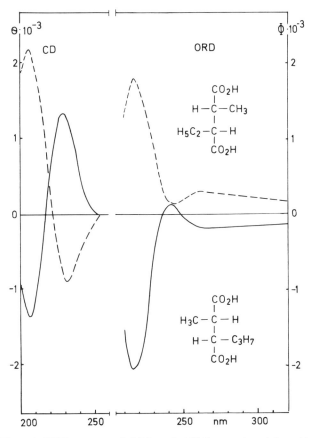

Fig. 7. CD and ORD spectra of 2(S)-methyl-3(S)-propylsuccinic acid (−60) and 2(R)-ethyl-3(R)-methylsuccinic acid (+47).

its dextrotatory di-p-bromophenacyl ester (61) were used for the direct comparison with samples of 60 and 61 prepared by hydrolysis and esterification of the 2-methyl-3-propyl succinimide (−40), which was obtained by chromic acid oxidation of 39 (see Section II, E).

59   60: R = H
     61: R = CH₂COC₄H₆Br   40

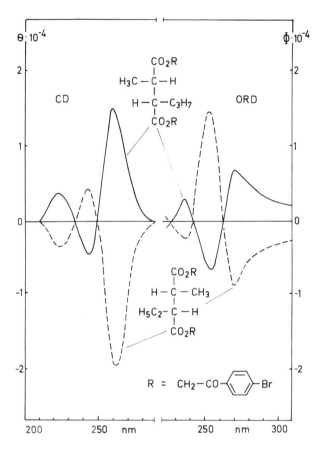

Fig. 8. CD and ORD spectra of the di-*p*-bromo-phenacyl esters of 2(*S*)-methyl-3(*S*)-propylsuccinic acid and 2(*R*)-ethyl-3(*R*)-methylsuccinic acid.

2(*S*),3(*S*) configuration of **40**, −**60**, and +**61** can also be derived from comparison of the ORD and CD spectra of these compounds with those of 2(*R*)-ethyl-3(*R*)-methylsuccinic acid and its derivatives. As can be seen from Figs. 6, 7, and 8, in all cases the curves are nearly mirror images.

III. BACTERIAL CHLOROPHYLLS

A. Bacteriochlorophyll *a*

The constitutional formula (**62**) of the long-known bacteriochlorophylls *a* from purple bacteria was determined by Mittenzwei[20] in 1942, and only recently it was found that bacteriochlorophyll (BChl) *a* with different terpene alcohol side chains (**62** and **63**) may occur in these organisms.[87-91] However, the bacteriomethylpheophorbides *a* (**64**) derived from BChl a$_P$ (**62**), as well as from BChl a$_{Gg}$ (**63**), have been shown to be identical also with respect to their stereochemistry.[87]

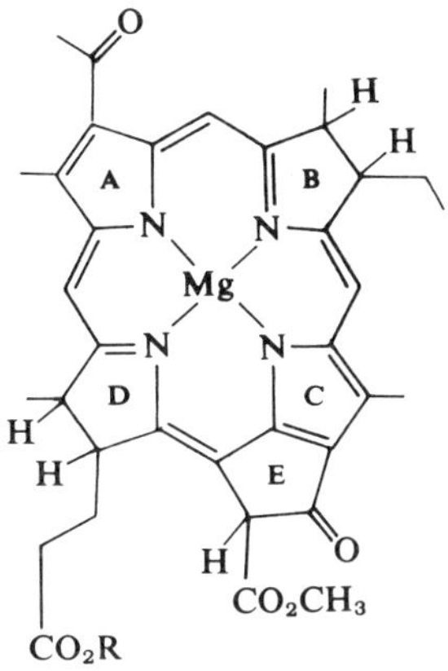

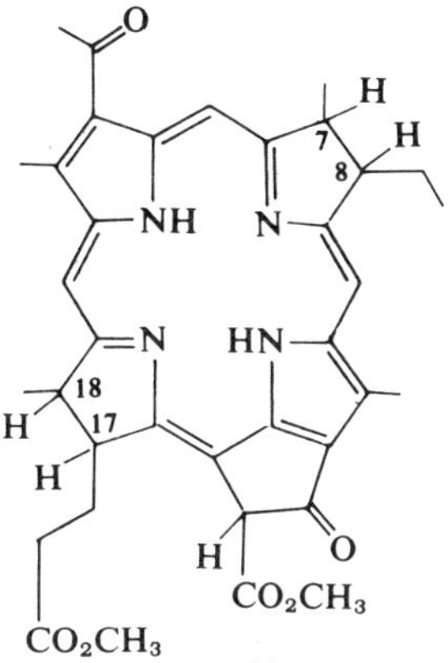

62: R =

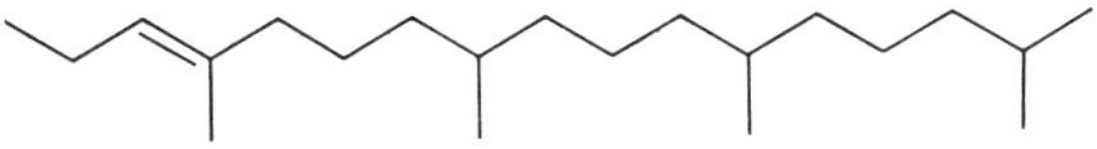

63: R =

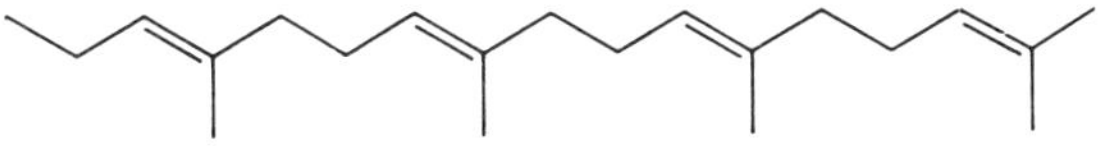

1. THE CONFIGURATION AT C-7 AND C-8

Already Mittenzwei has obtained, besides dihydrohematinimide (**33**), a second optically active compound by oxidative degradation of **64**. It is derived apparently from ring B and was believed to be 2-ethyl-3-methylsuccinic anhydride.[20] Mittenzwei even compared its optical rotation with that of the synthetic threo compound (**51**)[78] but did not draw any conclusion regarding its stereochemistry. With respect to modern knowledge about chromic acid

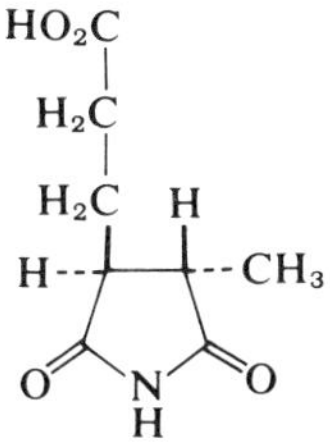

− 33 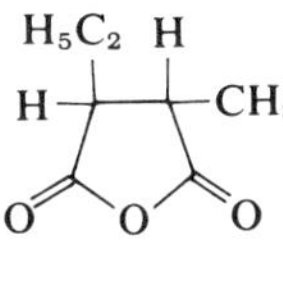   51 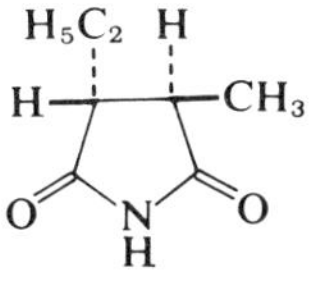   + 44

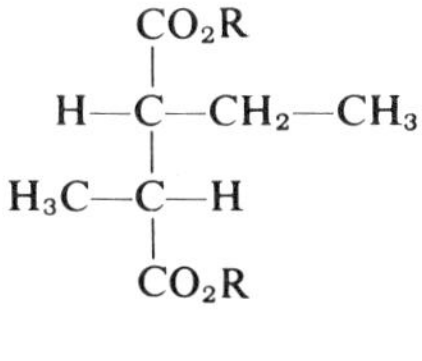

+47: R = H
−65: R = CH₂COC₆H₄Br

oxidation, it seems likely that Mittenzwei, in fact, had not obtained the anhydride but, rather, an impure sample of the corresponding imide (44).[71]

Golden et al.[23] repeated the degradation of a BChl a derivative and did obtain dextrorotatory 2-ethyl-3-methyl succinimide (+44), which was identified with a synthetic sample of 44 by paper chromatography. Hydrolysis of +44 and subsequent esterification with p-bromophenacyl bromide afforded the crystalline diester (−65), $[\alpha]_D = -37°$, the ir spectrum (KBr) of which was identical with that of synthetic racemic 65.

Because it is inadmissible to identify optically active compounds with racemic samples by ir spectra in solid phase,[32,71] final proof of the trans orientation of the hydrogens at C-7 and C-8 of bacteriochlorophyll a came only from the comparison of the degradation product (−65) with the synthetic levorotatory ester prepared from +48 of known 2(R),3(R) configuration.[28,29] By these experiments, also, the absolute configuration at C-7 and C-8 is determined.

2. CONFIGURATION AT C-17 AND C-18

That bacteriochlorophyll a (62) and chlorophyll a (1) have the same configuration at C-17 and C-18 follows from the indentity of 7,8-dehydro derivatives of 62 with compounds that could be prepared also from chlorophyll a. However, the proofs of identity used earlier by Fischer et al.[19,92,93] and by Mittenzwei[20] were mainly based on agreement of melting points, basicity, crystalline form, absorption spectra, and Debye–Scherrer photographs, whereas the values obtained for optical rotation agreed only qualitatively.[20] Moreover, the acidic, levorotatory oil obtained on oxidative degradation of 64 could only be insufficiently characterized as −33.[20]

Whereas dehydrogenation of 64 with tetrachloro-o-benzoquinone was

66　　　　　　　　　　67

accompanied by oxidation at C-13^2, Brockmann and Kleber[29] found that bacteriopyromethylpheophorbide a (**66**) in benzene reacted smoothly with this reagent to give [3-acetyl] pyromethylpheophorbide a (**67**) whose physical properties, and in particular the ORD spectrum, all agreed with those of a sample prepared from chlorophyll a.[29,94]

3. THE CONFIGURATION AT C-13^2

A priori, it could be expected that the 13^2-methoxycarbonyl group and the 17-propionic acid side chain of **64** are in the more stable trans configuration as they are in the methyl pheophorbides a and b. This assumption has been substantiated by ^{1}H nmr and ORD spectroscopy.[71] Removal of the 13^2-methoxycarbonyl group of **64** causes a slight upfield shift of the 12-CH$_3$ resonance in the ^{1}H nmr spectra, while the absorptions of the 17-hydrogen and the 17^3-ester methyl group are shifted downfield as in the chlorophyll a series.

In the ORD spectra, the loss of the 13^2-methoxycarbonyl group of **64** is accompanied by an enhancement of the $Q_{X(0-0)}$ Cotton effect and the disappearance of the Cotton effect (E) band as it is described by Wolf *et al.*[38,39,50] for other pheophorbide–pyropheophorbide pairs.

B. Bacteriochlorophyll b

Bacteriochlorophyll b, first isolated and characterized by Eimhjellen *et al.*,[95] occurs as a photosynthetically active pigment in a number of sulfur-containing and nonsulfur-containing purple bacteria, e.g., *Thiocapsa pfenigii* and *Rhodopseudomonas viridis*. It contains phytol as esterifying alcohol[91,96] and is easily converted to 7,8-dehydrobacteriochlorophyll a_P (**69**) in acetone solution. The structure and stereochemistry of the latter compound was established by preparation of the corresponding methyl pheophorbide (**70**) which, in turn, has been converted into **67**. The physical properties of the product thus obtained, in particular the ORD and CD spectra, were superimposable to those of authentic samples of **67** prepared either from chlorophyll a or from bacteriochlorophyll a.[96]

Recently, Scheer *et al.*[97] deduced structure **68** for bacteriochlorophyll b. Their evidence is based mainly on ^{1}H nmr and mass spectra, as well as on the results of a micro ozonolysis, which yielded acetaldehyde from the ethylidene side chain. Experiments to obtain ring B of **68** in the form of the corresponding imide by chromic acid degradation have been so far unsuccessful.[98] Thus, nothing is known about the configuration of the 8,8^1-double bond and the chirality at C-7.

CO$_2$phytyl

68

69: $\Delta^{7,8}$ instead of $\Delta^{8,8^1}$

CO$_2$CH$_3$

70: R = CO$_2$CH$_3$

67: R = H

C. Bacteriochlorophylls c, d, and e

In addition to bacteriochlorophyll a_P, which is probably located in the reaction center of the photosynthetic apparatus, green bacteria (Chlorobiaceae and Chloroflexaceae) contain a number of other photosynthetic tetrapyrroles, some structural features of which differ from all other known chlorophylls. These *Chlorobium* bacteriochlorophylls occur as mixtures of homologues,[99,100] are esterified with farnesol,[101,102] rather than with phytol or geranylgeraniol,* and lack a 13^2-methoxycarbonyl group.[102] They are dihydroporphyrin derivatives and possess a third chiral center in the 3-hydroxyethyl side chain in addition to those at C-17 and C-18.[102]

Until recently, only two types of *Chlorobium* bacteriochlorophylls were known and called the 660 and 650 series, according to their long-wavelength absorption maxima.[103] The designation bacteriochlorophyll c and d proposed for these pigments by Jensen *et al.*[104] is used so far mainly by biologists, but at least, since the discovery of a further type of *Chlorobium* bacteriochlorophyll (BChl e) isolated from *Chlorobium phaeovibrioides*,[105] the use of this more versatile nomenclature seems favorable and will be used here.

On the basis of chemical degradation studies and ^{1}H nmr investigations, Holt *et al.*[99,100,106–109] proposed structures **71** and **72** for the bacteriomethylpheophorbides *d* and *c*. For some time, the nature and position of the *meso*-alkyl group of **72** was doubted,[110–112] but the presence of a 20-methyl group has been proved unambiguously by reductive degradation experiments,[113]

---

* Recently a bacteriochlorophyll *c* with stearyl alcohol has been isolated from Chloroflexaceae, and minor components of a bacteriochlorophyll *e* mixture have been characterized as cetyl and phytyl esters.[102a]

$R^1 = C_2H_5$, $n\text{-}C_3H_7$ or $i\text{-}C_4H_9$
$R^2 = CH_3$ or C_2H_5

as well as by ^{1}H and ^{13}C nmr investigations.[114,115] The structure of bacterio-methylpheophorbide e (73) was determined by Brockmann et al.[105,116] mainly by spectrometric methods. The structural relationship between bacteriochlorophyll e and c is the same as that between chlorophyll b and a.

The ORD spectrum of bacteriochlorophyll c was first published by Ke.[117] It is similar to that of chlorophyll a. The ORD and CD spectra of 71 are virtually identical with those of [3-hydroxyethyl]pyromethylpheophorbide a (74).

The additional 20-methyl group in 72 and 73 gives rise to some distortion of the macrocycle as can be seen from absorption and ^{1}H nmr spectra but does not change the general pattern of the chiroptical properties. The conclusion drawn from these findings that 72 and 73 are 17(S),18(S) configurated has been proved by chromic acid oxidation. The levorotatory dihydrohematinimide obtained in all cases was identified as methyl or p-bromophenacyl ester with authentic samples from chlorophyll a.[114,118]

Addition of hydrogen bromide to the vinyl group of pyromethylpheophorbide a (11), followed by hydrolysis, affords a mixture of epimeric [2-hydroxyethyl]pyromethylpheophorbides a (74). Although these isomers have not been separated so far, their presence can be seen from the doubling of a number of signals in the ^{1}H nmr spectrum.[59,84,118a] As it was to be expected, the ^{1}H nmr spectra of 71, 72, and 73 do not show this behavior, revealing a definite configuration at C-3^1 in these compounds. The chirality of this carbon atom was first established by use of the modified Horeau analysis described by Brockmann and Risch.[119] All three compounds react preferentially with the levorotatory enantiomer of α-phenylbutyric imidazolide when esterified with an excess of this reagent as a racemate. According to Horeau's rule,[55]

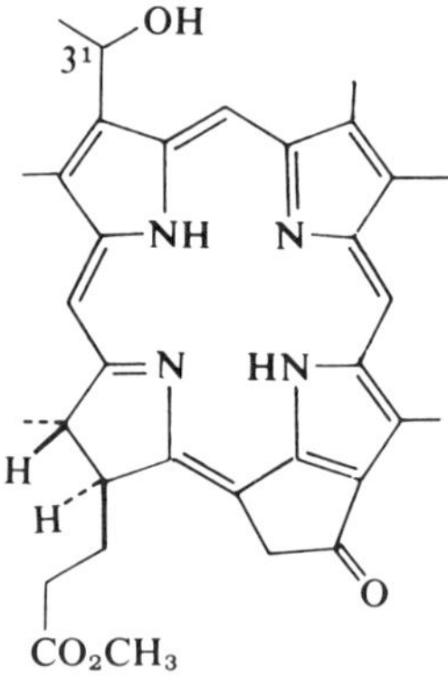

74

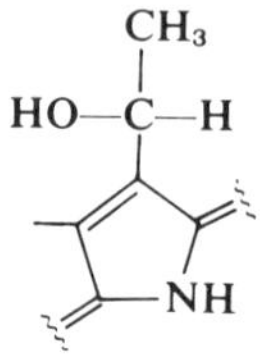

75

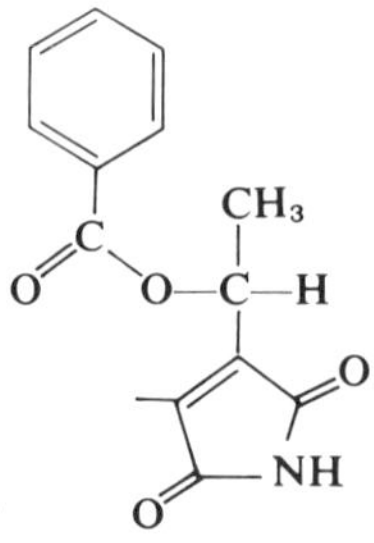

76

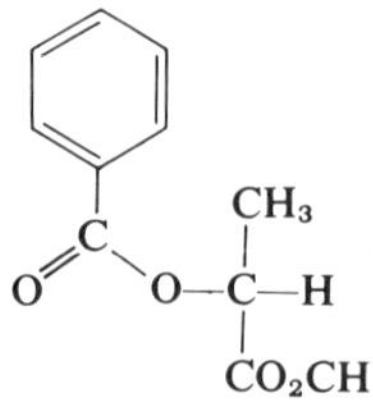

77

the alcohol must therefore have a configuration as depicted in **75**, i.e., (R) configuration.[114,120]

A more rigorous proof of this conclusion came from the following experiments.[114,118] If the bacteriomethylpheophorbides c, d, and e are benzoylated at the 3^1-hydroxy group prior to the chromic acid treatment, levorotatory 2(1-benzoyloxyethyl)-3-methylmaleimide (**76**) is obtained from ring A by oxidative degradation. Ozonolysis and subsequent esterification converted **76** to the derivative (**77**) of (−)-(R)-lactic acid.

IV. STEREOCHEMISTRY OF THE ESTERIFYING ALCOHOLS

Three terpene alcohols, i.e., phytol (**78**), geranylgeraniol (**79**), and farnesol (**80**), are so far found as esterifying components in chlorophylls. The stereochemical problems associated with these alcohols are of two types. Only phytol is a chiral compound due to two asymmetric carbon atoms. In addition to these, it contains an unsymmetrically substituted double bond, a structural

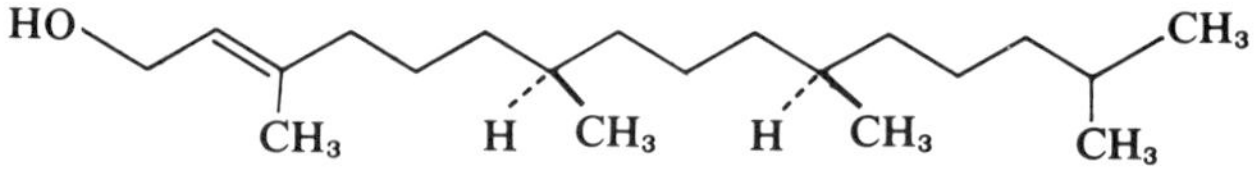

78

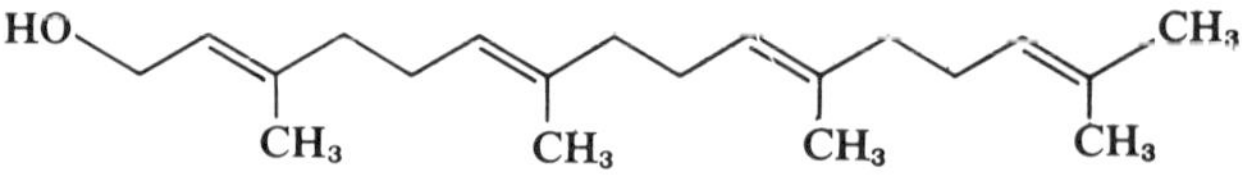

79

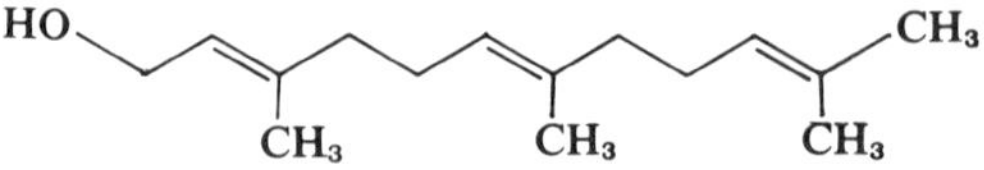

80

element that allows E, Z isomerism. Farnesol contains two double bonds of the same type and geranylgeraniol three.

A. Stereochemistry at the Double Bonds

The configuration at the double bonds of **78, 79,** and **80** has been established by ¹H nmr spectroscopy and by stereospecific syntheses of farnesol and geranylgeraniol from geraniol (**84**). Geraniol and its isomer nerol (**85**) are obtained by lithium aluminum hydride reduction of the isomeric methylgeranates (**82** and **83**), which can be synthesized from the methyl ketone (**81**) with methoxyacetylene, followed by acid-catalyzed isomerization of the resulting acetylenic alcohols.[121,122]

Upon chromatographic separation, the chemical shift for the β-CH$_3$ protons of **82** and **83** was found to be 1.73 ppm in one case and 1.98 ppm in the other. Because Jackman and Wiley[123,124] have found that, in proton magnetic resonance spectra of α,β-unsaturated esters, β-methyl groups trans to the ester group absorb at higher field than those in cis position, Burrel et al.[121,122] assigned E configuration (CH$_3$ and CO$_2$CH$_3$ cis to each other) to the isomer with the 1.73 ppm methyl resonance. This assignment is in harmony with the fact that nerol (**85**) can be cyclized in acids more rapidly to α-terpineol than geraniol (**84**)[125] and with the results of an X-ray analysis of geranylamine hydrochloride.[126]

Treatment of geraniol with PCl$_3$ yields the chloride (**86**), which was used for acetoacetic ester synthesis followed by ketonic hydrolysis.[127] The ketone (**87**) thus obtained was then used for the preparation of the 2(Z),6(E)-, and 2(E), 6(E)-methylfarnesoates, which could be resolved by gas-liquid chromatography (glc).[122] Again ¹H nmr allowed the configurational assignment of the double bond. Reduction of the 2(E),6(E) isomer (**88**) afforded the corresponding farnesol (**80**).

Recently, a new highly stereoselective synthesis of $2(E),6(E),11(E)$-geranylgeraniol (**79**) using geraniol (**84**) as starting material was reported.[128] Geranyl benzyl ether was oxidized with selenium dioxide to the allylic alcohol (**89**), and this reaction has been shown to be highly stereoselective.[129] The corresponding chloride (**90**) was then coupled in a Biellman-type reaction[130,131] with geranyl thiophenyl ether (**91**) to give **92**. Final reduction with lithium in ethylamine afforded compound **79** of high purity.

B. Absolute Configuration of Phytol

The optical rotation of phytol (**78**) from chlorophyll is extremely small, but that of the derived C_{18}-ketone (**93**) and of the phytadiene (**95**) clearly demonstrates optical activity of the natural alcohol.[132,133] The relative stereochemistry between the two asymmetric carbon atoms C-7 and C-11 has been deduced in the following way.[121,122] By treatment of **93** with methylenetriphenylphosphorane and reduction of the resulting olefin, optically inactive 2,6,10,14-tetramethylpentadecane (**94**) was obtained. That **94** really is the *meso*-hydrocarbon could be shown by the fact that a synthetic C_{18}-ketone with different configurations at the two asymmetric centers could be converted by the same reaction sequence to an optically active isomer of **94**.

Dehydration of **78** with phthalic acid anhydride in benzene solution affords

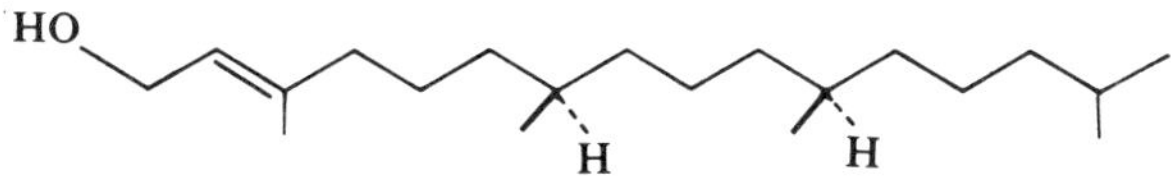

78

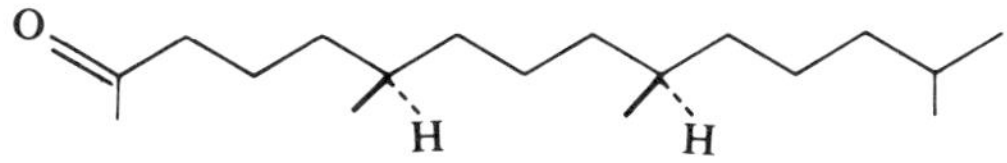

93

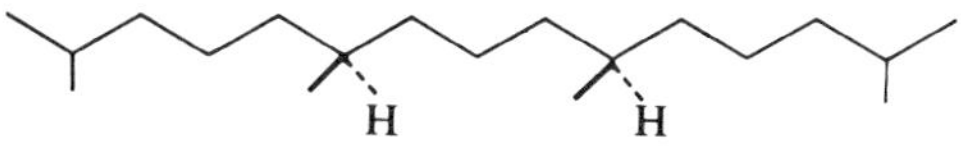

94

the phytadiene (**95**). Ozonolysis of the diene led to the C_{15} aldehyde (**97**) which exhibits a positive Cotton effect in the ORD spectrum. On the contrary, the C_{14} aldehyde (**100**) obtained by a second ozonolysis of the enolacetate (**98**) shows a negative Cotton effect. Because the aldehydes (**97** and **100**) give rise to opposite rotatory dispersion behavior as the aldehydes (**96** and **99**) of known (S) configuration do, Crabbé et al.[134] deduced (R) configuration at C-7 of phytol (**78**).

The configurational assignment based on ORD studies is in agreement with the results of a stereoselective synthesis of optically active **78**.[121,122] (+)-Citronellol (**101**), the absolute configuration (R) of which has been correlated

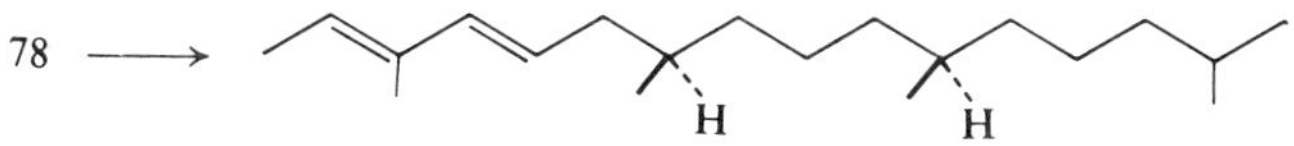

78 ⟶

95

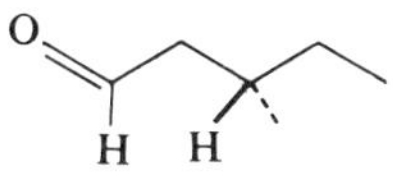

96
negative Cotton effect

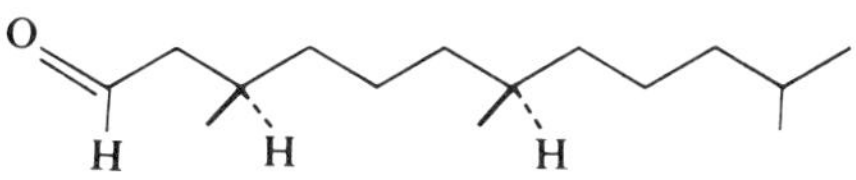

97

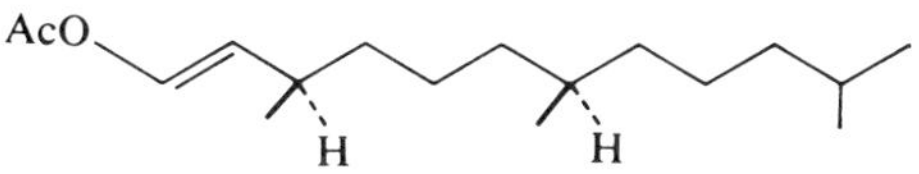

98

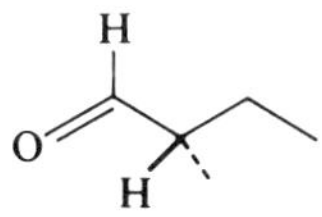

99
positive Cotton effect

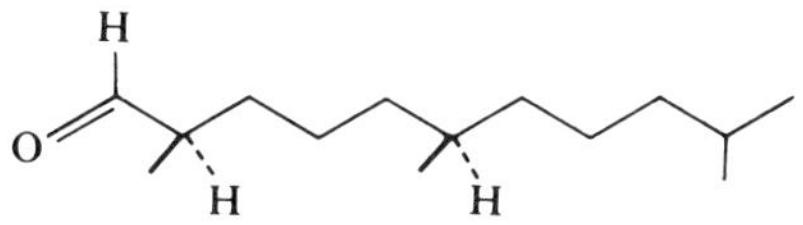

100

to that of D-glyceraldehyde,[135] on catalytic hydrogenation followed by chain expansion via the nitrile, gave the acid (104). This was then cross-coupled by Kolbe electrolysis with (+)-(R)-3-methylglutaric acid monomethyl ester (103). Upon hydrolysis of the resulting ester (106, R = CH₃), further anodic crossed coupling with levulinic acid (105) led to the 6(R),10(R)-6,10,14-trimethyl pentadecanone-2 (93), which was identical in optical rotation with that derived from natural phytol.

Reaction of 93 with methoxyacetylene, and subsequent isomerization, led to a mixture of E- and Z-phytenoates, which was separated by chromatography on alumina. ¹H nmr again was used to determine the configuration at the double bond as described in the previous section. Reduction of the E isomer (107) with lithium aluminum hydride afforded phytol, identical in all physical properties with that of natural sources.

V. LINEAR TETRAPYRROLES

A. Urobilinoids

The bile pigments, which are derived by biochemical degradation of hemo-proteins, were the first linear tetrapyrroles found in nature. Optical activity of natural (−)-stercobilin, first obtained by Watson [136] in crystalline form, was observed in 1935 by Fischer et al.[137] Closely related in structure to (−)-stercobilin (108) is the other optically active bile pigment (+)-urobilin (109).[138]

The constitutional formula (108) of (−)-stercobilin contains six centers of chirality.[139] Gray et al.[140] were able to demonstrate the relative configuration between pairs of these centers, e.g., the trans position of the hydrogens at C-2 and C-3, as well as at C-17 and C-18. By oxidative degradation, they obtained threo-2-ethyl-3-methyl succinimide (44) and mentioned in one of their publications that the mixture of the degradation imides was weakly dextrorotatory.[72] As described in Section II, F, the absolute configuration of

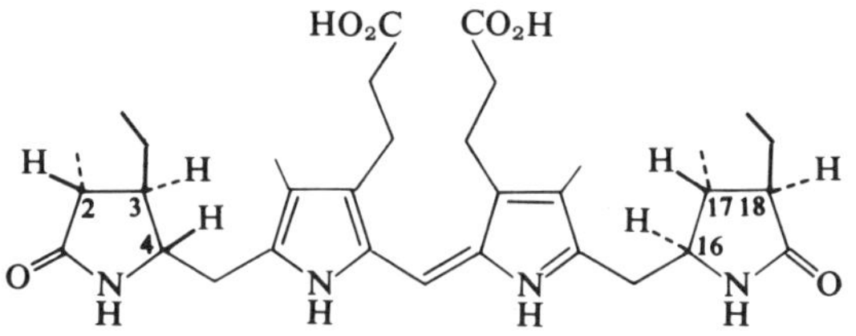

108

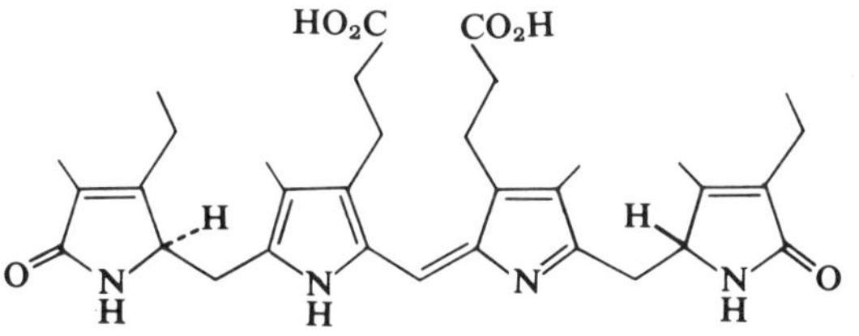

109

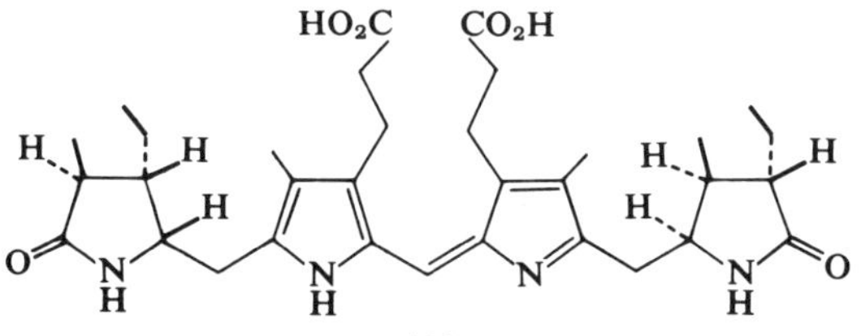

110

44 has been determined by stereochemical correlation of the 2-ethyl-3-methylsuccinic acids (**46**) and (**47**) with compounds of known chirality. In order to establish the absolute configuration of natural (−)-stercobilin, Brockmann et al.[73] repeated the chromic acid degradation of **108** and identified one of the degradation products which was obtained in high optical purity as (+)-2(R)-ethyl-3(R)-methyl succinimide (+**44**) by comparison with an authentic sample. Thus, the chirality of **108** at C-2, C-3, C-17, and C-18 was clearly established. It is opposite to that given in a recent review on bile pigments,[141] which does not mention at all the stereochemical investigations described in this section.

+44   −44

The relative configuration between the asymmetric β-carbons and the angular centers of chirality C-4 and C-16 in **108** could only be determined when a levorotatory stercobilin (**110**) was prepared by total synthesis[142] in which the steric relation of triads of chiral centers (C-2, C-3, C-4, and C-16, C-17, C-18) was known on the basis of the precursors used in the synthesis. Although a Debye–Scherrer diagram of **110** was undistinguishable from that of natural **108**, and the ir and ORD spectra of both compounds were almost superimposable, **110** turned out to be a diastereoisomer of **108**. Compound **110** failed to give a crystalline iron(III) chloride complex characteristic of **108**, and the ^{1}H nmr spectra measured in pyridine–water (9:1) were different for both compounds.

Chromic acid oxidation of the synthetic precursor of **110** yielded 2(S)-ethyl-3(S)-methyl succinimide, the enantiomer of +**44**, and it had to be concluded from this result that the chiroptical behavior of stercobilins is almost entirely dependent on the absolute configuration at C-4 and C-16. This is in agreement with a model developed by Moscowitz et al.,[143] according to which the helicity of the inherently dissymmetric dipyrromethene chromophore is only determined by the chirality of C-4 and C-16, and this helicity is responsible for the large amplitude of the Cotton effect of stercobilins in nonpolar solvents. Since cis configuration between C-3 and C-4, as well as between C-16 and C-17, was known for **110**, the hydrogen atoms at the corresponding asymmetric centers of natural **108** must be trans to each other, a conclusion that is in agreement with the chemical shift differences observed in the ^{1}H nmr spectra of **108** and **110**.[73]

(+)-Urobilin (**109**) has a visible absorption spectrum very similar to that of (−)-stercobilin, while the ORD spectra of the two compounds are like

Fig. 9. Helicity of the inherently dissymmetric chromophore of urobilinoids caused by intramolecular hydrogen bonds.

mirror images. Thus **109** must be 4(R),16(R) configurated according to the helicity model (Fig. 9).

B. Phycobilins

The chromoproteins phycoerythrin and phycocyanin occur in red and blue-green algae and act as light receptor pigments for the photosynthesis of these organisms. The prosthetic groups of phycoerythrin and phycocyanin are linear tetrapyrroles,[144] which can be liberated under different conditions.[145–149] Depending on the method used for solvolysis, phycoerythrin yields phycoerythrobilin or phycobiliviolin, while phycocyanobilin or phycobiliverdin is obtained from phycocyanin. The discussion of the contradictory nomenclatures used for phycobilins in the literature is beyond the scope of this chapter. In the following, the designations proposed by Rüdiger[150,151] are used.

The structures of phycobiliviolin (**111**) and of phycobiliverdin dimethyl ester (**113**) have been determined by chemical and spectrometrical methods.[75,152–156] Both compounds are marked by a reduced ring I with an asymmetric C-2 and an ethylidene side chain in position 3. The same partial structure has been demonstrated recently for bacteriochlorophyll *b* (see Section III, B). In addition to these elements of stereoisomerism, phycobiliviolin (**111**) contains a second center of chirality at C-16. Cole *et al.*[157] have shown that the configuration at this carbon atom is the same as that of natural (+)-urobilin (**109**) at C-4 and C-16. Catalytic hydrogenation of phycoerythrobilin—its structure probably differs only at ring I from that of **111**—followed by oxidation with iodine afforded the half-stercobilin (**112**), the ORD spectrum of which exhibits a Cotton effect with the same sign as **109** does. Due to the fact that **112** is a mixture of epimers at C-4, the maximum rotation of **112**, as it was to be expected, is only half of that observed for **109**.

The principal pigment in the purple defensive secretion of the sea hare *Aplysia* is aplysioviolin. It has been characterized as monomethyl ester of **111**[78] and is probably derived from the biliprotein of red algae, which the hare ingests with its food.

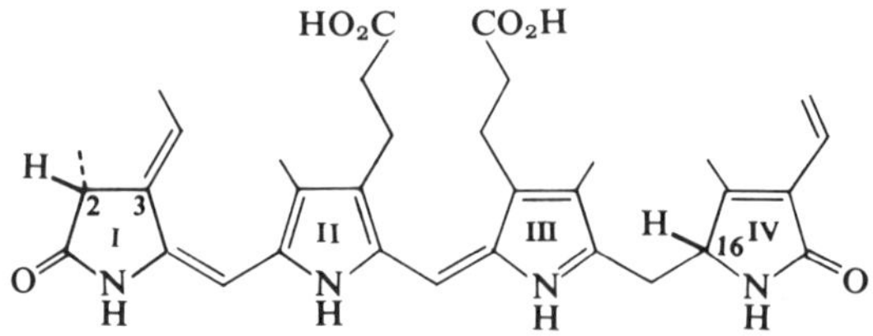

111

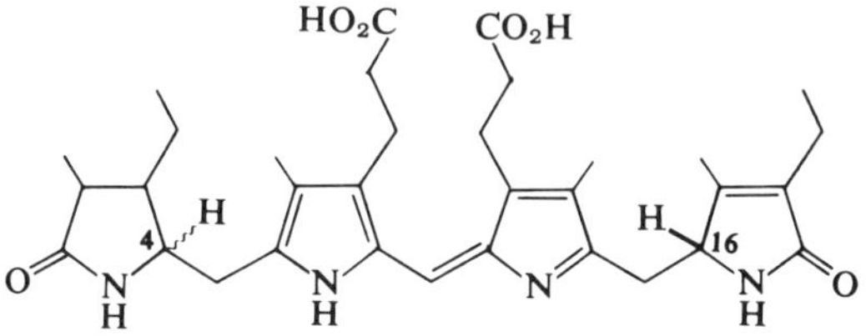

112

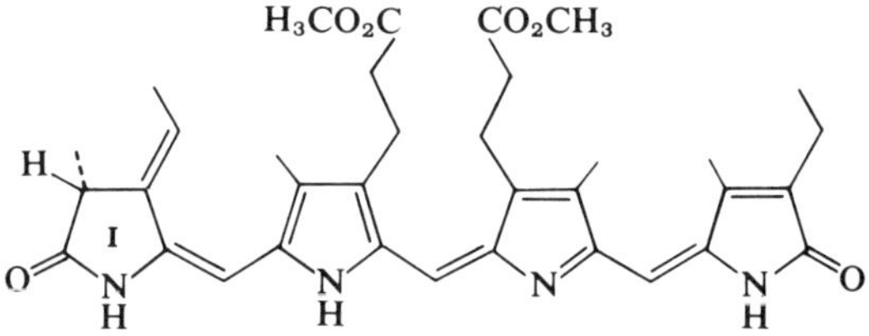

113

Aplysioviolin was the first tetrapyrrole that yielded 2-ethylidene-3-methyl succinimide (114) upon oxidative degradation.[74] The same imide was then found to be also a degradation product of denatured phycoerythrin and phycocyanin[75] as well as of the corresponding chromophores, regardless of the mode of their preparation.[75] Furthermore, 114 has been obtained from phytochrome.[76]

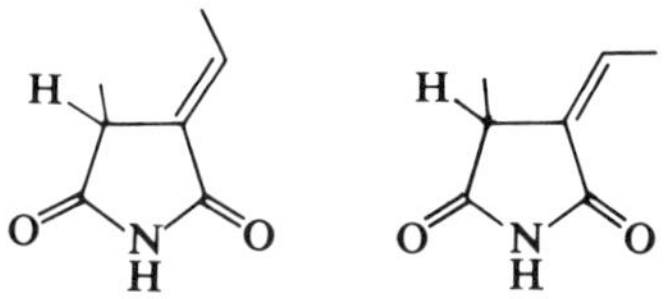

114   115

Several syntheses of 114 are known.[77,158–160] The method used by Brockmann and Knobloch, in addition to 114, afforded also its isomer (115), and the stereochemistry at the exocyclic double bonds of both compounds could be established by comparison of their ^1H nmr spectra.[77,159]

Partial resolution of synthetic racemic (114) was achieved by chromatography on acetylated cellulose powder. Compound (+114), when hydrogena-

$+114$ -44 116

ted catalytically, yielded *erythro*-2-ethyl-3-methyl succinimide (116) and (−)-*threo*-2(S)-ethyl-3(S)-methyl succinimide (−44); thus +114 is 3(S) configurated.[77]

The 2-ethylidene-3-methyl succinimide obtained by degradation of aplysio-violin is reported to be slightly levorotatory,[157] and the imide (114) derived from phycobiliverdin dimethyl ester (113) has been identified unequivocally with −114 by ORD and ^{1}H nmr spectra.[98]

In the course of his preparation of 113 by methanolysis of denatured C-phycocyanin, Knobloch[98] isolated a second violet pigment, which was characterized as 117. This pigment, on chromic acid oxidation, afforded the imide (118), which should be 2(R),3(R) configurated according to its chiropti-cal properties.[98]

117   118

Chromic acid oxidation of phycoerythrin was reported recently to yield also small amounts of 115, besides 114. This fact has been interpreted by the assumption that the exocyclic double bond probably does not exist in the native chromoprotein but is generated only during the degradation.[161]

REFERENCES

1. D. C. Hodgkin, J. Pickworth, J. H. Robertson, K. N. Trueblood, R. J. Prosen, and J. G. White, *Nature (London)* 176, 325 (1955).
2. D. C. Hodgkin, J. Kamper, J. Lindsey, M. McKay, J. Pickworth, J. H. Robertson, C. B. Shoemaker, J. G. White, R. J. Prosen, and K. N. Trueblood, *Proc. R. Soc. London, Ser. A* 242, 228 (1957).
3. K.-G. Paul, *Acta Chem. Scand.* 5, 389 (1951).
4. F. C. Yong and T. E. King, *Biochem. Biophys. Res. Commun.* 27, 59 (1967); *J. Biol. Chem.* 244, 509 and 515 (1969).
5. J. Barret, *Biochem. J.* 64, 626 (1956).
6. M. J. Murphy, L. M. Siegel, and H. Kamin, *J. Biol. Chem.* 248, 2801 (1973).
7. M. J. Murphy and L. M. Siegel, *J. Biol. Chem.* 248. 6911 (1973).

8. A. Stoll and E. Wiedemann, *Fortschr. Chem. Org. Naturst.* **1**, 159 (1938).
9. R. C. Dougherty, H. H. Strain, W. A. Svec, R. A. Uphaus, and J. J. Katz, *J. Am. Chem. Soc.* **92**, 2826 (1970).
10. H. Budzikiewicz and K. Taras, *Tetrahedron* **27**, 1447 (1971).
11. H. H. Strain, *Carnegie Inst. Washington, Yearb.* **42**, 79 (1943).
12. H. H. Strain, *in* "Manual of Phycology" (G. M. Smith, ed.), p. 243. Chronica Botanica, Waltham, Massachusetts, 1951.
13. H. Fischer, O. Süs, and G. Klebs, *Justus Liebigs Ann. Chem.* **490**, 38 (1931).
14. H. Fischer and H. Siebel, *Justus Liebigs Ann. Chem.* **499**, 84 (1932).
15. A. Stoll and E. Wiedemann, *Helv. Chim. Acta* **16**, 307 (1933).
16. H. Fischer and A. Stern, *Justus Liebigs Ann. Chem.* **519**, 58 (1935); **520**, 88 (1935).
17. H. Fischer and H. Wenderoth, *Justus Liebigs Ann. Chem.* **545**, 140 (1940).
18. H. Fischer and H. Gibian, *Justus Liebigs Ann. Chem.* **552**, 153 (1942).
19. H. Fischer, R. Lambrecht, and H. Mittenzwei, *Hoppe-Seyler's Z. Physiol. Chem.* **253**, 1 (1938).
20. H. Mittenzwei, *Hoppe-Seyler's Z. Physiol. Chem.* **275**, 93 (1942).
21. G. E. Ficken, R. B. Johns, and R. P. Linstead, *J. Chem. Soc.* p. 2272 (1956).
22. G. E. Ficken, R. B. Johns, and R. P. Linstead, *J. Chem. Soc.* p. 2280 (1956).
23. J. H. Golden, R. P. Linstead, and G. H. Whitham, *J. Chem. Soc.* p. 1725 (1958).
24. J. H. Golden and R. P. Linstead, *J. Chem. Soc.* p. 1732 (1958).
25. J. Fleming, *Nature (London)* **216**, 151 (1967).
26. J. Fleming, *J. Chem. Soc. C.* p. 2765 (1968).
27. H. Brockmann, Jr., *Angew. Chem.* **80**, 233 (1968); *Angew. Chem., Int. Ed. Engl.* **7**, 221 (1968).
28. H. Brockmann, Jr., *Angew. Chem.* **80**, 234 (1968); *Angew. Chem., Int. Ed. Engl.* **7**, 222 (1968).
29. H. Brockmann, Jr. and I. Kleber, *Angew. Chem.* **81**, 626 (1969); *Angew. Chem., Int. Ed. Engl.* **8**, 610 (1969).
30. H. Brockmann, Jr., *Justus Liebigs Ann. Chem.* **754**, 139 (1971).
31. R. Thomas, Doctoral Thesis, Technical University, Braunschweig (1967).
31a. H. Fischer and H. Gibian, *Justus Liebigs Ann. Chem.* **550**, 208 (1942).
32. R. W. Rickards and R. M. Smith, *Tetrahedron Lett.* p. 1025, footnote p. 1027 (1970).
33. R. B. Woodward, W. A. Ayer, J. M. Beaton, F. Bickelhaupt, R. Bonnett, P. Buchschacher, G. L. Closs, H. Duttler, J. Hannach, F. P. Hauck, S. Itô, A. Langemann, E. Le Goff, W. Leimgruber, W. Lwowski, J. Sauer, Z. Valenta, and H. Volz, *J. Am. Chem. Soc.* **82**, 3800 (1960).
34. R. B. Woodward, *Angew. Chem.* **72**, 651 (1960).
35. G. L. Closs, J. J. Katz, F. C. Pennington, M. R. Thomas, and H. H. Strain, *J. Am. Chem. Soc.* **85**, 3809 (1963).
36. G. Jeckel, Doctoral Thesis, Technical University, Braunschweig (1967).
37. G. Brockmann and H. Brockmann, Jr., *IIT-NMR-Newslett.* **117**, 62 (1968).
38. H. Wolf, H. Brockmann, Jr., H. Biere, and H. H. Inhoffen, *Justus Liebigs Ann. Chem.* **704**, 208 (1967).
39. H. Wolf, I. Richter, and H. H. Inhoffen, *Justus Liebigs Ann. Chem.* **725**, 177 (1969).
40. C. Houssier and K. Sauer, *Biochim. Biophys. Acta* **172**, 492 (1969).
41. C. Houssier and K. Sauer, *J. Am. Chem. Soc.* **92**, 779 (1970).
42. H. H. Strain and W. M. Manning, *J. Biol. Chem.* **146**, 275 (1942).
43. J. J. Katz, G. D. Norman, W. A. Svec, and H. H. Strain, *J. Am. Chem. Soc.* **90**, 6841 (1968).

44. W. Hoppe, *Z. Kristallogr. Kristallgeom., Kristallphys., Kristallchem.* **108**, 335 (1957).
45. W. Hoppe and G. Will, *Z. Kristallogr. Kristallgeom., Kristallphys., Kristallchem.* **113**, 104 (1960).
46. W. Hoppe, G. Will, J. Gassmann, and H. Weichselgartner, *Z. Kristallogr., Kristallgeom., Kristallphys., Kristallchem.*, **128**, 18 (1967).
47. J. Gassmann, I. Strell, F. Brandl, M. Sturm, and W. Hoppe, *Tetrahedron Lett.* p. 4609 (1971).
48. M. S. Fischer, D. H. Templeton, A. Zalkin, and M. Calvin, *J. Am. Chem. Soc.* **94**, 3613 (1972).
49. C. E. Strouse, *Proc. Natl. Acad. Sci. U.S.A.* **71**, 325 (1974),
49a. H.-C. Chow, R. Serlin, and C. E. Strouse, *J. Am. Chem. Soc.* **97**, 7230 (1975).
49b. C. Kratky and J. D. Dunitz, *Acta Cryst.* B **31**, 1586 (1975).
49c. R. Serlin, H.-C. Chow, and C. E. Strouse, *J. Am. Chem. Soc.* **97**, 7237 (1975).
50. H. Wolf, H. Brockmann, Jr., I. Richter, C.-D. Mengler, and H. H. Inhoffen, *Justus Liebigs Ann. Chem.* **718**, 162 (1968).
51. I. Richter, Doctoral Thesis, Technical University, Braunschweig (1969).
52. H. Wolf and H. Scheer, *Justus Liebigs Ann. Chem.* **745**, 87 (1971).
53. H. Brockmann, Jr. and J. Bode, *Justus Liebigs Ann. Chem.* **1974**, 1017 (1974).
54. A. Horeau, *Tetrahedron Lett.* p. 506 (1961).
55. A. Horeau and H. Kagan, *Tetrahedron* **20**, 2431 (1964).
56. J. Bode, Doctoral Thesis, Technical University, Braunschweig (1971).
57. W. Küster, *Hoppe-Seyler's Z. Physiol. Chem.* **28**, 1 (1899).
58. H. Fischer and H. Wenderoth, *Justus Liebigs Ann. Chem.* **537**, 170 (1939).
59. G. E. McCasland and S. Proskow, *J. Am. Chem. Soc.* **78**, 5646 (1956).
60. B. Carnmalm, *Chem. Ind. (London)* p. 1093 (1956).
61. K. Nagarajan, C. Weissmann, H. Schmied, and P. Karrer, *Helv. Chim. Acta* **46**, 1212 (1963).
62. A. Fredga and L. Terenius, *Acta Chem. Scand.* **18**, 2081 (1964).
63. H. Brockmann, Jr. and D. Müller-Enoch, unpublished.
64. D. Müller-Enoch, Doctoral Thesis, Technical University, Braunschweig (1971).
65. A. Werner and M. Basyrin, *Ber. Dtsch. Chem. Ges.* **46**, 3229 (1913).
66. J. D. M. Asher and G. A. Sim, *J. Chem. Soc.* pp. 1584 and 6041 (1965).
67. A. T. McPhail, B. Rimmer, J. M. Robertson, and G. A. Sim, *J. Chem. Soc.* B p. 101 (1967).
68. M. Nakazaki and H. Arakawa, *Proc. Chem. Soc., London* p. 151 (1962).
69. M. Nakazaki, *Bull. Chem. Soc. Jpn.* **35**, 1904 (1962).
70. M. Nakazaki and K. Ikematsu, *Bull. Chem. Soc. Jpn.* **37**, 459 (1964).
71. H. Brockmann, Jr., Habilitationsschrift, Technical University, Braunschweig (1969).
72. C. H. Gray and D. C. Nicholson, *J. Chem. Soc.* p. 3085 (1958).
73. H. Brockmann, Jr., G. Knobloch, H. Plieninger, K. Ehl, J. Ruppert, A. Moscowitz, and C. J. Watson, *Proc. Natl. Acad. Sci. U.S.A.* **68**, 2141 (1971).
74. W. Rüdiger, *Hoppe-Seyler's Z. Physiol Chem.* **348**, 129 (1967).
75. W. Rüdiger and P. O'Carra, *Eur. J. Biochem.* **7**, 509 (1969).
76. W. Rüdiger and D. L. Correl, *Justus Liebigs Ann. Chem.* **723**, 208 (1969).
77. H. Brockmann, Jr. and G. Knobloch, *Chem. Ber.* **106**, 803 (1973).
78. E. Berner and R. Leonardsen, *Justus Liebigs Ann. Chem.* **583**, 1 (1939).
79. H. Brockmann, Jr. and D. Müller-Enoch, *Chem. Ber.* **104**, 3704 (1971).
80. J. Bode and H. Brockmann, Jr., *Chem. Ber.* **105**, 34 (1972).

81. K. Freudenberg and W. Lwowski, *Justus Liebigs Ann. Chem.* **594**, 76 (1955).
82. H. Brockmann, Jr. and D. Müller-Enoch, *Angew. Chem.* **80**, 562 (1968); *Angew. Chem., Int. Ed. Engl.* **7**, 543 (1968).
83. H. Brockmann, Jr. and W. Trowitzsch, unpublished.
84. W. Trowitzsch, Doctoral Thesis, Technical University, Braunschweig (1974).
85. V. Prelog and E. Zalán, *Helv. Chim. Acta* **28**, 545 (1944).
86. H. Brockmann, Jr. and J. Bode, *Justus Liebigs Ann. Chem.* **748**, 20 (1971).
87. H. Brockmann, Jr. and G. Knobloch, *Arch. Mikrobiol.* **85**, 123 (1972).
88. J. J. Katz, H. H. Strain, A. L. Harkness, M. H. Studier, W. A. Svec, T. R. Janson, and B. T. Cope, *J. Am. Chem. Soc.* **94**, 7938 (1972).
89. H. Brockmann, Jr., G. Knobloch, I. Schweer, and W. Trowitzsch, *Arch. Mikrobiol.* **90**, 161 (1973).
90. A. Künzler and N. Pfennig, *Arch. Mikrobiol.* **91**, 83 (1973).
91. A. Gloe and N. Pfennig, *Arch. Microbiol.* **96**, 93 (1974).
92. H. Fischer, W. Lautsch, and K.-H. Lin, *Justus Liebigs Ann. Chem.* **534**, 1 (1938).
93. H. Fischer, H. Mittenzwei, and D. B. Hevèr, *Justus Liebigs Ann. Chem.* **545**, 145 (1940).
94. H. H. Inhoffen, P, Jäger, R. Mählhop, and C.-D. Mengler, *Justus Liebigs Ann. Chem.* **704**, 188 (1967).
95. K. E. Eimhjellen, O. Aasmundrud, and A. Jensen, *Biochem. Biophys. Res. Commun.* **10**, 232 (1963).
96. H. Brockmann, Jr. and I. Kleber, *Tetrahedron Lett.* p. 2195 (1970).
97. H. Scheer, W. A. Svec, B. T. Cope, M. H. Studier, R. G. Scott, and J. J. Katz, *J. Am. Chem. Soc.* **96**, 3714 (1974).
98. G. Knobloch, Doctoral Thesis, Technical University, Braunschweig (1972).
99. D. W. Hughes and A. S. Holt, *Can. J. Chem.* **40**, 171 (1962).
100. A. S. Holt, D. W. Hughes, H. J. Kende, and J. W. Purdie, *J. Am. Chem. Soc.* **84**, 2835 (1962).
101. H. Rapoport and H. P. Hamlow, *Biochem. Biophys. Res. Commun.* **6**, 134 (1961).
102. A. S. Holt, D. W. Hughes, H. J. Kende, and J. W. Purdie, *Plant Cell Physiol.* **4**, 49 (1963).
102a. H. Brockmann, Jr., A. Gloe, and N. Risch, unpublished results.
103. R. Y. Stanier, *in* "Comparative Biochemistry of Photoreactive Systems" (M. B. Allen, ed.), p. 69. Academic Press, New York, 1960.
104. A. Jensen, O. Aasmundrud, and K. E. Eimhjellen, *Biochim. Biophys. Acta* **88**, 466 (1964).
105. A. Gloe, N. Pfennig, H. Brockmann, Jr., and W. Trowitzsch, *Arch. Microbiol.* **102**, 103 (1975).
106. A. S. Holt and D. W. Hughes, *J. Am. Chem. Soc.* **83**, 499 (1961).
107. H. V. Morley and A. S. Holt, *Can. J. Chem.* **39**, 755 (1961).
108. J. W. Purdie and A. S. Holt, *Can, J. Chem.* **43**, 3347 (1965).
109. A. S. Holt, J. W. Purdie, and J. W. F. Wasley, *Can. J. Chem.* **44**, 88 (1966).
110. J. W. Mathewson, W. R, Richards, and H. Rapoport, *J. Am. Chem. Soc.* **85**, 364 (1963).
111. J. W. Mathewson, W. R. Richards, and H. Rapoport, *Biochem. Biophys. Res. Commun.* **13**, 1 (1963).
112. J. J. Katz, R. C. Dougherty, and L. J. Boucher, *in* "The Chlorophylls" (L. P. Vernon and G. R. Seely, eds.), p. 185. Academic Press, New York, 1966.
113. R. A. Chapman, M. W. Roomi, J. C. Norton, D. T. Krajcarski, and S. F. MacDonald, *Can. J. Chem.* **49**, 3544 (1971).

114. H. Brockmann, Jr., *Philos. Trans. R. Soc. London, Ser. B* **273**, 277 (1976).
115. K. M. Smith and J. F. Unsworth, *Tetrahedron* **31**, 367 (1975).
116. H. Brockmann, Jr., A. Gloe, N. Risch, and W. Trowitzsch, *Justus Liebigs Ann. Chem.* **1976**, 546 (1976).
117. B. Ke, *in* "The Chlorophylls" (L. P. Vernon and G. R. Seely, eds.), p. 427. Academic Press, New York, 1966.
118. R. Tacke, Doctoral Thesis, Technical University, Braunschweig (1975).
118a. H. Brockmann, Jr., W. Trowitzsch, and V. Wray, *Org. Mag. Reson.* **8**, 380 (1976).
119. H. Brockmann, Jr. and N. Risch, *Angew. Chem.* **86**, 707 (1974); *Angew. Chem., Int. Ed. Engl.* **13**, 664 (1974).
120. N. Risch and H. Brockmann, Jr., *Justus Liebigs Ann. Chem.* **1976**, 578 (1976).
121. J. W. K. Burrel, L. M. Jackman, and B. C. L. Weedon, *Proc. Chem. Soc., London* p. 263 (1959).
122. J. W. K. Burrel, R. F. Garwood, L. M. Jackman, E. Oskay, and B. C. L. Weedon, *J. Chem. Soc. C.* p. 2144 (1966).
123. L. M. Jackman and R. H. Wiley, *Proc. Chem. Soc., London* p. 196 (1958).
124. L. M. Jackman and R. H. Wiley, *J. Chem. Soc.* pp. 2881 and 2886 (1960).
125. O. Zeitschel, *Ber. Dtsch. Chem. Ges.* **39**, 1780 (1906).
126. G. A. Jeffrey, *Proc. R. Soc. London, Ser. A* **183**, 388 (1945).
127. R. B. Bates, D. M. Gale, B. J. Gunner, and P. P. Nicholas, *Chem. Ind. (London)* p. 1907 (1961).
128. L. J. Altman, L. Ash, and S. Marson, *Synthesis* p. 129 (1974).
129. U. T. Bhalerao and H. Rapoport, *J. Am. Chem. Soc.* **93**, 4835 (1971).
130. J. F. Biellmann and J. H. Ducep, *Tetrahedron Lett.* p. 3703 (1969).
131. J. F. Biellmann and J. H. Ducep, *Tetrahedron* **27**, 5861 (1971).
132. P. Karrer, A. Geiger, H. Rentschler, E. Zbinden, and A. Kugler, *Helv. Chim Acta* **26**, 1741 (1943).
133. P. Karrer, H. Simon, and E. Zbinden, *Helv. Chim. Acta* **27**, 313 (1944).
134. P. Crabbé, C. Djerassi, E. J. Eisenbaum, and S. Liu, *Proc. Chem. Soc., London* p. 263 (1959).
135. J. A. Mills and W. Klyne, *Prog. Stereochem.* **2**, 177 (1954).
136. C. J. Watson, *Hoppe-Seyler's Z. Physiol. Chem.* **208**, 101 (1932).
137. H. Fischer, H. Halbach, and A. Stern, *Justus Liebigs Ann. Chem.* **579**, 254 (1935).
138. S. Schwartz and C. J. Watson, *Proc. Soc. Exp. Biol. Med.* **49**, 643 (1942).
139. A. J. Birch, *Chem. Ind. (London)* p. 625 (1955).
140. C. H. Gray, G. A. Lemmon, and D. C. Nicholson, *J. Chem. Soc. C* p. 178 (1967).
141. M. F. Hudson and K. M. Smith, *Chem. Soc. Rev.* **4**, 363 (1975).
142. H. Plieninger and J. Ruppert, *Justus Liebigs Ann. Chem.* **736**, 43 (1970).
143. A. Moscowitz, I. T. Kay, W. C. Krueger, G. Skewes, and S. Bruckenstein, *Proc. Natl. Acad. Sci. U.S.A.* **52**, 1190 (1964).
144. R. Lemberg, *Justus Liebigs Ann. Chem.* **461**, 46 (1928); **477**, 195 (1930).
145. Y. Fujita and A. Hattori, *J. Biochem. (Tokyo)* **51**, 89 (1962).
146. Y. Fujita and A. Hattori, *J. Gen. Appl. Microbiol.* **9**, 253 (1963).
147. C. O'hEocha, *Biochemistry* **2**, 375 (1963).
148. P. O'Carra, C. O'hEocha, and D. M. Carrol, *Biochemistry* **3**, 1343 (1964).
149. P. O'Carra and C. O'hEocha, *Phytochemistry* **5**, 993 (1966).
150. W. Rüdiger, *Angew. Chem.* **82**, 527 (1970); *Angew. Chem., Int. Ed. Engl.* **9**, 437 (1970).
151. W. Rüdiger, *Fortschr. Chem. Org. Naturst.* **29**, 60 (1971).
152. H. L. Crespi, L. J. Boucher, G. D. Norman, J. J. Katz, and R. C. Dougherty, *J. Am. Chem. Soc.* **89**, 3642 (1967).

153. H. L. Crespi, U. Smith, and J. J. Katz, *Biochemistry* **7**, 2232 (1968).
154. W. J. Cole, D. J. Chapman, and H. W. Siegelman, *J. Am. Chem. Soc.* **89**, 3643 (1967).
155. W. J. Cole, D. J. Chapman, and H. W. Siegelman, *Biochemistry* **7**, 2929 (1968).
156. W. Rüdiger, P. O'Carra, and C. O'hEocha, *Nature (London)* **215**, 1477 (1967).
157. W. J. Cole, C. O'hEocha, A. Moscowitz, and W. C. Krueger, *Eur. J. Biochem.* **3**, 202 (1967).
158. W. Rüdiger and W. Klose, *Tetrahedron Lett.* p. 1177 (1967).
159. H. Brockmann, Jr. and G. Knobloch, *Tetrahedron Lett.* p. 267 (1970).
160. A. Gossauer and W. Hirsch, *Justus Liebigs Ann. Chem.* **1974**, 1496 (1974).
161. H. P. Köst and W. Rüdiger, *Tetrahedron Lett.* p. 3417 (1974).

10

Pyrrolic Macrocycles Other than Porphyrins

R. GRIGG

The interest in polypyrrole macrocycles, other than porphyrins, has been fostered by the fascinating chemistry of the porphyrins and the vitamin B_{12} coenzyme. A further influence has been the interest in aromaticity and the extent to which modification and expansion of the aromatic porphyrin nucleus is possible while still retaining aromatic character. The synthetic efforts in this field have resulted in a range of new and intriguing macrocycles, largely developed since 1965. These macrocycles include aromatic (corroles, sapphyrins, dioxonorsapphyrins) and nonaromatic (1-alkyl- and 1,19-dialkyltetradehydrocorrins, corrins) systems, and the chemistry of these macrocycles has, in many cases, been little studied as yet. All the macrocycles considered in this chapter have a direct link between two of the pyrrole or, where appropriate, furan rings.

In developing synthetic strategies for macrocycles related to corrins (i.e., containing a direct link), one initial choice that must be made is whether to incorporate the direct link at the outset or to generate it in the final cyclization step. Examples of both approaches have been reported, but syntheses in which the direct link is generated in the final cyclization are more numerous.

I. CORROLES

A. Synthesis

These were the first of the pyrrolic macrocycles containing a direct link to be synthesized. The Fischer–Stachel compounds (1), in which the direct link is already incorporated, were originally thought[1] to cyclize with formaldehyde in acidic ethanol to pentadehydrocorrins. However, further work[2] showed the product to be the palladium oxacorrole (2). Treatment of 1 with ammonia, methylamine, or sodium sulfide[3] gave the hetero analogues (3; X = NH, NMe, and S). The rigidity and stereochemistry imposed on 1 by coordination to palladium is beneficial to the final nucleophilic attack required to form the macrocycles 2 and 3, but is not mandatory for cyclizations generating corroles, provided the direct link is generated last.

The most successful general synthesis of corroles involves construction of a 1,19-dideoxybiladiene-a,c (6) by a 2 + 1 + 1 approach[4] of condensing, under acidic conditions, either (a) 2 moles of a 3,4-dialkylpyrrole (4; R = H) with a 5,5′-diformyldipyrromethane (5; R = CHO) or (b) dipyrromethane-

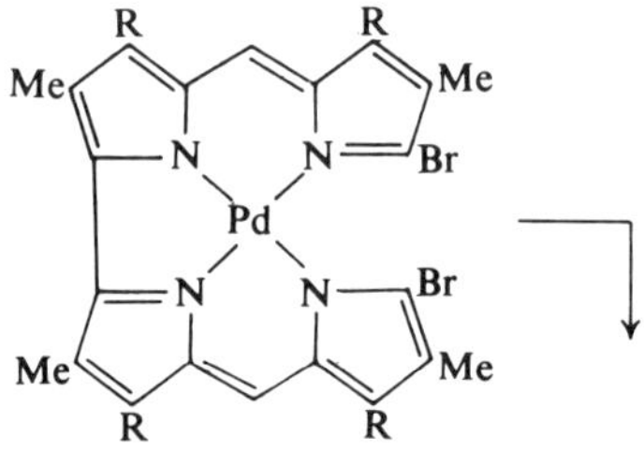

1

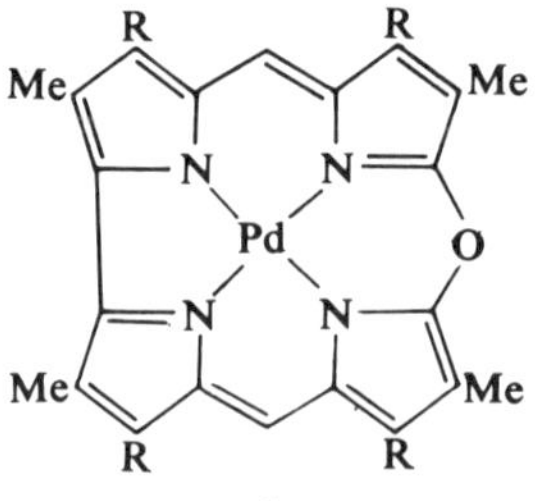

2

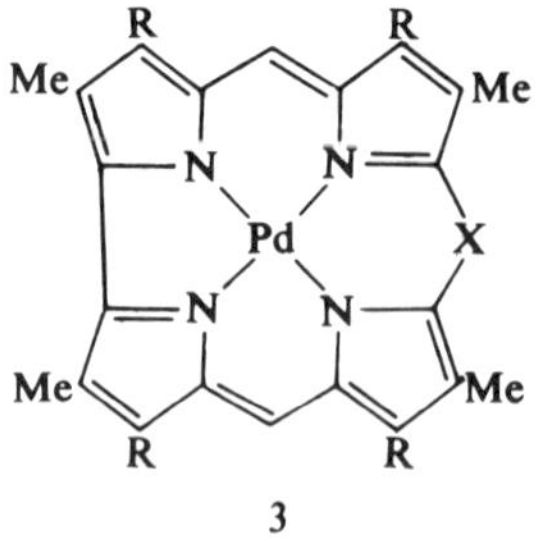

3

5,5'-dicarboxylic acids (**5**; R = CO_2H) with 2 moles of **4** (R = CHO). The biladiene salts (**6**) undergo smooth oxidative cyclization under basic conditions to corroles (**7**; 20–70%), with the direct link being generated by the final closure. Small amounts of azaporphyrins ($\sim 5\%$) are also produced when ammonia is used as the base.

The mechanism of this cyclization has been the subject of several studies.[5,6] The first step involves formation of the green bilatriene-free base (**8**) under the basic conditions of the reaction (Fig. 1). The next step of the reaction is more problematical and, in particular, it is not known whether oxidation precedes or follows ring closure. Initial studies[5] were interpreted in favor of a free radical mechanism. Typically, the cyclization is effected by irradiation of **6** in methanolic ammonia with visible light (200 W tungsten lamp) or by treating basic solutions with various oxidizing agents [ceric sulfate, ferric chloride, potassium ferricyanide, di(*tert*-butyl)peroxide, benzoyl peroxide, hydrogen peroxide]. However, cyclizations proceed normally in the presence of a large excess of either hydroquinone or *p-tert*-butylcatechol,[6] both of which might be expected[7] to intercept any free radicals produced during the cyclizations. The Woodward–Hoffmann rules governing pericyclic processes[8] suggest an interpretation of the biladiene cyclization.

The bilatriene (**8**) seems unlikely to be the species undergoing cyclization since it contains a through conjugated 18π-electron system, which would be

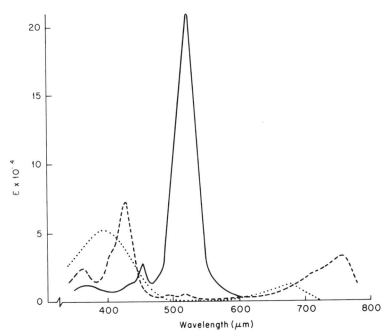

Fig. 1. Electronic spectra of 1,19-dideoxy-2,18-diethoxycarbonyl-8,12-diethyl-1,3,7,13,17,19-hexamethylbiladiene-ac hydrobromide: (——) in chloroform, (–––) in ethanol (i.e., bilatriene-abc salt), and (· · ·) in chloroform containing piperidine (i.e., bilatriene-abc free base).

expected to undergo a thermally allowed disrotatory cyclization producing the sterically unfavorable cis geometry (**9**) at the direct link. A one-electron oxidation of the bilatriene would generate a 17π-electron system, and the geometry of electrocyclic reactions of radicals is the subject of unresolved controversy.[9] However, while the negative trapping experiments would seem to rule this out, we cannot make a final judgement on the present evidence. A two-electron oxidation would generate a 16π-electron cation (**10**), which would be expected to undergo conrotatory cyclization giving trans geometry (**11**) at the direct link. Molecular models indicate **11** is less sterically strained than **9**. Finally, prototropy would give the corrole (**7**).

Studies on the mechanism of the related cyclization of 1,19-dideoxy-1,19-dimethylbiladienes-a,c to nickel 1,19-dimethyl tetradehydrocorrin salts provide evidence for the involvement of a 16π-electron cation analogous to **10** (see Section III, A).

The addition of metal ions (e.g., Cu, Ni, Co) to solutions of 1,19-dideoxy-biladienes-a,c undergoing cyclization results in the formation of the corresponding metallocorroles in good yield (60–70%), although a larger range of

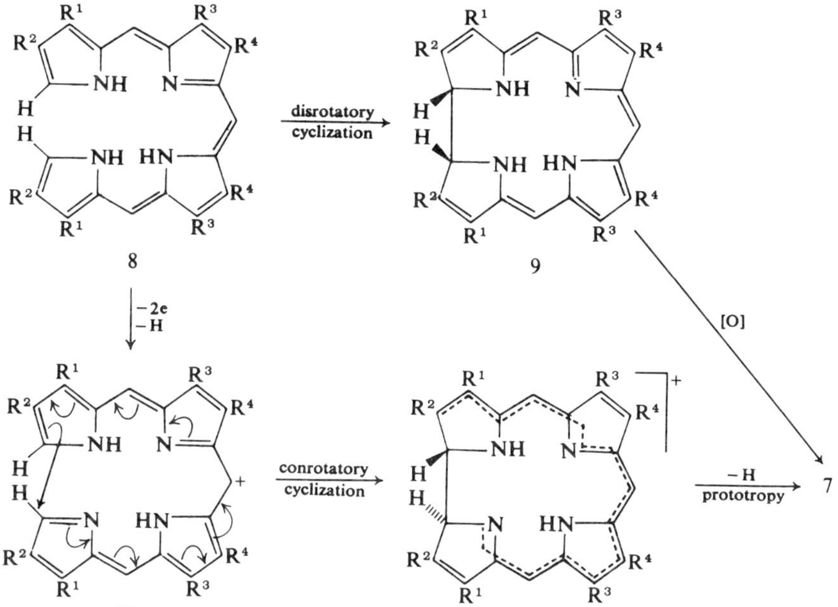

8   9   10   11

metals have been introduced into the preformed corrole [e.g., Cu(II), Ni(II), Pd(II), Fe(III), Co(III)]. The template effect provided by cobalt proved crucial to the cyclization of the labile fully β-unsubstituted 1,19-dideoxybiladiene-a,c to the parent fully β-unsubstituted Co(III) corrole (5%).[10] Corroles are also produced by heating 1,19-dideoxy-1,19-dibromobiladiene-a,c salts in refluxing o-dichlorobenzene (e.g., 12 → 13; 26%).[11] This reaction is little used because the intermediate pyrrole bromoaldehydes required for the synthesis of 12 are less accessible than (4; R = CHO or H). However, it has been used to synthesize the corroles with the uroporphyrin III and vitamin B_{12} β-substitution pattern.[11a]

Attempts to prepare biladienes incorporating furan or thiophen rings (e.g., 14; X = O or S) as precursors to oxa- and thiacorroles were unsuccessful.[12]

Schemes patterned on the acid catalyzed MacDonald 2 + 2 porphyrin

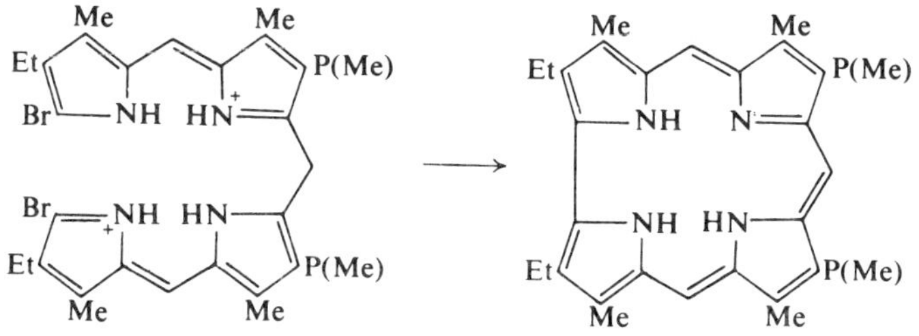

12   13

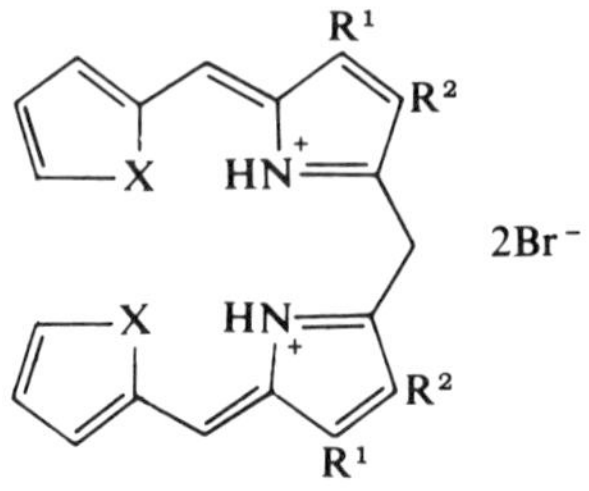

14

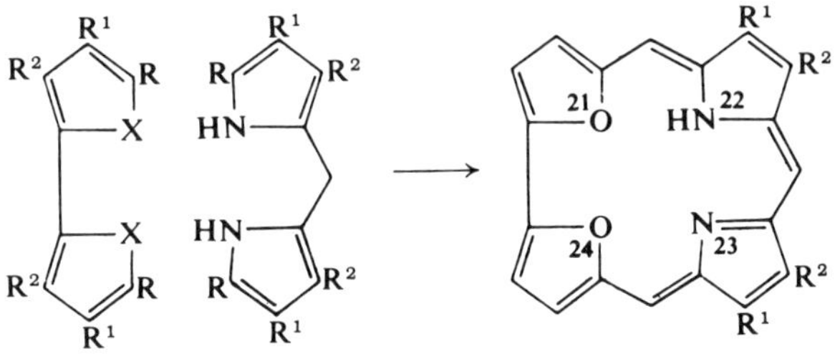

15   16: (a) $R^1 = R^2 = $ Me   17
               (b) $R^1 = $ Me, $R^2 = $ Et
               (c) $R^1 = R^2 = $ Et

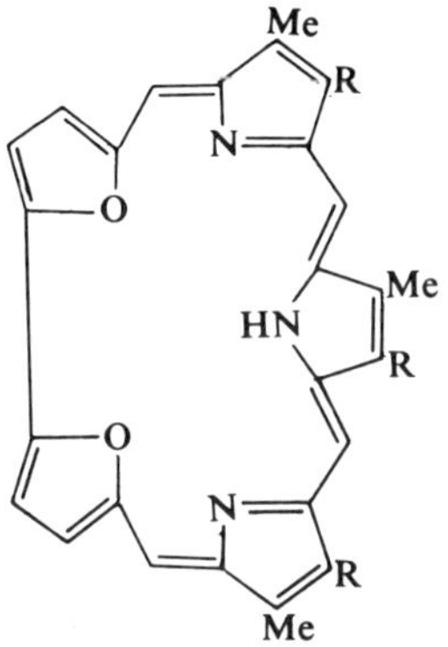

18: (a) R = Me
    (b) R = Et

synthesis are in general unsuccessful. Thus, the bipyrroles (15; X = NH, R = H or CHO) failed to give corroles when reacted with the appropriate dipyrromethane (16; R = CHO or CO_2H).[13,14] However, cobalt (III) corroles can be prepared (21%) by a 2 + 2 approach involving bipyrroles by combining either 15 (X = NH, R = CHO) and 16 (R = CO_2H) or 15 (X = NH, R = CO_2H) and 16 (R = CHO) and heating the initial condensation product, presumably an a-norbilene-b, with cobalt(II) acetate and triphenylphosphine in methanol.[10]

Reactions designed to produce dithiacorroles involving **15** (X = S, R = CHO, $R^1 = R^2 = H$) and **16** (R = CO_2H) also failed.[12] The corresponding reaction with the almost insoluble bifuran (**15**; X = O, R = CHO, $R^1 = R^2 = H$) and **16** (R = CO_2H) in dilute chloroform solutions does produce the corresponding 21,24-dioxacorroles (**17**) but in very low yield (1–5%). The red dioxacorroles are admixed with a green macrocyclic product, also produced in low yield (7%), which was shown to be the 26,30-dioxasapphyrin (**18**).[12] The dioxacorroles (**17**) are readily separated from the more polar **18** by chromatography on alumina. The 26,30-dioxasapphyrins clearly result via acid-catalyzed cleavage–recombination reactions of the dipyrromethanes (**16a,b**; R = CO_2H) prior to condensation. Cleavage reactions of this type are well known[15] and result in isomer mixtures in some porphyrin syntheses. Rational 26,30-dioxasapphyrin syntheses are discussed later.

A more successful 21,24-dioxacorrole synthesis has been devised involving a sulfur extrusion reaction.[12] The diformyldifuryl sulfide (**19**; X = O) is readily prepared by reaction of 2-bromo-5-formylfuran with aqueous sodium sulfide. Acid catalyzed condensation of **19** (X = O) with the dipyrromethane diacids (**16a–c**; R = CO_2H) in methanol was expected to lead to the nonaromatic *meso*-thiadioxaphlorins (**20**). It was anticipated that **20** would undergo consecutive electrocyclic and cheletropic reactions resulting in loss of sulfur and formation of a dioxacorrole (**20** → **21** → **17**). The π-electron complement of **20** favors the $(4n + 2)$ π-electron disrotatory closure (18-electron process) to **21** and cheletropic loss of sulfur.[16] In particular, the disrotatory process results in a sterically favorable cis-fused intermediate thiirane

19

20   21

(21), whereas a $4n$ π-electron system would undergo a conrotatory closure resulting in an energetically unfavorable trans-fused thiirane.

Monitoring the reaction by uv showed the gradual development of a Soret-like absorption at approximately 390 nm, suggesting that an aromatic product was forming. The major reaction products were found to be the 21,24-dioxacorroles (**17a–c**; 27–30%) together with small amounts (1–2%) of by-products (**22**; R = Me or Et). The structure of the by-products was assigned on the basis of spectral data (uv, nmr). The alternative formulation (**23**; R = Me or Et) for the by-products cannot be ruled out, but it is considered less likely on the grounds of the presumed mode of formation, via cleavage–recombinations reactions of the dipyrromethane, generating an intermediate tripyrrane as found in the reaction producing **18** as a by-product.

Nucleophilic attack of the tripyrrane at the furan α-position (**24**) would produce **25** in which the electron-withdrawing aldehyde group facilitates sulfide cleavage (**25**; arrows).

24: X = tripyrrane   25

This facile sulfur extrusion (**20** → **17**) could not be extended to the thiophen case; **19** (X = S) and **16a–c** failed to give macrocyclic products. However, several types of sulfur-containing macrocycle are isolable from reactions involving only pyrrolic precursors. Acid-catalyzed condensation of **26** (R = R^1 = Me) and **16b** or **c** (R = CO_2H) at $-10°C$ gives a mixture of two macrocycles in low yield.[12] The minor green product (5%) was conveniently isolated as its stable zinc complex (**27**; R = Me or Et) and can be isolated in higher yield (10–15%) if zinc acetate is added to the reaction mixture. *Meso*-thiaporphyrin-free bases are known to be unstable.[17] The major purple red product (10–27%) had an nmr consistent with the unsymmetrical, nonaromatic formulation (**28**) [i.e., **28**; R^1 = Et; nmr (CDCl$_3$), low field signals at τ -0.54 (1H, NH, exchangeable with D$_2$O), 0.7 (1H angular C-1 proton), 3.22,

3.53, and 5.18 (3 × 1H, *meso* protons)]. Chemical evidence also accords with structure **28** since with Zn(OAc)$_2$, the purple red products give **27**, and heating **28** (R = Me) with excess triphenylphosphine in dichlorobenzene gives the corrole (**29**, 15%).

A new type of macrocycle, the desired *meso*-thiaphlorin (**30**; R = H), was obtained (52%) when the dialdehyde (**26**; R = CO$_2$Et, R^1 = Me) was condensed with the dipyrromethane (**16b**; R = CO$_2$H) in chloroform at 0°C, using dry hydrogen chloride as catalyst. The nmr spectrum (CDCl$_3$) of the deep blue macrocycle (**30**; R = H) demonstrated the absence of an aromatic ring current in the molecule with meso proton signals at τ 2.43 (2H) and 3.7 (1H) and the imino proton signals at τ5.44. The visible spectrum of **30** (R = H) closely resembles that of a phlorin (Fig. 2). The *meso*-thiaphlorin (**30**; R = H) is smoothly oxidized by high potential quinones to the unstable green thiaporphyrin, which is conveniently isolated as its stable zinc complex.[12] The mechanism postulated for the extrusion of sulfur in the formation of the dioxacorroles (**20** → **21** → **17**) is equally applicable to **30**. Indeed, heating **30** (R = H) in boiling *o*-dichlorobenzene gave the corresponding corrole (**31**; R = R^1 = H, 40%). An improved yield (60%) of **31** (R = R^1 = H) was obtained when excess triphenylphosphine was present. A concerted mechanism analagous to (**20** → **21** → **17**) is favored for this process, since reactions carried out in the presence of radical scavengers, hydroquinone or

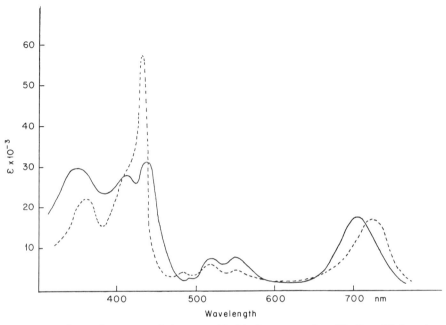

Fig. 2. Electronic spectra of the *meso*-thiaphlorin monocation (**30**; R = H) (———)
and a phlorin monocation (- - -).

tert-butylcatechol, caused no significant difference in yield.[12] N-Alkylation
of **30** (R = H) with methyl iodide/diisopropylethylamine occurred regio-
specifically giving the *N*-methyl derivative (**30**; R = Me), which showed
characteristic upfield shifts in the nmr spectrum for the protons of the methyl
and ethyl ester substituents attached to the alkylated ring (Fig. 3).[4,18] This
regiospecific alkylation is intriguing and is clearly related to the presence of
the highly polarizable sulfur atom adjacent to the ring A nitrogen atom. It
is instructive that in the N methylation of the corresponding corrole (**31**;
R = R^1 = H), the deactivating influence of the ester substituents results in a
predominance of the N(22)-methylcorrole (**31**; R = H, R^1 = Me) (46%) over
the N(21)-methylcorrole (**31**; R = Me, R^1 = H) (26%), in contrast to the
N-alkylation of β-octaalkylcorroles, where the N-21 isomer predominates.
The *N*-methylthiaphlorin (**30**, R = Me) underwent extrusion of sulfur at
213°C in the presence of triphenylphosphine, giving the corresponding
corrole (**31**; R = Me, R^1 = H) in excellent yield (85%). The slower rate of
sulfur extrusion from (**30**; R = Me) reflects the steric influence of the *N*-
methyl group.

A remarkable increase in the rate of the extrusion reaction occurred when
(**30**; R = Me) was heated in boiling acetic acid in the presence of palladium

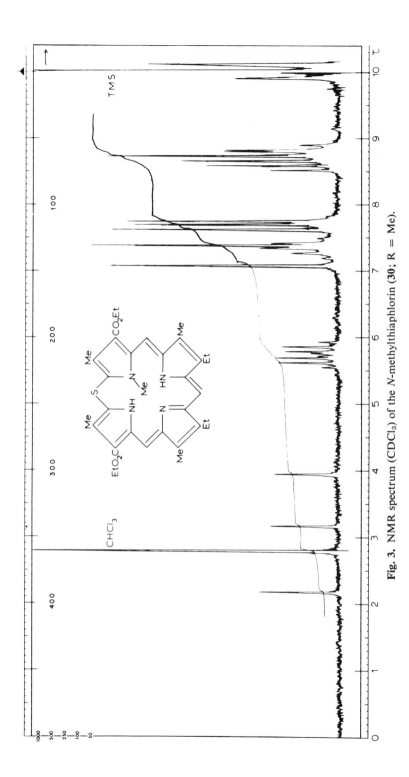

Fig. 3. NMR spectrum (CDCl₃) of the *N*-methylthiaphlorin (**30**; R = Me).

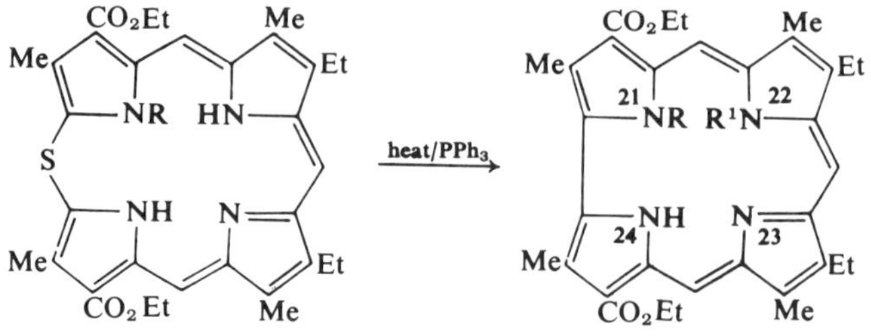

30   31

acetate. After 2 minutes, the corresponding Pd(II) N(21)-methylcorrole (32; M = Pd, R = Me) (27%) could be isolated. A related remarkably easy expulsion of a nitrogen atom, at room temperature, from the *meso*-homoazaporphyrin (33) in the presence of metal ions [Cu(II), Zn(II)] has been observed (33 → 34).[19] These rate enhancements are thought to arise from the efforts of the complexed metal ion to form four equivalent bonds to the ligand nitrogen atoms. Molecular models suggest that such metal ion complexation of 30 and 33 results in a twisting (35; X = S or NCO$_2$Et) of the *p*-orbitals on the carbon atoms flanking the heteroatom in a manner beneficial to a disrotatory cyclization (analogous to 20 → 21), which is considered to be the first step in both 30 → 31 and 33 → 34. Studies with a range of metal ions on the thiaphlorin (30; R = H)[20] show that copper nickel, palladium, and

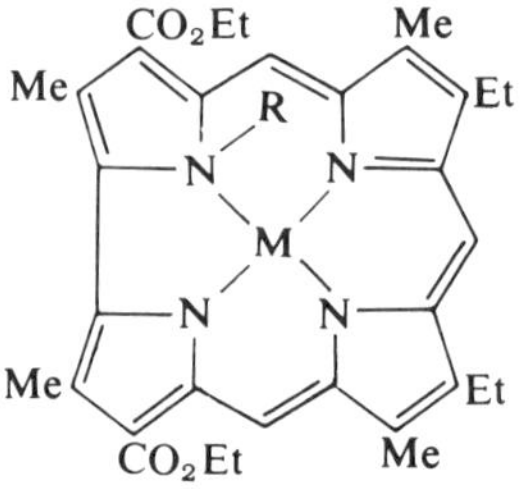

32 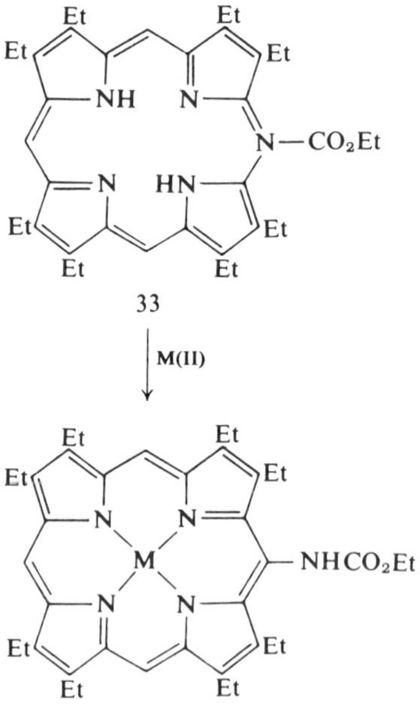   33

35   34

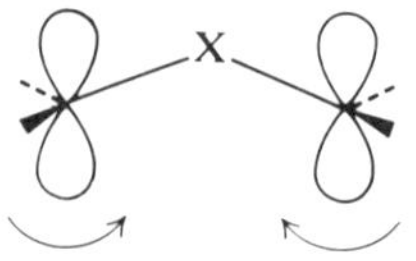

cobalt all give the corresponding metallocorrole (e.g., **32**; M = Cu, R = H) (52%)* under analogous conditions, with Cu(II) producing the best yield of metallocorrole.

The *meso*-dithiamacrocycle (**37**), prepared (44%) by condensation of the dialdehyde (**26**; R = CO_2Et, R^1 = Me) with the diacid (**36**), undergoes slow extrusion of sulfur (10 hours at 213°C) to give an approximately 50:50 mixture (42%) of the two *meso*-thiacorroles (**39** and **40**).[12] The same combination of a disrotatory cyclization and cheletropic sulfur loss is involved, but in this case the intermediate is a mesomeric zwitterion (e.g., **38**). Formation of neutral metal complexes [e.g., Zn(II) and Pd(II)] assists the extrusion of sulfur and results in a higher yield (e.g., the Zn complex of **37** gives 66% of the zinc *meso*-thiacorroles), but the isomer distribution is unaffected. A related case of sulfur extrusion from the unsymmetrical *p*-dithiin (**41**) shows selective loss of one sulfur atom giving the corresponding 2-nitrothiophen (**43**), due to the stabilizing influence of the nitro group favoring one of the possible zwitterions (**42**).[21]

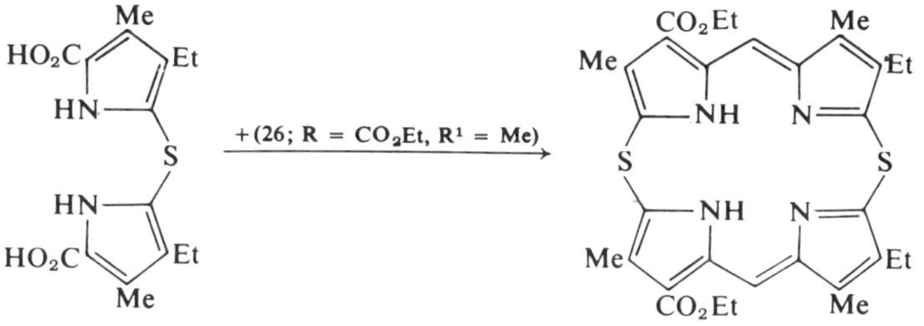

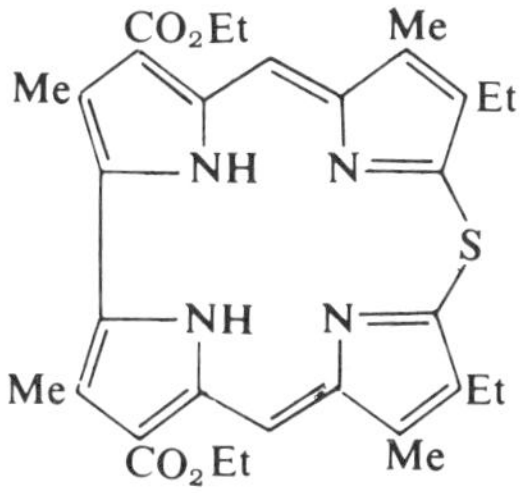

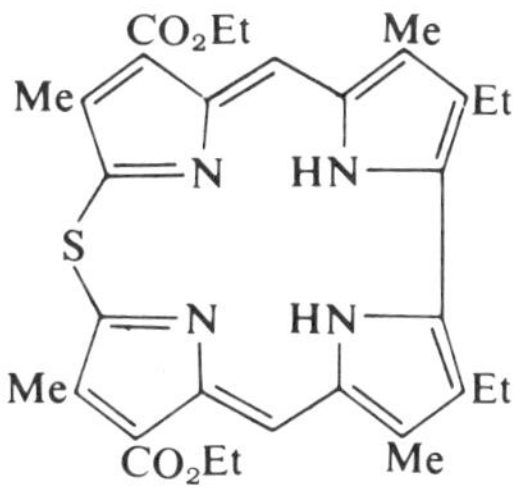

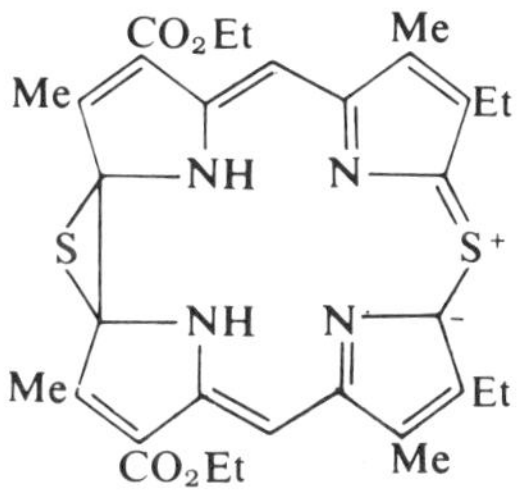

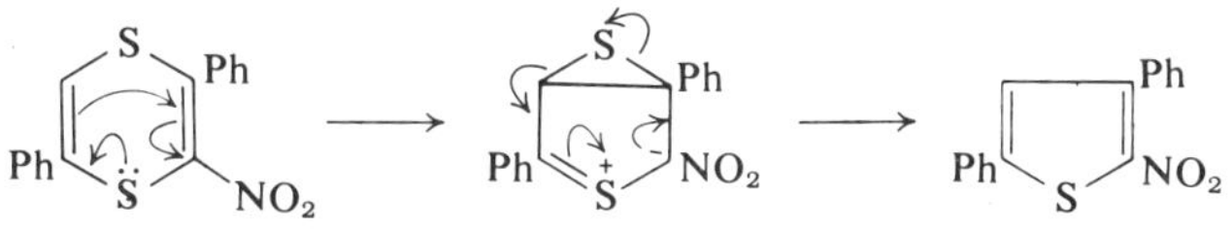

* The position of the extra hydrogen atom shown here on *N*-21 has not been established with certainty.

B. Properties and Reactions of Corroles and Their Hetero Analogues

The corroles and their hetero analogues are all delocalized aromatic systems containing 18π-electron chromophores like the closely related porphyrins. X-Ray structures of both a corrole[22] and an N-methyl copper corrole[23] have been reported and show the peripheral bonds of the corrole to be intermediate in length between normal carbon–carbon single and double bonds. Ring A of the corrole is twisted by about 8° out of the general molecular plane, presumably as a result of crystal packing forces, and, within the five-membered rings, the α,β-bonds are longer than the β,β-bonds. This is also the case in porphyrins[24] but is the reverse of the situation in simple pyrroles.[25] Calculations of the bond lengths and π-electron distribution in corroles and corrole anions have been reported.[26] These calculations indicate that a proton hole is localized at N-22 and that the $N(24)$-pyrrole ring is tilted out of the mean plane of the other three rings.

The electronic spectra of these macrocycles contain, as expected, an intense Soret band (Fig. 4). However, corroles differ from porphyrins in that they possess three central "pyrrole type" nitrogen atoms and one "pyridine type" nitrogen atom, whereas the porphyrins possess two of each type. This difference has important effects on the chemistry of the corroles.

Corroles are stronger acids than porphyrins but weaker bases.[4,27] They readily form stable aromatic monoanions (e.g., **44**) by N deprotonation with dilute alkali.[4] The acidity of 21,24-dioxacorroles, however, more closely resembles the porphyrins than the corroles. Thus the visible spectra of dioxacorroles are virtually unchanged in dimethylformamide solution after the

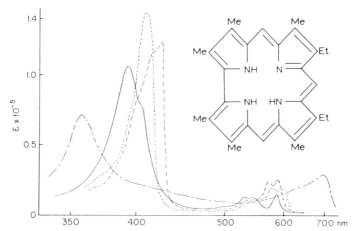

Fig. 4. Electronic spectra of neutral (——), anion (---), monocation (·· · ·), and dication (—·—·—) of a corrole.

addition of aqueous sodium hydroxide. Protonation studies with N(21)- and N(22)-methylcorroles show that mono-, di-, and triprotonation occurs (Fig. 4).[27] The mono- and diprotonated forms retain a Soret band in their electronic spectra, which accords with N protonation. The addition of a third proton causes the disappearance of the Soret band from the spectrum and this, taken in conjunction with nmr studies in fluorosulfonic acid,[28] supports meso protonation. Two meso-protonated structures (45)[28] and the analogous C-10 protonated form[4] have been suggested, but both were based on the mistaken belief that a dicationic species was involved. Thus, further protonation of the monoprotonated corrole (46) apparently involves addition of a proton at either N-21 or N-22, with N-21 being intuitively favored. The third and final protonation then occurs at one of the three meso positions.

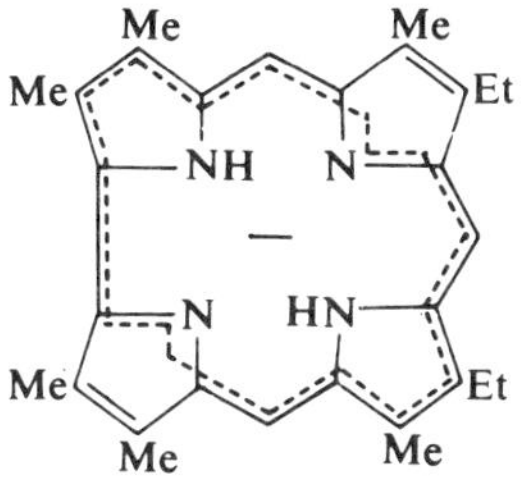

44

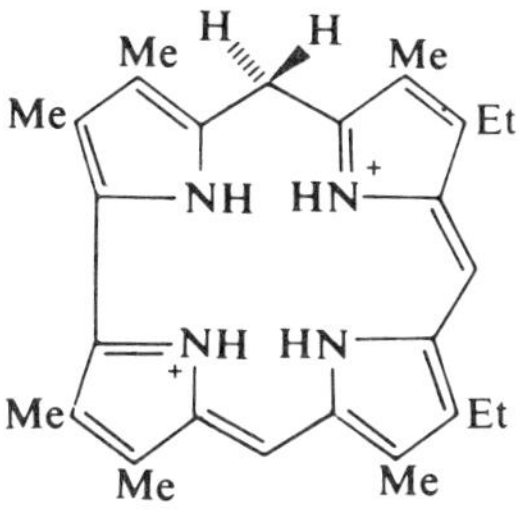

45

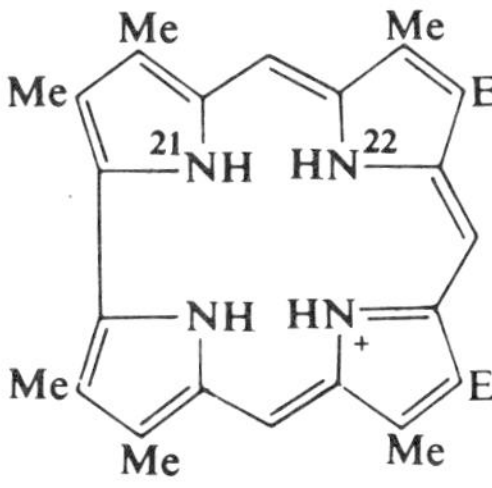

46

Virtually nothing has been published on the chemical reactivity of the hetero analogues of corroles apart from the dioxacorroles (47a, b; R^1 = H). Corroles undergo rapid exchange of the meso protons for deuterium in deuteriotrifluoroacetic acid (DTFA) at room temperature. Exchange is complete in less than 15 minutes.[28] In contrast, no detectable exchange of the meso protons of the dioxacorrole (47b; R^1 = H) occurs in DTFA solution at 35°C, but after 1 hour at 100°C virtually complete exchange of the meso protons had occurred, while no exchange of the β-protons of the furan rings was observed after 100 hours at 100°C.[12] Porphyrins and their furan and thiophen analogues showed no exchange of meso protons or the β-protons of furan or thiophen rings under analogous conditions.[29]

Attempts to effect Friedel–Crafts carbon acylation of corroles were un-successful, and Lewis acid catalysts invariably cause decomposition of the macrocycle. Heating corroles under reflux in acetic anhydride solution gives the N(21)-acetyl derivatives (**48a, b**)[27] which, as might be expected, are prone to hydrolysis. In contrast, the less reactive dioxacorrole (**47a**; $R^1 = H$) reacts with acetyl chloride/$AlCl_3$ to give a mixture of **47a** ($R^1 = Ac$; 31%) and the 2,18(3,17?)-diacetyl derivative (27%).[28]

47 (a) R = Me   48 (a) R = Me, R^1 = Et
   (b) R = Et   (b) R = Et, R^1 = Me

Alkylation of the ambident corrole anions (RI/acetone/K_2CO_3) is a kinetic-ally controlled process and produces a mixture of N(21)- and N(22)-alkyl-corroles with the N-21 isomer predominating.[4,28] More forcing conditions (100°C) give N,N'-dialkylcorrole salts, for which structure **49** is favored. The dioxacorrole (**47a**, $R^1 = H$) methylates (MeI/iPr$_2$NEt/100°C) to give a mixture of the *N*-methyl (**50**; R = Me, R^1 = H) and *trans-N,N^1*-dimethyl (**50**; R = R^1 = Me) salts. Aromaticity is retained in all the *N*-alkylcorroles and *N*-alkyldioxacorroles as evidenced by the retention of a Soret band in the electronic spectra (Fig. 5) and the chemical shift of the *N*-alkyl groups in their nmr spectra (e.g., Fig. 6).

In common with N(22)-alkylcorroles and *N*-alkylporphyrins, the Soret band in the electronic spectrum showed a characteristic shift (~ 11 nm) to longer wavelength, after the addition of each alkyl group. The *N*-alkyl-corroles are both more basic and better nucleophiles in alkylation reactions than the parent corroles. The *N*-alkyldioxacorroles also show enhanced basicity, compared to the parent system. This enhanced basicity presumably reflects the relief of steric strain upon further protonation. Thermal isomeriza-tion of N(22)-allylcorroles to the N-21 isomers occurs at 110°C and is accom-panied by substantial amounts of cleavage to the parent corrole. The forma-tion of cleavage products, together with trapping experiments in cumene, and the observation that the dimethylallyl group in **51** migrates without inversion to give **52** suggest a free radical process is involved, rather than an allowed [3,11]-sigmatropic process.[28] *N*-Methylcorroles are stable up to 200°C but decompose at 300–350°C.

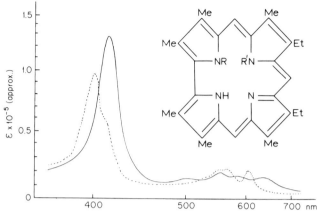

Fig. 5. Electronic spectra of an N(21)-methylcorrole (R = Me, R^1 = H; ---) and an N(22)-methylcorrole (R = H, R^1 = Me; ——).

Corroles and their C-10 meso hetero analogues form a range of metal complexes. Complexes of corroles with Ni(II)[4,30] and Pd(II)[31] are non-aromatic, with one of the potentially tautomeric hydrogen atoms being displaced from nitrogen to carbon in such a way as to interrupt the chromophore. The location of this displaced proton is still unresolved. PMR studies of the Ni(II) and Pd(II) complexes, which were expected to be diamagnetic, were

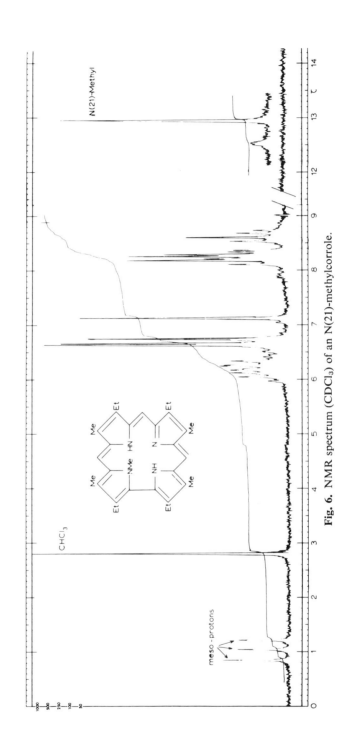

Fig. 6. NMR spectrum (CDCl₃) of an N(21)-methylcorrole.

not possible since they were found to be paramagnetic. This paramagnetism is due to the presence of radical species in the complexes which arise from partial loss of the displaced "blocking" hydrogen atom, generating a delocalized metallocorrole radical. The "displaced" proton is arbitrarily placed at C-10 (i.e., **53**; M = Ni or Pd). In contrast to the Ni(II) and Pd(II) corroles, the paramagnetic Cu(II) corroles have spectra similar to those of the fully conjugated N(21)-methyl Cu(II) corroles (**54**; R^1 = Me) (see later) and show similar color changes on protonation. The parent Cu(II) corroles are thus formulated as **54**; R^1 = H.[31]

Recently a new type of metallocorrole complex has been prepared in which the metal atom lies above the plane of the corrole ring.[31a] Thus corrole, N(21)-methylcorrole, and N(22)-methylcorrole react with $[Rh(CO)_2Cl]_2$ to give **54a** (R = R^1 = H; R = Me, R^1 = H and R = H, R^1 = Me). The N(22)-methylcorrole complex (**54a**; R = H, R^1 = Me) equilibrates in boiling acetone with an isomeric complex thought to be **54b**.[31b] The nmr spectra of **54a** and **54b** show that the complexes, though obviously distorted, are still delocalized aromatic systems.

21-Oxacorroles (**22**) readily form stable, neutral, aromatic (Fig. 7), transition metal complexes [Ni(II), Co(II), Cu(II)], but these have not been studied further. In contrast, the 21,24-dioxocorroles (**17**) failed to form metal complexes (Co, Ni, Cu, Zn).[12]

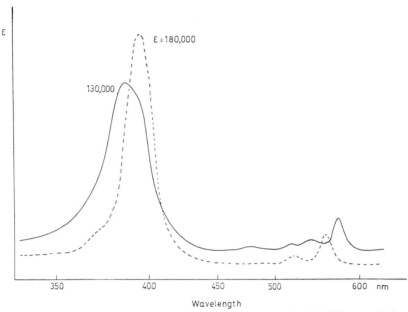

Fig. 7. Electronic spectra of a Ni(II) 21-oxacorrole (———) and Ni(II) octaethylporphyrin (---).

The metallocorroles [**53**; M = Ni(II) or Pd(II) and **54**; R^1 = H] are readily deprotonated to give the corresponding aromatic ambident anions [**55**; M = Cu(II), Ni(II), or Pd(II)]. The palladium corroles are the least stable and most acidic of the three metallocorroles and are best isolated and handled as their stable pyridinium salts.[31] The order of acidity of the metallocorroles is Pd > Ni > Cu. Copper corroles are the weakest acids, since they are themselves aromatic and acquire less stabilization on forming the aromatic metallocorrole anion than do the nonaromatic Ni(II) and Pd(II) corroles. The metallocorrole ambident anions [**55**; M = Ni(II), Cu(II), or Pd(II)] undergo kinetically controlled alkylation to give the corresponding N(21)-alkylmetallocorroles (**56**), together with a preponderance of C-3 alkylation [**57**; M = Pd(II)] in the palladium case. Use of allyl halides or bulky alkylating agents, e.g., isobutyl iodide, results in C-3 alkylation of Ni(II) corroles [**57**; M = Ni(II)].[30,31] There is no evidence of alkylation at other sites, e.g., N-22, meso, or metal centers, in contrast to the metal-free macrocycle where substantial amounts of N-22 alkylation are found. The ratio of N-21:C-3 alkylation of palladium corrole anions shows some solvent dependence[31] with 2,2,2-trifluoroethanol suppressing N-21 alkylation by hydrogen bonding with the nitrogen centers and hexamethylphosphoric triamide producing most N alkylation (N-21:C-3, ~1:2). Acetone gave an intermediate result (N-21:C-3, ~1:4).

X-Ray crystal structures of an N(21)-methyl Cu(II) corrole[23] and an N(21)-ethylnickel corrole (56; $R^1 = R^2 = Me$, $R^3 = R^4 = Et$)[32] show the N-21 nitrogen atom to be sp^3 hybridized, with ring A markedly twisted out of the general molecular plane. The aromaticity of the macrocycle is not destroyed by these distortions as evidenced by nmr (e.g., 56; M = Ni, $R^1 = R^2 = Me$, $R^3 = R^4 = Et$) (Fig. 8) and electronic spectra (presence of a Soret band). The methylene protons of the N-ethyl substituent in 56 (M = Ni, $R^1 = R^2 = Me$, $R^3 = Et$) are magnetically nonequivalent and give rise to an ABX_3 pattern (Fig. 8). The N(21)-alkyl metallocorroles can also be prepared by insertion of Ni(II), Pd(II), or Cu(II) into the preformed N(21)-alkylcorrole.[30,31] Nickel acetate is unsatisfactory for this purpose and $(CH_3CN)_6NiClO_4$ is the preferred reagent. Insertion of metal ions into N(22)-alkylcorroles has not proved possible with Ni(II), but Pd(II) gives unstable complexes. In contrast, the insertion of Cu(II) into an N(22)-ethylcorrole gave 58.[31] Attempts to insert Cu(II) into N(21)- and N(22)-allylcorroles results in loss of the allyl group and formation of Cu(II) corrole. Protonation of the N(21)-alkyl complexes (56) causes a reversible red to green color change and nmr spectra of the Ni(II) and Pd(II) complexes showed that the aromaticity had been destroyed by angular protonation. The precise site of protonation has not been determined, but there is a marked similarity

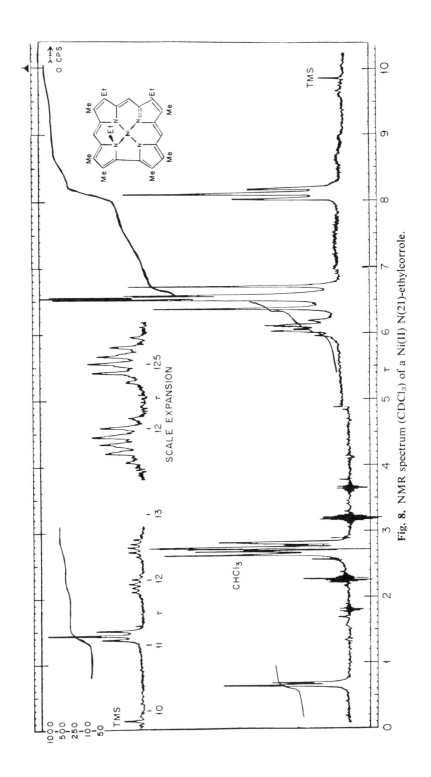

Fig. 8. NMR spectrum (CDCl₃) of a Ni(II) N(21)-ethylcorrole.

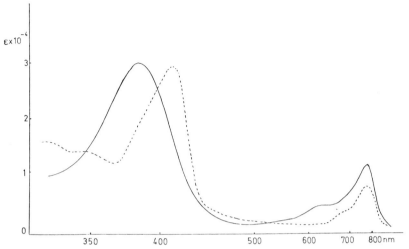

Fig. 9. Electronic spectra of the protonated form of **56** [M = Ni(II), R^1 = R^2 = R^4 = Me, R^3 = Et] (——) and a Ni(II) 1-methyltetradehydrocorrin [**97**; M = Ni(II), R = R^2 = R^3 = Me, R^1 = R^4 = Et] (- - -).

(Fig. 9) between the visible spectra of the protonated species and those of nickel 1-alkyltetradehydrocorrins.

The N(21)-alkylmetallocorroles show interesting differences in their thermal behavior. The Ni(II) and Pd(II) complexes (**56**) undergo thermal rearrangement to the corresponding 3,3-dialkylmetallocorroles [**57**; M = Ni(II) or Pd(II)] at 130°C and 180°C, respectively. Appropriate substituent-labeling experiments show that the corresponding metallo-1-alkyltetradehydrocorrin (**59**) is not an intermediate.[30] Thermal rearrangements of N(21) alkyl Ni(II) corroles in the presence of a radical trapping agent (hydroquinone) or a radical trapping solvent (cumene) did not alter the yields of rearrangement product [**57**; M = Ni(II)], and thermal rearrangement of mixed N(21)-alkyl Ni(II) corroles did not produce "crossed" products. These observations are in accord with a process involving two concerted [1,5]-sigmatropic shifts of the N(21)-alkyl group via **60** as an intermediate. In contrast to the Ni(II) and Pd(II) series, the N(21)-alkyl Cu(II) corroles undergo thermal cleavage to the parent Cu(II) corroles. Presumably, the presence of an unpaired electron in the d^9 Cu(II) complex encourages uncoupling of the electron pair in the N-alkyl bond by d_π–p_π bonding between copper and the developing framework radical resulting in homolytic cleavage of the N-alkyl bond. Subsequent abstraction of a hydrogen atom from the solvent would then lead to the observed products. A similar process occurring under mild conditions may account for the observation that attempts to insert Co(II) into an N(21)-methylcorrole gave only Co(III) corrole in low yield.[31] The 3,3-dialkyl-metallocorroles undergo protonation (TFA) and allylation at C-17 [61; M = Ni(II) or Pd(II) R^5 = H or $CH_2CH = CH_2$]. The site of protonation was delineated by appropriate substituent labeling.[31,33] When there are two different alkyl substituents at C-3, i.e., a chiral center, there is a slight stereoselectivity (6:5) in the addition of the proton at C-17. This stereoselectivity is enhanced (7:3) in the C-17 allylation of nickel 3-ethyl-3-methyl-corroles, presumably in favor of allylation from the least hindered face of the molecule.[30]

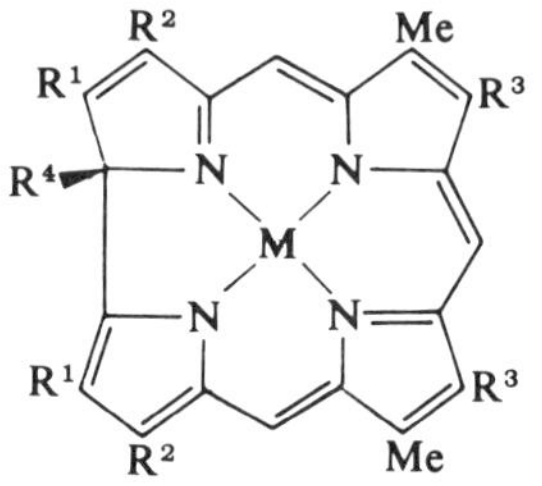

59

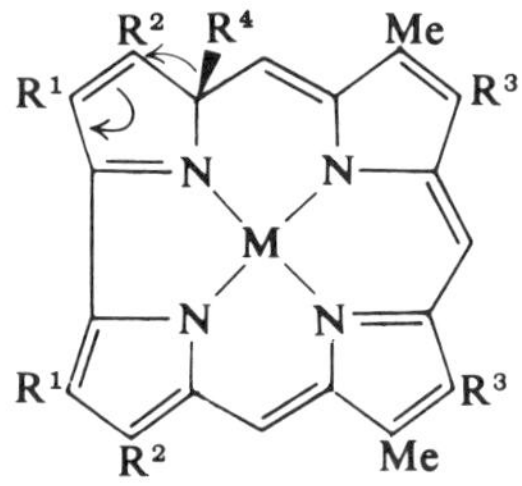

60

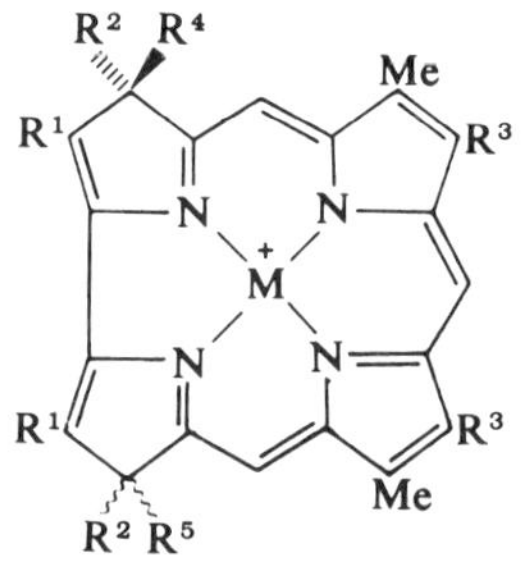

61

Some doubts regarding the valency of cobalt in cobalt corroles have been resolved using electrochemical techniques. The existence of square planar, pyramidal, and octahedral complexes was established, and all species were found to contain Co(III).[10] Treatment of the pentacoordinate monopyridine derivative (62; R = C_5H_5N) with aryllithium reagents gave aromatic, paramagnetic, arylated products formulated as cobalt(II) derivatives (63). The aryl group of 63 dissociated on heating the complex and regenerated the cobalt(III) corrole. Oxidation of 3,17-dimethylcobalt corroles such as 62 (R = PPh$_3$, R^1 = Et, R^2 = Me) with an excess of 2,3-dichloro-5,6-dicyano-1,4-benzoquinone gives the corresponding 3,17-diformylcobalt corroles (e.g., 64).

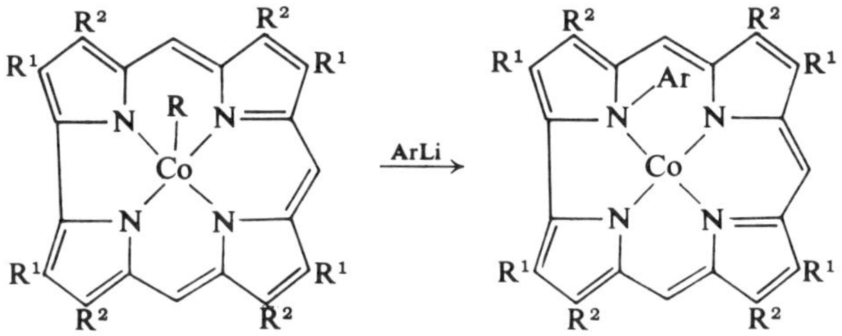

62   63

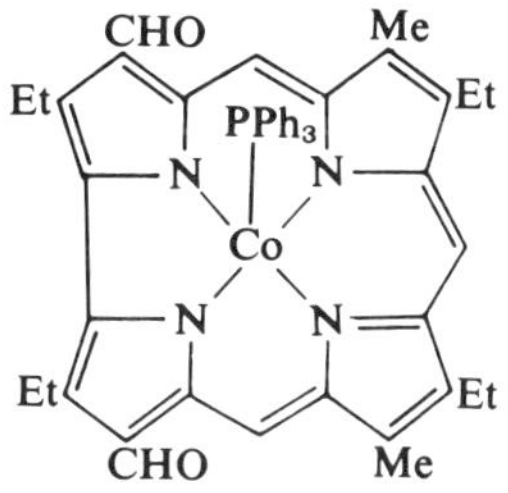

64

II. 22π-ELECTRON MACROCYCLES: SAPPHYRINS AND RELATED COMPOUNDS

A. Synthesis

Syntheses of 22π-electron macrocycles were prompted by an interest in the ability of Huckel systems with a larger complement of π-electrons than the 18π-electron porphyrins to retain aromatic character.

The first reported member of this series was the all-pyrrole system, sapphyrin (**67**; X = NH),[34] and this was followed by the isolation of small amounts of 26,30-dioxasapphyrins as by-products in the synthesis of 21,24-dioxacorroles (see Section I, A).[12]

Two general 2 + 3 synthetic schemes have been devised for the 22π-electron macrocycles.[35] One method involves combination of two-ring dialdehyde component (**65**), which already possesses a direct link, with a bis(pyrrolylmethyl)pyrrole diacid (**66**).

Thus, acid-catalyzed condensation of **65** (X = O, R = R^1 = H) with **66** followed by oxidation with either air or iodine gave the 26,30-dioxasapphyrin (**67**; X = O, R = R^1 = H) isolated as its dihydroperchlorate salt (40%). The tripyrrolic fragment (**66**) is acid sensitive, but, under the reaction conditions (1.5 hours at room temperature), there is no randomization of substituents in the product. The 2,2'-bipyrrole (**65**; X = NH, R = Me, R^1 = Et) undergoes an analogous condensation with **66** to give the sapphyrin (**67**;

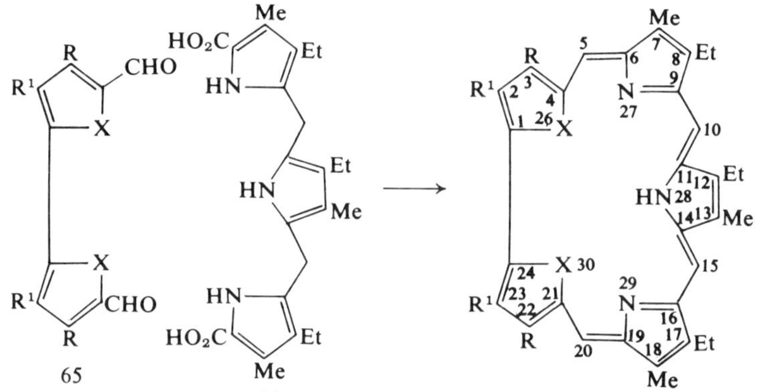

65   66   67

X = NH, R = Me, R^1 = Et; 46%). A further example of this type of reaction is the condensation of (65; X = NH, R = Me, R^1 = Et) with the
dipyrrolylthiophene (68)[29] to give (69; 19%). These syntheses and the alternative approaches discussed below are remarkable for the facility with which
they occur, when one considers the size of ring being generated. Presumably,
intramolecular hydrogen bonding between the heterocyclic rings involving
pyrrolic N—H groups results in conformations beneficial to the cyclization
process.

An interesting modification of the sapphyrin synthesis leads to the dioxanor-
sapphyrins (71), which contain two direct links. Molecular models of 71

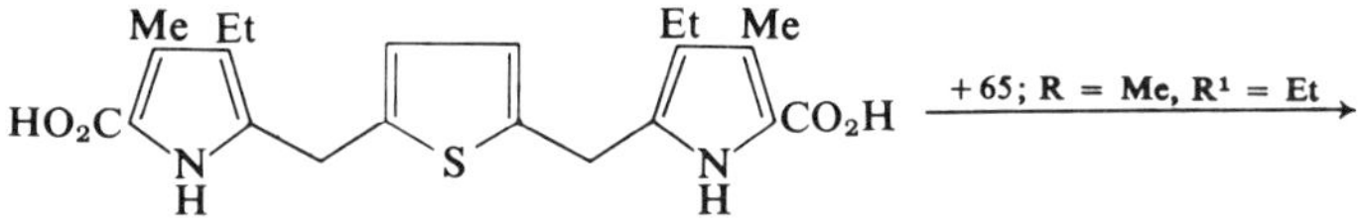

68

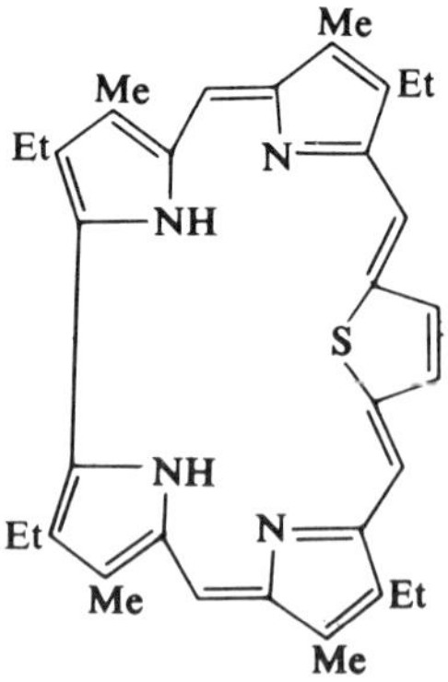

69

Me R   R   Me   Me R¹
(pyrrole structures) 70

(a) R = H, R¹ = Me
(b) R = R¹ = Et

+65; X = O, R = R¹ = H →

71
(a) R = H, R¹ = Me
(b) R = R¹ = Et

Me R   R   Me   Me R¹
(pyrrole structures)
Br⁻
72

revealed it could exist in a relatively planar, strain-free conformation. The synthesis involves condensation of the bifuran dialdehyde (65; X = O, R = R¹ = H) with the pyrrolyldipyrromethanes (70a, b).[35]

The pyrrolyldipyrromethanes (70a, b) are generated *in situ* by borohydride reduction of the corresponding pyrrolyldipyrromethenes (72) which, in turn, are prepared by hydrogen bromide-catalyzed condensation of 2-formyl-pyrroles with 5,5'-unsubstituted bipyrroles.[13] Attempts to prepare the norsapphyrins by condensation of 70 with 5,5'-diformylbipyrroles gave reaction mixtures with the expected visible spectra (λ_{max}, ~450 nm), but attempts to isolate the macrocycles resulted in their decomposition. However, addition of suitable metal ions might allow isolation of the corresponding metal complexes.

The second general method developed for the synthesis of 22π-electron macrocycles is an extension of the sulfur extrusion process developed for dioxacorroles and corroles (see Section I, A), and involves reacting the bis(formylfuryl)sulfide (19; X = O) with the tripyrrane (66) in the presence of hydrogen bromide. The initial product from this reaction (73) can then tautomerize to 74. The π-electron complement of 74, 24π-electrons (4n-π-system), now favors a sterically unfavorable conrotatory cyclization to a trans-fused thiirane intermediate. However, oxidation of 74 to 75 would generate a (4n + 2) π-electron system (22π-electrons excluding the sulfur lone pair electrons) which can undergo a disrotatory cyclization to a sterically favorable cis-fused thiirane intermediate (76). Collapse of the thiirane to the dioxasapphyrin should then occur. In accord with the foregoing analysis, no macrocyclic product is isolated unless iodine (as oxidant) is added after the reaction has been allowed to proceed for 1 hour. Addition of the iodine is

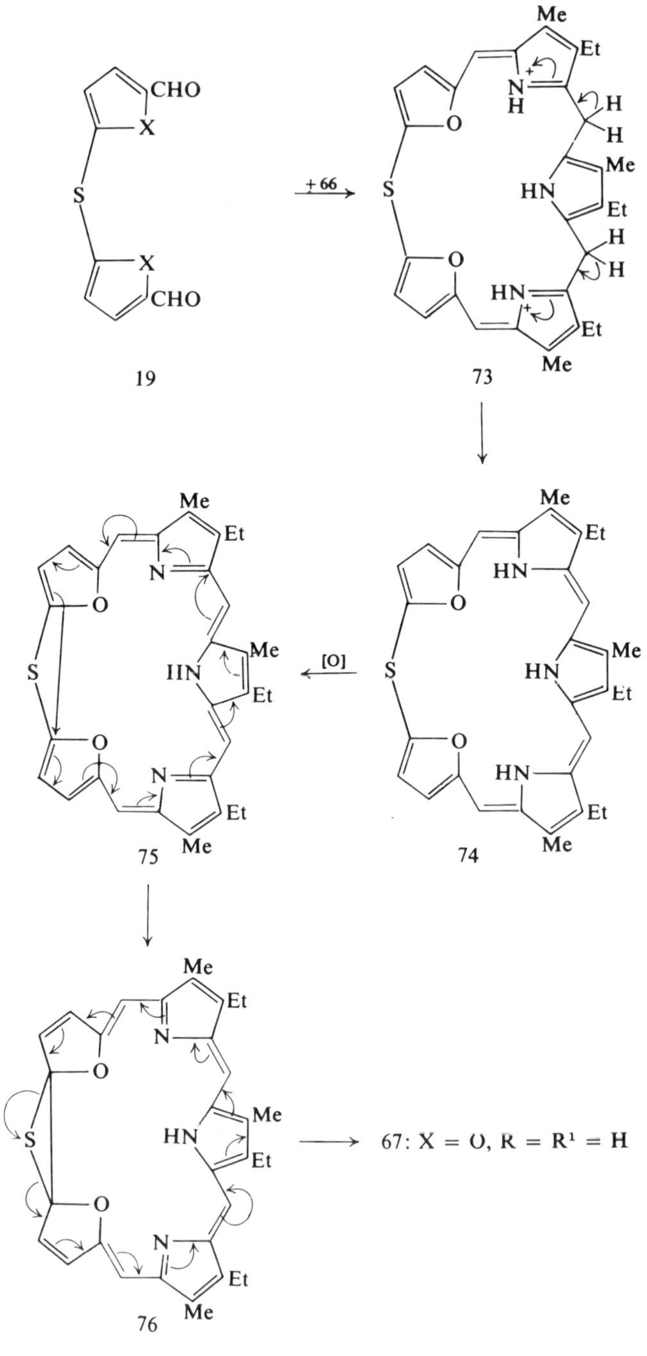

19

+66 →

73

[O] ←

74

75

76

→ 67: X = O, R = R¹ = H

followed by rapid development of an intense band at 435 nm in the visible spectrum of the mixture and leads to the dioxasapphyrin (**67**; 22%).[35]

B. Properties of Sapphyrins and Related Compounds

Theoretical predictions place the upper limit of aromaticity of [4n + 2] annulenes at the 22π-electron system.[36] The 22π-electron sapphyrins and related compounds can be viewed as heteroatom-bridged annulenes, and both nmr (Table 1) and electronic (Table 2) spectra show them to be strongly aromatic.[35] A series of hetero-bridged annulenes[36] containing furan and thiophen rings, but not pyrrole rings, is outside the scope of this article.[37]

TABLE 1

NMR Spectra (τ, CDCl₃) of 22π-Electron Macrocycles

Compound	NH	β-H	meso-H
71	16.55	−0.04 (2H)	−0.48, −0.45, −0.38,
		0.28 (2H)	−0.34
67 (X = NH, R = Me, R¹ = Et)	13.9	—	−0.5 (2H), −0.32 (2H)
69	12.2	−0.12 (2H)	−0.87 (2H), −0.31 (2H)
71b	14.85	0.42 (AB, 4H)	−0.52 (1H), −0.06 (2H)

TABLE 2

Electronic Spectra (Me₂CO-HClO₄) of 22π-Electron Macrocycles (Cations)

Compound	Soret band (λ/nm)	ε
71	435.5	754,600
67 (X NH, R = Me, R¹ = Et)	450	530,200
69	461	472,800
71b⁺	448 and 459	184,000 and 212,000

⁺ in CHCl₃–HBr

The Soret band (Table 2) of the 22π-electron macrocycles occurs at a significantly longer wavelength than that of the porphyrins (∼ 400 nm) and is much more intense. The Soret band of the dioxasapphyrin (**71**) occurs at a shorter wavelength than that of the all-nitrogen analogue (**67**; X = NH, R = Me, R¹ = Et) and a similar trend was found for oxaporphyrins/porphyrins.[29] Dioxanorsapphyrin (**71**) and dioxasapphyrin-(**67**; X = O, R = R¹ = H) free bases have similar visible spectra, but their spectra in acid solution differ in that **71** exhibits a split Soret band.

The basicity of 26,30-dioxasapphyrins is unusually high and mixtures of the mono- and dihydrobromide are eluted from alumina columns. In contrast, the 26,30-dioxanorsapphyrins are markedly less basic, and the free bases are readily obtainable. The 26,30-dioxasapphyrins do not appear to form metal complexes, and no studies of the metal-complexing ability of the other 22π-electron macrocycles have yet appeared. Electrophilic deuteration studies of the dioxasapphyrin (67; X = O, R = R^1 = H) in DTFA solution at 100°C showed marked differences in reactivity of the four meso positions. Two meso protons exchanged with a half-life of 2.2 hours, whereas the remaining two meso protons had exchanged to only a small extent after 100 hours at 100°C. The precise assignment of the meso protons of (67; X = O, R = R^1 = H) has not yet proved possible. No exchange of the β-protons of the furan rings was observed, and a similar lack of exchange was observed for the β-protons of the furan rings in oxacorroles[12] and oxaporphyrins.[29]

III. 1-SUBSTITUTED AND 1,19-DISUBSTITUTED METALLOTETRADEHYDROCORRINS

A. Synthesis

The synthesis of these two classes of macrocycles are conveniently discussed together, since the methods and precursors employed are very similar, and there are reactions for converting one into the other. The template effect of metal ions is required for the synthesis of both classes.

The successful cyclization of 1,19-dideoxybiladienes-a,c to corroles focused attention on the previously synthesized 1,19-disubstituted-1,19-dideoxybiladienes-a,c as possible precursors of 1-substituted and 1,19-disubstituted tetradehydrocorrins. The 1-bromo-1,19-dideoxy-19-methylbiladienes-a,c (79) are prepared by Friedel–Crafts alkylation of a 5-methyl-5'-unsubstituted dipyrromethene hydrobromide (77) with a 5-bromo-5'-bromomethyldipyrromethene hydrobromide (78) in the presence of stannic chloride.[11] The intermediate tin complexes of 79 are not isolated but converted directly to 79 by treatment with hydrogen bromide.

The biladienes (**79**) give porphyrins when heated in *o*-dichlorobenzene or on keeping a dimethylsulfoxide solution at room temperature in the presence of pyridine.[11,38] The reaction is diverted into a new course when **79** is treated with base in the presence of nickel ions.[39] A rapid cyclization to nickel 1-methyl tetradehydrocorrins (**82**; 45–75%) occurs.

The intermediate nickel complexes (**80**) have not been isolated, but analogous compounds have been isolated during cyclizations of 1,19-dimethyl-1,19-dideoxybiladienes-a,c in the presence of nickel ions (see later in this section). The mechanism proposed for this cyclization[39] involves deprotonation at C-10 followed by intramolecular nucleophilic displacement of bromide (**80 → 81 → 82**). Nickel complexation ensures the favorable juxtaposition of C-1 and C-19. Later work on the cyclization of 1,19-dimethyl-1,19-dideoxybiladienes-a,c (below) suggests another possibility involving prior oxidation of the deprotonated complex (**80**) to the 16π-electron cationic species (**83**) followed by an allowed conrotatory cyclization to **84**. However, formation of the product **84 → 82** would involve loss of Br$^+$ (which could function as the oxidant required to generate **83**) but no brominated by-products have been observed in these reactions. Small amounts of lactam (**85**) are, however, formed in some reactions, and there is evidence showing the lactam results, at least in part, from oxidative degradation of **82**.

The biladiene cyclization can accommodate larger alkyl groups[39] at C(1) and both Ni(II) 1-ethyl- and 1-isopropyltetradehydrocorrins have been

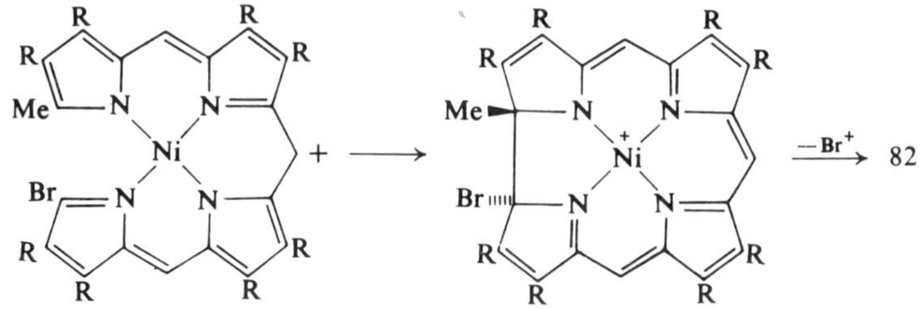

83   84   82

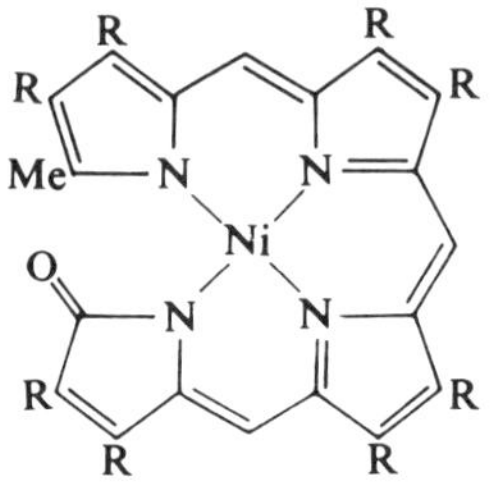

85

prepared. The cyclization of 1-iodo-1,19-dideoxy-19-methylbiladienes-a,c with a uroporphyrin III and a vitamin B_{12} β-substitution pattern to **82** ($R = R^2 = CH_2CO_2Me$, $R^1 = CH_2CH_2CO_2Me$) and **82** ($R = CH_2CO_2Me$, $R^1 = CH_2CH_2CO_2Me$, $R^2 = Me$) has been reported.[11a] Attempts to vary the metal ion template by using cupric acetate, cupric sulfate or cobalt (II) acetate under a variety of conditions were largely unsuccessful. The copper salts gave low yields (8–25%) of copper porphyrins, while cobalt salts produced complex mixtures under the usual conditions. However, it is possible to generate the Co(II) 1-substituted tetradehydrocorrins under anerobic conditions and trap them by alkylation as the stable Co(II) 1,19-disubstituted tetradehydrocorrin salts (see Section III, B). The precise reason for the instability of the cobalt 1-substituted tetradehydrocorrins is not, at present, understood, although it is clearly related to the variable valency of cobalt. Palladium(II) salts have not been studied, but it appears likely that they would prove successful templates.

Biladienes-a,c, in which both the terminal 1- and 19-positions are occupied by alkyl groups, also cyclize in the presence of Ni(II) and Co(II) ions to the corresponding metal 1,19-dialkyltetradehydrocorrin salts in high yield (60–85%).[40] The terminally symmetric biladienes are more readily prepared than the unsymmetrical biladienes (**79**) required for the synthesis of the Ni(II) 1-alkyltetradehydrocorrins. Thus, condensation of 5,5′-unsubstituted dipyrromethanes (generated *in situ* by borohydride reduction of the corresponding dipyrromethene) or dipyrromethane-5,5′-dicarboxylic acids with 2 moles of a 2-formyl-5-methylpyrrole, in presence of hydrogen bromide, gives the required terminally symmetric biladienes-a,c (**86**; R = Me). Biladienes in which the 1- and 19-substituents differ (e.g., **86**; R = CO_2Et) can be synthesized by a 2 + 1 + 1 approach involving condensation of a 5-formyldipyrromethane-5'-carboxylic acid first with a 2,3,4-trialkylpyrrole and then with a 2-ethoxycarbonyl-5-formylpyrrole.[40] The oxidative cyclization of

1,19-dideoxy-1,19-dimethylbiladiene-a,c dihydrobromides (**86**; R = Me) in the presence of cupric salts to Cu(II) porphyrins (**87**) had been known for some years[41,42] before the successful corrole synthesis prompted a study of the behavior of **86** (R = Me) in the presence of Ni and Co salts. The biladienes (**86**; R = Me or CO_2Et) cyclize to the corresponding Ni(II) 1,19-disubstituted tetradehydrocorrin salts (**88**; M = Ni, R = Me or CO_2Et) when heated in methanol, with aeration, in the presence of nickel and sodium acetates. When the reaction is run for a shorter time without aeration the intermediate complex (e.g., **89**) can usually be isolated.[40] Analogous cyclizations to **88** [M = Pd(II) or Co(II)] can be effected.[40,43] In the cobalt case, a rapid cyclization prevents isolation of the intermediate complex analogous to **89**. Treatment of the paramagnetic cobaltous complexes [**88**; M = Co(II)] with sodium cyanide gives the diamagnetic Co(III) complexes (**90**).

The trans arrangement of the 1,19-substituents has been established for nickel 1,19-dimethyltetradehydrocorrin salts by partial resolution of the (+)-D-camphorsulfonate salt [**88**; M = Ni(II), $R^1 = R^2 = R^3$ = Me, R^4 = Et, X = (+)-D-camphorsulfonate],[44] and is assumed, by analogy, for the corresponding cobalt and palladium complexes. The cis-1,19-disubstituted tetradehydrocorrin would possess a plane of symmetry.

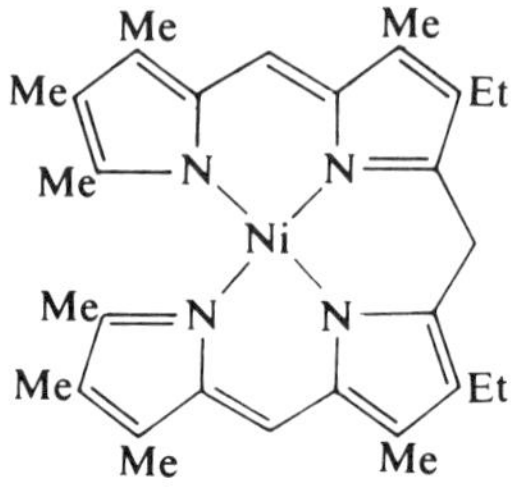

89

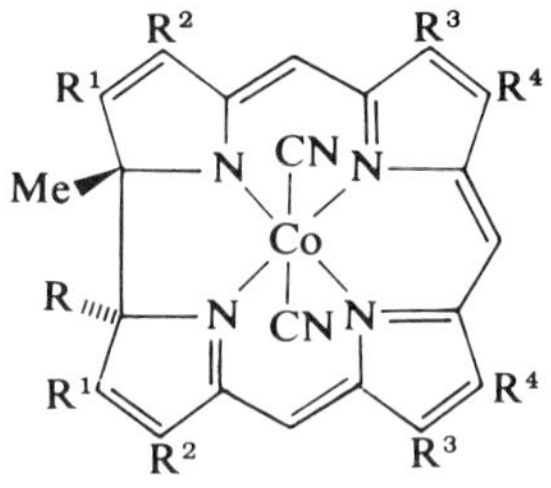

90

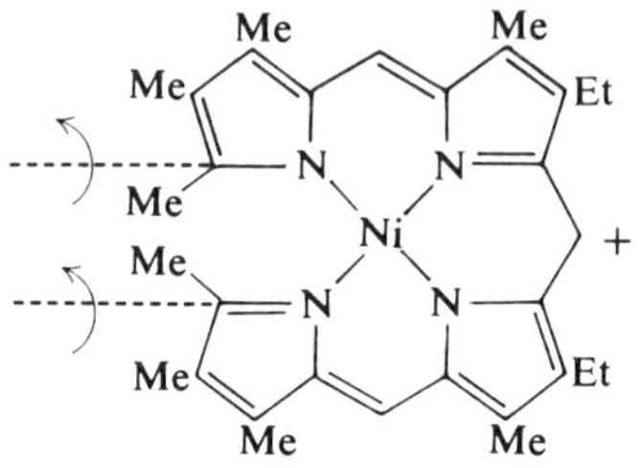

91

The usual mechanistic problem associated with the oxidative cyclization of 1,19-dideoxybiladienes-a,c to tetradehydrocorrins—i.e., does oxidation precede or follow cyclization—is more amenable to study in this series, since the intermediate nickel complexes (e.g., **89**) can be isolated. Thus, **89** when treated with the hydride-abstracting reagent, trityl perchlorate, in the absence of oxygen, gave the corresponding nickel *trans*-1,19-dimethyltetradehydrocorrin [**88**; M = Ni(II), R = R^1 = R^2 = R^3 = Me, R^4 = Et, X = ClO_4].[44] This result suggests that the species undergoing cyclization is the 16π-electron cation (**91**) and that oxidation precedes cyclization. The trans geometry is also in accord with this scheme since a conrotatory cyclization is the expected mode for a 16π-electron species, although the stereochemistry of the intermediate nickel biladiene complex (**89**) would make a cis geometry unlikely, if not impossible.

Studies with the 10-mono- and 5,15-di-*meso*-methyl derivatives of 1,19-dideoxy-1,19-dimethylbiladienes-a,c dihydrobromides show that, under standard conditions ($NiOAc_2/NaOAc_2/MeOH$), the relative cyclization rates are 5,15-dimethyl ≈ unsubstituted > 10-methyl.[45] These rates reflect the lower acidity of the C-10 proton when C-10 is monomethylated.[5] As expected, the 10-di-*meso*-methyl derivative (**92**) does not cyclize,[45] and the 1,19-dideoxy-1,19-dimethylbiladienes-a,c do not cyclize in the absence of metal ions.[40]

The synthesis of metal 1,19-dideoxy-1,19-disubstituted tetradehydrocorrin salts from the corresponding biladienes-a,c tolerates a wide range of substituents. The parent nickel and cobalt 1,19-dideoxy-1,19-dimethyltetradehydro-corrin perchlorates have been recently synthesized ($\sim 25\%$).[46] The requisite biladiene was constructed from 2,2'-dipyrromethane and 2-formyl-5-methylpyrrole, but the acid lability of both 2,2'-dipyrromethane and the intermediate 1,19-dideoxy-1,19-dimethylbiladiene-ac necessitates careful control of the reaction conditions. A 1,19-diethoxycarbonyl derivative has also been prepared,[40] and this links the metallocorroles and metal 1,19-disubstituted tetradehydrocorrin salts, since hydrolysis and decarboxylation of latter gives the former.[40]

Bilirubin (**93**) provided a convenient starting material for the synthesis of the interesting 1,19-diethoxytetradehydrocorrin derivatives [**95**; M = Ni(II) or Co(II)].[47] Alkylation of **93** with triethyloxonium fluoroborate gives the di(imino ester) (**94**) which cyclized to (**95**; M = Ni or Co) under the usual conditions in the presence of nickel acetate or cobaltous perchlorate. β-Alkoxy derivatives [**96**; M = Ni(II) or Co(II), R = R^2 = R^3 = R^4 = Me, R^1 = OBz][48] and analogues with the vitamin B$_{12}$ substitution pattern [**96**; M = Ni(II) or Co(II), R = CH$_2$CO$_2$Me, R^1 = R^2 = R^4 = CH$_2$CH$_2$CO$_2$Me, R^3 = Me][49] have been prepared from the appropriate biladienes under standard conditions.

Ni(II) and Co(II), 1,19-disubstituted tetradehydrocorrin salts with different

96

1- and 19-substituents are conveniently prepared by protonation or alkylation of the corresponding metallo 1-substituted tetradehydrocorrins (see Section III, B).

B. Properties of 1-Substituted and 1,19-Disubstituted Metallotetradehydrocorrins

These macrocycles contain interrupted nonaromatic chromophores and have characteristic nmr (Figs. 10 and 11) and visible (Fig. 12) spectra. In particular, the angular 1- and 1,19-substituents are attached to sp^3 carbon centers and are well separated, in the nmr spectrum, from the other signals (Fig. 10, angular 1-methyl signal at τ 8.52; Fig. 11, angular 1- and 19-methyl signals at τ 9.35). The high position of the signals corresponding to the 1- and 19-methyl substituents (Fig. 11) is due to shielding by the adjacent π-system of the macrocycle arising from deformations engendered by the sp^3 hybridization of two centers. The nonaromatic nature of the macrocycles is emphasized by the comparatively high τ-values of the signals due to the meso protons [Fig. 10; τ 2.80, 3.23, and 4.12 (the τ 2.80 signal is enhanced by superposition with the $CHCl_3$ signal); Fig. 11; τ 2.45 (1H) and 2.53 (2H)].

The two types of macrocycle can be chemically interrelated with each other and with the corroles and porphyrins (Scheme 1).

The alkylation of 97 [M = Ni(II)] with alkyl halides is both regiospecific and stereospecific and gives the corresponding metallo $trans$-1,19-disubstituted tetradehydrocorrin salts [98; M = Ni(II); 47–92%].[50] The cobalt complexes [97; M = Co(II)] are unstable and cannot be isolated. They are alkylated $in\ situ$, in the absence of air. Protonation of 97 also occurs at C-19 with strong acids giving 98 (R^5 = H).[39] These alkylations are remarkably facile suggesting the central metal atom assists the electrophilic alkylation by back donation from the filled d_{yz} and d_{xz} orbitals of the metal to the π-system of the ligand (Fig. 13). Alkylation at C-19 results from the transition state involving the least perturbation of the conjugated system of 97, and trans alkylation minimizes steric interactions.

Nickel 1,19-diethoxycarbonyl tetradehydrocorrin salts can be hydrolyzed,

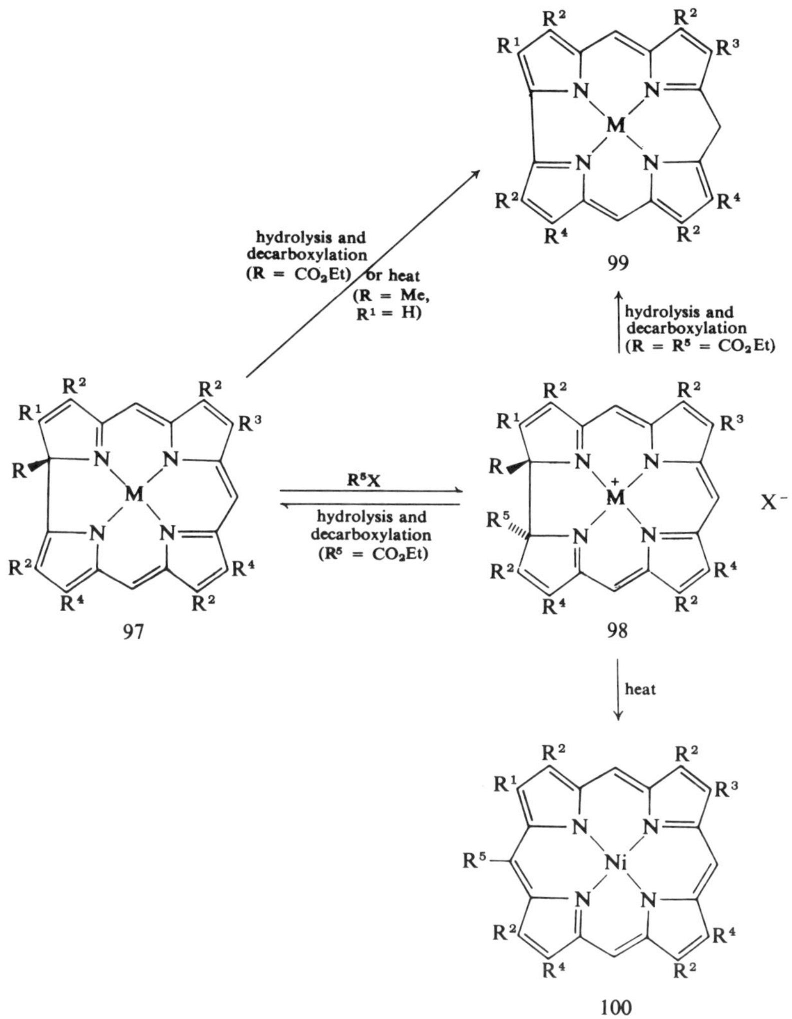

Scheme 1

under appropriate conditions, to either **97** [M = Ni(II), R = CO_2Et][51] or to corroles (**99**).[40] Similar hydrolyses of **98** [M = Ni(II), R = alkyl, R^5 = CO_2Et] gives **97** [M = Ni(II), R = alkyl].

The Ni(II) salts [**98**; M = Ni(II), R = Me, R^5 = alkyl] undergo an interesting thermal rearrangement at 180°C to Ni(II) porphyrins (**100**).[52] The yields of **100** are, however, low and both yield and product structure are dependent upon the anion present in **98** [M = Ni(II)]. Perchlorate salts of **98** [M = Ni(II); R = Me] give the meso substituted porphyrins (**100**; 7–21%) in which

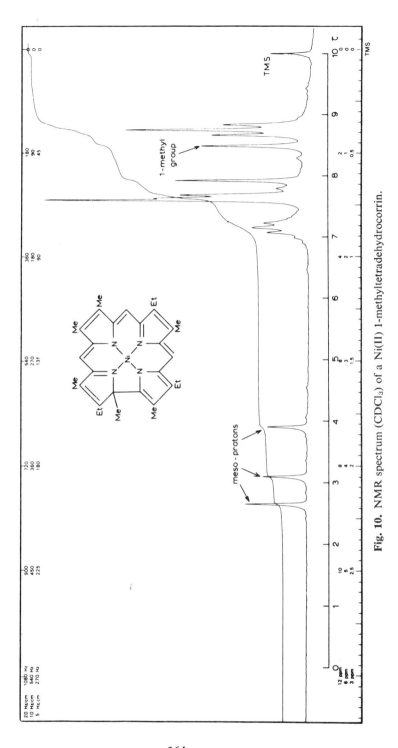

Fig. 10. NMR spectrum (CDCl₃) of a Ni(II) 1-methyltetradehydrocorrin.

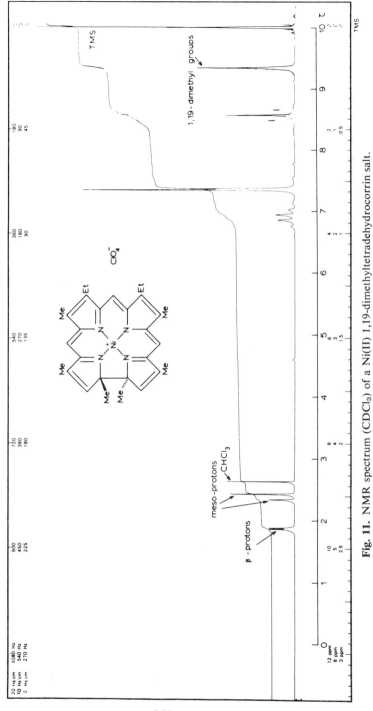

Fig. 11. NMR spectrum (CDCl₃) of a Ni(II) 1,19-dimethyltetradehydrocorrin salt.

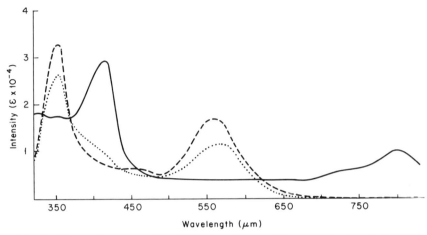

Fig. 12. Electronic spectra of a Ni(II) 1-methyltetradehydrocorrin (——), its C-19 protonated form (· · · ·), and a Ni(II) 1,19-dimethyltetradehydrocorrin salt (– – –).

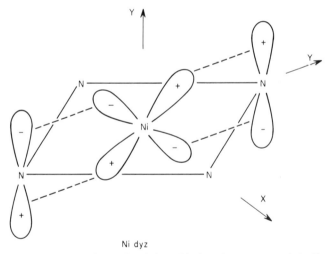

Ni d$_{yz}$

Fig. 13. Back donation from nickel d$_{yz}$ orbital to the π-system of the ligand.

the alkyl group R^5 is retained intact as a meso substituent. Small amounts of meso unsubstituted porphyrins (**100**; R^5 = H, 1–8%) are also formed. When salts other than perchlorates are thermolyzed [i.e., **98**: M = Ni(II), R = Me, R^5 = alkyl, X = F, Cl, I, or NO_3], a faster reaction ensues, and the products are the meso unsubstituted porphyrins (**100**; R^5 = H, 19–48%). A number of mechanisms were considered for this rearrangement which is considered to involve an initial Hofmann-type elimination (**98**; partial structure, arrows) to **101** followed by an 18π-electron electrocyclic process giving the angularly substituted porphyrin (**102**). Oxidation of **102** to **103** is then followed by nucleophilic attack of the anion (when X = F, Cl, I, or NO_3) at the angular group R^5 generating the meso-unsubstituted porphyrin (**100**; R^5 = H). Methyl iodide, but not methane, was detected in the volatile products from the thermolysis of a Ni(II) 1,19-dimethyltetradehydrocorrin iodide. The very poor nucleophilicity of the perchlorate ion permits the intermediate (**103**) to undergo a 1,17-sigmatropic shift (**103**; arrows) followed by deprotonation to give the meso-substituted porphyrin (**100**; R^5 = alkyl). When the chloride

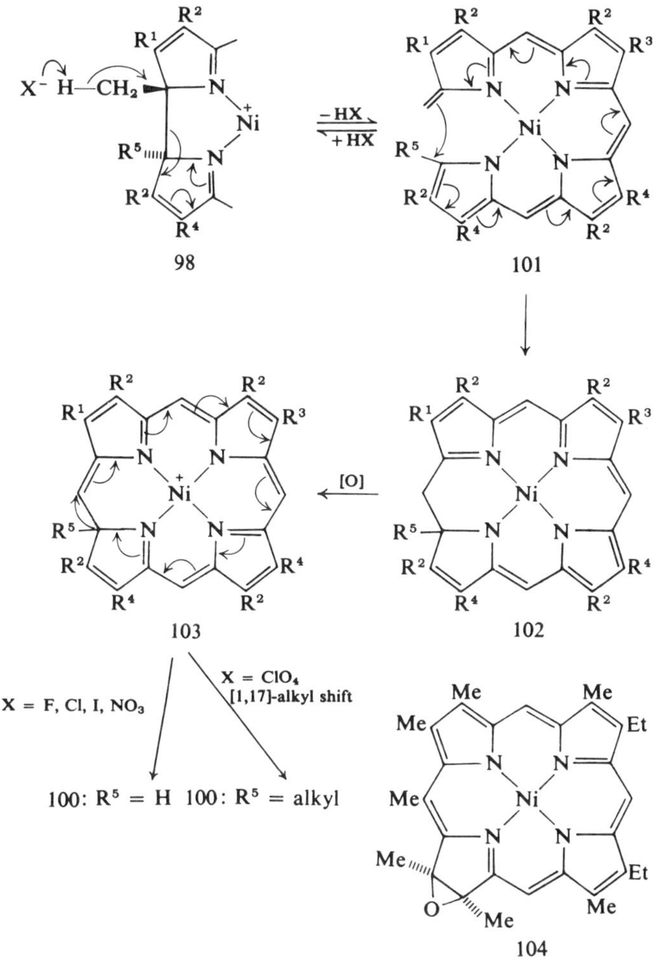

Scheme 2

salts (**98**; X = Cl, R^5 = Me) are thermolyzed, substantial amounts of *meso*-methyl epoxychlorins (e.g., **104**) are obtained as well as the meso-unsubstituted porphyrins.[52,53] Good evidence for the reversibility of the step (**98** ⇌ **101**) was obtained by studying the thermal racemization of an optically active compound [**98**; M = Ni(II)] which was shown to be anion dependent (Scheme 2).[44]

Apart from C-19 angular protonation and alkylation, the most interesting reaction of the Ni(II) 1-substituted tetradehydrocorrins is their thermal rearrangement at 213°C to the corresponding 3,3-dialkylcorroles (**97** → **105** → **106**, 20–40%).[51] The use of appropriate substituent labeling confirms the intermediacy of **105**, and the order of migratory aptitude of the C-1 substituent was found to be allyl > ethoxycarbonyl > ethyl > methyl. The migrations of C-1-alkyl substituents are examples of [1,5]-sigmatropic rearrangements. The migration of the ester group occurs in good yield (73.5%) and is one of comparatively few examples of migration of an ester group.[54] However, other alternative mechanisms are possible for the ester and allyl group migration. Thus, the ester group could migrate via cyclopropyl intermediates (e.g., **107** → **108** → **106**; R = CO_2Et) possibly involving nickel participation. It did not prove possible to obtain evidence to distinguish between a [1,5]- or [3,3]-sigmatropic mechanism for the allyl group migration.[51] The reaction of the 3,3-disubstituted corroles (**106**) with electrophiles (H^+, R^+) at C-17 to give 3,3,17,17-tetrasubstituted corroles has been discussed previously (see Section I, B).

The presence of a positive charge on the metallo-1,19-disubstituted tetradehydrocorrins renders them sensitive to nucleophilic attack.[53,55] However, while the Ni(II) complexes undergo nucleophilic attack at *meso*-carbon, in the Co(II) complexes the metal is the site of attack,[55] and only on subsequent thermal rearrangement are the meso-substituted derivatives obtained. In reactions of both nickel and cobalt complexes generating these meso-substituted derivatives, there is a tendency to form radical species, but, in the cobalt case, this can be accommodated by a redox reaction at the metal, generating a Co(I) species, while no such route is open to the nickel derivatives. Thus, in the presence of hydroxide ions and air 109 (R = R¹ = H) gives the 5-keto derivative (110; R = H; 60%). Reaction of 109 (R = R¹ = H) with aqueous NaOH/CHCl₃ gives 110 (R = CHO and CHCl₂), while methanolic KCN gives 110 (R = CN), together with two other products that are formulated as stable free radicals with the unpaired electron localized on the ligand. These latter two products, on treatment with strong acids, give the 5-cyano (109; R = CN, R¹ = H; 50% overall) and the 5,15-dicyano (109; R = R¹ = CN) derivatives which undergo the standard thermal rearrangement to Ni(II) *meso*-cyanoporphyrins.[53] In contrast, the Co(II), 1,19-dimethyltetradehydrocorrin salts react with methanolic KCN, NaN₃ or NaNO₂/CH₃CO₂H by attack at the cobalt atom giving the Co(III) complexes (111; L = CN, N₃, or ONO).[55]

Thermolysis of Co(II) 1,19-dimethyltetradehydrocorrin salts gives the analogous cobalt porphyrins as in the Ni(II) series. However, thermolysis at

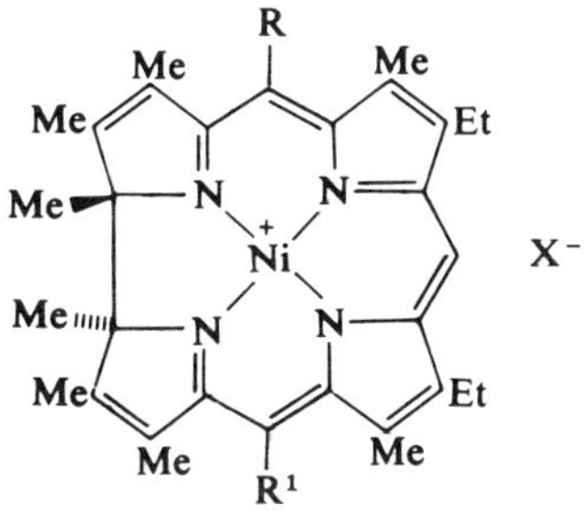

109

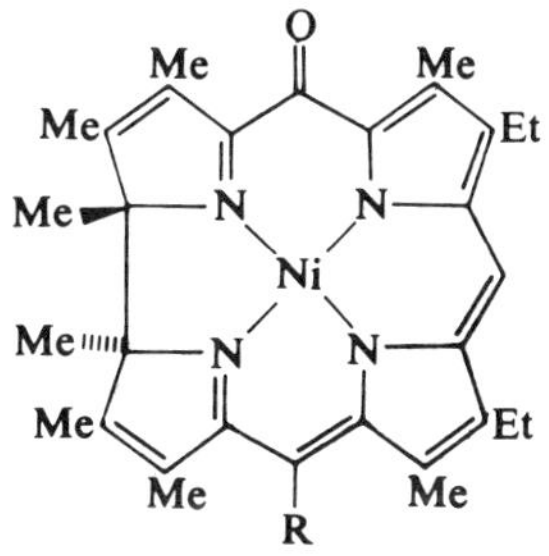

110

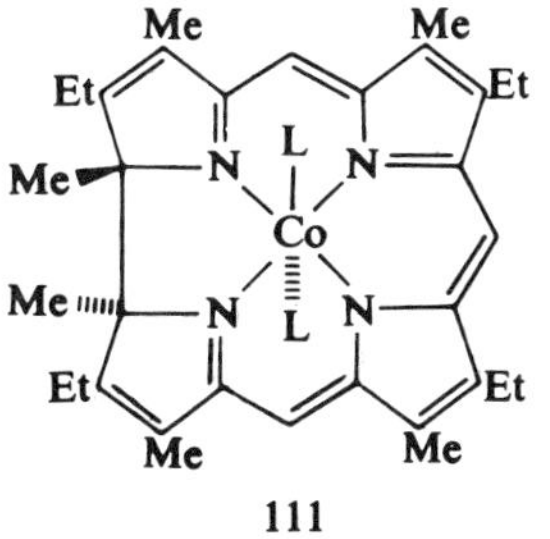

111

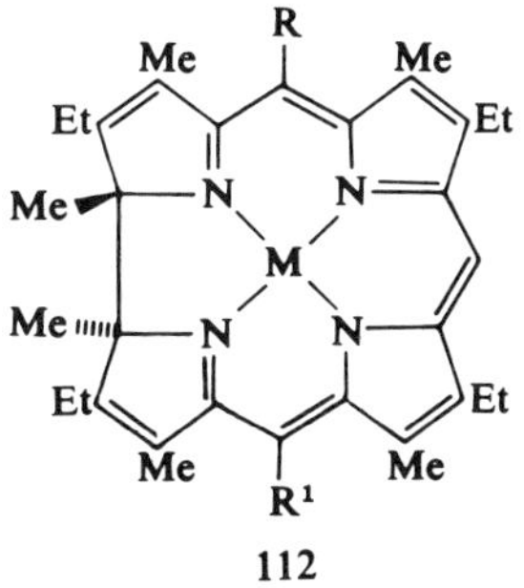

112

100°C of the Co(III) complexes (**111**; L = CN) induces rearrangement of cyanide from cobalt to the chromophore giving the 5-cyano Co(II) complex [**112**; M = Co(II), R = CN, R^1 = H; 43%] as the major product together with a small amount of the 5,15-dicyano Co(I) complex [**112**; M = Co(I), R = R^1 = CN; 6%]. The dicyano complex [**112**; M = Co(I), R = R^1 = CN] could be obtained in better yield by recycling the monocyano Co(II) complex [**112**; M = Co(II), R = CN, R^1 = H] through the KCN/thermolysis treatment. The rearrangement of the cyano groups from cobalt to carbon is probably intramolecular.[55]

Electrochemical studies support the suggestion that electron-attracting meso substituents stabilize the Co(I) oxidation state.[55] Electron transfer to Ni(II) and Co(II) 1,19-disubstituted tetradehydrocorrin salts using sodium anthracenide,[56] sodium in THF,[57] and electrochemical techniques[58] has been reported. Reductive methylation of Ni(II) decamethyltetradehydrocorrin iodide with sodium anthracenide and methyl iodide caused ring opening to the corresponding biladiene methylated at C-10 (**113**; R = H). The biladiene (**113**; R = H) can be recycled to the corresponding Ni(II) *meso*-methyl-tetradehydrocorrin salt and reductively methylated to (**113**; R = Me).[56] The Ni(II) and Co(II) 1,19-disubstituted tetradehydrocorrin salts both form unusually stable one-electron reduction products,[57,58] and both two- and three-electron transfer products are detectable. The one-electron reduction product of the Ni(II) complexes is an unusually stable free radical analogous to those isolated from the nucleophilic substitution reactions (see before). The Co(II) complexes show a remarkable fluctuation of oxidation state of the central cobalt atom as successive electrons are added. Thus the one-electron reduction product is a neutral Co(I) 1,19-disubstituted tetradehydrocorrin while the two-electron reduction product is best formulated as a Co(II) complex with the two additional electrons located on the organic ligand. Further studies show that the Co(II) ions in Co(II) corrole anion and the Co(II) 1,19-disubstituted tetradehydrocorrin cation have different electronic ground states for the central cobalt ion.[59]

Electrophilic substitution reactions of Ni(II) and Co(II) 1,19-disubstituted

113   114

tetradehydrocorrin salts are facile despite the positive charge on the macro-cycles. However, the fully β-unsubstituted Co(II) macrocycle is unstable to acids and undergoes cleavage probably to the corresponding biladiene-a,c.[46] The Ni(II) complexes [114; M = Ni(II), R = R^1 = R^2 = Me and R = R^2 = Me, R^1 = H] and the Co(II) complex [114; M = Co(II), R = R^1 = R^2 = Me] are reported to undergo N protonation in concentrated sulfuric acid (but not TFA) to give bright green solutions of cations formulated as 115 [M = Ni(II) or Co(II)]. The nmr and electronic spectra of these green cations are interpreted in terms of their being quasi-aromatic.[60]

115

116

Deuterium exchange of the meso protons of 114 [M = Ni(II), R = H, R^1 = Me, R^2 = Et] occurs in DTFA at 35°C. The half-life for the C-10 proton is 15 minutes while the C-5/C-15 protons have a half-life of 11 minutes, assuming they both exchange at the same rate.[39] In accord with these data, halogenation (Cl$_2$, Br$_2$) gives the unstable 5,15-dihalogeno derivatives [e.g., 116; M = Ni(II), R = Me, R^1 = Cl, or Br, R^2 = H] that react with lithium copper dimethyl to give the *meso*-methylated compound [116; M = Ni(II), R = R^1 = Me, R^2 = H][53] in good yield. The 5,15-dibromo derivative is readily converted to the meso-unsubstituted compound by traces of acid.[53] Nitration of both Co(II)[55] and Ni(II)[53,61] 1,19-dimethyltetradehydrocorrins has been reported. Mono-, di-, and tri-*meso*-nitro derivatives have been reported for Ni(II) complexes, and di- and tri-*meso*-nitro derivatives for Co(II) complexes. Formylation, in contrast to halogenation and nitration, occurs at the C-10 meso position giving the 8,12-diethyl analogue of 116 [M = Ni(II), R = R^1 = H, R^2 = CHO].[62] No attack at the unsubstituted 2,18-β-positions of 116 [M = Ni(II), R = R^1 = R^2 = H] was observed. Chloromethylation and hydroxymethylation have also been studied.[63] The nmr spectra of a range of meso-substituted Ni(II) 1,19-dimethyltetradehydro-corrins have been correlated with σ-substituent constants.[64]

Hydroxylation of Ni(II) and Co(II) 1,19-disubstituted tetradehydrocorrin salts with osmium tetroxide gives a mixture of 7,8-dihydroxy-[117; M = Ni(II) or Co(II)] and 7,8,12,13-tetrahydroxy [118; M = Ni(II) or Co(II)] derivatives.[47,63] Further oxidation of the Ni(II) series with osmium tetroxide gives the Ni(II) hexahydroxy and octahydroxy derivatives.[65] Treatment of **117** [M = Ni(II) or Co(II)] with sulfuric acid causes pinacol rearrangement to a mixture of **119** and **120** (partial formulae). The tetraol (**118**) rearranges under similar conditions to a mixture of **121** and **122** (partial formulae).[63] Reductive ring opening of Ni(II) 1,19-dicarbethoxytetradehydrocorrin perchlorate, the Ni(II) perchlorate salts of **117** [M = Ni(II), R = CO_2Et] and **118** [M = Ni(II), R = CO_2Et], and their corresponding pinacol rearrangement products, with hydrogen sulfide or sodium dithionite, leads to the corresponding bilatrienes, which can be recycled with Co(II) to effect an overall substitution of Co(II) for Ni(II).[66]

117   118

119   120   121   122

IV. 1-SUBSTITUTED AND 1,19-DISUBSTITUTED CORRINS

A. Synthesis

The elucidation of the structure[67] of vitamin B_{12} led to ever growing interest in solving the tremendous synthetic challenge presented by its complex structure with nine asymmetric centers. This challenge was taken up by a number of groups of workers and culminated in the total synthesis of cobyric acid (**123**)[68] by a group led jointly by Eschenmoser and Woodward. The

$$A = CH_2CONH_2$$
$$P = CH_2CH_2CONH_2$$

123

efforts expended by this group and others, notably groups led by Cornforth,[69] Johnson,[70] and Todd,[71] in the quest for a synthesis of corrins have led to a truly remarkable harvest of new synthetic methods, intriguing new macrocycles, and preeminently to the concept of conservation of orbital symmetry in concerted reactions.[8]

The synthesis of corrins has been approached in two main ways: (a) the utilization of precursors at the correct oxidation level and the assembling of these to form the corrin; (b) using pyrroles as precursors and adjusting the oxidation level by a final reduction stage. The latter approach of necessity leads to an amorphous mixture of stereoisomers except in the case of the β-unsubstituted macrocycle.[46]

The first successful corrin synthesis was a 2 + 2 approach (124 + 125) in which the direct link was incorporated in a two-ring precursor (124).[72] Of particular note is the activation of lactam carbonyl groups by conversion to their imino esters using Meerwein's triethyloxonium fluoroborate reagent. Selectivity in the mixed condensation (124 + 125) is engineered by positioning the more reactive conjugated imino ester group in the component containing the less reactive enamine center. Formation of metal complexes (126) ensures the close proximity of the sites of final ring closure in rings A and B and, incidentally, also activates the iminoester center in ring A. Deprotonation of 126 [M = Ni(II) or Co(III)] with excess potassium tert-butoxide then effects ring closure to 127 and acid work up affords the corrin complexes [128; M = Ni(II) or Co(III), R = CN] in 80–90% yield. The electronic spectrum of [128; M = Ni(II), R = CN] shows the expected bathochromic shift of the long wavelength band (Fig. 14) and resembles the characteristic electronic spectrum of the vitamin B_{12} chromophore. Acid hydrolysis–decarboxylation of 128 (R = CN) at 210–220°C removes the C-15 cyano group giving 128 (R = H). Alternative approaches to corrins employing isoxazoles, originally pioneered by Cornforth, have been reported,[73] and Stevens has recently reviewed his work in this area.[73a]

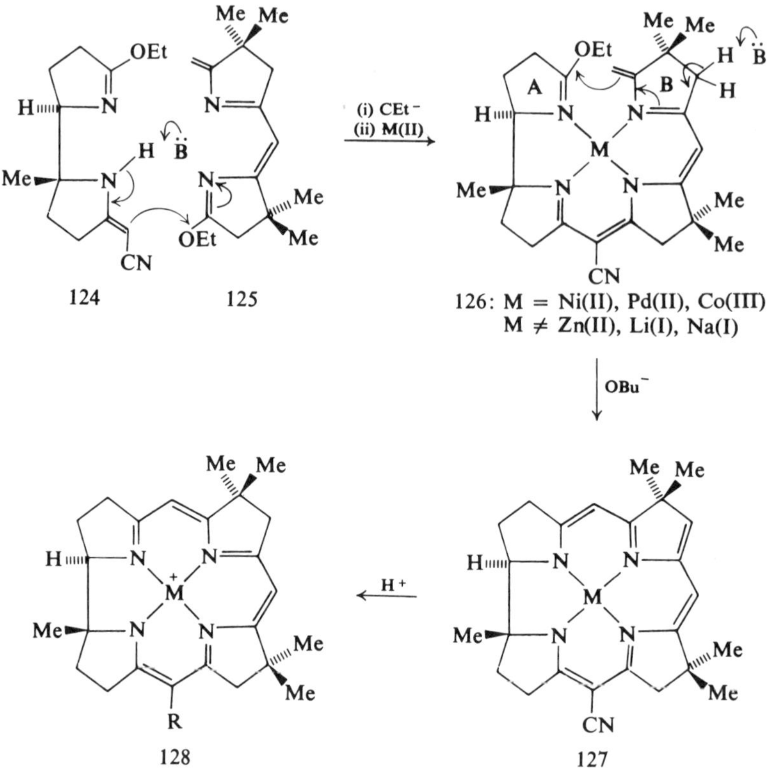

124   125

(i) CEt⁻
(ii) M(II)

126: M = Ni(II), Pd(II), Co(III)
     M ≠ Zn(II), Li(I), Na(I)

OBu⁻

128   ← H⁺   127

Isoxazoles are masked vinylogous amides and therein lies their attraction as corrin precursors. The vinylogous amides are readily released by hydrogenolysis of the isoxazoles. Thus the isoxazole **128a** on hydrogenolysis gave the vinylogous amide **128b** which underwent spontaneous ring closure to **128c**. Treatment of **128c** with methanolic ammonia followed by dehydration with potassium *t*-butoxide gave the semicorrin **128d**, a precursor of **125**.[73a]

A reaction sequence for *meso*-methylation with some degree of selectivity has been developed.[74] Chloromethylphenyl sulfide, in the presence of silver tetraborate at room temperature, alkylates **129** at C-15 giving **130** (R = CH₂SPh), the product of steric control. Repetition of the process on the product of Raney nickel desulfurization (**130**; R = Me) gives both the 5,15-dimethyl- (**131**) and the 10,15-dimethyl-corrin complex in a ratio of about 5:1. The 5- and 15-positions are, thus, the favored sites for electrophilic substitution, and similar trends are found in deuteration and cyanidation studies.[74,75] HMO calculations also support a preference for attack at 5- and 15-positions.[76]

Severe problems were encountered in adapting the iminoether route (**124 + 125 → 128**) to a synthesis of vitamin B₁₂. In particular, the steric

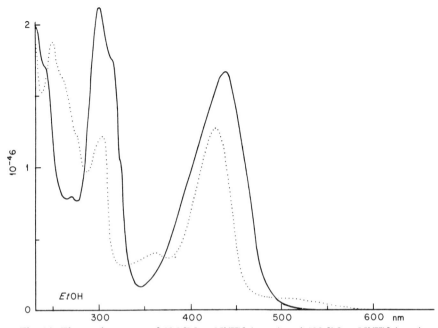

Fig. 14. Electronic spectra of **126** [M = Ni(II)] (· · · ·) and **128** [M = Ni(II)] (——).

encumbrances and increased lability of multiply substituted intermediates led to a search for alternative coupling methods and gave rise to a sulfur extrusion process, which proved to be the key method for constructing the methine bridges in the synthesis of vitamin B_{12}. Thus, the thiolactam (**132**) on oxidation with one equivalent of benzoylperoxide in the presence of the enamide (**134**) and a trace of hydrogen chloride gives, via the bisimidoyl-disulfide (**133**), the coupled product (**135**) which is desulfurized by triethyl-phosphite without isolation to a ring B epimeric mixture of the B/C precursor (**136**)[77] utilized in the synthesis of vitamin B_{12}. Analogous sequences were developed for constructing the A/B and C/D bridging moieties in vitamin B_{12}, but a full discussion of this magnificent synthesis is outside the scope of this review.

The discovery of naturally occurring cobalt-free corrinoids[78] and the inability of chemists to remove the cobalt atom from vitamin B_{12} and its derivatives without disrupting the corrin chromophore[79] focused attention on the synthesis of metal-free corrins. The sulfur extrusion sequence proved the key to this class of corrins. The problem with the iminoester route is that the A/B cyclization step (**126** → **127** → **128**) does not occur for cases with easily removable metal atoms, i.e., **126** [M = Zn(II), Li(I) or Na(I)]. However, treatment of the precorrinoid sodium salt (**137**) with H_2S/CF_3CO_2H

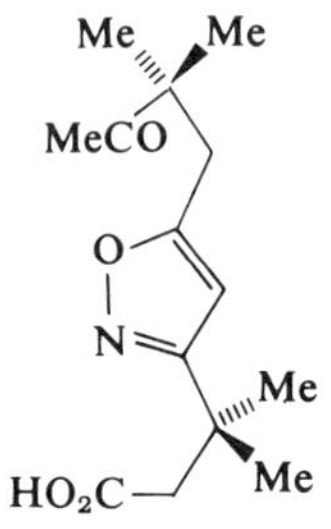

128a 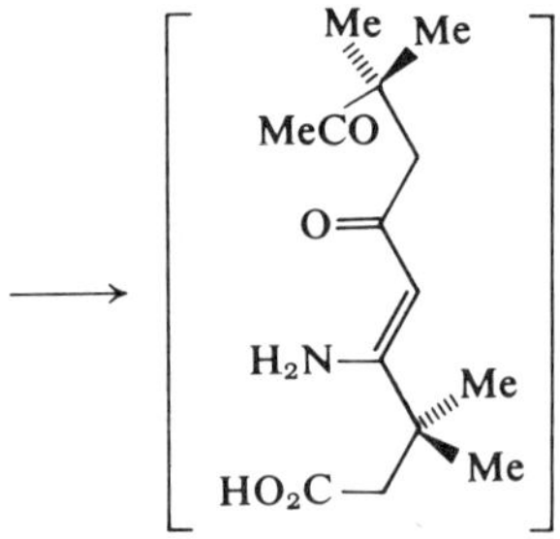 → 128b 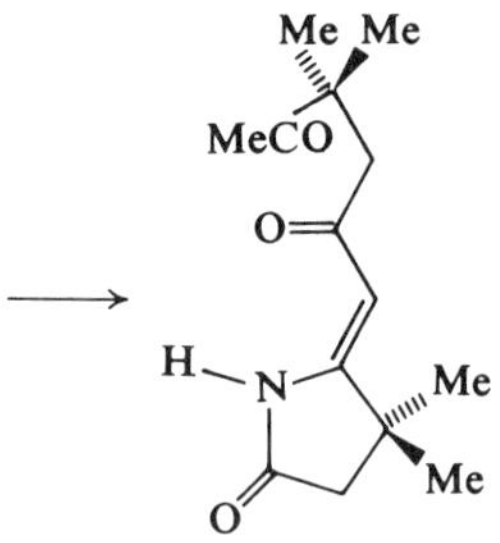 → 128c 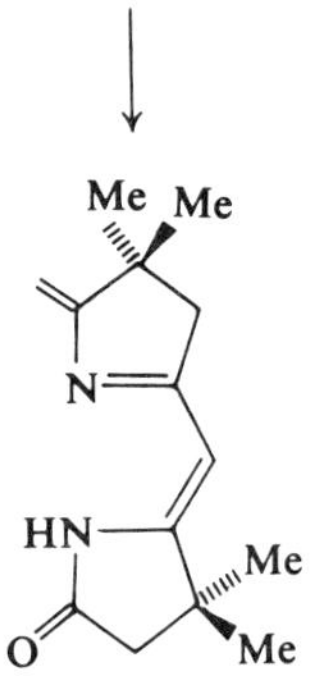 → 128d

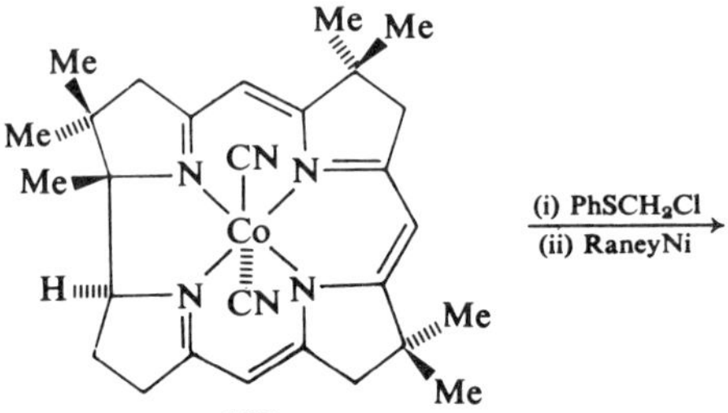

(i) PhSCH$_2$Cl
(ii) RaneyNi

129 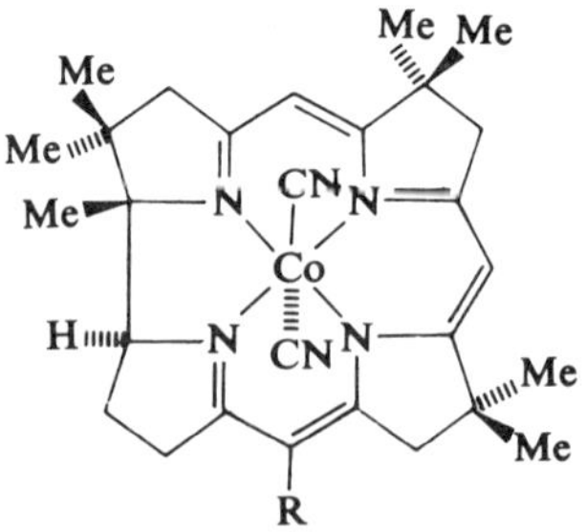   130 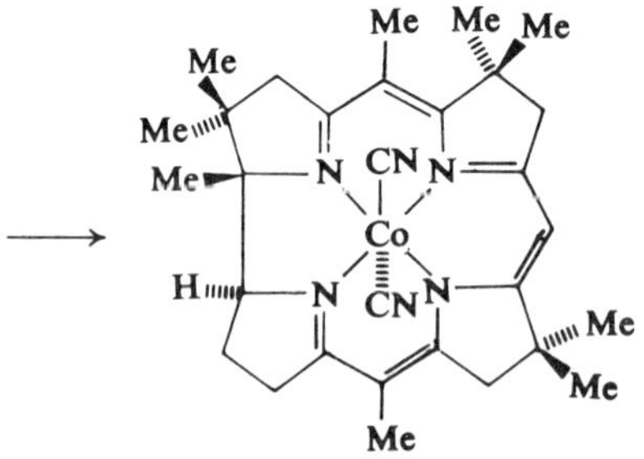   131

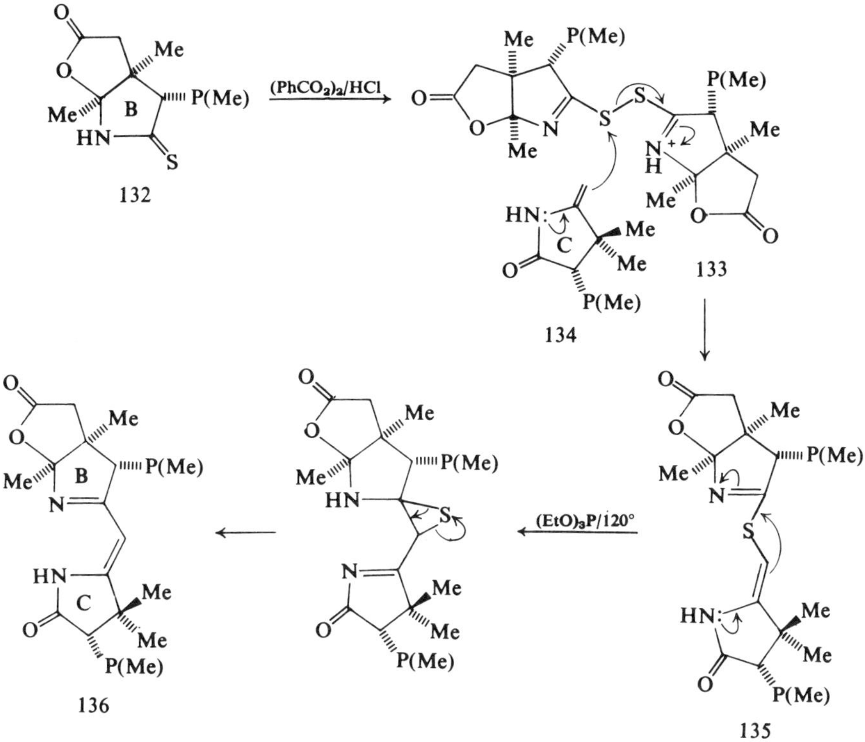

132

(PhCO₂)₂/HCl

133

134

(EtO)₃P/120°

136

135

followed by insertion of Zn(II) gives the *meso*-thia macrocycle (**138**). Although not the desired product, **138** is converted to **139** by treatment with benzoyl-peroxide–trifluoroacetic acid followed by methanol. The conversion of **139** to the corrin complexes (**140a, b**) is achieved in high yield ($\sim 80\%$) by treatment of **139** with trifluoroacetic acid in DMF at 70°C.[80]

The conversion of **138** to **139** is believed to involve acid-catalyzed formation of the thiolactam (**141**), followed by peroxide attack giving **142**. Nucleophilic displacement (**142** → **143**) then occurs, followed by addition of methanol giving **139**. Sulfide extrusion (**139** → **140**) apparently involves decomplexation of zinc, which allows double bond isomerization to **144**, episulfide formation (**144**; arrows) and rearrangement to **140** with reinsertion of zinc.

The *meso*-mercapto complex (**140a**) is desulfurized to **140b** by triphenyl-phosphine in the presence of trifluoroacetic acid in chloroform. Further treatment of **140b** with trifluoroacetic acid in acetonitrile gives the metal free corrin salt (**145**).[80] X-Ray structure analysis of the bromide (**145**; X = Br) shows that rings B, C, and D are essentially coplanar, while ring A is tilted out of the plane of the other three rings.[81] A range of metal ions have been incorporated into the corrin ligand [**146**; M = Li(I), Zn(II), Ni(II), Pd(II),

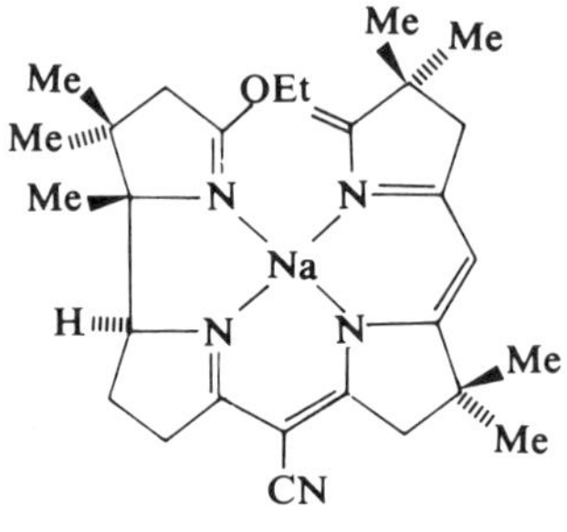

137

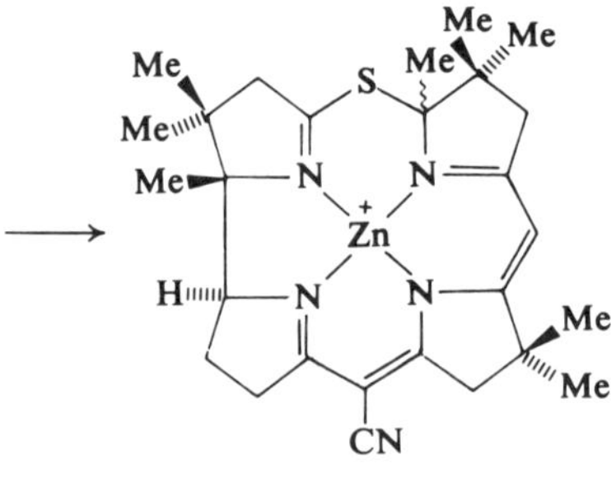

138

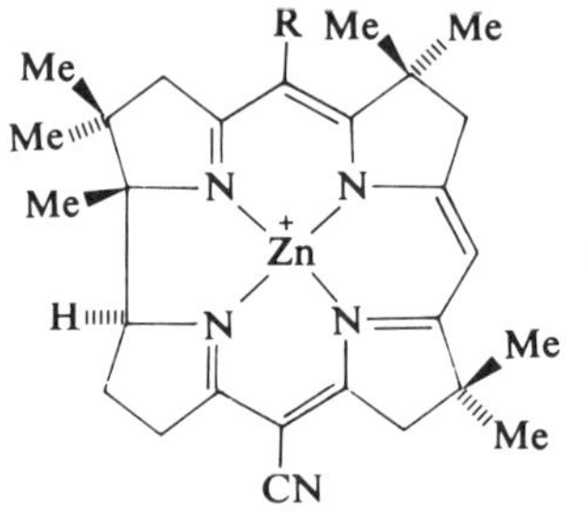

140: (a) R = SH
(b) R = H

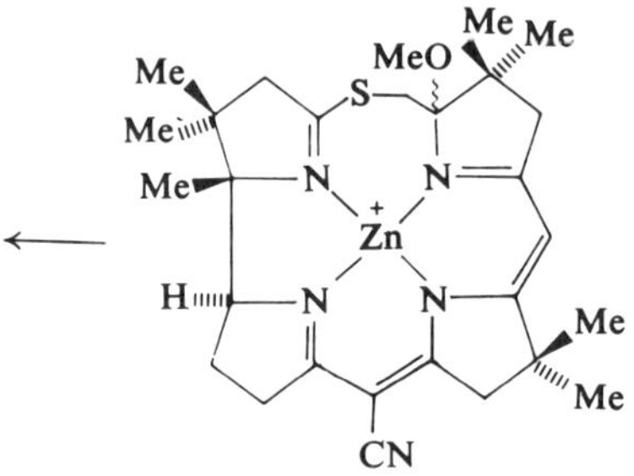

139

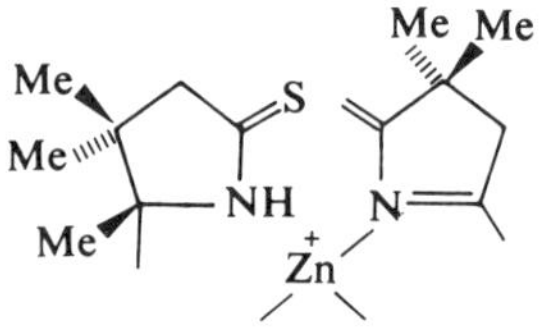

141

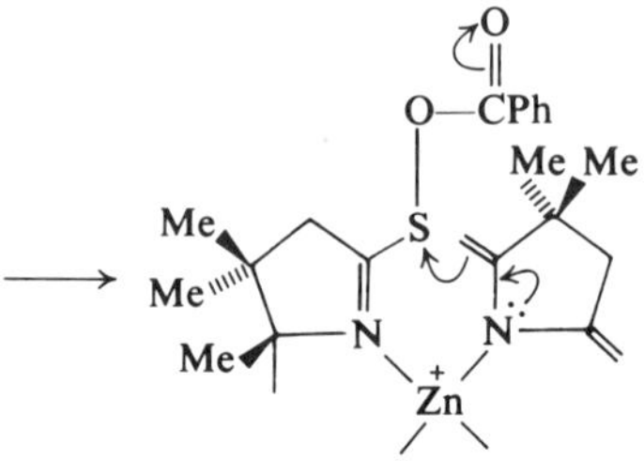

142

140

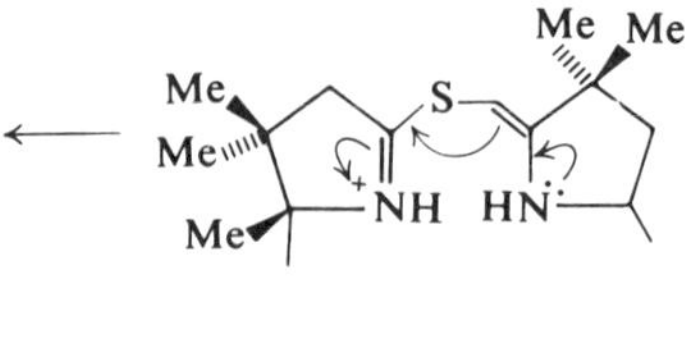

144

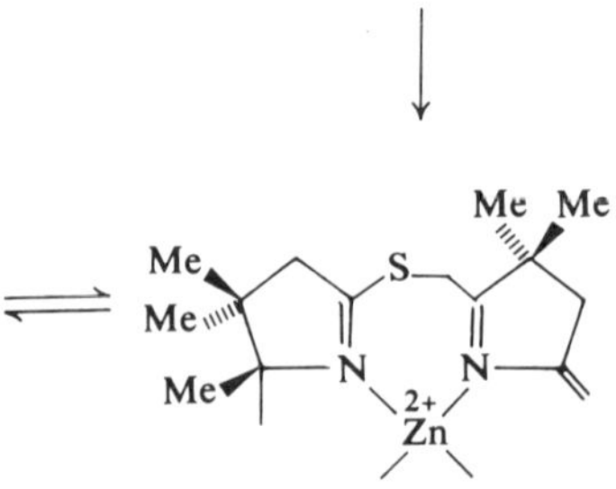

143

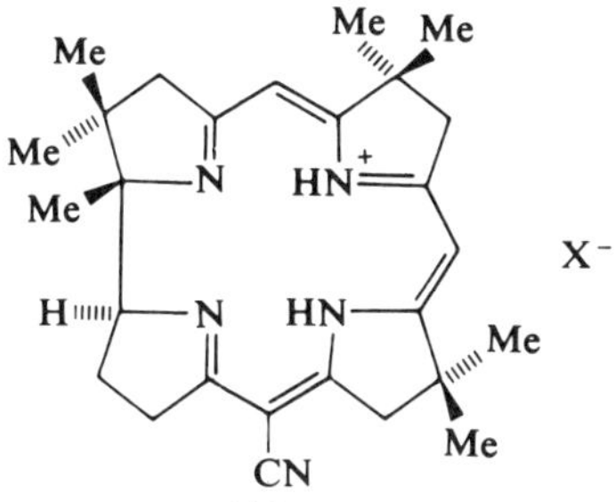

145

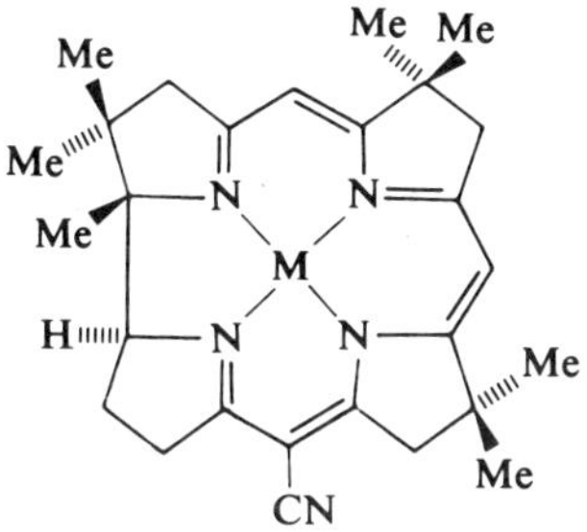

146

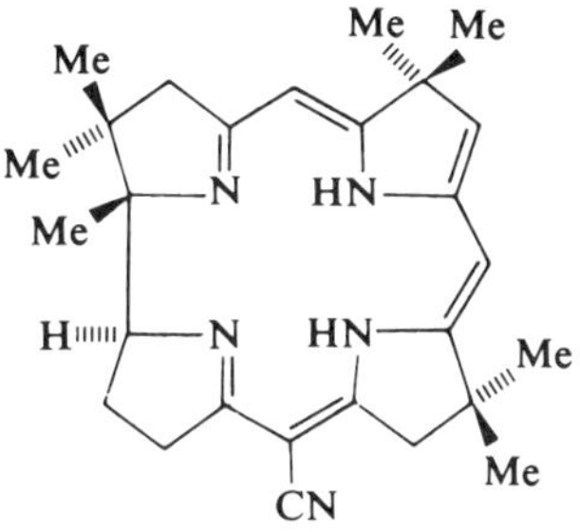

147

Co(III), Rh(III)] and will provide the basis for a study of metal–ligand interactions. Attempts to liberate the corrin free base from **145** lead to the cross-conjugated macrocycle (**147**) (Fig. 15).

The two previous approaches to corrins involving assembling rings with the correct oxidation level both incorporated the A/D direct link from the outset. The later corrin syntheses differ from the earlier ones in that the final step, a photochemical[82] or electrochemical[83] closure, forms the A/D direct link. In the photochemical route, the penultimate complex (**150**) is prepared by elimination of hydrogen cyanide from **149** using potassium *tert*-butoxide. The synthesis of **149** involved both iminoester and sulfur extrusion methods for linking up the rings, with **148** serving as precursor for ring A.

A range of metal complexes (**149** and **150**) can be prepared. Degassed solutions of the complexes (**150**; M = Li, MgCl, ZnCl, CdCl) undergo rapid photochemical cyclization to the corresponding *trans*-metallocorrins (**151**; R = CN) in essentially quantitative yields. The complexes [**150**; M = Pt(II) or Pd(II)] cyclize more slowly while the Co(III), Ni(II), and Cu(II) complexes fail to cyclize. This striking metal ion effect appears to be due to the very effective quenching of the excited state by metal ions with unfilled *d*-shells. The photoactive species of **150** is the triplet state and the product, **151**, is an efficient sensitizer. Moreover, careful studies suggest that the presence of trace amounts of contaminants such as **149** are required as sensitizers to initiate

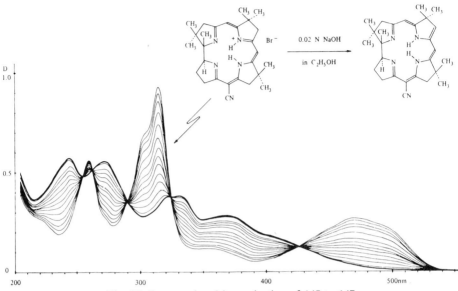

Fig. 15. Base-catalyzed isomerization of **145** to **147**.

the photocyclization.[83a] The rationale of, and impetus for, this synthesis came from orbital symmetry considerations[8] which "allow" an antarafacial photochemical sigmatropic 1,16-H transfer (**150** → **152**) followed by an antarafacial thermal 16π-electron electrocyclic process (**152** → **151**; R = CN). The geometry of the secocorrin complex (**150**) results in the favorable juxtaposition of the exocyclic methylene on ring A and the migrating hydrogen on ring D and ensures collapse of **152** to the *trans*-corrin. This intriguing cyclization has been extended to **153**, which cyclizes to the 1-hydroxycorrin (**154**; 89%) on irradiation at ∼ 40°C in benzene.[84] An improved sulfide contraction procedure utilizing iodinated enamides (e.g., **155**) extends the range of enamides/thiolactams that can be successfully coupled.[85] The isoxazole route has also been used to produce **156**.[73a]

The discovery that the secocorrin complex (**156**) undergoes electrochemical reductive cyclization (50–60%) to **151** [M = Ni(II), R = H] in a protonating medium (acetonitrile/TFA; 9:1)[83] directed studies to the fruitful field of redox stimulated secocorrin ring closure.[83a] Thus the Ni(II) perchlorate complex [**150**; M = Ni(II)] undergoes an electrochemical two electron oxidation in moist acetonitrile, with incorporation of an oxygen atom from the water, to give the secocorrin oxide (**156a**) in almost quantitative yield.[83b] In the absence of water, a low yield of the nickel corrin [**151**; M = Ni(II)] is obtained. The oxide bridge of **156a**, though inert to acid catalyzed cleavage, undergoes thermal conversion to **156b** (175°C in presence of NaCl). This

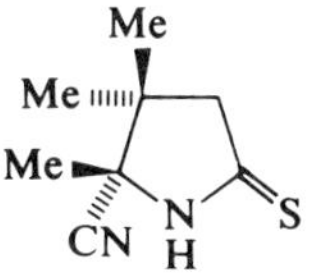

148

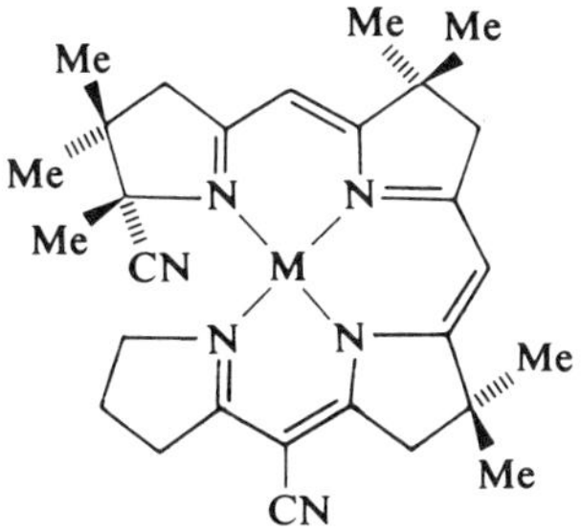

149

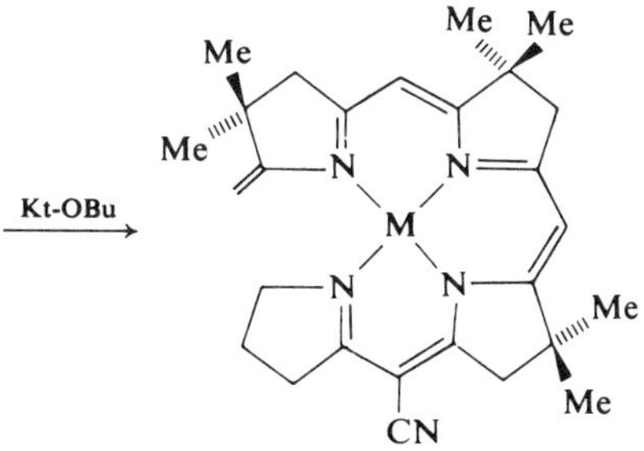

Kt-OBu

150

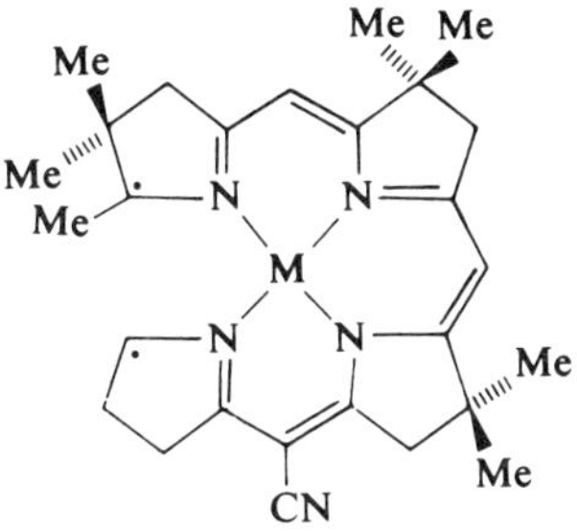

152

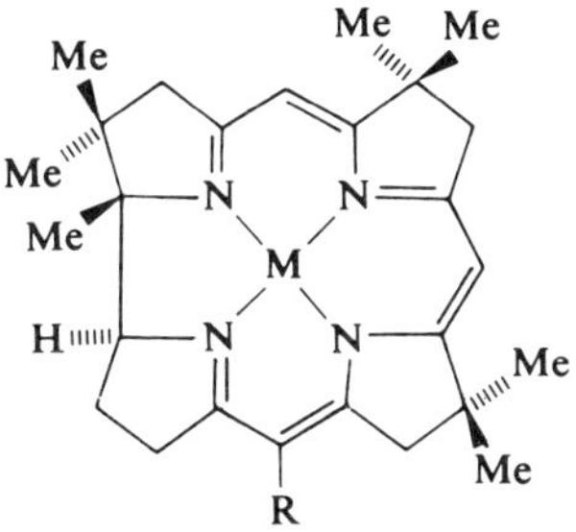

151

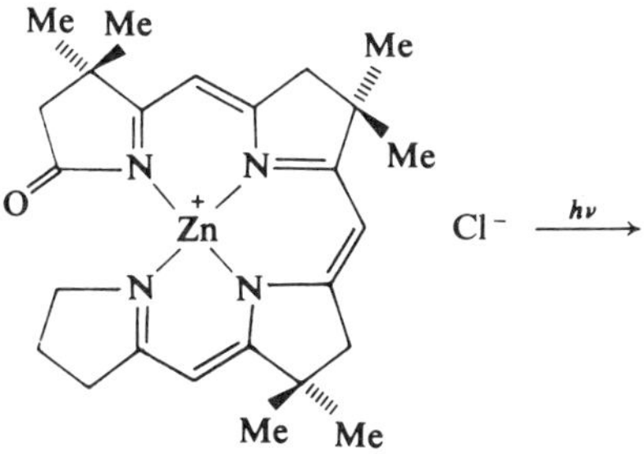

153

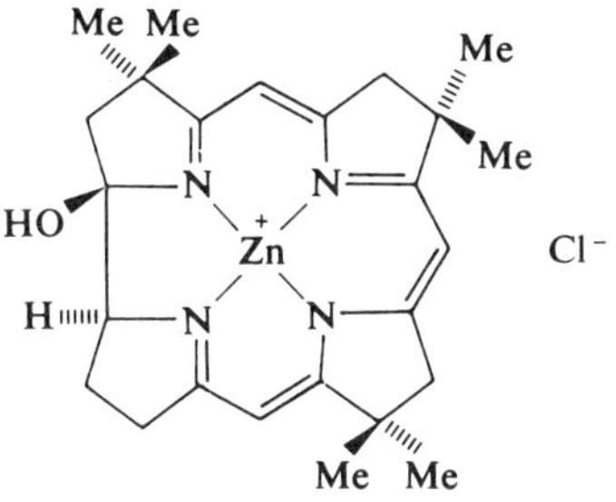

$h\nu$

154

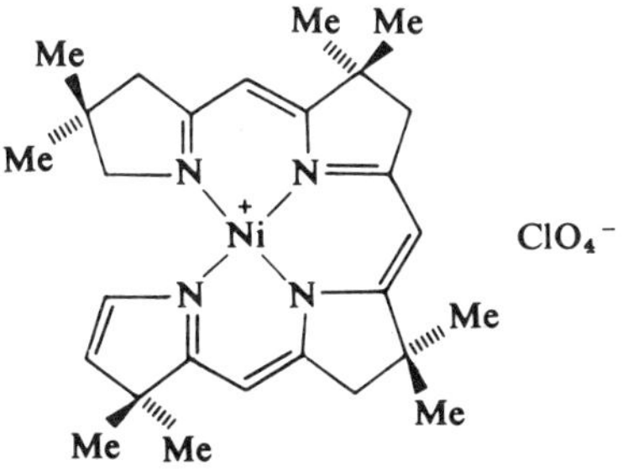

156

ClO_4^-

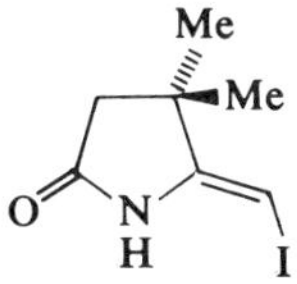

155

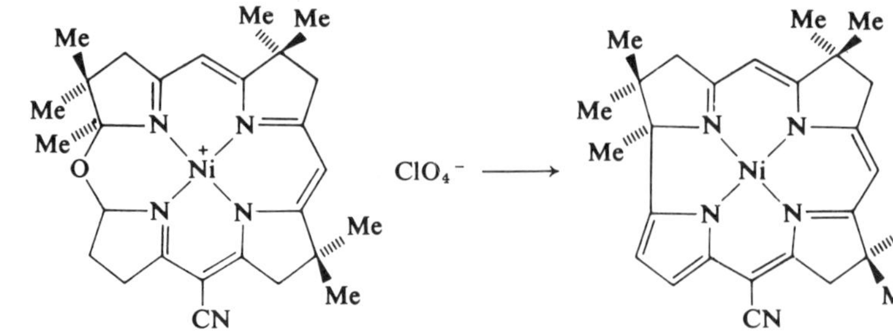

156a

156b

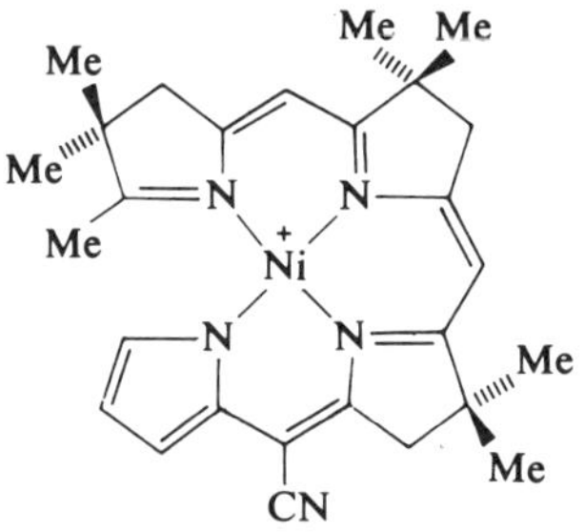

156c

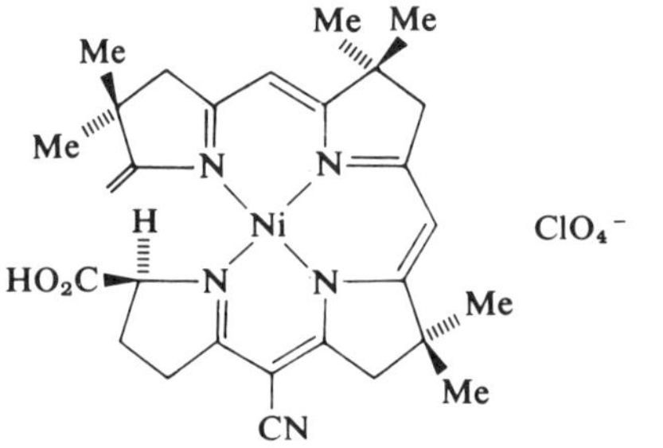

156d

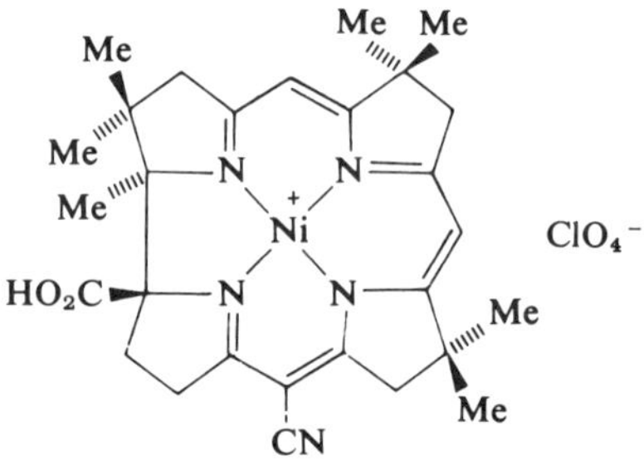

156e

transformation probably proceeds via 16π-electron conrotatory 1,16-($\pi \to \sigma$) cycloisomerization of **156c**. This type of process is analogous to the cyclization of **91** (See Section III, A).

The latest reported series of experiments in secocorrin–corrin cyclizations have involved the 19-carboxy derivative (**156d**)[83a] which undergoes thermal cyclization to **151** [M = Ni(II)]. Preliminary labeling studies suggest deprotonation at C-19 is followed by cyclization to **156e** which then decarboxylates to **151** [(M = Ni(II)]. The 19-carboxycorrin **156e** has been prepared and shown to undergo decarboxylation to **151** [M = Ni(II)] under mild conditions.

This particular cyclization was prompted by the biosynthetic studies of Shemin and Scott [83c,d] which indicated that the 1-methyl group in vitamin B_{12} was not derived from the Urogen III *meso*-methylene group. Secocorrin→ corrin cyclizations are clearly an area where further exciting developments can be expected.

The "pyrrole" approach to corrins involves the construction of nickel and cobalt tetradehydrocorrin salts (see Section III, A) and their catalytic reduction with various catalysts and conditions to a range of reduced tetradehydrocorrins including corrins.The course of the reduction is profoundly affected by the β-alkyl substitution pattern. The nickel and cobalt β-unsubstituted 1,19-dimethyltetradehydrocorrin salts [157; M = Ni(II) or Co(III)] are reduced by W2-Raney nickel and hydrogen at room temperature to the corresponding crystalline, metallocorrin [158; M = Ni(II) or Co(III)CN$_2$] and 4,5-dihydrocorrin salts [159; M = Ni(II) or Co(III)CN$_2$].[46]

157   158   159

The metallocorrin is the major product in both the nickel (44%) and cobalt (72%) cases. More 4,5-dihydrocorrin is obtained from the reduction of the nickel (28%) than the cobalt (1–11%) tetradehydrocorrin salt. The Ni(II) and dicyanocobalt (III) corrin complexes (158) exhibit nmr signals for the 1,19-dimethyl groups at 8.53 and 8.58 τ, respectively, compared to 9.23 τ for 157 [M = Ni(II)]. Identification of the nickel and cobalt corrins (158) was facilitated by spectral comparisons with vitamin B_{12} derivatives and Eschenmoser's synthetic corrins. The structure of the 4,5-dihydrocorrin (159; M = Ni) is based on spectral studies, e.g., the appearance of two signals for the angular methyl groups of 159 (M = Ni) at 8.54 and 8.82 τ and the formation of 159 (M = Ni; 57%) from 158 (M = Ni) by reduction with W2 Raney nickel at 200°C under 120 atmospheres of hydrogen. Hydrogenation of 158 (M = Ni), using conditions under which both 158 (M = Ni) and 159 (M = Ni) were formed (25 atmospheres H$_2$/room temperature), caused no change, demonstrating that 158 (M = Ni) is not a precursor of 159 (M = Ni) under these mild conditions. A fairly constant ratio of corrin to dihydrocorrin derivatives was obtained under a wide range of reaction conditions.

β-Alkyl substituted tetradehydrocorrins are more difficult to reduce, and the products are amorphous. However, by altering the substitution pattern, various partially reduced products can be obtained.[86,87] The tetradehydro-corrins (**160**; R = H or Me, R^1 = Me or Et) can be reduced in two stages, first to the BC-didehydrocorrin (**161**; R = H or Me; R^1 = Me or Et) and, under more forcing conditions, to the corresponding corrin.[87] When rings A and D have their full complement of β-substituents (e.g., **162**), there is severe steric crowding around rings A and D, and reduction to the corrin level is not observed. Products in which one (**163**)[86] or both (**164**)[86,87] of the peripheral double bonds in rings B and C have been reduced are obtained. More vigorous reduction (W2 Raney Ni/160°C/100 atmospheres H_2) leads to the formation of the isocorrin (**165**).[86]

B. Properties and Reactions of Corrins

The reduced tetradehydrocorrins (**163** and **164**) and the Ni(II) corrins are all susceptible to dehydrogenation at the 7,8- and 12,13-positions.[87] Thus, **164** is dehydrogenated to the corresponding nickel tetradehydrocorrin salt (67%) on keeping in ethanolic sodium hydroxide overnight. The corrin (**166**) is selectively dehydrogenated to the BC-didehydrocorrin (**167**) under similar conditions. Careful exclusion of air allows an intermediate enamine to be trapped with methyl iodide in the base-induced dehydrogenation (**168** → **169**; 50%).[87] The site of deprotonation of **168**, i.e., C-8 ($\equiv$ C-12), is analogous to that of the Ni(II) and Co(III) corrins[75,77] and metal-free corrins are most stable in the cross-conjugated form (**147**). The parent fully β-unsubstituted Ni(II) and Co(II), 1,19-dimethylcorrins [**158**; M = Ni(II) or Co(II)], are unstable in the presence of bases.[46] Vitamin B_{12} undergoes an acid-catalyzed equilibration with its C-13 epimer neovitamin B_{12}.[88]

Electrophilic substitution reactions of the synthetic metallocorrins have been little studied as yet, but a preference for C-5/C-15 substitution, rather than C-10, has been demonstrated in several cases. Thus, although all three meso protons of **158** [M = Ni(II)] are exchanged in DTFA in less than 1 minute,[46] the meso protons of the Co(III) complex [**170**; M = Co(III)CN$_2$, R = R^1 = H] underwent deuterium exchange over 5 days under less severe conditions (*tert*-BuOD/D$_2$O/DTFA), and evidence was obtained showing exchange occurred most rapidly at C-15.[75] The corresponding β-unsubstituted Co(III) corrin [**158**; M = Co(III)CN$_2$] decomposed in DTFA, but slow exchange of the meso protons was observed in CDCl$_3$-CD$_3$CO$_2$D. The Ni(II) 4,5-dihydrocorrin [**159**; M = Ni(II)] showed rapid exchange of the C-10- and C-15-meso protons in DTFA,[46] followed by a further much slower exchange of β-protons. Thus two β-protons exchanged after 24 hours and a further two after 175 hours. The precise position of β-deuteration is not known but positions C-2 and C-18 are ruled out on nmr evidence. Eschenmoser and co-workers have achieved specific C-15 monocyanation with chlorosulfonyl

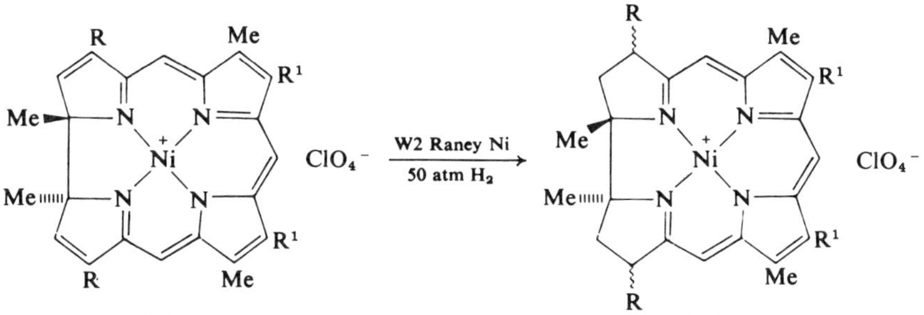

160   161

162

163 + 164

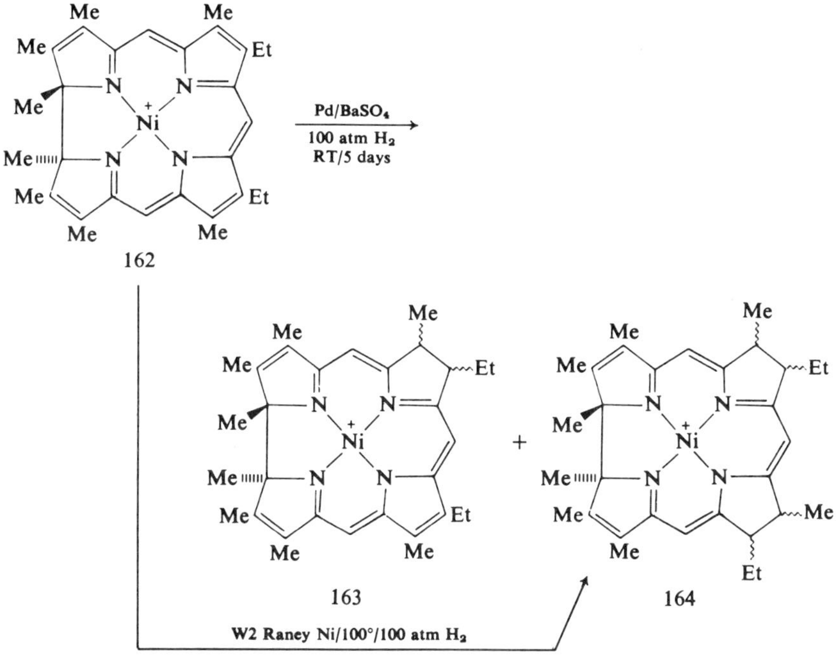

W2 Raney Ni/100°/100 atm H$_2$

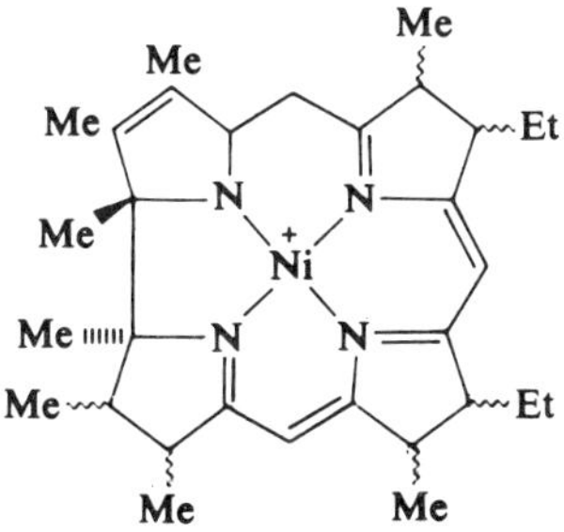

165

isocyanate in both Ni(II) [**170**; M = Ni(II), R = Me, R^1 = CN] and Co (III) [**170**; M = Co(III)CN$_2$, R = H, R^1 = CN] corrins in 80% yield.[75] The dominance of reaction at C-15 as opposed to C-5 is due to steric shielding of C-5 by the adjacent *gem*-dimethyl group. *Meso*-methylation of corrins[74] via reaction with chloromethylphenyl sulfide has been discussed previously (see Section IV, A). Selective meso substitution can also be achieved by treating corrins (e.g., **171**; R = H) with excess CH$_2$=$\overset{+}{\mathrm{N}}$Me$_2$I$^-$ to give **171** (R = CH$_2$NMe$_2$).[89]

Relatively few studies of the chemistry of the corrin chromophore in vitamin B$_{12}$ have appeared apart from ring B lactam and lactone formation.[79] Early chlorination studies[79] have recently been followed up and chlorination of cobyrinic esters and the corresponding ring B lactone has been shown to give adducts formulated as **172**, on the basis of analytical data and chemical behavior.[79a] Interestingly **172** is dehalogenated by [Rh(CO$_2$)Cl]$_2$ with regeneration of cobyrinic ester.

The chemistry of the cobalt-carbon bond in vitamin B$_{12}$ and model compounds has attracted great interest and these studies have been reviewed recently.[79b]

HMO calculations on the corrin system are in accord with the experimental observations of preferred electrophilic attack at C-5/C-15. The predicted

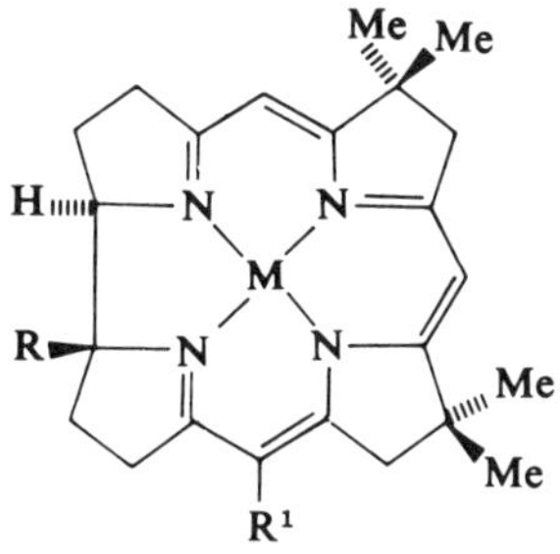

170

relative rates for deprotonation are C-8 > C-13 > C-3.[76] Other calculations concerned with the cobalt atom in corrins and its effects on the spectra of corrins have been reported.[90] The mass spectra,[91] magnetic circular dichroism,[92] and hplc[93] of various corrin derivatives have also been studied.

The synthesis and properties of corphins (e.g., **173**) and metallocorphins[73,94] are beyond the scope of this review.

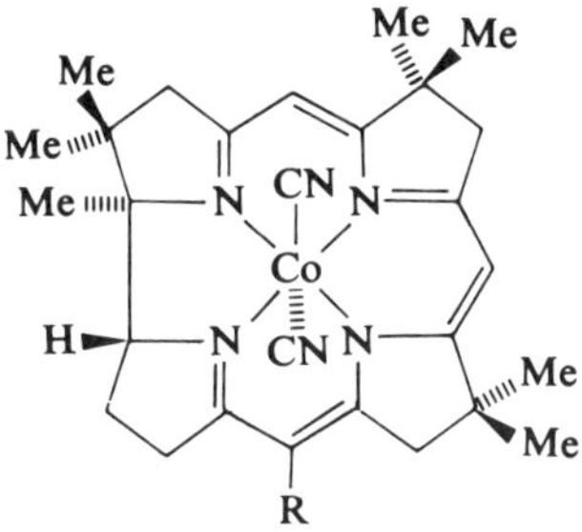

171

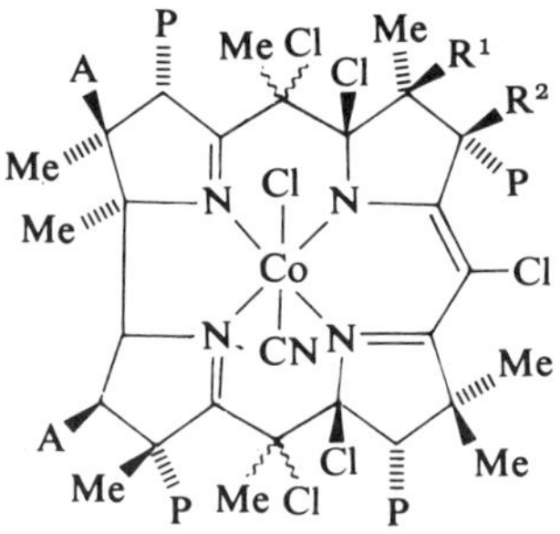

A = CH$_2$CO$_2$Me
P = CH$_2$CH$_2$CO$_2$Me

172

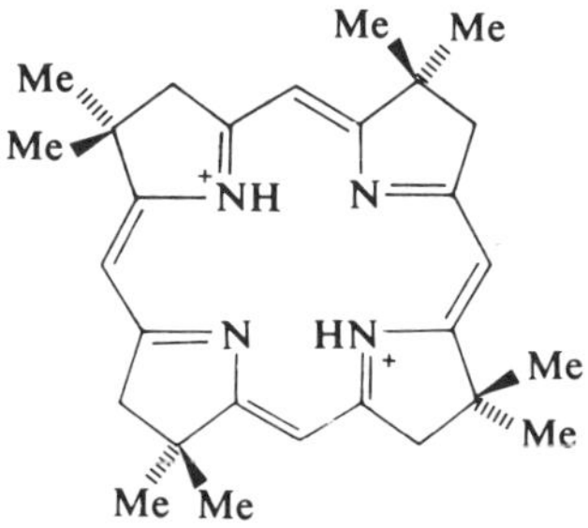

173

REFERENCES

1. A. W. Johnson and R. Price, *J. Chem. Soc.* p. 1649 (1960).
2. A. W. Johnson and I. T. Kay, *Proc. Chem. Soc., London* p. 168 (1961).
3. A. W. Johnson, I. T. Kay, and R. Rodrigo, *J. Chem. Soc.* p. 2336 (1963).
4. A. W. Johnson and I. T. Kay, *J. Chem. Soc.* p. 1620 (1965).
5. D. Dolphin, A. W. Johnson, J. Leng, and P. van den Broek, *J. Chem. Soc. C* p. 880 (1966).
6. G. Shelton, Ph.D. Thesis, Nottingham University (1970); R. Grigg, A. W. Johnson, and G. Shelton, *J. Chem. Soc. C* p. 2287 (1971) (experimental section of paper).
7. C. Walling, "Free Radicals in Solution." Wiley, New York, 1957.
8. R. B. Woodward and R. Hoffmann, *Angew. Chem., Int. Edn. Engl.* **8**, 781 (1969).
9. M. J. S. Dewar and S. Kirschner, *J. Am. Chem. Soc.* **93**, 4290 (1971).
10. M. Conlon, C. M. Elson, A. W. Johnson, W. R. Overend, and D. Rajapaksa, *J. Chem. Soc., Perkin Trans.* 2 p. 2281 (1973).
11. R. L. N. Harris, A. W. Johnson, and I. T. Kay, *J. Chem. Soc. C* p. 22 (1966).
11a. J. Engel and A. Gossauer, *Chem. Commun.* p. 713 (1975).
12. M. J. Broadhurst, R. Grigg, and A. W. Johnson, *J. Chem. Soc., Perkin Trans. 1* p. 1124 (1972).
13. E. Bullock, R. Grigg, A. W. Johnson, and J. W. F. Wasley, *J. Chem. Soc.* p. 2326 (1963).
14. M. J. Broadhurst, Ph.D. Thesis, Nottingham University (1970).
15. A. H. Jackson, G. W. Kenner, and G. S. Sach, *J. Chem. Soc.* p. 2045 (1967).
16. R. Grigg, R. Hayes, and J. L. Jackson, *Chem. Commun.* p. 1167 (1969); B. P. Stark and A. J. Duke, "Extrusion Reactions." Pergamon, Oxford, 1967.
17. R. L. N. Harris, *Tetrahedron Lett.* p. 3689 (1969).
18. M. J. Broadhurst, R. Grigg, G. Shelton, and A. W. Johnson, *Chem. Commun.* p. 231 (1970).
19. R. Grigg, *J. Chem. Soc. C* p. 3664 (1971).
20. R. Grigg and V. Viswanatha, unpublished results.
21. W. E. Parham and V. J. Traynelis, *J. Am. Chem. Soc.* **77**, 68 (1955); D. S. Breslow and H. Skolnik, "Multi-Sulphur and Sulphur and Oxygen Five- and Six-Membered Heterocycles," Part 2, Chapter 12. Wiley (Interscience), New York, 1966.
22. H. R. Harrison, O. J. R. Hodder, and D. C. Hodgkin, *J. Chem. Soc. B* p. 640 (1971).
23. R. Grigg, T. J. King, and G. Shelton, *Chem. Commun.* p. 56 (1970).
24. J. L. Hoard, *Ann. N.Y. Acad. Sci.* **206**, 18 (1973); A. Tulinsky, *ibid.* p. 47, and references therein.
25. B. Bak, D. Christensen, L. Hansen, and J. Rastrup-Anderson, *J. Chem. Phys.* **24**, 720 (1956).
26. N. S. Hush, J. M. Dyke, M. L. Williams, and I. S. Woolsey, *Mol. Phys.* **17**, 559 (1969); **20**, 1149 (1971).
27. R. Grigg, R. J. Hamilton, M. L. Josefowicz, C. H. Rochester, R. J. Terrell, and H. Wickwar, *J. Chem. Soc., Perkin Trans.* 2 p. 407 (1973).
28. M. J. Broadhurst, R. Grigg, A. W. Johnson and G. Shelton, *J. Chem. Soc., Perkin, Trans. 1* p. 143 (1972).
29. M. J. Broadhurst, R. Grigg, and A. W. Johnson, *J. Chem. Soc. C* p. 3681 (1971).
30. R. Grigg, A. W. Johnson, and G. Shelton, *Justus Liebigs Ann.* **746**, 32 (1971).
31. R. Grigg, A. W. Johnson, and G. Shelton, *J. Chem. Soc. C* p. 2287 (1971).
31a. A. M. Abeysekera, R. Grigg, J. Trocha-Grimshaw, and V. Viswanatha, *Tetrahedron Lett.* p. 3189 (1976).

31b. A. M. Abeysekera and R. Grigg, unpublished observations.
32. G. Shelton, Ph.D. Thesis, Nottingham University (1970).
33. R. Grigg, A. W. Johnson, K. Richardson, and M. J. Smith, *J. Chem. Soc. C* p. 1289 (1970).
34. Reported by Professor R. B. Woodward, *Aromaticity Conf.*, *Sheffield* (1966).
35. M. J. Broadhurst, R. Grigg, and A. W. Johnson, *J. Chem. Soc., Perkin Trans. 1* p. 2111 (1972).
36. M. J. S. Dewar and G. J. Gleicher, *J. Am. Chem. Soc.* **87**, 685 (1965); H. C. Longuet-Higgins and L. Salem, *Proc. R. Soc. London, Ser. A* **251**, 172 (1959).
37. K. P. C. Vollhardt, *Synthesis* **12**, 765 (1975).
38. P. Bamfield, R. L. N. Harris, A. W. Johnson, I. T. Kay, and K. W. Shelton, *J. Chem. Soc. C* p. 1436 (1966).
39. D. A. Clarke, R. Grigg, R. L. N. Harris, A. W. Johnson, I. T. Kay, and K. W. Shelton, *J. Chem. Soc. C* p. 1648 (1967).
40. D. Dolphin, R. L. N. Harris, J. L. Huppatz, A. W. Johnson, and I. T. Kay, *J. Chem. Soc. C* p. 30 (1966).
41. A. H. Corwin and G. C. Kleinspehn, *J. Am. Chem. Soc.* **82** 2750 (1960).
42. A. W. Johnson and I. T. Kay, *J. Chem. Soc.* p. 2418 (1961); G. M. Badger, R. L. N. Harris, and R. A. Jones, *Aust. J. Chem.* **17**, 1013 (1964).
43. A. Gossauer, H. Maschler, and H. H. Inhoffen, *Tetrahedron Lett.* p. 1277 (1974).
44. R. Grigg, A. P. Johnson, A. W. Johnson, and M. J. Smith, *J. Chem. Soc. C* p. 2457 (1971).
45. D. Dolphin, R. L. N. Harris, J. L. Huppatz, A. W. Johnson, I. T. Kay, and J. Leng, *J. Chem. Soc. C* p. 98 (1966).
46. A. W. Johnson and W. R. Overend, *J. Chem. Soc., Perkin Trans 1* p. 2681 (1972).
47. A. Gossauer, H. H. Inhoffen, and H. Maschler, *Justus Liebigs Ann. Chem.* p. 141 (1973).
48. K. P. Heise, H. H. Inhoffen, and N. Schwartz, *Justus Liebigs Ann. Chem.* 146 (1973).
49. J. Engel and A. Gossauer, *Chem. Commun.* p. 570 (1975).
50. R. Grigg, A. W. Johnson, and K. W. Shelton, *J. Chem. Soc. C* p. 1291 (1968).
51. R. Grigg, A. W. Johnson, K. Richardson, and M. J. Smith, *J. Chem. Soc. C* p. 1289 (1970).
52. R. Grigg, A. W. Johnson, K. Richardson, and K. W. Shelton, *J. Chem. Soc. C* p. 655 (1969).
53. A. Hamilton and A. W. Johnson, *J. Chem. Soc. C* p. 3879 (1971).
54. R. K. Bramley, R. Grigg, G. Guilford, and P. Milner, *Tetrahedron* **29**, 4159 (1973); R. M. Acheson, *Acc. Chem. Res.* **4**, 177 (1971).
55. C. M. Elson, A. Hamilton, and A. W. Johnson, *J. Chem. Soc., Perkin Trans. 1* p. 775 (1973).
56. H. H. Inhoffen, J. W. Buchler, L. Puppe, and K. Rohbock, *Justus Liebigs Ann. Chem.* **747**, 133 (1971).
57. N. S. Hush and I. S. Woolsey, *J. Am. Chem. Soc.*, **94**, 4107 (1972).
58. N. S. Hush and J. M. Dyke, *J. Inorg. Nucl. Chem.* **35**, 4341 (1973).
59. N. S. Hush and I. S. Woolsey, *J. Chem. Soc., Dalton Trans.* p. 24 (1974).
60. T. A. Melenteva, N. D. Pekel, and V. M. Berezovskii, *Zh. Obshch. Khim.* **42**, 183 (1972).
61. T. A. Melenteva, N. D. Pekel, N. S. Genokhova, and V. M. Berezovskii, *Dokl. Akad. Nauk SSSR* **194**, 591 (1970).
62. T. A. Melenteva, N. D. Pekel, and V. M. Berezovskii, *Zh. Obshch. Khim.* **42**, 180 (1972).

63. T. A. Melenteva, N. S. Genokhova, and V. M. Berezovskii, *Dokl. Akad. Nauk SSSR* **201**, 366 (1971).

64. T. A. Melenteva, N. S. Genokhova, N. D. Pekel, and V. M. Berezovskii, *Zh. Obshch. Khim.* **43**, 1337 (1973).

65. H. H. Inhoffen, J. Ullrich, H. A. Albrecht, G. Klinzmann, and R. Scheu, *Justus Liebigs Ann. Chem.* **738**, 1 (1970); H. H. Inhoffen, F. Fattinger, and N. Schwartz, *ibid.* p. 412 (1974).

66. H. H. Inhoffen and H. Maschler, *Justus Liebigs Ann. Chem.* p. 1269 (1974).

67. D. Crowfoot-Hodgkin, A. W. Johnson, and A. R. Todd, *Chem. Soc., Spec. Publ.* **3**, 109 (1955); D. Crowfoot-Hodgkin, J. Kamper, J. Lindsey, M. Mackay, J. Pickworth, J. H. Robertson, C. B. Shoemaker, J. G. White, R. J. Prosen, and K. N. Trueblood, *Proc. R. Soc. London, Ser. A* **242**, 288 (1957).

68. R. B. Woodward, *Pure Appl. Chem.* **33**, 145 (1973); W. Fuhrer, P. Schneider, W. Schilling, H.-J. Wild, H. Maag, N. Obata, A. Holmes, J. Schreiber, and A. Eschenmoser, *Chimia* **26**, 320 (1972).

69. J. W. Cornforth, reported by P. B. de la Mare, *Nature* (*London*) **195**, 441 (1962); J. W. Cornforth, "Discussions of Recent Experiments on Corrin Chemistry." Royal Society, London, 1964.

70. A. W. Johnson, *Chem. Br.* p. 253 (1967).

71. R. Bonnett, V. M. Clark, A. Giddey, and A. R. Todd, *J. Chem. Soc.* p. 2087 (1959), and subsequent papers.

72. E. Bertele, H. Boos, J. D. Dunitz, F. Elsinger, A. Eschenmoser, I. Felner, H. P. Gribi, H. Gschwend, E. F. Meyer, M. Pesaro, and R. Scheffold, *Angew. Chem., Int. Ed. Engl.* **3**, 490 (1964); E. Bertele, A. Eschenmoser, H. Gschwend, M. Pesaro, and R. Scheffold, *Proc. R. Soc. London, Ser. A* **288**, 306 (1965).

73. G. Traverso, G. P. Pollini, A. Barco, and G. De Guli, *Gazz. Chim. Ital.* **102**, 243 (1972); R. V. Stevens, L. E. Du Pree, W. L. Edmonson, L. L. Magid, and M. P. Wentland, *J. Am. Chem. Soc.* **93**, 6637 (1971); R. V. Stevens, C. G. Christensen, R. M. Cory, and E. Thorsett, *ibid.* **97**, 5940 (1975).

73a. R. V. Stevens, *Tetrahedron* **32**, 1599 (1976).

74. E. L. Winnacker, Doctoral Thesis, ETH Zurich, Switzerland (1968).

75. D. Bormann, A. Eschenmoser, A. Fischli, and R. Keese, *Angew. Chem., Int. Ed. Engl.* **6**, 868 (1967).

76. R. Keese, *Tetrahedron Lett.* p. 149 (1969).

77. A. Eschenmoser, *Q. Rev., Chem. Soc.* **24**, 366 (1970).

78. J. I. Toohey, *Proc. Natl. Acad. Sci. U.S.A.* **54**, 934 (1965); *Fed. Proc., Fed. Am. Soc. Exp. Biol.* **25**, 1628 (1966).

79. R. Bonnett, *Chem. Rev.* **63**, 573 (1963).

79a. A. Gossauer, K. P. Heise, H. Laas, and H. H. Inhoffen, *Justus Liebigs Ann. Chem.* 1150 (1976); *Phil. Trans. Roy. Soc.* B273, 327 (1976).

79b. R. H. Abeles and D. Dolphin, *Acc. Chem. Res.* **9**, 114 (1976); G. N. Schrauzer, *Angew. Chem., Int. Ed. Engl.* **15**, 417 (1976).

80 A. Eschenmoser and A. Fischli, *Angew. Chem. Int. Ed. Engl.* **6**, 866 (1967).

81. E. Edmond and D. Crowfoot-Hodgkin, unpublished results, quoted by Eschenmoser, in ref. 77, p. 391.

82. A. Eschenmoser, B. Golding, R. Keese, P. Loliger, D. Miljkovic, K. Muller, P. Wehrli, and Y. Yamada, *Angew. Chem., Int. Ed. Engl.* **8**, 343 (1969).

83. A. Pfaltz, B. Hardegger, P. M. Muller, S. Farooq, B. Kraeutler, and A. Eschenmoser, *Helv. Chim. Acta* **58**, 1444 (1975).

83a. A. Eschenmoser, *Q. Rev., Chem. Soc.* **5**, 377 (1976).

83b. B. Kraeutler, A. Pfaltz, R. Nordmann, K. O. Hodgson, J. D. Dunitz, and A. Eschenmoser, *Helv. Chim. Acta* **59**, 924 (1976).

83c. A. I. Scott, *Heterocycles* **2**, 125 (1974); *Science* **184**, 760 (1974); *Tetrahedron* **31**, 2639 (1975).

83d. C. E. Brown, J. J. Katz, and D. Shemin, *Proc, Natl. Acad. Sci. U.S.A.* **69**, 2585 (1972); *J. Biol. Chem.* **248**, 8015 (1973); A. I. Scott, C. A. Townsend, K. Okada, M. Kajiwara, P. J. Whitman, and R. J. Cushley, *J. Am. Chem. Soc.* **94**, 8267 (1972); **96**, 8054 (1974).

84. A. Eschenmoser and E. Goetschi, *Angew. Chem., Int. Ed. Engl.* **12**, 912 (1973).

85. E. Goetschi, W. Hunkeler, H.-J. Wild, W. Schneider, W. Fuhrer, J. Gleason, and A. Eschenmoser, *Angew. Chem., Int. Ed. Engl.* **12**, 910 (1973).

86. A. L. Hamilton, A. W. Johnson, and W. R. Overend, *J. Chem. Soc., Perkin Trans. 1* p. 991 (1973).

87. I. D. Dicker, R. Grigg, A. W. Johnson, H. Pinnock, K. Richardson, and P. van den Broek, *J. Chem. Soc. C* p. 536 (1971).

88. R. Bonnett, J. M. Godfrey, V. E. Math, E. Edmond, H. Evans, and O. J. R. Hodder, *Nature (London)* **229**, 473 (1971); R. Bonnett, *Phil. Trans. Roy. Soc.* **B273**, 295 (1976).

89. J. Schreiber, H. Maag, N. Hashimoto, and A. Eschenmoser, *Angew. Chem., Int. Ed. Engl.* **10**, 330 (1971).

90. P. O. Offenhartz, B. H. Offenhartz, and M. M. Mayme, *J. Am. Chem. Soc.* **92**, 2966 (1970); A. Veillard and B. Pullman, *J. Theor. Biol.* **8**, 307 (1965).

91. J. Seibl, *Org. Mass Spectrom.* **1**, 215 (1968); *Adv. Mass Spectrom.* **4**, 317 (1968).

92. B. Briat and C. Djerassi, *Bull. Soc. Chim. Fr.* p. 135 (1969).

93. J. Schreiber, *Chimia* **25**, 405 (1971).

94. P. M. Muller, S. Farooq, B. Hardegger, W. S. Salmond, and A. Eschenmoser, *Angew. Chem., Int. Ed. Engl.* **12**, 914 (1973); N. Muller and H. H. Inhoffen, *Tetrahedron Lett.* p. 3209 (1969).

Author Index

228(224), 229(111, 112), 230(6, 224), 231(224), 237(245), 239(252), 240(111, 224, 258), 242(111, 112, 252, 266), *246, 248, 249, 250, 252, 253*

Miroshnichenko, L. D., 211(137), 219(173), 220(173, 178, 179), 223(179), *250, 251*

Misaki, S., 209(127), *249*

Mittenzwei, H., 20(111), *41,* 288(19, 20), 298, 302(20), 306, 307, 308, *322, 324*

Molvig, H., 222(208), *251*

Moore, T. A., 23(127), *42*

Morell, D. B., 126(68a), 127(75), *130*

Morley, H. V., 35(177), *43,* 310(107), *324*

Moron, J., 200(31), *247*

Morrell, D. B., 259(15), *283*

Morsingh, F., 202(54), *247*

Moscowitz, A., 256(4), 262(4), *283,* 302(72), 318, 319(157), 321(157), *323, 325, 326*

Moser, F. H., 136(28), *157*

Moslov, V. G., 186(211, 212), *195*

Motekaitis, R. J., 203(85), *248*

Müller, H., 166(62), *191*

Müller, K., 140(44), *157, 390*

Müller, N., 4(5), *37,* 78, 82(170), *89,* 141(53), 144(65), *157, 158,* 387(94), *391*

Müller, P. M., 3(2), *37,* 6 6(96), *88,* 379(83), 380(83), 387(94), *390, 391*

Müller-Enoch, D., 36(183), *44,* 298, 302(79), 303(79, 82), *323, 324*

Mukhin, E. N., 165(45), 168(45), *191*

Muljiani, Z., 48(29), *85,* 167(91), *192,* 108(24), 109(24), *128*

Murakami, Y., 256(8), 257(8, 9), 263(8), 264(8), 265(8), *283*

Murphy, M. J., 3(1b), 20(1b), *37,* 288(6, 7), *321*

Murphy, R. F., 127(83), *130*

Nasralla, S. M., 215(151), 216(151), 242(151), *250*

Negelein, E., 72(147), *89*

Neidle, S., 24(140), *42,* 210(133), *250*

Neiman, R., 265(36), *284*

Neri, B. P., 22(119), *42,* 52(64), 53(64, 68), 59(68), *86, 87*

Neuberger, A., 5(34), *39,* 232(234), *252*

Neumann, W., 154(95a), *159*

Newman, D. J., 214(140), *250*

Nichol, A. W., 48(25), *85,* 105(8, 9), 106(8, 9, 15, 17), 107(8), 108(8, 17), 111(8, 17), 115(8, 17), 116(17), 118(8), 127(75, 87), *128, 130,* 146(72), *158,* 200(24), 201(24), 202(42), 207(24, 42), 208(24), 210(130), 215(145), 222(24, 145), *247, 250, 259(15), *283*

Nicholas, P. P., 313(127), *325*

Nicholaus, R., 126(70), *130*

Nicholson, D. C., 72(149), *89,* 123, 124(59), 126(70), 127(72), *130,* 302(72), 317(72, 140), *323, 325*

Nickon, A., 166(66), *191*

Nolte, W., 22(118), *42,* 64(94), 77(94, 167), 78(94), 79(94), 80(94), *88, 89,* 141(52), *157*

Nordmann, R., 380(83b), *391*

Norman, G. D., 33(172), 37(172), *43,* 71(141), *89,* 291(43), 319(152), *322, 325*

Norman, R. O. C., 127(78, 79), *130*

Norrie, M., 72(150), *89,* 104(3), 118(3), 121(50), 123(3), *128, 129*

Norris, J. R., 4(12a), 11(60), 32(12a), 36(12a), 37(12a), *38, 40,* 47(15), 51(55a), 64(15, 89c), 69(121a), *85, 86, 87, 88*

Norton, J. C., 310(113), *324*

Nüssler, L., 32(166), *43*

N

Nagarajan, K., 298(61), 301(61), 302(61), *323*

Nagarkatti, J. P., 202(46), *247*

Nagata, C., 132(2, 3), *156*

Nakajima, H., 124, 127(62, 80), *130*

Nakajima, O., 127(80, 81), *130*

Nakata, B. T., 217(159), 219(159), 223(211), *250, 252*

Nakazaki, M., 300(68, 69, 70), *323*

Nandi, J. C., 211(135), *250*

O

Obata, N., 372(68), *390*

O'Brien, T. P., 162(23), *190*

O'Carra, P., 126(68a, 69), 127(74, 83), 128, *130,* 258(10), *283,* 319(75, 148, 149, 156), 320(75), *323, 325, 326*

Oester, M. Y., 5(19), 6(19), 30(19), 32(169), *38, 43,* 49(38), 50(38), 59(79), *86, 87*

Oettmeier, W., *44*

Offenhartz, B. H., 387(90), *391*

Subject Index

E

432 SUBJECT INDEX